Frédéric Mangolte

Real Algebraic Varieties

Frédéric Mangolte
LAREMA
Université d'Angers
Angers, France

Translated by
Catriona Maclean
Institut Fourier
Grenoble Alpes University
Saint-Martin-d'Hères, France

ISSN 1439-7382 ISSN 2196-9922 (electronic)
Springer Monographs in Mathematics
ISBN 978-3-030-43106-8 ISBN 978-3-030-43104-4 (eBook)
https://doi.org/10.1007/978-3-030-43104-4

Mathematics Subject Classification (2020): 14P25, 14P05, 14Pxx, 14E05, 14E07, 14R05, 14Jxx, 14Hxx, 26C15, 32Jxx, 57M50, 57Mxx, 57N10

Translation from the French language edition: *Variétés algébriques réelles* by Frédéric Mangolte, 2017. Published by SMF. All Rights Reserved.

This Springer imprint is published by the registered company Springer Nature Switzerland AG
The registered company address is: Gewerbestrasse 11, 6330 Cham, Switzerland

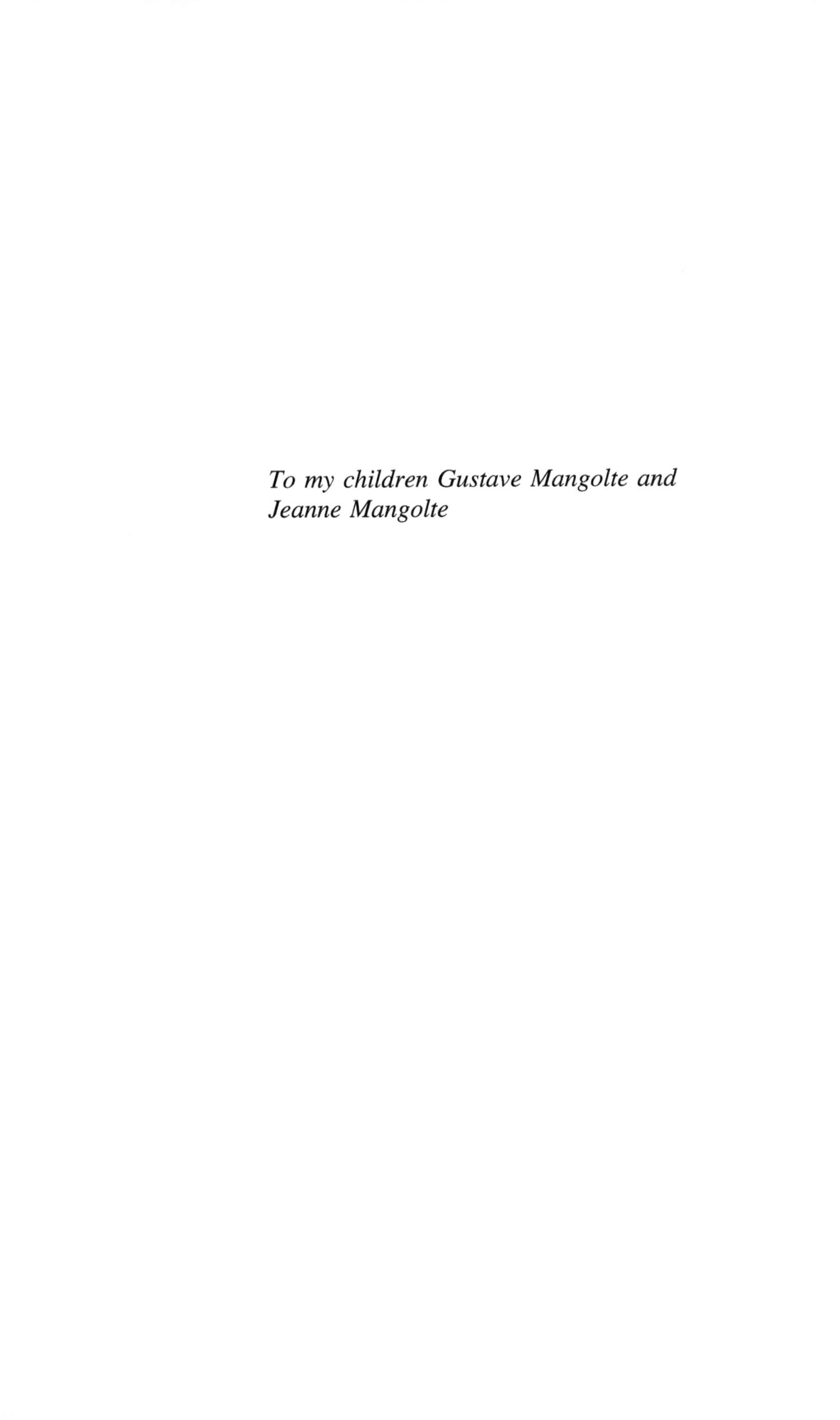

To my children Gustave Mangolte and Jeanne Mangolte

Preface

The present volume is a translation of my book *Variétés algébriques réelles* originally published in French, [Man17b].

Apart from corrections and incorporation of several new bibliographical references, this translated version is not substantially different from the original. In particular, any statement from the original French edition to the new English one has the same number: Théorème 5.4.16 in [Man17b] is now Theorem 5.4.16 in this English version.

I wish to thank Catriona MacLean for the quality of her careful translation of the text which allowed to improve it in several places.

Angers, France
December 2019

Frédéric Mangolte

Introduction

My work always tried to unite the truth with the beautiful, but when I had to choose one or the other, I usually chose the beautiful.

Attributed to Hermann Weyl (Elmshorn, Allemagne, 1885–Zurich, 1955).

Pessimism of the intelligence, optimism of the will.

Antonio Gramsci. (Ales, Sardaigne, 1891–Rome, 1937). Taken from a letter written in prison to his brother Carlo the 19th December 1929 (Selections from the Prison Notebooks, International Publishers, 1971).

Mathematicians often consider the set of real roots of a polynomial with real coefficients: it is just as natural to consider the set of its complex roots. In this book we will adopt the point of view that a real variety is *also* a complex variety.

When I was a doctoral student in the 90s there were essentially three reference books in real algebraic geometry. As well as Benedetti and Risler [BR90], there was **the** general reference, Bochnak Coste and Roy [BCR87][1] and Silhol's book [Sil89] for the classification of real algebraic surfaces. Since then [DIK00] by Degtyarev, Itenberg and Kharlamov has appeared, containing the classification of surfaces of special type summarised in [Sil89] plus the major progress made in the following decade.

The natural first port of call for a mathematician looking for a reference for real algebraic geometry is Bochnak, Coste and Roy, but for more information on surfaces or higher dimensional varieties he or she will need to look elsewhere. Silhol's book contains an overview of surfaces which was complete at the time of publication (1989): more up-to-date information can be found in Degtyarev, Itenberg and Kharlamov (2000).

[1]English translation: [BCR98].

It is my belief that a reader discovering real algebraic geometry can compare, contrast and link the different points of view of these three texts only if they have significant mathematical maturity, and the result is that many foundational results are not easily available. For example, [BCR87] uses germs of real varieties rather than schemes defined over $\mathbb{R}$: these germs appear in our text as the *real algebraic varieties* of Chapter 1. Meanwhile, [Sil89] adopts the point of view that real algebraic varieties are *schemes over* $\mathbb{R}$ - a steep learning curve for readers interested in topological applications, since the link between a scheme over $\mathbb{R}$ and a scheme over $\mathbb{C}$ with real structure is not obvious for inexperienced readers. On the other hand, [DIK00] considers "complex varieties with an anti-holomorphic involution", which appear in this text in Chapter 2 under the name of $\mathbb{R}$-*varieties*: this choice makes topological applications more accessible but is difficult to link to scheme-theoretic results.

My goal in this book is to present the foundations of real algebraic varieties, including their topological, geometric and algebraic structures and singularities, from each of the three points of view described above simultaneously. The first few chapters are intended to be accessible to PhD students and specialists of other areas. Compared with the three texts mentioned above, this work presents the proofs of the main theorems of the area in a uniform language, supplementary material on the topology and birational geometry of real algebraic surfaces, and some new work on three-dimensional varieties due to Kollár et al.

Before getting to the heart of the matter, we begin with a motivating discussion of the Nash conjectures.

Algebraic Models of Smooth Manifolds

Any closed smooth curve M is diffeomorphic as a differential manifold to the circle $\mathbb{S}^1 := \{(x,y) \in \mathbb{R}^2,\ x^2+y^2 = 1\}$. We say that the manifold M has a *real algebraic model*[2]—namely, the algebraic curve $\mathbb{S}^1$. In arbitrary dimension, Nash and Tognoli showed that any smooth compact manifold has a real algebraic model. More precisely, for any such manifold M there are real polynomials $P_1(x_1,\ldots,x_m),\ldots,P_r(x_1,\ldots,x_m)$ such that their locus of common zeros

$$X(\mathbb{R}) := \{x \in \mathbb{R}^m \text{ such that } \mathrm{P}_1(\mathrm{x}) = \ldots = \mathrm{P}_r(\mathrm{x}) = 0\}$$

[2]We will come back to this notion in Chapter 5, see Definition 5.1.1.

is smooth[3] and diffeomorphic to M (Nash and Tognoli theorem, below.)

Given that the existence of such a model is guaranteed for any M, does there exist a model which is “simpler[4]” than the others? Let us go back to the example of a smooth compact curve M. We have seen that $\mathbb{S}^1$ is a possible model for this curve but for any even non-zero integer d the set $X_d := \{(x,y) \in \mathbb{R}^2,\ x^d + y^d = 1\}$ is obviously another model of M.

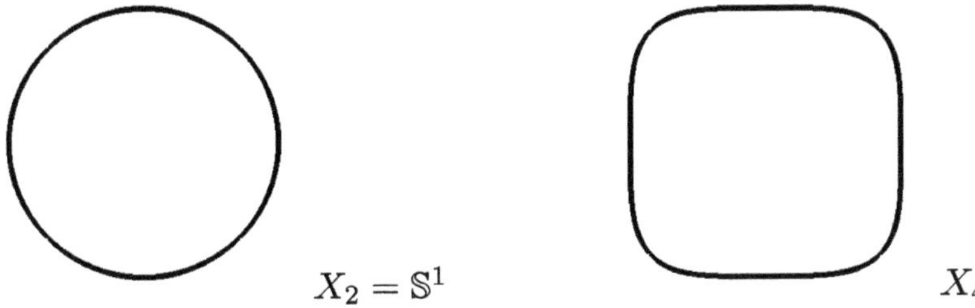

Similarly, the ellipse Y of equation $\frac{1}{2}x^2 + y^2 = 1$ and the curve Z of quartic equation $-x(1-x-y)^3 + y^2(1-x-y)^2 - \frac{1}{2}xy^3 = 0$ are models of M.

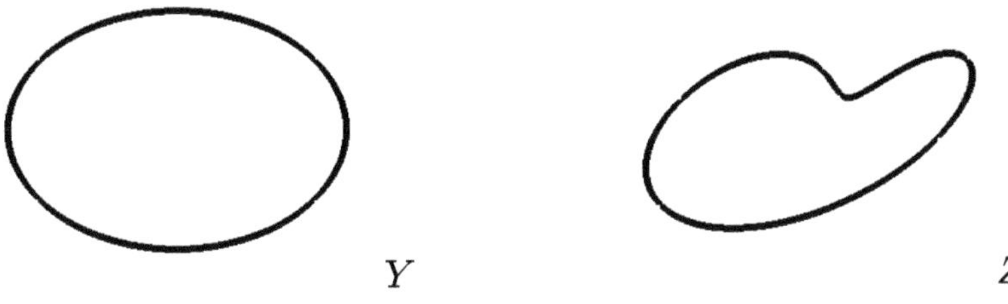

We could argue that $\mathbb{S}^1$ and Y are “simple” real algebraic models of M because their equations are of minimal degree—but an abstract algebraic variety does not have a well-defined degree. (All the above examples are plane curves.) We will consider the topology of the associated complex variety instead. For example, the associated complex variety of $\mathbb{S}^1$ is the set $\{(x,y) \in \mathbb{C}^2,\ x^2 + y^2 = 1\}$. This complex curve is not compact but can be made compact by adding two points at infinity corresponding to the asymptotic directions $\pm i$. The curve then becomes isomorphic to the Riemann sphere $\mathbb{P}^1(\mathbb{C})$ whose underlying smooth manifold is an orientable surface of genus 0. Since the models Y, X_d and Z are also (up to addition of a few points) irreducible plane curves, the genus formula $g = (d-1)(d-2)/2$ implies that the genus of Y is 0, X_4 and Z are of genus 3 and the genus of X_d grows

---

[3]By Hironaka's theorem on the resolution of singularities Theorem 1.5.54, we may also suppose that the set of complex zeros—real or not—is smooth.

[4]See also [Kol01c, LV06].

quadratically with d. In dimension 1 we will, as in this example, consider that a model is "simple" if its genus is 0. In higher dimensions manifold topology is more complicated and there is no longer a single numerical invariant that detects simple real algebraic models. For curves we have just seen that simple models are "close to" $\mathbb{P}^1(\mathbb{C})$. In dimension n the class of *rational* varieties (defined below and further studied in Chapter 1, see Definition 1.3.37), which are in a certain sense "close" to $\mathbb{P}^n$, is a useful generalisation to higher dimensions of the class of algebraic curves isomorphic to $\mathbb{P}^1$ minus a few points.

Nash and Tognoli Theorems

When an algebraic variety $X \subset \mathbb{P}^m(\mathbb{C})$, defined by homogeneous polynomials with real coefficients, has at least one non-singular real point then $\dim_{\mathbb{R}} X = 2 \dim_{\mathbb{R}} X(\mathbb{R})$. In particular, when the complex algebraic variety X is smooth and the real locus $X(\mathbb{R}) = X \cap \mathbb{P}^m(\mathbb{R})$ is non-empty the algebraic subsets $X \subset \mathbb{P}^m(\mathbb{C})$ and $X(\mathbb{R}) \subset \mathbb{P}^m(\mathbb{R})$ come equipped with the structure of a smooth compact differentiable sub-variety. (See Section 1.5 for the definitions of a non-singular point and the dimension of an algebraic variety.)

Conversely, given a smooth manifold, can it be realised as the set of points of a smooth algebraic variety? It is quite clear that a general smooth manifold is not necessarily diffeomorphic to any *complex* algebraic variety, if only because any such variety is always orientable and even dimensional. There are many more sophisticated obstructions to such a realisation—see [FM94], for example. On the other hand, Nash proved there is no obstruction to the realisation of a compact manifold as a real algebraic variety.

Theorem ([Nas52]) If M is a smooth connected compact manifold without boundary, then there is a projective algebraic variety[5] X whose real locus has a connected component $A \subset X(\mathbb{R})$ which is diffeomorphic to M,

$$M \approx A \hookrightarrow X(\mathbb{R}).$$

The reader will find a proof of this theorem in [Nas52, BCR98, Theorem 14.1.8] or [Kol17, Theorem 2].

Following this theorem Nash proposed two conjectures strengthening its conclusion. The first of these, proved by A. Tognoli in the early seventies, stated that there is a variety X such that $X(\mathbb{R}) \approx M$.

[5]I.e. defined by homogeneous real polynomials.

Theorem ([Tog73]) In the statement of Nash's theorem, we may require $X(\mathbb{R})$ to be connected.

The proof ([Tog73] or [BCR98, Theorem 14.1.10]) uses a deep result from cobordism theory which states that any compact smooth manifold is cobordant to a compact smooth real algebraic set.

It is then easy to construct a real algebraic variety whose real locus is the union of the real loci of a set of given varieties.

Corollary (Nash–Tognoli) If M is a smooth compact manifold without boundary then there is a projective algebraic variety X whose real locus is diffeomorphic to M:

$$M \approx X(\mathbb{R}).$$

In fact any given manifold has not one but an infinity of possible different algebraic models. The theorem below is taken from [BK89, Theorem 1.1] supplemented by [Bal91], see also [BK91].

Anticipating Definition 1.3.27, algebraic sub-varieties $X \subset \mathbb{P}^n(K)$ and $Y \subset \mathbb{P}^N(K)$ are said to be birationally equivalent if there exist Zariski dense open subsets $U \subset X$, $V \subset Y$ and an isomorphism $U \xrightarrow{\simeq} V$ defined by rational functions with coefficients in K.

Theorem Let M be a smooth compact manifold without boundary of strictly positive dimension. There is an uncountably infinite set of real algebraic models for M which are pairwise non-birationally equivalent.

An algebraic variety $X \subset \mathbb{P}^N(K)$ of dimension n is *rational over K* if and only if it is birationally equivalent to projective space $\mathbb{P}^n(K)$, or in other words, if there are dense open Zariski subsets $U \subset X$, $V \subset \mathbb{P}^n(K)$ and an isomorphism $U \xrightarrow{\simeq} V$ defined by rational functions with coefficients in K.

Examples

1. Blowing up a variety along a subvariety (see Appendix F) is a birational morphism.
2. The varieties $\mathbb{P}^n(K)$ and K^n are rational over K.
3. The surface $\mathbb{P}^1 \times \mathbb{P}^1$ with the product real structure such that $(\mathbb{P}^1 \times \mathbb{P}^1)(\mathbb{R}) \approx \mathbb{S}^1 \times \mathbb{S}^1$ and the Hirzebruch surfaces $\mathbb{F}_k$ with their canonical real structure (see Definition 4.2.1) are rational surfaces over $\mathbb{R}$.

Nash's second conjecture is as follows.

Conjecture ([Nas52]) For any smooth compact connected manifold M without boundary there is a *rational* algebraic variety X whose real locus is diffeomorphic to M.

We will see in Chapter 1 that when an algebraic variety is irreducible its ring of rational functions is a field called the *function field* of the variety and moreover the function field of a (reduced and irreducible) algebraic variety of dimension n over K is a finite degree extension of the field of rational functions in n unknowns $K(X_1, \ldots, X_n)$. The variety X is then rational if and only if its field of functions is *isomorphic* to $K(X_1, \ldots, X_n)$. Nash's conjecture is therefore much stronger than Nash's theorem, since it claims that we can choose X with a function field of degree 1 over $\mathbb{R}(X_1, \ldots, X_n)$ for any M.

This conjecture is wrong. We will see a counter-example for surfaces in Chapter 4 and for higher dimensional varieties, in Chapter 6. There exist smooth manifolds for which no real algebraic model is rational. In other words, there is no universal answer to the question 'what is the the "simplest" real algebraic model of a given smooth manifold?' One of the leitmotifs of this book is a sort of converse to this question: for some given class of real algebraic varieties, which are the smooth manifolds which can be realised by models in this class?

Such a class of models may be characterised, as in Nash's conjecture, by birational constraints on the abstract real algebraic model. More classically, we might require that this model be defined by a single equation of given "small" degree, particularly for planar curves or surfaces in $\mathbb{P}^3$. Here are some examples of specific questions dealt with in this text. What are the possible topological types of the real locus of a rational surface? Of a degree 4 surface in $\mathbb{P}^3$? Conversely, what is the smallest possible degree of a real algebraic model in $\mathbb{P}^3$ of an orientable surface of genus 11? Of a disjoint union of 23 compact connected surfaces? In a similar vein, in Chapter 3 we discuss the first part of the Hilbert's famous sixteenth problem which rounds off the sections on plane curves in Chapters 1 and 2.

Thanks Many people have encouraged me over the 4 years of writing this text. Amongst them I would particularly like to thank (in alphabetical order) for their rereading, corrections, improvement and support : Mouadh Akriche, Mohamed Benzerga, Jérémy Blanc, Erwan Brugallé, Fabrizio Catanese, Michel Coste, Julie Déserti, Adrien Dubouloz, Denis Eckert, Éric Edo, Marianne Fabre, José Fabre, Goulwen Fichou, Michel Granger, Lucy Halliday, Ilia Itenberg, Viatcheslav Kharlamov, János Kollár, Wojciech Kucharz, Jacques Lafontaine, Stéphane Lamy, Gustave Mangolte, Jeanne Mangolte, Jean-Philippe Monnier, Delphine Pol, Ronan Terpereau, Olivier Wittenberg, Mikhal Zaidenberg, Susanna Zimmermann, and the three anonymous referees who proposed many improvements on the initial text.

Contents

List of Figures

Chapter 1
Algebraic Varieties

1.1 Algebraic Varieties: Points or Spectra?

In this chapter we have chosen the *naive* (as opposed to scheme theoretic) point of view in which an algebraic variety is a topological space equipped with a sheaf of functions, called regular functions. The scheme theoretic point of view starts with a ring (which turns out *later* to be the ring of regular functions) and constructs from it a topological space called the spectrum. Many fundamental differences result from this change of perspective. First of all, the spectrum has more points than the naive space—for example, when the base field is algebraically closed, the naive space is the set of closed points of the spectrum (also called the maximal spectrum). When the base field is not algebraically closed, the situation is even more complicated. For example, consider the algebraic set $V := \{(x, y) \in \mathbb{R}^2 \mid x^2 + y^2 = 0\}$ which *naively* consists of a single point $(0, 0)$. In scheme theory a variety "is" its defining equation. More precisely, the scheme-theoretic variety would be the union of two complex lines L and $\overline{L}$ with equations $x - iy = 0$ and $x + iy = 0$ in $\mathbb{C}^2$. In the naive point of view V "is" the point $(0, 0)$ (i.e. the intersection of the lines L and $\overline{L}$), which leads us to consider that the real equations of V are $x = 0$, $y = 0$, losing all the information contained in the lines L and $\overline{L}$ in the process (see Example 1.5.20).

We will stick with the naive point of view, but from Chapter 2 onwards we will address its most glaring weaknesses by associating complexifications to real varieties. We will not, however, bypass sheaf theory, which is both necessary and reasonably accessible. For the reader's convenience, a summary of the necessary results from sheaf theory in provided in Appendix C.

We have chosen to avoid the arduous "rite of passage" of scheme theory since we do not believe its refinement to be necessary over $\mathbb{C}$ and $\mathbb{R}$, especially since varieties over these fields have a natural topology which is stronger than the Zariski topology, and we believe it is helpful to the reader's intuition to consider these two topologies on the same naive set. We will come back to the scheme-theoretic point of view at the end of Chapter 2.

F. Mangolte, *Real Algebraic Varieties*, Springer Monographs in Mathematics,
https://doi.org/10.1007/978-3-030-43104-4_1

In short, this book is entirely accessible to a reader who does not want to invest the time and energy required to learn scheme theory. This being said, any reader intending to make a career in algebraic geometry will need to understand schemes. We recommend Antoine Ducros's lecture notes [Duc14] and Qing Liu's book [Liu02] for readers looking for an introduction to scheme theory. We have included many remarks aimed at readers experienced with schemes.

In this first chapter we review the standard results of algebraic geometry over an arbitrary field, particularly the real and complex numbers. Many elementary textbooks on algebraic geometry only deal with algebraically closed base fields and when non-algebraically closed fields are discussed they are almost always arithmetic. When the base field $\mathbb{R}$ is introduced the reader is typically assumed to be already familiar with complex algebraic geometry. In short, our aim in writing this book is to provide the reader with all the tools needed for algebraic geometry over a non-algebraically closed base field of characteristic zero, including both the Euclidean topology and the powerful results of complex algebraic geometry.

We end this introduction with an important remark: over $\mathbb{R}$ or $\mathbb{C}$ algebraic varieties are naturally also analytic varieties and it is unsurprising that such varieties appear throughout this book. We have chosen to summarise the theory of analytic varieties in Appendix D to avoid swamping the inexperienced reader—the field of study generated by the various different definitions of a "real algebraic variety" is rich and complicated enough as it is. As with schemes, the first-time reader can skip the references to analytic varieties, which are only needed in a handful of proofs in this book. Most of the time we deal only with smooth projective varieties and the link between projective algebraic varieties and projective analytic varieties is well understood (see Section D.5) despite the radical change in topology, since the algebraic variety is equipped with the Zariski topology (Definition 1.2.3) and the analytic variety with the Euclidean topology (Definition 1.4.1). Switching from one topology to the other causes no problems for smooth varieties (a slippery concept, see Section 1.5) but for singular varieties it becomes important to distinguish the algebraic and analytic structures. For example, there are algebraic singularities that are smooth from an analytic point of view, (see Example 1.5.1).

1.2 Affine and Projective Algebraic Sets

[The results from commutative algebra on which the discussion below relies are summarised in Appendix A.]

Affine algebraic sets are local models of abstract algebraic varieties. We will formalise this idea by fixing a base field K and giving a local definition of the functions (i.e. morphisms to K) authorised in the category of algebraic varieties over K. Such functions will be called regular functions (Definition 1.2.33). When the base field is algebraically closed the regular functions on an algebraic subset of a affine space are simply restrictions of polynomial functions (Theorem 1.2.50). When the base field is $\mathbb{R}$, however, this correspondence is no longer valid and regular functions are restrictions of rational functions without real poles (Theorem 1.2.52).

Affine Space

Let K be a field and let n be a natural number. As usual, we denote by K^n the set of n-tuples of elements of K with its natural K-vector space structure. (By convention $K^0 = \{0\}$ is the trivial vector space and when $n = 0$ the notation $K[X_1, \ldots, X_n]$ means the ring K of constant polynomials). When $K = \mathbb{R}$ or $\mathbb{C}$, we will mostly consider K^n as a finite-dimensional topological vector space, all norm-induced topologies being equivalent. The Zariski topology defined immediately below is not induced by a norm, since it is not even Hausdorff (see Appendix B.1).

Definition 1.2.1 Let K be a field and let n be a natural number. A subset F of K^n is an *affine algebraic set* if F is the zero locus of a set of polynomials with coefficients in K. In other words a set F is algebraic if and only if there exist polynomials $P_1, \ldots, P_l \in K[X_1, \ldots, X_n]$ such that[1]

$$F = \left\{(x_1, \ldots, x_n) \in K^n \mid P_1(x_1, \ldots, x_n) = \cdots = P_l(x_1, \ldots, x_n) = 0\right\}.$$

It is easy to check that algebraic sets are the closed subsets of a topology called the *Zariski topology*. The *affine space* $\mathbb{A}^n(K)$ of dimension n over K is the set K^n equipped with the Zariski topology. For any $F \subset \mathbb{A}^n(K)$ we define the Zariski topology on F to be the topology induced by the Zariski topology on $\mathbb{A}^n(K)$.

Exercise 1.2.2 Let K be an infinite field. Show that the Zariski topology on $\mathbb{A}^2(K)$ is strictly finer than the product of the Zariski topology on $\mathbb{A}^1(K)$ with itself. See Appendix B.1, particularly Exercise B.1.4, for a deeper exploration of this subject.

For any subset U in $\mathbb{A}^n(K)$ we let $\mathcal{I}(U)$ be the ideal in $K[X_1, \ldots, X_n]$ of polynomials that vanish on U. If F is a closed subset of $\mathbb{A}^n(K)$ then the quotient K-algebra $\mathcal{A}(F) := K[X_1, \ldots, X_n]/\mathcal{I}(F)$ is called the the *K-algebra of affine coordinates of F*.

Projective Space

Let K be a field and let n be a natural number. Projective space $\mathbb{P}(K^{n+1})$ is the set of orbits of the action of the multiplicative group K^* on the set $K^{n+1} \setminus \{0\}$ given by $(x_0, \ldots, x_n) \mapsto (\lambda x_0, \ldots, \lambda x_n)$. The orbit of $(x_0, \ldots, x_n)$ under this action is

[1] As the ring $K[X_1, \ldots, X_n]$ is Noetherian we may assume that this family of polynomials is finite, see Exercise A.3.14.

denoted by $(x_0 : \cdots : x_n)$. A polynomial in $n+1$ variables does not define a function on $\mathbb{P}(K^{n+1})$ but the vanishing locus of a *homogeneous* polynomial[2] is well defined.

Definition 1.2.3 Let K be a field and let n be a natural number. A subset F in $\mathbb{P}(K^{n+1})$ is said to be a *projective algebraic set* if F is the zero locus of a set of homogeneous polynomials with coefficients in K. In other words the set F is algebraic if and only if there are homogeneous polynomials $P_1, \ldots, P_l \in K[X_0, \ldots, X_n]$ such that

$$F = \big\{(x_0 : \cdots : x_n) \in \mathbb{P}(K^{n+1}) \mid P_1(x_0, \ldots, x_n) = \cdots = P_l(x_0, \ldots, x_n) = 0\big\}.$$

As in the affine case, the *Zariski topology* on $\mathbb{P}(K^{n+1})$ is the topology whose closed sets are the zero loci of families of homogeneous polynomials. The set $\mathbb{P}(K^{n+1})$ with this topology is called the *projective space* of dimension n over K and is denoted by $\mathbb{P}^n(K)$.

Definition 1.2.4 If U is a subset of $\mathbb{P}^n(K)$ we denote by $\mathcal{I}(U)$ the *homogeneous* ideal in $K[X_0, \ldots, X_n]$ of polynomials vanishing on U. If F is a Zariski closed subset in $\mathbb{P}^n(K)$ then the quotient K-algebra

$$\mathcal{S}(F) := K[X_0, \ldots, X_n]/\mathcal{I}(F)$$

is called the *K-algebra of homogeneous coordinates* of F.

Exercise 1.2.5 Let $F \in \mathbb{P}^n(K)$ be a projective algebraic set and let $I \subset \mathcal{S}(F)$ be a homogeneous ideal. Let $\sqrt{I}$ be the radical of I (see Definition A.2.3). We then have that $\mathcal{Z}(I) = \varnothing$ (see Definition 1.2.12) if and only if $\sqrt{I} = \mathcal{S}(F)$ or $\sqrt{I}$ is the homogeneous ideal $\bigoplus_{d>0} S_d$.

Remark 1.2.6 The word *dimension* appears in Definitions 1.2.1 and 1.2.3. We will define the dimension of an algebraic set further on (see Definition 1.5.9 for the dimension of an affine algebraic set) and we will check (see Exercises 1.5.16 and 1.5.46) that affine space $\mathbb{A}^n(K)$ and projective space $\mathbb{P}^n(K)$ really are of (algebraic) dimension n. For the moment we simply note that when $K = \mathbb{R}$ the affine and projective spaces $\mathbb{A}^n(\mathbb{R})$ and $\mathbb{P}^n(\mathbb{R})$ with their affine (resp. projective) topology are topological (or differentiable) manifolds of dimension n. When $K = \mathbb{C}$, on the other hand, the algebraic dimension is half the topological dimension of the associated manifold. See also Remark 1.5.4 on the finiteness of the dimension.

[2] A homogeneous polynomial does not define a map to K but rather a section of a certain "K-bundle", see Exercise 2.6.15.

Algebraic and Quasi-algebraic Sets

Definition 1.2.7 A set F is an *algebraic set* over K if it is a Zariski closed subset in $\mathbb{A}^n(K)$ or in $\mathbb{P}^n(K)$ for some integer n. A set U is a *quasi-algebraic set* over K if it is a Zariski open subset of an algebraic set over K. We will say that an open subset of an affine algebraic set is *quasi-affine* and that an open subset of a projective algebraic set is *quasi-projective*.

Remark 1.2.8 In other words, an algebraic set F over K is a closed subset of either affine space $\mathbb{A}^n(K)$ or projective space $\mathbb{P}^n(K)$. We will emphasise the unusual topology being used by saying that F is *Zariski closed* in $\mathbb{A}^n(K)$ (or $\mathbb{P}^n(K)$).

Exercise 1.2.9 A set U is quasi-algebraic over K if and only if it is a subset of $\mathbb{A}^n(K)$ or $\mathbb{P}^n(K)$ satisfying one of the following equivalent conditions:

1. U is *locally closed* in the Zariski topology, i.e. U is the intersection of a closed subset and an open subset,
2. U is an open subset of its Zariski closure.

Definition 1.2.10 Let $f\colon K^n \to K$ be a function. The *vanishing locus* (or *zero-set*) of f is defined by

$$\mathcal{Z}(f) := \{x \in K^n \mid f(x) = 0\}$$

and the *non-vanishing locus* of f is defined by

$$\mathcal{D}(f) := \{x \in K^n \mid f(x) \neq 0\} .$$

Remark 1.2.11 If the function f in the above definition is polynomial then $\mathcal{Z}(f)$ is Zariski-closed in $\mathbb{A}^n(K)$ and $\mathcal{D}(f)$ is Zariski-open in $\mathbb{A}^n(K)$.

Definition 1.2.12 Let K be a field and let $I \subset K[X_1, \ldots, X_n]$ be an ideal. The *zero-set* of I is denoted by

$$\mathcal{Z}(I) := \mathcal{Z}_K(I) = \{x \in K^n \mid \forall f \in I, f(x) = 0\} .$$

More generally, if L is an extension of K then the zero-set in L^n of the ideal I is denoted by

$$\mathcal{Z}_L(I) := \{x \in L^n \mid \forall f \in I, f(x) = 0\}$$

and the ideal of $L[X_1, \ldots, X_n]$ generated by I is denoted by I_L.

In particular, if I is an ideal in $\mathbb{R}[X_1, \ldots, X_n]$ then $\mathcal{Z}(I) = \mathcal{Z}_{\mathbb{R}}(I)$ is the set of real zeros of the ideal I and $\mathcal{Z}_{\mathbb{C}}(I) = \mathcal{Z}_{\mathbb{C}}(I_{\mathbb{C}})$ is the set of its complex zeros.

Remarks 1.2.13 (*Zero sets of ideals*)

1. The ideal in $L[X_1, \ldots, X_n]$ generated by I turns out to be isomorphic to the tensor product
$$I_L = I \otimes_{K[X_1,\ldots,X_n]} L[X_1, \ldots, X_n] \,.$$
(see Proposition A.4.1 defining the tensor product $\otimes$)
2. As in Remark 1.2.11, $\mathcal{Z}(I)$ is obviously Zariski closed in $\mathbb{A}^n(K)$ and $\mathcal{Z}_L(I)$ is Zariski closed in $\mathbb{A}^n(L)$.
3. We have that $Z_L(I) = Z(I_L)$. In particular, if I is an ideal in $\mathbb{R}[X_1, \ldots, X_n]$ then $I_{\mathbb{C}}$ is an ideal in $\mathbb{C}[X_1, \ldots, X_n]$ and $\mathcal{Z}_{\mathbb{C}}(I) = \mathcal{Z}_{\mathbb{C}}(I_{\mathbb{C}}) = \mathcal{Z}(I_{\mathbb{C}})$. The notation $I_{\mathbb{C}}$ denotes an ideal in $\mathbb{C}[X_1, \ldots, X_n]$ generated by a family of real polynomials.

Exercise 1.2.14 Let K be a field.

1. Let F be a closed subset of $\mathbb{A}^n(K)$. Prove that
$$F = \mathcal{Z}(\mathcal{I}(F)) \,.$$
2. Let $I \subset K[X_1, \ldots, X_n]$ be an ideal. Prove that
$$I \subseteq \mathcal{I}(\mathcal{Z}(I)) \,.$$
3. Let $I \subset K[X_1, \ldots, X_n]$ be an ideal. Prove that if I is not radical (see Definition A.2.3) then
$$I \subsetneq \mathcal{I}(\mathcal{Z}(I)) \,.$$
4. Find an example where $I \subset K[X_1, \ldots, X_n]$ is radical but
$$I \subsetneq \mathcal{I}(\mathcal{Z}(I)) \,.$$
[Hint: the Nullstellensatz (see Corollary A.5.13) tells us that any such example will involve a non-algebraically closed field K.]

Irreducible Algebraic Sets

Definition 1.2.15 We say that a non-empty subset U of a topological space X is *irreducible* if for any pair of closed sets F_1 and F_2 in X such that $U \subset F_1 \cup F_2$ we have that $U \subset F_1$ or $U \subset F_2$. A subset that is not irreducible is said to be *reducible*.

Remark 1.2.16 Requiring an irreducible subspace to be non-empty is a convention which corresponds in commutative algebra to the convention that the zero ring is not an integral domain. See Remark 1.2.31(4).

Exercise 1.2.17 As an exercise, the reader may wish to prove the following statements.

1. A subspace $U \subset X$ is irreducible if and only if it is non-empty and is not the union of two non-empty closed sets (in the induced topology) which are strict subsets of U. In particular, X itself is irreducible if and only if it is non-empty and cannot be written as the union of two closed strict subsets.
2. If U is irreducible then any non-empty open subset of U is dense in U.

Lemma 1.2.18 *Let $\varphi\colon X \to Y$ be a continuous map. The image under φ of any irreducible subspace of X is an irreducible subspace of Y.*

Proof Let $U \subset X$ be irreducible and let $Y_1 \cup Y_2 \supset \varphi(U)$ be a union of two closed sets in Y. The set $\varphi^{-1}(Y_i)$ is then closed in X for $i = 1, 2$ and $\varphi^{-1}(Y_1) \cup \varphi^{-1}(Y_2) \supset U$. As U is irreducible we may assume that $\varphi^{-1}(Y_1) \supset U$ and hence $Y_1 \supset \varphi(U)$. It follows that $\varphi(U)$ is irreducible. □

Remark 1.2.19 Irreducibility is only relevant for relatively coarse topologies such as the Zariski topology. We invite the reader to check that in a Hausdorff topological space the only irreducible subspaces are isolated points.

Definition 1.2.20 The maximal irreducible closed subsets of a topological space U are called its *irreducible components*.

Exercise 1.2.21 (Note that this exercise is immediate once Proposition 1.2.30 has been established.) Let n be a non-zero natural number.

1. Let K be an infinite field. Prove the following statements.
   (a) Affine space $\mathbb{A}^n(K)$ is irreducible.
   (b) Projective space $\mathbb{P}^n(K)$ is irreducible.
2. Suppose that K is finite. Prove that the above spaces are reducible.

Definition 1.2.22 A topological space X is said to be *Noetherian* (or has a *Noetherian* topology) if any decreasing sequence of closed sets stabilises (or alternatively "if any decreasing sequence of closed sets is stationary"). This means that for any sequence $F_1 \supset F_2 \supset \dots$ of closed subspaces there is an integer r such that $F_r = F_{r+1} = \dots$

Example 1.2.23 For any field K the affine space $\mathbb{A}^n(K)$ is Noetherian. Suppose that $F_1 \supset F_2 \supset \dots$ is a decreasing sequence of closed subsets of $\mathbb{A}^n(K)$. We then have that $\mathcal{I}(F_1) \subset \mathcal{I}(F_2) \subset \dots$ is an increasing sequence of ideals of $K[X_1, \dots, X_n]$ which is Noetherian. This sequence of ideals therefore stabilises and it follows that $F_1 \supset F_2 \supset \dots$ stabilises because for every i we have that $F_i = \mathcal{Z}(\mathcal{I}(F_i))$.

Proposition 1.2.24 *Any non-empty quasi-algebraic set U admits a decomposition into a finite number of irreducible components, i.e.*

$$U = U_1 \cup \dots \cup U_m$$

where U_i is irreducible for every i and $U_i \not\subset U_j$ whenever $i \neq j$. This decomposition is unique up to permutation of the components.

Proof This follows from the fact that the Zariski topology is Noetherian. □

Regular Functions

Definition 1.2.25 Let K be a field and let $U \subset \mathbb{A}^n(K)$ be a quasi-algebraic set. A function $f\colon U \to K$ is *polynomial* if there is a $g \in K[X_1, \ldots, X_n]$ such that for every $x \in U$, $f(x) = g(x)$. We denote the K-algebra of polynomial functions on U by $\mathcal{P}(U)$.

Remark 1.2.26 The polynomial g is only determined by the polynomial function f up to addition of an element in $\mathcal{I}(U)$. In particular, if K is finite the ideal of $\mathbb{A}^n(K)$ is not trivial and a polynomial on K^n is not uniquely determined by a polynomial function $K^n \to K$.

Let $F \subset \mathbb{A}^n(K)$ be an algebraic set and let $\mathcal{I}(F)$ be the ideal of elements of $K[X_1, \ldots, X_n]$ which vanish on F. The following proposition enables us to identify $\mathcal{P}(F)$ with the K-algebra of affine coordinates $\mathcal{A}(F) := K[X_1, \ldots, X_n]/\mathcal{I}(F)$.

Proposition 1.2.27 *For any field K and any algebraic set $F \subset \mathbb{A}^n(K)$ the restriction morphism $g \mapsto g|_F$ induces an isomorphism*

$$\mathcal{A}(F) \xrightarrow{\simeq} \mathcal{P}(F).$$

Proof The proof is immediate and is left as an exercise. □

Remark 1.2.28 In particular, if K is infinite then the ring of polynomial functions on $\mathbb{A}^n(K)$ is the ring of polynomials in n variables

$$\mathcal{P}(\mathbb{A}^n(K)) = K[X_1, \ldots, X_n]\,.$$

Remark 1.2.29 Note that for any subset $U \subset K^n$ the quotient ring $K[X_1, \ldots, X_n]/\mathcal{I}(U)$ is reduced because the ideal $\mathcal{I}(U)$ is radical (see Definition A.2.3 and Exercise A.2.4). This "vanishing" multiplicity—which identifies the algebraic sets $V := \{x \in K^n \mid f(x) = 0\}$ and $W := \{x \in K^n \mid f^2(x) = 0\}$[3] for example—is reflected in the differential geometric definition of a manifold as the zero set of a submersion. If K is $\mathbb{C}$ or $\mathbb{R}$ and if f is a submersion at every point of V—and hence at every point of W—then W is a differentiable submanifold of K^n, despite the fact that f^2 is not a submersion at any point of W.

One of our motivations for scheme theory is that it allows us to distinguish V and W by including nilpotent elements in the associated ring. Another illustration of the weakness of the naive point of view is provided by the algebraic set $V := \{(x, y) \in \mathbb{R}^2 \mid x^2 + y^2 = 0\}$, whose ideal $\mathcal{I}(V) = (x, y)$ is strictly larger than the ideal generated by its defining equation $(x^2 + y^2)$. Compare this with Remark 1.2.31(3).

---

[3]Here f^2 is the function whose value at a point is the square of the value of f at that point.

Proposition 1.2.30 *Let K be a field.*

An algebraic set $F \subset \mathbb{A}^n(K)$ is irreducible if and only if its ideal $\mathcal{I}(F)$ is a prime ideal of $K[X_1, \ldots, X_n]$ or in other words if and only if its ring of affine coordinates $\mathcal{A}(F)$ is an integral domain.

An algebraic set $F \subset \mathbb{P}^n(K)$ is irreducible if and only if its homogeneous ideal $\mathcal{I}(F)$ is a prime ideal of the graded ring $K[X_0, \ldots, X_n]$.

Proof This is left as an exercise for the reader. □

Remark 1.2.31 ($\mathcal{I}(\mathcal{Z}(I))$ *vs.* I)

1. If K is algebraically closed and $I \subset K[X_1, \ldots, X_n]$ is a prime ideal then $\mathcal{Z}(I) \subset \mathbb{A}^n(K)$ is an irreducible space.
2. The polynomial $P(x, y) = (x^2 - 1)^2 + y^2 = x^4 - 2x^2 + 1 + y^2$ is irreducible in $\mathbb{R}[x, y]$. Indeed, the rings $\mathbb{R}[x, y] \subset \mathbb{C}[x, y]$ are both factorial and since $P(x, y) = (x^2 - 1 + iy)(x^2 - 1 - iy)$ in $\mathbb{C}[x, y]$ and the polynomials $x^2 - 1 \pm iy$ are irreducible in $\mathbb{C}$ the polynomial P is irreducible in $\mathbb{R}[x, y]$. On the other hand $\mathcal{Z}(P)$ is a reducible subspace of $\mathbb{A}^2(\mathbb{R})$ since
$$\mathcal{Z}(P) = \{(1, 0), (-1, 0)\} = \mathcal{Z}(x - 1, y) \cup \mathcal{Z}(x + 1, y) .$$
3. (See Exercise 1.2.14(4).) The ideal $(x^2 + y^2)$ is prime in $\mathbb{R}[x, y]$ but $\mathcal{I}(\mathcal{Z}(x^2 + y^2)) = (x, y)$. In $\mathbb{C}[x, y]$ the ideal $(x^2 + y^2) = ((x - iy)(x + iy))$ is not prime.
4. The ideal $I = (x^2 + y^2 + 1)$ is prime in $\mathbb{R}[x, y]$ but $\mathcal{Z}(I) = \varnothing \subset \mathbb{A}^n(\mathbb{R})$ is not irreducible since $\mathcal{I}(\mathcal{Z}(I)) = \mathbb{R}[x, y]$ is not a prime ideal in $\mathbb{R}[x, y]$.

Exercise 1.2.32 Following on from Exercise 1.2.14(1): let K be a field. Show that a subset U in $\mathbb{A}^n(K)$ or $\mathbb{P}^n(K)$ is Zariski closed if and only if $U = \mathcal{Z}(\mathcal{I}(U))$.

Definition 1.2.33 Let K be a field and let $U \subset \mathbb{A}^n(K)$ be a quasi-algebraic set. A function $f : U \to K$ is said to be *regular at* a point $x \in U$ if there is a neighbourhood V of x in U and two polynomials $g, h \in K[X_1, \ldots, X_n]$ such that for any $y \in V$, $h(y) \neq 0$ and $f(y) = \frac{g(y)}{h(y)}$.

In the following definition, note that the homogeneous polynomials g and h do not define functions on U but as they have the same degree their quotient $\frac{g}{h}$ is a well-defined function on U.

Definition 1.2.34 Let K be a field and let $U \subset \mathbb{P}^n(K)$ be a quasi-algebraic set. A function $f : U \to K$ is said to be *regular at* a point $x \in U$ if there is a neighbourhood V of x in U and two homogeneous polynomials $g, h \in K[X_0, \ldots, X_n]$ of the same degree such that for any $y \in V$, $h(y) \neq 0$ and $f(y) = \frac{g(y)}{h(y)}$.

Definition 1.2.35 Let U be a quasi-algebraic set over K. A function $f : U \to K$ is said to be *regular* if it is regular at every point in U. We denote by $\mathcal{R}(U)$ the K-algebra of regular functions on U.

Remark 1.2.36 Of course, any polynomial function on an affine algebraic set F over a field K is regular

$$\mathcal{A}(F) \stackrel{\simeq}{\to} \mathcal{P}(F) \hookrightarrow \mathcal{R}(F)$$

but the converse is false if $K = \mathbb{R}$ (see Proposition 1.2.38(1)–(3) below and Exercise 1.2.51(2)).

Exercise 1.2.37 Check that any regular function $f : U \to K$ is continuous with respect to the Zariski topology on U and $K = \mathbb{A}^1(K)$.

Proposition 1.2.38 (Is the algebra $\mathcal{R}$ finitely generated?)

1. *Let K be a field. For any Zariski-closed subset $F \subset \mathbb{A}^n(K)$ the algebra $\mathcal{A}(F)$ is a finitely generated K-algebra*
2. *Similarly, the $\mathbb{C}$-algebra $\mathcal{R}(\mathbb{C}^n) = \mathcal{R}(\mathbb{A}^n(\mathbb{C}))$ is finitely generated*
3. *On the other hand, the $\mathbb{R}$-algebra $\mathcal{R}(\mathbb{R}^n) = \mathcal{R}(\mathbb{A}^n(\mathbb{R}))$ is not finitely generated.*

Proof 1. By definition $\mathcal{A}(F)$ is a quotient ring of the polynomial algebra $K[X_1, \ldots, X_n]$ so it is generated by a finite number of elements, namely the classes of the elements $X_1, \ldots, X_n$.
2. By Theorem 1.2.50, for example, $\mathcal{R}(\mathbb{C}^n)$ is isomorphic to $\mathbb{C}[X_1, \ldots, X_n]$.
3. Let $A_1, \ldots, A_l$ be elements of $\mathcal{R}(\mathbb{R}^n)$. For any $i = 1 \ldots l$ the function $x \mapsto \frac{1}{1+(x-A_i)^2}$ is a regular function on $\mathbb{R}^n$ but does not belong to the algebra generated by $A_1, \ldots, A_l$.

□

Germs of Regular Functions

We refer the reader to Appendix C for basic sheaf theory. The notions of inductive limit (A.1.2) and the stalk of a sheaf (C.3.1) will be particularly important.

Definition 1.2.39 Let K be a field and let U be a quasi-algebraic set over K. The *sheaf of regular functions* $\mathcal{O}_U$ on U is the sheaf of K-algebras whose set of sections over an open set $V \subset U$ is the K-algebra $\mathcal{R}(V)$.

$$\Gamma(V, \mathcal{O}_U) = \mathcal{O}_U(V) := \mathcal{R}(V)\,.$$

Definition 1.2.40 Let K be a field and let U be a quasi-algebraic set over K. The *germ of a K-valued function* on U, regular at x, is an equivalence class of pairs (V, f) where V is an open set of U containing x and f is a regular function on V. Two pairs (V, f) and (W, g) are equivalent if and only if $f = g$ on some neighbourhood of x contained in $V \cap W$.

We denote by $\mathcal{O}_x := \mathcal{O}_{U,x}$ the K-algebra of germs of K-valued functions on U which are regular at the point x. The K-algebras $\mathcal{O}_{U,x}$ are then the stalks of the sheaf $\mathcal{O}_U$:

$$\mathcal{O}_{U,x} = \varinjlim_{V \ni x} \mathcal{O}_U(V)$$

where the inductive limit is taken over all open neighbourhoods V of x contained in U (see Definition C.3.1 and Examples A.1.3 and A.1.4.)

Our definition of germs of regular functions is more "local" than the definition given in [Har77, Section I.3, page 16]. When U is irreducible—which is assumed in [*Ibid.*]—the two definitions coincide.

Definition 1.2.41 Let x be a point in a quasi-algebraic set U defined over a field K and let A be a ring of K-valued functions on U. We denote by $\mathfrak{m}_x := \mathfrak{m}_x^A$ the maximal ideal in A of functions vanishing at x.

We recall that, as in Definition A.3.1, for any prime ideal $\mathfrak{p} \subset A$ we denote by $A_{\mathfrak{p}}$ the localisation of A with respect to the multiplicative set $A \setminus \mathfrak{p}$.

Exercise 1.2.42 Let K be a field. Prove that for any $x \in \mathbb{A}^n(K)$ the algebra $\mathcal{O}_{\mathbb{A}^n(K),x} \subset K(X_1, \ldots, X_n)$ is isomorphic to the algebra $K[X_1, \ldots, X_n]_{\mathfrak{m}_x}$ of fractions $\frac{g}{h}$ such that $h(x) \neq 0$.

Lemma 1.2.43 *A quasi-affine set $U \subset \mathbb{A}^n(K)$ can be seen as a quasi-projective set $j(U) \subset \mathbb{P}^n(K)$ where $j\colon \mathbb{A}^n(K) \hookrightarrow \mathbb{P}^n(K)$ is the inclusion*

$$(x_1, \ldots, x_n) \mapsto (1 : x_1 : \cdots : x_n).$$

The morphism of K-algebras $j^*\colon K[X_0, \ldots, X_n] \to K[X_1, \ldots, X_n]$ which sends $(X_0, X_1, \ldots, X_n)$ to $(1, \ldots, X_n)$ then induces a sheaf isomorphism $j_*\mathcal{O}_U \simeq \mathcal{O}_{j(U)}$.

Proof Left as an exercise for the reader □

Exercise 1.2.44 Let $P \in K[X_1, X_2]$ be a polynomial of degree $d \geqslant 1$ and let $C = \mathcal{Z}(P) \subset \mathbb{A}^2(K)$ be the corresponding plane curve. The *projective completion* $\widehat{C} := \overline{j(C)}^{Zar} \subset \mathbb{P}^2(K)$ is the Zariski closure of $j(C)$ in the projective plane. Prove that $\widehat{C}$ is the set of zeros of the homogenised polynomial $\widehat{P} \in K[X_0, X_1, X_2]$. Here, if $a_{ij}X_1^iX_2^j$ is a monomial in P then $a_{ij}X_0^{d-i-j}X_1^iX_2^j$ is the corresponding monomial in $\widehat{P}$ and $\widehat{C} = \mathcal{Z}(\widehat{P}) \subset \mathbb{P}^2(K)$.

Proposition 1.2.45 *Let $U \subset \mathbb{P}^n(K)$ be a quasi-projective set, let x be a point in U and let $H \subset \mathbb{P}^n(K)$ be a hyperplane not containing x. The open set $V := \mathbb{P}^n(K) \setminus H$ is then a neighbourhood of x isomorphic to affine space. (We will see further on that V is a principal open set, see Exercise 1.2.60). We denote this isomorphism by $j\colon \mathbb{A}^n(K) \xrightarrow{\simeq} V$. The K-algebras $\mathcal{O}_{U,x}$ and $\mathcal{O}_{j^{-1}(U\cap V), j^{-1}(x)}$ are then isomorphic.*

Proof Let (W, f) be a pair representing an element of $\mathcal{O}_{U,x}$. There are then homogeneous polynomials of same degree $g, h \in K[X_0, \ldots X_n]$ such that h does not vanish on the open set W and $f = \frac{g}{h}$ on W. We consider the subset $W' = j^{-1}(W \cap V)$ in $j^{-1}(U \cap V)$. This is a neighbourhood of $j^{-1}(x)$ contained in $j^{-1}(U \cap V)$ and we can assume that j is written in coordinates as $(x_1, \ldots, x_n) \mapsto (1 : x_1 : \cdots : x_n)$. As the polynomials g and h are homogeneous and of same degree the rational function $\frac{g(1,X_1,\ldots,X_n)}{h(1,X_1,\ldots,X_n)}$ is in fact a regular function $j^*(f)$ on W'. The pair $(W', j^*(f))$ represents an element in $\mathcal{O}_{j^{-1}(U\cap V), j^{-1}(x)}$.

Conversely, let $(W, f = g/h)$ be a pair representing an element in

$$\mathcal{O}_{j^{-1}(U\cap V), j^{-1}(x)} .$$

Set $d = \max(\deg g, \deg h)$: denoting by $\widehat{p}$ the degree d homogenisation with respect to X_0 of a polynomial $p \in K[X_1, \ldots X_n]$, the fraction $\widehat{g}/\widehat{h}$ represents an element of $\mathcal{O}_{U,x}$. □

When F is an algebraic set then for any neighbourhood U of x in F we can identify $\mathcal{O}_{F,x}$ with the localisation of $\mathcal{R}(U)$ with respect to the maximal ideal $\mathfrak{m}_x^{\mathcal{R}}$ of regular functions vanishing at x by the following proposition.

Proposition 1.2.46 *Let K be a field and let F be an algebraic set over K. Let x be a point in F and let U be a neighbourhood of x in F. There is then a natural isomorphism*

$$\mathcal{O}_{F,x} \simeq \mathcal{R}(U)_{\mathfrak{m}_x^{\mathcal{R}}} .$$

If moreover F is affine then we also have a natural isomorphism

$$\mathcal{O}_{F,x} \simeq \mathcal{P}(U)_{\mathfrak{m}_x^{\mathcal{P}}} .$$

Proof By Proposition 1.2.45 we can assume in the proof of the first part of this proposition that $F \subset \mathbb{A}^n(K)$ is affine. By the natural map $\mathcal{R}(U) \to \mathcal{O}_{F,x}$ sending f to the class of the pair $(U \cap \mathcal{D}(f), f)$ the image of a function which is non zero at x is invertible in $\mathcal{O}_{F,x}$. Indeed, if $f \notin \mathfrak{m}_x^{\mathcal{R}}$ then $\frac{1}{f}$ is regular on the neighbourhood $U \cap \mathcal{D}(f)$ of x. By the universal property of localisations (Proposition A.3.2) this map induces a surjective map $\beta\colon \mathcal{R}(U)_{\mathfrak{m}_x^{\mathcal{R}}} \to \mathcal{O}_{F,x}$. Indeed, consider an element in $\mathcal{O}_{F,x}$ represented by (U, f). There is a neighbourhood $V \subset U$ de x and polynomials $g, h \in K[X_1, \ldots, X_n]$ such that h does not vanish on V and $\frac{g}{h} = f$ on V. The rational function $\frac{g}{h}$ represents an element of $\mathcal{R}(U)_{\mathfrak{m}_x^{\mathcal{R}(U)}}$ whose image under β is equivalent to f (see the solution to Exercise 1.2.42). We now prove that β is injective. If the image of $\frac{f}{g}$ vanishes then f vanishes in a neighbourhood $V \subset U$ of x. If U is irreducible this implies that f vanishes on U. Otherwise, decompose $U = \cup U_j$ into irreducible components. For each component of U_j containing x the function f vanishes on $V \cap U_j$ and therefore vanishes on U_j. Let W be the union of all components of U not containing x. There is a $h \in \mathcal{I}(W)$ such that $h(x) \neq 0$. It follows that the function hf vanishes on U and since $h \notin \mathfrak{m}_x^{\mathcal{R}}$ it is an invertible element of $\mathcal{R}(U)_{\mathfrak{m}_x^{\mathcal{R}}}$. It follows that $\frac{f}{1}$ vanishes in $\mathcal{R}(U)_{\mathfrak{m}_x^{\mathcal{R}}}$.[4]

The proof when F is affine is identical, except that we replace $\mathcal{R}(U)_{\mathfrak{m}_x^{\mathcal{R}}}$ by $\mathcal{P}(U)_{\mathfrak{m}_x^{\mathcal{P}}}$. □

---

[4]If K is infinite, this can also be proved using the avoidance Lemma A.3.12.

Corollary 1.2.47 *Under the hypotheses of Proposition 1.2.46 we have that*

$$\mathcal{O}_{F,x} \simeq \mathcal{O}_{U,x} \simeq \mathcal{R}(U)_{\mathfrak{m}_x^{\mathcal{R}(U)}} \simeq \mathcal{R}(F)_{\mathfrak{m}_x^{\mathcal{R}(F)}}$$

and if F is affine,

$$\mathcal{O}_{F,x} \simeq \mathcal{P}(U)_{\mathfrak{m}_x^{\mathcal{P}(U)}} \simeq \mathcal{P}(F)_{\mathfrak{m}_x^{\mathcal{P}(F)}}.$$

Example 1.2.48 If $F \subset \mathbb{A}^n(K)$ is an affine algebraic set then any element of $\mathcal{O}_{F,x}$ is represented by a fraction $\frac{g}{h}$ where $g \in K[X_1, \ldots, X_n]/\mathcal{I}(F)$, $h \in K[X_1, \ldots, X_n]/\mathcal{I}(F)$ and $h(x) \neq 0$.

Proposition 1.2.49 *Let U be a quasi-algebraic set and let x be a point of U. The point x belongs to a unique component of U if and only if the local ring $\mathcal{O}_x$ is an integral domain. More generally, $\mathcal{O}_x$ is a reduced ring whose minimal prime ideals can be identified with the irreducible components of U passing through x.*

Proof Let V be a neighbourhood of x in U. The ring $\mathcal{R}(V)$ is reduced so its local rings are also reduced. The prime ideals of $\mathcal{R}(V)_{\mathfrak{m}_x}$ correspond to prime ideals of $\mathcal{R}(V)$ contained in $\mathfrak{m}_x$. The minimal prime ideals of $\mathcal{R}(V)_{\mathfrak{m}_x}$ then correspond to irreducible components of V containing x. Indeed, the maps $I \mapsto \mathcal{Z}(I)$ and $F \mapsto \mathcal{I}(F)$ provide a bijection between prime ideals and irreducible subvarieties and are strictly decreasing for inclusion. The conclusion follows because any reduced ring with only one minimal prime ideal is an integral domain. This can be proved as follows—if a ring A contains only one minimal prime ideal then it must be equal to the intersection I of all the prime ideals of A. It follows that the ideal I is prime, but I is the radical $\sqrt{(0)}$ of the zero ideal in A, see [Eis95, Corollary 2.12], which is equal to (0) if A is reduced. The zero ideal of A is therefore prime, or in other words A is integral. □

When F is an affine algebraic set and K is an algebraically closed field we usually identify regular functions (locally defined) and polynomial functions (globally defined) using the following proposition.

Theorem 1.2.50 (K algebraically closed) *If K is algebraically closed and F is Zariski closed in $\mathbb{A}^n(K)$ then the injection from $\mathcal{P}(F)$ to $\mathcal{R}(F)$ is a bijection.*

$$\mathcal{A}(F) \simeq \mathcal{P}(F) \simeq \mathcal{R}(F).$$

Proof By hypothesis F is algebraic and there is a canonical morphism

$$\iota\colon K[X_1, \ldots, X_n]/\mathcal{I}(F) \to \mathcal{O}_F(F) = \mathcal{R}(F)$$

which is injective by definition of $\mathcal{I}(F)$. By Proposition 1.2.46 we can identify $\mathcal{O}_{F,x}$ with the ring of fractions of $\mathcal{P}(F)$ with respect to the ideal $\mathfrak{m}_x$ of polynomials which vanish at x.

We will assume that F is irreducible—see [Ser55a, Cor. 3, page 237] for the general case. In this case, $\mathcal{P}(F)$ is an integral domain and the rings $\mathcal{O}_{F,x}$ can be considered as subrings of the field of fractions Frac $\mathcal{P}(F)$, see Definition A.3.8. We then have that

$$\Gamma(F, \mathcal{O}_F) = \bigcap_{x \in F} \mathcal{O}_{F,x} \,. \tag{1.2.1}$$

Any maximal ideal of $\mathcal{P}(F)$ is equal to some $\mathfrak{m}_x$ by Hilbert's Nullstellensatz (Theorem A.5.12). It follows immediately that $\mathcal{P}(F) = \bigcap_{x \in F} \mathcal{O}_{F,x} = \Gamma(F, \mathcal{O}_F)$ by Proposition A.3.11 and Equation (1.2.1). □

Exercise 1.2.51 1. The hypothesis that F is Zariski closed is necessary.
The function $f\colon K^2 \setminus \mathcal{Z}(x^2 + y^2 + 1) \to K,\ (x, y) \mapsto \frac{1}{x^2+y^2+1}$ is a regular function that is not the restriction of any polynomial function on K^2.
2. The hypothesis that K is algebraically closed is necessary.
The function $g\colon \mathbb{R}^2 \to \mathbb{R},\ (x, y) \mapsto \frac{1}{x^2+y^2+1}$ is a regular function not given by a polynomial function on $\mathbb{R}^2$. This follows from the fact that it does not define a regular complex function on $\mathbb{C}^2$.

This last example shows that when $K = \mathbb{R}$ the correspondence given in Theorem 1.2.50 between polynomial and regular functions no longer holds. In this example, $(x^2 + y^2 + 1)$ is a maximal ideal in $\mathbb{R}[x, y]$ which does not correspond to any point in $\mathbb{R}^2$.

We cannot develop the theory of algebraic varieties over a non algebraically closed field exactly as in [Ser55a] because of the non surjectivity of the map $\mathcal{P}(F) \hookrightarrow \mathcal{R}(F)$. However, over $\mathbb{R}$ (and more generally over any real closed field, see Definition A.5.18) we do still have a global characterisation of regular functions on quasi-affine algebraic sets. The algebra in question is now an algebra of rational functions, rather than an algebra of polynomials, and is no longer finitely generated—see Proposition 1.2.38—but the result has the advantage of applying to the quasi-algebraic case. This is a very useful result, notably when applied to a principal open set of the form $U = \mathcal{D}(f)$ for some $f \in \mathcal{P}(F)$—see Exercise 1.2.60(1) and Definition 1.3.14, especially since over $\mathbb{R}$ any Zariski open subset is principal—see Proposition 1.2.61.

Theorem 1.2.52 (K a real closed field) *Let U be a quasi-algebraic set which is an open subset of a closed set $F \subset \mathbb{A}^n(\mathbb{R})$. The injection from the localisation of $\mathcal{P}(F)$ with respect to the multiplicative system $\mathcal{S}_U := \{h \in \mathcal{P}(F) \mid \forall x \in U, h(x) \neq 0\}$ into $\mathcal{R}(U)$ is a bijection*

$$\mathcal{S}_U^{-1}\mathcal{P}(F) \simeq \mathcal{R}(U).$$

In particular, any regular function $f : U \to \mathbb{R}$ is the restriction of a global rational function defined at any point of U. In other words, there are polynomial functions $g, h \in \mathcal{P}(F)$ such that h does not vanish at any point of U and for all $x \in U$, $f(x) = g(x)/h(x)$.

Proof Let $g \in \mathcal{P}(F)$ and $h \in \mathcal{P}(F)$ be functions defined on U and assume that h does not vanish at any point of U. The function g/h is then clearly regular everywhere on U. We now show that any regular function $U \to \mathbb{R}$ is of this form. Consider an element $f \in \mathcal{R}(U)$. As U is quasi-compact (see Definition B.1.5) in the Zariski topology, there is a finite covering of U by Zariski open sets $\bigcup_{i=1}^{l} U_i = U$ and polynomial functions $g_i, h_i \in \mathcal{P}(F), h_i(x) \neq 0 \, \forall x \in U_i$ such that $f|_{U_i} = g_i/h_i$. Let $s_i \in \mathcal{P}(F)$ be such that $F \setminus U_i = \{x \in F \mid s_i(x) = 0\}$. In other words $U_i = \mathcal{D}(s_i)$. The function $h := \sum_{i=1}^{l} s_i^2 h_i^2$ is then a polynomial function on F which does not vanish on U. It will now be enough to prove that

$$f = \frac{\sum_{i=1}^{l} s_i^2 g_i h_i}{h}$$

Consider a point $x \in U$ and let $J_x \subset \{1, \dots l\}$ be the set of points such that $x \in U_i$. For any $i \in J_x$ we then have that $f(x) = g_i(x)/h_i(x)$ and for any $i \notin J_x$ we then have that $s_i(x) = 0$. In particular

$$\left(\frac{\sum_{i=1}^{l} s_i^2 g_i h_i}{h}\right)(x) = \frac{\sum_{i \in J_x} s_i^2(x) g_i(x) h_i(x)}{\sum_{i \in J_x} s_i^2(x) h_i^2(x)} = \frac{\sum_{i \in J_x} s_i^2(x) h_i^2(x) \frac{g_i(x)}{h_i(x)}}{\sum_{i \in J_x} s_i^2(x) h_i^2(x)} .$$

We choose an index $i_0 \in J_x$. For any $i \in J_x$ we then have that $f(x) = g_i(x)/h_i(x) = g_{i_0}(x)/h_{i_0}(x)$ and

$$\left(\frac{\sum_{i=1}^{l} s_i^2 g_i h_i}{h}\right)(x) = \frac{\sum_{i \in J_x} s_i^2(x) h_i^2(x) \frac{g_{i_0}(x)}{h_{i_0}(x)}}{\sum_{i \in J_x} s_i^2(x) h_i^2(x)} = g_{i_0}(x)/h_{i_0}(x) = f(x) .$$

□

In Corollary 1.2.66 below we will see the extent to which the ring of regular functions of an affine algebraic set over $\mathbb{R}$ or $\mathbb{C}$ characterises the algebraic set. Over $\mathbb{R}$ this result also holds for a projective algebraic set since by Proposition 1.2.63 any such set is affine. On the other hand, the following theorem shows that the ring of regular functions of a complex projective set is as simple as possible.

Theorem 1.2.53 *Let F be a projective algebraic set over a base field K. If K is algebraically closed and F is irreducible then the only regular functions on F are the constants, i.e.*

$$\mathcal{R}(F) \simeq K .$$

Proof Assume that $F \subset \mathbb{P}^n(K)$ and consider the K-algebra of homogeneous coordinates $\mathcal{S}(F) := K[X_0, \dots, X_n]/\mathcal{I}(F)$ (see Definition 1.2.4) which is an integral domain because F is irreducible. For any $N \geqslant 0$ let $\mathcal{S}(F)_N$ be the K-vector space of homogeneous polynomials of degree N and for any $i = 0, \dots, n$ let x_i be the image in $\mathcal{S}(F)$ of X_i. Using Proposition 1.2.45 we freely identify the complement of a hyperplane in $\mathbb{P}^n(K)$ with affine space $\mathbb{A}^n(K)$. Let $f \in \mathcal{R}(F)$ be a regular function on the whole of F. For any $U_i := \mathcal{D}(x_i)$ the function f is regular on $U_i \cap F$ and since K is algebraically closed, $f \in \mathcal{A}(U_i \cap F) \simeq \mathcal{S}(F)_{(x_i)}$ by Theorem 1.2.50. It follows that there is a natural number N_i and a homogeneous polynomial function $g_i \in \mathcal{S}(F)_{N_i}$ such that

$$f = \frac{g_i}{x_i^{N_i}} .$$

Considering the rings $\mathcal{R}(F)$ and $\mathcal{S}(F)$ as subrings of the field of fractions $\operatorname{Frac} \mathcal{S}(F)$ (not to be confused with its subfield $K(F)$—see Definition 1.2.69), we see that for all i the element $x_i^{N_i} f$ is homogeneous of degree N_i. Set $N \geqslant \sum_{i=0}^{n} N_i$. The K-vector space $\mathcal{S}(F)_N$ is generated by monomials of degree N in the variables $x_0, \ldots, x_n$ and in each of these monomials at least one of the variables x_i appears with an exponent that is larger than N_i. In particular, for any homogeneous polynomial $h \in \mathcal{S}(F)_N$ we have that $hf \in \mathcal{S}(F)_N$. Iterating we get that for any $q > 0$, $f^q \mathcal{S}(F)_N \subset \mathcal{S}(F)_N$. In particular $x_0^N f^q \in \mathcal{S}(F)_N$ for any $q > 0$. The subring $\mathcal{S}(F)[f]$ of $\operatorname{Frac} \mathcal{S}(F)$ is therefore contained in $x_0^{-N} \mathcal{S}(F)$ which is a finitely generated $\mathcal{S}(F)$-module. It follows that f is integral over $\mathcal{S}(F)$ (see Definition A.5.1) or in other words there exist elements $a_1, \ldots, a_m \in \mathcal{S}(F)$ such that

$$f^m + a_1 f^{m-1} + \cdots + a_m = 0 . \tag{1.2.2}$$

Since f is of degree 0, equation (1.2.2) still holds if we replace each of the a_is by their homogeneous degree 0 components. But now $\mathcal{S}(F)_0 = K$ so for every $i = 1 \ldots m$ we have that $a_i \in K$ and f is algebraic over K. Since K is algebraically closed, $f \in K$. □

Regular Maps and Morphisms of Algebraic Sets

Definition 1.2.54 Let V and W be quasi-algebraic sets over K. A *morphism of quasi-algebraic sets* (or *regular map*) $\varphi \colon V \to W$ is a continuous map (with respect to the Zariski topologies) such that for any open set $U \subset W$ and any regular function $f \colon U \to K$ the function $f \circ \varphi \colon \varphi^{-1}(U) \to K$ is regular.

A map $\varphi \colon V \to W$ is an *isomorphism of quasi-algebraic sets* if φ is a homeomorphism and φ and φ^{-1} are regular maps.

Remark 1.2.55 From the sheaf theoretic point of view (see Example C.5.3 in Appendix C) the map φ is regular if and only if it is continuous and the image of the pull back map $\varphi^{\#} \colon \mathcal{O}_W \to \varphi_* \mathcal{F}_V$ (a morphism of sheaves on W) is contained in $\varphi_* \mathcal{O}_V$. A map $\varphi \colon V \to W$ is an isomorphism of quasi-algebraic sets if and only if φ is a homeomorphism and the induced maps of sheaves $\varphi^{\#} \colon \mathcal{O}_W \xrightarrow{\simeq} \varphi_* \mathcal{O}_V$ is an isomorphism of K-algebra sheaves.

Exercise 1.2.56 (*Morphisms and polynomial functions*)

1. Let $F_1 \subset \mathbb{A}^n(K)$ and $F_2 \subset \mathbb{A}^m(K)$ be two algebraic sets over the same algebraically closed field K. Using Theorem 1.2.50, prove that a map $\varphi \colon F_1 \to F_2$

is a morphism if and only if there are polynomial functions $f_1, \ldots, f_m \in K[x_1, \ldots, x_n]$ such that for every point $(x_1, \ldots, x_n) \in F_1$,

$$\varphi(x_1, \ldots, x_n) = (f_1(x_1, \ldots, x_n), \ldots, f_m(x_1, \ldots, x_n)) .$$

2. Let $F_1 \subset \mathbb{A}^n(\mathbb{R})$ and $F_2 \subset \mathbb{A}^m(\mathbb{R})$ be algebraic sets over the same real closed field. Using Theorem 1.2.52 prove that a map $\varphi\colon F_1 \to F_2$ is a morphism if and only if there are polynomial functions $g_1, \ldots, g_m \in \mathbb{R}[x_1, \ldots, x_n]$ and $h_1, \ldots, h_m \in \mathbb{R}[x_1, \ldots, x_n]$ such that for every point $(x_1, \ldots, x_n) \in F_1$, $h_1(x_1, \ldots, x_n) \neq 0, \ldots, h_m(x_1, \ldots, x_n) \neq 0$ and

$$\varphi(x_1, \ldots, x_n) = \left(\frac{g_1(x_1, \ldots, x_n)}{h_1(x_1, \ldots, x_n)}, \ldots, \frac{g_m(x_1, \ldots, x_n)}{h_m(x_1, \ldots, x_n)} \right) .$$

3. Let $F_1 \subset \mathbb{P}^n(K)$ and $F_2 \subset \mathbb{P}^m(K)$ be algebraic sets over the same algebraically closed base field K. A map $\varphi\colon F_1 \to F_2$ is a morphism if and only if there exist homogeneous polynomials $f_0, \ldots, f_m \in K[x_0, \ldots, x_n]$ without common zeros such that for all $x = (x_0 : \cdots : x_n) \in F_1$,

$$\varphi(x) = (f_0(x_0, \ldots, x_n) : \cdots : f_m(x_0, \ldots, x_n)) .$$

4. The projective real case follows from the affine real case using Proposition 1.2.63.

Example 1.2.57 Note that the image of a quasi-algebraic set under an algebraic morphism is not necessarily quasi-algebraic. The image $B \subset \mathbb{A}^2(K)$ of the affine plane under the map $\mathbb{A}^2(K) \to \mathbb{A}^2(K), (x, y) \mapsto (xy, y)$ is a union of the point $(0, 0)$ and the complement of the line $y = 0$. It is neither open nor closed in $\mathbb{A}^2(K)$. The set B, which is the image under an algebraic map of an affine algebraic set is neither algebraic nor quasi-algebraic, but only *constructible*. See [Har77, Exercise II.3.18 & 3.19] for an introduction to this notion.

Exercise 1.2.58 (*Quasi-algebraic sets*)

1. Prove that the quasi-algebraic set $\mathbb{C}^* \subset \mathbb{C}$ is isomorphic to an algebraic set.
2. Similarly, prove that the groups $\mathbf{GL}_n(\mathbb{C})$, which are open subsets of the spaces $\mathcal{M}_n(\mathbb{C})$, are affine algebraic sets. These groups are *algebraic groups*.
3. Prove the same results over an arbitrary base field K.

Exercise 1.2.59 (*Affine algebraic sets*)

1. Let K be a field and let $H \subset \mathbb{P}^n(K)$ be a hypersurface. Prove that the complement $\mathbb{P}^n(K) \setminus H$ is isomorphic to an affine space.
2. Let K be an algebraically closed field. Prove that the quasi-affine set $U := \mathbb{A}^2(K) \setminus \{(0, 0)\}$ is not isomorphic to an affine algebraic set.
[Hint: prove that any regular function on U extends to a regular function on the whole of $\mathbb{A}^2(K)$. If the base field is $\mathbb{C}$ this result is a corollary of Hartog's theorem [GH78, page 7] on extending holomorphic functions of two variables.]

3. Let K be an algebraically closed field. Prove that the only irreducible algebraic set which is both affine over K and isomorphic to a projective algebraic set over K is a point.

Exercise 1.2.60 (*Principal open sets*) Let K be an algebraically closed field (such as $\mathbb{C}$) or a real closed field—see Definition A.5.18—(such as $\mathbb{R}$). Let n be a non-zero natural number and let $F \subset \mathbb{A}^n(K)$ be an algebraic set.

1. Using Theorems 1.2.50 and 1.2.52, prove that if $f \in \mathcal{P}(F)$ is a polynomial function then the set $\mathcal{D}(f)$ of points where f does not vanish is isomorphic to an affine algebraic set and there is an isomorphism of K-algebras.

$$\mathcal{O}_F(\mathcal{D}(f)) \simeq \mathcal{R}(F)[\frac{1}{f}]\,.$$

When K is algebraically closed there is in fact an isomorphism

$$\mathcal{O}_F(\mathcal{D}(f)) \simeq \mathcal{A}(F)[\frac{1}{f}]\,.$$

2. Consider the function $f\colon \mathbb{A}^2(\mathbb{R}) \to \mathbb{R}$, $(x, y) \mapsto x^2 + y^2$. Give an affine algebraic set which is isomorphic to $\mathcal{D}(f)$. Compare with Exercise 1.2.59(2).

The following three results illustrate important differences between real and complex algebraic varieties. We invite the reader to compare them with (2) and (3) of Exercise 1.2.59. The first property, very different from complex case, follows from the fact that any real algebraic set can be defined by a principal ideal.

Proposition 1.2.61 *Any open set in a real affine algebraic set is principal.*

Corollary 1.2.62 *Any open set in a real affine algebraic set is isomorphic to a real affine algebraic set.*

Proof For any real algebraic set $F \subset \mathbb{A}^n(\mathbb{R})$ and any open set $U \subset F$ consider $F' = F \setminus U$ and take a set $\{P_1, \ldots, P_l\}$ of generators of $\mathcal{I}(F')$. We have that $F' = \mathcal{Z}(P)$ where $P \in \mathbb{R}[X_1, \ldots, X_n]$ is the polynomial $P = P_1^2 + P_2^2 + \cdots + P_l^2$. It follows that the open set U in $\mathbb{A}^n(\mathbb{R})$ is isomorphic to the following closed set in $\mathbb{A}^{n+1}(\mathbb{R})$

$$\left\{(x_1, \ldots, x_n, y) \in \mathbb{A}^{n+1}(\mathbb{R}) \mid (x_1, \ldots, x_n) \in F \text{ et } yP(x_1, \ldots, x_n) = 1\right\}\,.$$

□

Proposition 1.2.63 *Any real projective algebraic variety is isomorphic to a real affine algebraic set.*

Proof Simply note that $\mathbb{P}^n(\mathbb{R})$ is isomorphic to the real algebraic variety $\mathbb{P}^n(\mathbb{R}) \setminus H$, where H is the hypersurface in $\mathbb{P}^n(\mathbb{R})$ whose equation is $x_0^2 + \cdots + x_n^2 = 0$. The space $\mathbb{P}^n(\mathbb{R})$ is therefore an affine algebraic set by Exercise 1.2.59(1). □

Theorem 1.2.64 *Let F and F' be algebraic sets over the same base field K. If F' is affine then there is a natural bijection*

$$\alpha\colon \operatorname{Hom}(F, F') \xrightarrow{\cong} \operatorname{Hom}(\mathcal{A}(F'), \mathcal{R}(F))$$

where on the left hand side Hom *represents the set of morphisms of algebraic sets, and on the right hand side* Hom *represents the set of morphisms of K-algebras.*

If the field K is real closed or algebraically closed then there is a natural bijection

$$\beta\colon \operatorname{Hom}(\mathcal{R}(F'), \mathcal{R}(F)) \xrightarrow{\cong} \operatorname{Hom}(\mathcal{A}(F'), \mathcal{R}(F))\ .$$

We will see a generalisation of this result to abstract varieties in Theorem 1.3.18.

Proof We start by defining the map α. Any morphism of algebraic sets $\varphi\colon F \to F'$ induces a morphism of K-algebras $\varphi^*\colon \mathcal{R}(F') \to \mathcal{R}(F), \varphi^*(f) = f \circ \varphi$. We set $\alpha(\varphi) := \varphi^*|_{\mathcal{A}(F')}$. Conversely let $h\colon \mathcal{A}(F') \to \mathcal{R}(F)$ be a morphism of K-algebras. We can assume that F' is a closed set in $\mathbb{A}^N(K)$ and that $\mathcal{A}(F') = K[y_1, \ldots, y_N]/\mathcal{I}(F')$. Let the element $\xi_i \in \mathcal{R}(F)$ be the image under h of the class in $\mathcal{A}(F')$ of the polynomial function y_i. For any i, this is a globally defined functions on F. Using these we can construct a function $\psi\colon F \to \mathbb{A}^N(K)$ given by $\psi(x) = (\xi_1(x), \ldots, \xi_N(x))$ for any $x \in F$. Since $F' = \mathcal{Z}(\mathcal{I}(F'))$ and h is a morphism of K-algebras it is immediate that $\psi(x) \in F'$ for any $x \in F$. Since each of the components ξ_i of ψ is regular on F the map ψ is easily seen to be a morphism and by construction $h \mapsto \psi$ is the inverse of the bijection α.

We define the map β using the injection $\mathcal{A}(F') \hookrightarrow \mathcal{R}(F')$ (see Remark 1.2.36) sending the K-algebra of affine coordinates into the K-algebra of regular functions. Let $h\colon \mathcal{R}(F') \to \mathcal{R}(F)$ be a morphism of K-algebras and set $\beta(h) := h|_{\mathcal{A}(F')}$. By Theorem 1.2.50 (if K is algebraically closed) or Theorem 1.2.52 (if K is real closed) every element of $\mathcal{R}(F')$ is represented by a global rational function and it easily follows that the map h is determined by its values on $\mathcal{A}(F')$. □

The following corollaries are immediate. The first of them forms the basis for the dictionary between algebraic geometry and commutative algebra. We will see similar results for abstract varieties in Corollary 1.3.19.

Corollary 1.2.65 *Let F and F' be affine algebraic sets over the same base field K. If K is algebraically closed then F is isomorphic to F' if and only if the K-algebras $\mathcal{A}(F)$ and $\mathcal{A}(F')$ are isomorphic.*

Corollary 1.2.66 *Let F and F' be affine algebraic sets over the same base field K. If K is either real closed or algebraically closed then F is isomorphic to F' if and only if the K-algebras $\mathcal{R}(F)$ and $\mathcal{R}(F')$ are isomorphic.*

Remark 1.2.67 If we consider polynomial morphisms instead of regular morphisms then we obtain the same result over $\mathbb{R}$ as over $\mathbb{C}$—the set F is polynomially isomorphic to F' if and only if the $\mathbb{R}$-algebras $\mathcal{A}(F)$ and $\mathcal{A}(F')$ are isomorphic. See [CLO15, Chapter V, Section 4, Proposition 8].

Exercise 1.2.68 (*Conics*) The full definition of an affine or projective plane curve will be given in Definition 1.6.1. For now we define a *conic* to be a degree two plane curve.[5] An *affine conic* is thus given by a degree 2 polynomial in two variables. A *projective conic* is given by a degree 2 homogeneous polynomial in three variables. If the zero locus of a *conic* is non singular and non empty then it is a 1-dimensional variety (see Definitions 1.5.9 and 1.5.43). Abusing notation, the zero set of a conic will also be called a conic. We will be careful to keep track of information contained in the equation but lost on passing to the zero locus. For example, the equation $(x + y - 1)^2$ is a conic called the *double line* but it has the same zero set as the equation $(x + y - 1)$ which is not a conic.

1. Let P be an irreducible polynomial of degree 2. The set $\mathcal{Z}(P)$ is then either empty or irreducible. (See Definition 1.2.15).
2. Calculate the rings of affine coordinates of the conics given by the equations $y = x^2$ and $xy = 1$ and show that they are not isomorphic.
3. Assume the base field K is *algebraically closed.*
   (a) Prove that any irreducible conic in $\mathbb{P}^2(K)$ is isomorphic to $\mathbb{P}^1(K)$.
   (b) Prove that $\mathbb{A}^1(K)$ is not isomorphic to $\mathbb{A}^1(K) \setminus \{0\}$.
   (c) Prove that any irreducible conic in $\mathbb{A}^2(K)$ is isomorphic to $\mathbb{A}^1(K)$ or $\mathbb{A}^1(K) \setminus \{0\}$.
4. Assume that $K = \mathbb{C}$.
   (a) Classify up to isomorphism the (possibly reducible) conics in $\mathbb{P}^2(\mathbb{C})$.
   (b) Classify up to isomorphism the (possibly reducible) conics in $\mathbb{A}^2(\mathbb{C})$.
5. Assume that $K = \mathbb{R}$.
   (a) Construct two degree 2 irreducible polynomials defining non-isomorphic conics in $\mathbb{P}^2(\mathbb{R})$ and prove that any conic in $\mathbb{P}^2(\mathbb{R})$ defined by an irreducible polynomial is isomorphic to one of them.
   (b) Construct four degree 2 irreducible polynomials defining pairwise non-isomorphic conics in $\mathbb{A}^2(\mathbb{R})$ and prove that any conic in $\mathbb{A}^2(\mathbb{R})$ defined by an irreducible polynomial is isomorphic to one of them.
   (c) Classify up to isomorphism the (possibly reducible) conics in $\mathbb{P}^2(\mathbb{R})$.
   (d) Classify up to isomorphism the (possibly reducible) conics in $\mathbb{A}^2(\mathbb{R})$.

Rational Functions

Definition 1.2.69 Let K be a field and let U be a quasi-algebraic set over K. The *K-algebra of rational functions* of U is the K-algebra

$$K(U) = \varinjlim_{\overline{V}=U} \mathcal{O}_U(V)$$

[5] More generally, a *quadric* over a base field K is an equivalence classes of degree 2 polynomials with coefficients in K, where polynomials P and Q are declared to be *equivalent* if there is a $\lambda \in K^*$ such that $P = \lambda Q$.

where the limit is taken over all dense open sets in U. An element of $K(U)$ is therefore an equivalence class of pairs (V, f), where V is a dense open set in U, f is a regular function on V, and we identify two pairs (V, f) and (W, g) if and only if $f = g$ on some dense open set contained in $V \cap W$. Elements of $K(U)$ are called *rational functions* on U.

Remark 1.2.70 As $K(U)$ is an inductive limit of K-algebras it is a K-algebra.

Proposition 1.2.71 *For any dense open set V in U the natural map*

$$K(U) \xrightarrow{\simeq} K(V)$$

is an isomorphism.

Proof By definition of the inductive limit we can take the limit over dense open sets contained in V. □

Proposition 1.2.72 *Let U be a quasi-algebraic set over a field K. If U is irreducible then $K(U)$ is a field.*

Proof If (V, f) represents an element of $K(U)$ with $f \neq 0$ then we can restrict f to the non-empty open set $W = V \setminus \mathcal{Z}(f)$ which is dense because U est irreducible. It follows that $\frac{1}{f}$ is regular on W, and the class of $(W, \frac{1}{f})$ is the inverse of the class of (V, f) in $K(U)$. □

Definition 1.2.73 Let K be a field and let U be a quasi-algebraic set over K. If U is irreducible then $K(U)$ is called the *field of rational functions* or *function field* of U.

Remark 1.2.74 The field of rational functions of V is often denoted by $K(V)$ even when the base field is not denoted by K. For example, if V is an irreducible algebraic set over $\mathbb{C}$ then its function field will be denoted either $K(V)$ or $\mathbb{C}(V)$.

Proposition 1.2.75 *Let F be an irreducible algebraic set over a field K. For any point $x \in F$ and any neighbourhood U of x in F the canonical morphisms*

$$\mathcal{R}(U) \hookrightarrow \mathcal{O}_{U,x} \hookrightarrow K(U)$$

are injective. Moreover the restriction morphisms

$$\mathcal{O}_{F,x} \xrightarrow{\simeq} \mathcal{O}_{U,x} \quad \text{and} \quad K(F) \xrightarrow{\simeq} K(U)$$

are isomorphisms.

Proof Since the open set U is non-empty it is dense in the irreducible set F and we have an isomorphism $K(U) \simeq K(F)$ by Proposition 1.2.71. The three injectivity results follow from the fact that any regular function is continuous for the Zariski topology, so if such a function vanishes on a non-empty open set in an irreducible space then it vanishes everywhere. □

Remark 1.2.76 If F is an irreducible affine algebraic set over a field K and x is a point of F then the natural maps $\mathcal{O}_F(V) \to K(F)$, where V is any open set in F containing x, are injective and the rings $\mathcal{O}_F(V)$ can be thought of as subrings of the field of functions $K(F)$. By Remark C.3.2 we then have that

$$\mathcal{O}_{F,x} = \bigcup_{V \ni x} \mathcal{O}_F(V) .$$

Proposition 1.2.77 *Let F be an algebraic set over a field K. For any dense open set $U \subset F$ we then have that*

$$K(F) \simeq \operatorname{Frac} \mathcal{R}(U)$$

where Frac A *denotes the total ring of fractions of some ring A (see Definition A.3.8).*

Remark 1.2.78 Note that the analogous statement in complex analytic geometry is false. For example, the function $z \mapsto \exp \frac{1}{z}$ is holomorphic on $\mathbb{C} \setminus \{0\}$ but cannot be written as a quotient of two holomorphic functions on $\mathbb{C}$. (See Appendix D).

Proof Since the open set U is dense in F, Proposition 1.2.71 applies and the map $\mathcal{R}(U) \to K(U) \simeq K(F)$ is injective because a regular function is continuous in the Zariski topology, so if it vanishes on a dense open subset it vanishes everywhere. The induced morphism on the total ring of fractions $\operatorname{Frac} \mathcal{R}(U) \to K(F)$ is therefore also injective. As this map is surjective by definition of a regular function the result follows. □

Corollary 1.2.79 *Let F be an affine algebraic set over a field K which is either real closed or algebraically closed. We then have that*

$$K(F) \simeq \operatorname{Frac} \mathcal{P}(F).$$

Proof If K is algebraically closed then the above isomorphism follows from Proposition 1.2.77 and Theorem 1.2.50. If K is real closed then any regular function $f \in \mathcal{R}(F)$ is the restriction of a rational function defined on F by Theorem 1.2.52. In particular $\operatorname{Frac} \mathcal{R}(F) \simeq \operatorname{Frac} \mathcal{P}(F)$. □

Exercise 1.2.80 Let K be a field which is algebraically closed or real closed.

1. For any non-zero natural number n we have that

$$K(\mathbb{A}^n(K)) = K(X_1, \ldots, X_n).$$

2. For any non-zero natural number n we have that

$$K(\mathbb{P}^n(K)) = K(X_1, \ldots, X_n) = K(\mathbb{A}^n(K)) .$$

3. Let $C = \mathcal{Z}(P) \subset \mathbb{A}^2(K)$ be a plane curve of equation $P(x, y) = 0$ where $P \in K[x, y]$ is a polynomial of non-zero degree.
   (a) If $f_1 = x|_C$ and $f_2 = y|_C$ then $P(f_1, f_2) = 0$ in $K(f_1, f_2)$ and

   $$K(C) = K(f_1, f_2)$$

   is a finite degree extension of $K(X)$.
   (b) If K is algebraically closed then $K(C) \simeq \operatorname{Frac} K[x, y]/\sqrt{(P)}$.

1.3 Abstract Algebraic Varieties

We started by defining algebraic and quasi-algebraic sets as *sub*-sets of affine and projective spaces. Once we had introduced the sheaf of regular functions and defined isomorphisms of algebraic sets we could give a definition of an algebraic variety which was independent of the surrounding space. As in differential geometry, we can go further and define algebraic varieties without any reference to an embedding: an abstract algebraic variety is a ringed space[6] covered by a finite number[7] of open sets that are isomorphic to affine varieties.[8] The class of spaces thus defined is larger than the class of quasi-projective varieties considered so far. This is an important difference with differential geometry, where every abstract real differential manifold can be smoothly embedded in $\mathbb{R}^n$ for some n. (See [Hir76, Theorem I.3.4], for example, for a proof of this classical result).

Most of the time we will only consider varieties that are isomorphic to quasi-projective varieties. This class includes all projective varieties, affine varieties and quasi-affine varieties by Lemma 1.2.43. Even when dealing with quasi-projective varieties, the notion of an abstract variety is useful for varieties which are not naturally described by a set of homogeneous defining equations, such as quotient varieties or fibre spaces.

Definition 1.3.1 An (*abstract*) *algebraic variety* over a field K is a pair $(X, \mathcal{O}_X)$ where X is a topological space (which will turn out to be *quasi-compact*, see Definition B.1.5) and $\mathcal{O}_X$ is a sub-sheaf of the sheaf of K-valued functions on X such that there is a covering of the space X by a finite collection of open sets U_i such that

---

[6]Or in other words, a topological space equipped with a sheaf of rings satisfying certain properties, see Definition C.5.1.

[7]This is the algebraic analogue of the requirement that the topology on a differential manifold should have a countable basis.

[8]These open sets play the role of local charts in differential geometry.

$(U_i, \mathcal{O}_X|_{U_i})$ is isomorphic as a ringed space to an affine algebraic subset of some K^n with its ring of regular functions (see Definitions 1.2.1 and 1.2.33). By analogy with the theory of affine and projective varieties the topology on X is called the *Zariski topology* and the sheaf $\mathcal{O}_X$ is called the *sheaf of regular functions* or *structural sheaf* of X.

Definition 1.3.2 An open set $U \subset X$ such that $(U, \mathcal{O}_X|_U)$ is isomorphic as a ringed space to an affine algebraic set is called an *affine open set* of X and a covering of X by such open sets is called an *affine covering* of X.

Remark 1.3.3 We emphasise the fact that our algebraic varieties are not assumed irreducible, which is often required, especially in America.

Definition 1.3.4 If X and Y are algebraic varieties over K, a *morphism* (or *regular map*) $\varphi\colon X \to Y$ is a continuous map such that for any open set $V \subset Y$ and any regular function $f\colon V \to K$ the function $f \circ \varphi\colon \varphi^{-1}(V) \to K$ is regular.

Remark 1.3.5 (*See Example* C.5.3 *in Appendix* C) A morphism of algebraic varieties is a morphism of ringed spaces which induces a morphism of sheaves of K-algebras. In other words, a map $\varphi\colon X \to Y$ is a morphism of algebraic varieties if and only if it is continuous and the image of the pullback morphism on sheaves $\varphi^{\#}\colon \mathcal{O}_Y \to \varphi_*\mathcal{F}_X$ is contained in $\varphi_*\mathcal{O}_X$.

Let X be an algebraic variety over K and let $Y \subset X$ be a locally closed subset of X. We want to define a structural sheaf on Y such that the inclusion map is a morphism of varieties. We cannot simply define this sheaf on Y using only restrictions of regular functions on open sets of X, since this construction will generally yield a presheaf rather than a sheaf, see [Per95, III.4.8] for example. We use the fact that $\mathcal{O}_X$ is a sheaf of functions on X and we set $\mathcal{O}_Y := (\mathcal{O}_X)_Y$, which is a sheaf on Y, see Definition C.1.6. We recall the definition of this sheaf in terms of local sections below.

If Y is open in X we have that $\mathcal{O}_Y = \mathcal{O}_X|_Y$. Indeed, the pair $(Y, \mathcal{O}_X|_Y)$ is clearly an algebraic variety over K since the fact that any open set in U contained in Y is also an open set in X implies that $\mathcal{O}_X|_Y(U) = \mathcal{O}_X(U)$, see Example C.4.9. In general the sections of $\mathcal{O}_Y$ over an open set U in Y are

$$\begin{aligned} &\mathcal{O}_Y(U) = \{f\colon U \to K \mid \forall x \in U, \\ &\exists V \text{ neighbourhood of } x \text{ in } X \text{ and } \exists g \in \mathcal{O}_X(V) \mid g|_{V\cap U} = f|_{V\cap U}\}\,. \end{aligned}$$

Proposition 1.3.6 *Let X be an algebraic variety over a field K and let Y be a locally closed subset of X. The pair $(Y, \mathcal{O}_Y)$ is an algebraic variety and $Y \hookrightarrow X$ is a morphism of algebraic varieties.*

Proof Immediate by definition. □

Definition 1.3.7 Let X be an algebraic variety over K and let Y be a locally closed subset of X. The pair $(Y, \mathcal{O}_Y)$ is said to be an *algebraic subvariety* of $(X, \mathcal{O}_X)$. If

Y is Zariski-closed (resp. open) in X then we will say that Y is a *closed algebraic subvariety* (resp. *open algebraic subvariety*) of X.

Example 1.3.8 Consider the case where $X = \mathbb{A}^n(K)$ and $Y = F$ is a closed subset of $\mathbb{A}^n(K)$. The K-algebra $\mathcal{O}_F(F)$ is then the quotient of the ring $\mathcal{O}_{\mathbb{A}^n(K)}(F)$ by the ideal $\mathcal{I}(F)$.

Definition 1.3.9 Let X be an algebraic variety over K. If $K = \mathbb{C}$ we call X a *complex algebraic variety* (or *complex variety* if it is clear from the context that X is algebraic). If $K = \mathbb{R}$ we say that X is a *real algebraic variety*.

Definition 1.3.1 taken from [Per95, Section III.4] was first used by Serre in [Ser55a, Chapitre II], where it is used for algebraically closed K.[9] As in [BCR87] our Definition 1.3.9 of real algebraic varieties is the same as Serre's definition for algebraically closed fields. With this definition, any locally closed subset U of affine space $\mathbb{A}^n(\mathbb{R})$ is real algebraic variety with induced topology and sheaf $\mathcal{O}_U$ as defined in Definition 1.2.39. Similarly, any algebraic projective set over $\mathbb{R}$ (see Definition 1.2.3) is a real algebraic variety. More generally, for any base field K any algebraic affine set with its sheaf of regular functions (Definitions 1.2.33 and 1.2.39) is an algebraic variety, as is any algebraic projective set with its sheaf of regular functions (Definitions 1.2.34 and 1.2.39). This inspires the following definitions.

Definition 1.3.10 An algebraic variety $(Y, \mathcal{O}_Y)$ over a field K is said to be

1. *affine* if it is isomorphic as a ringed space to an algebraic affine set with its sheaf of regular functions (see Definitions 1.2.33 and 1.2.39);
2. *projective* if it is isomorphic as a ringed space to an algebraic projective set with its sheaf of regular functions (see Definitions 1.2.34 and 1.2.39);
3. *quasi-affine* (resp. *quasi-projective*) if Y is a Zariski-open subset of an affine (resp. projective) variety X and $\mathcal{O}_Y = \mathcal{O}_X|_Y$ is the restriction to Y of the sheaf $\mathcal{O}_X$.

The following result illustrates an important difference between real and complex algebraic varieties. We invite the reader to compare it with Propositions 1.2.61, 1.2.63 and Corollary 1.2.62.

Proposition 1.3.11 (Real affine algebraic varieties) *A real algebraic variety is affine if and only if it is quasi-projective.*

Proof Whatever the field, any affine algebraic variety X is quasi-affine by definition and any quasi-affine variety is quasi-projective by Lemma 1.2.43. Conversely, if the base field is $\mathbb{R}$, Proposition 1.2.63 implies the remarkable fact (which does not hold for complex varieties) that any real projective algebraic variety is affine. It follows from Corollary 1.2.62 that any open set in an affine real algebraic variety is a real affine algebraic variety. □

---

[9]Note that Serre adds the technical condition that the space should be *separated* (see Appendix B.1) which is not used in [BCR87] and [Per95] because all quasi-projective spaces are separated.

By definition, an algebraic variety X over K is affine (resp. projective) if and only if there exists an integer n and a morphism of algebraic varieties

$$\varphi\colon X \to \mathbb{A}^n(K) \quad (\text{resp.}\, \mathbb{P}^n(K))$$

such that $\varphi(X)$ is locally closed in the Zariski topology on the target space and φ induces an isomorphism of algebraic varieties between $(X, \mathcal{O}_X)$ and the subvariety $(\varphi(X), \mathcal{O}_{\varphi(X)})$ in $(\mathbb{A}^n(K), \mathcal{O}_{\mathbb{A}^n(K)})$ (resp. $(\mathbb{P}^n(K), \mathcal{O}_{\mathbb{P}^n(K)})$). Such a morphism is called an embedding.

Definition 1.3.12 Let X and Y be algebraic varieties over the same base field K. A morphism $\varphi\colon X \to Y$ is an *embedding* of X in Y si $\varphi(X)$ is locally closed in Y and φ induces an isomorphism between X and $\varphi(X)$.

Exercise 1.3.13 (*Segre embedding*) Consider the map

$$\varphi\colon \begin{cases} \mathbb{P}^r(K)_{a_0:\cdots:a_r} \times \mathbb{P}^s(K)_{b_0:\cdots:b_s} & \longrightarrow \qquad \mathbb{P}^N(K) \\ (a_0 : \cdots : a_r) \times (b_0 : \cdots : b_s) & \longmapsto \left(\cdots : a_i b_j : \ldots\right) \end{cases}$$

where we set $N = rs + r + s$ and the set $\left(\cdots : a_i b_j : \ldots\right)$ is ordered lexicographically. Check that φ is well-defined and injective and prove that its image $X := \varphi\left(\mathbb{P}^r(K) \times \mathbb{P}^s(K)\right)$ is a subvariety of $\mathbb{P}^N(K)$ or in other words that φ is an embedding. [Hint—see [Har77, Exercise I.2.14].]

In particular, check that the image of $\mathbb{P}^1(K) \times \mathbb{P}^1(K)$ under the Segre embedding is a quadric surface in $\mathbb{P}^3(K)$.

A projective algebraic variety cannot always be embedded in a projective space of given dimension. In particular, there are smooth irreducible curves that cannot be embedded in the projective plane. See Section 1.6 for more details.

Note that over $\mathbb{R}$ a fibre space whose base and fibre are both real affine algebraic varieties is not necessarily a real algebraic affine variety. See Example 2.5.6 for more details. This is a major obstacle in the theory of real algebraic varieties.

Definition 1.3.14 Let $I, J \subset K[X_0, \ldots, X_n]$ be two homogeneous ideals and let $X \subset \mathbb{P}^n(K)$ be the quasi-projective variety defined by

$$X := \mathcal{Z}(J) \setminus \mathcal{Z}(I) = \mathcal{D}(I) \cap \mathcal{Z}(J)\ .$$

An open set U in X is a *principal open set* of X if there is a function $f \in \mathcal{I}(\mathcal{Z}(I))$ such that U is the non-vanishing locus of f in X,

$$U = \mathcal{D}(f) := \mathcal{D}(f) \cap \mathcal{Z}(J)\ .$$

Exercise 1.3.15 (*Affine and principal open sets*)

1. All open affine sets in a real quasi-projective algebraic variety are principal.

2. Prove that the intersection of two principal open sets in a quasi-projective variety remains principal.
3. Deduce that in a quasi-projective algebraic variety the principal open sets are a basis for the Zariski topology (see Exercise 1.2.60).
4. Prove that the intersection of two affine open sets in a quasi-projective variety remains affine.
5. Let X be a quasi-projective algebraic variety over a base field K and let $f : X \to K$ be a function. Prove that f is regular if and only if there is an affine covering of X such that the restriction of f to each open affine set in the covering is regular.
6. Let K be an algebraically closed field and let $X = \mathcal{Z}(J) \setminus \mathcal{Z}(I) \subset \mathbb{P}^n(K)$ be a quasi-projective variety defined by two homogeneous ideals $I, J \subset K[X_0, \dots, X_n]$. Let $h \in I$ be a homogeneous polynomial function. Every regular function on the principal open set $\mathcal{D}(h) := \mathcal{D}(h) \cap X$ can be written in the form $\frac{g}{h^k}$ where g is a homogeneous function and $\deg g = k \deg h$.

Let f be an element of a ring A and recall that as in Definition A.3.1 we denote by A_f the localisation of the ring A with respect to the multiplicative system of powers of f. The following statement follows from Exercise 1.2.60(1).

Proposition 1.3.16 *Let K be an algebraically closed field (such as $\mathbb{C}$) or a real closed field (such as $\mathbb{R}$). Let F be an affine algebraic set and let $f \in \mathcal{P}(F)$ be a polynomial function on F. There is then an isomorphism of K-algebras*

$$\mathcal{O}_F(\mathcal{D}(f)) \simeq \mathcal{O}_F(F)_f \ .$$

By definition, if Y is isomorphic to X then the K-algebras $\mathcal{O}_X(X)$ and $\mathcal{O}_Y(Y)$ are isomorphic, so the algebra of global regular functions is an *invariant* of the variety X. Similarly, the algebra of global rational functions is an invariant of X. Moreover, for any point $x \in X$ the algebra $\mathcal{O}_{X,x}$ is an invariant of the pair (X, x) in the following sense: if $\varphi \colon X \to Y$ is an isomorphism then $\mathcal{O}_{X,x}$ and $\mathcal{O}_{Y,\varphi(y)}$ are isomorphic as algebras.

The algebra of affine coordinates $\mathcal{A}$, the algebra of polynomials $\mathcal{P}$ and the algebra of homogeneous coordinates $\mathcal{S}$ may depend on the embedding. For any affine variety X over an algebraically closed field, the coordinate algebra $\mathcal{A}(X)$ is an invariant by Theorem 1.2.50 whereas for a projective variety X the ring of homogeneous coordinates $\mathcal{S}(X)$ depends on the projective embedding, see Example 1.3.17 below.

Example 1.3.17 Consider the embedding (Definition 1.3.12) of $\mathbb{P}^1(K)$ as a plane conic (also called the degree 2 Veronese embedding of $\mathbb{P}^1(K)$):

$$\varphi \colon \mathbb{P}^1(K)_{x:y} \to \mathbb{P}^2(K)_{X:Y:Z} \quad , \qquad (x : y) \mapsto (x^2 : y^2 : xy) \ .$$

The image of this map $C := \varphi(\mathbb{P}^1(K))$ is a conic of equation $XY = Z^2$. The ring of homogeneous coordinates of $\mathbb{P}^1(K)$ is $K[x, y]$, but the ring of homogeneous coordinates of C, $K[C] = K[X, Y, Z]/(XY - Z^2)$ is not isomorphic to $K[x, y]$ because the space of degree 1 elements in $K[C]$ is of dimension 3.

Theorem 1.3.18 *Let X and Y be algebraic varieties over the same base field K. If Y is affine there is a natural bijection*

$$\alpha\colon \operatorname{Hom}(X, Y) \longrightarrow \operatorname{Hom}(\mathcal{A}(Y), \mathcal{R}(X))$$

where on the left hand side Hom *denotes the set of morphisms of algebraic varieties and on the right hand side* Hom *denotes the set of morphisms of K-algebras.*

If moreover the field K is algebraically closed or real closed then there is a natural bijection

$$\beta\colon \operatorname{Hom}(\mathcal{R}(Y), \mathcal{R}(X)) \longrightarrow \operatorname{Hom}(\mathcal{A}(Y), \mathcal{R}(X)) .$$

Proof The proof of Theorem 1.2.64 applies on considering an affine algebraic set F' isomorphic to Y. □

Corollary 1.3.19 *Let X and Y be affine varieties over the same base field K. If K is algebraically closed then X is isomorphic to Y if and only if the K-algebras $\mathcal{A}(X)$ and $\mathcal{A}(Y)$ are isomorphic.*

Corollary 1.3.20 *Let X and Y be affine varieties over the same base field K. If K is algebraically closed or real closed then X is isomorphic to Y if and only if the K-algebras $\mathcal{R}(X)$ and $\mathcal{R}(Y)$ are isomorphic.*

We now adapt Definition 1.2.40 for abstract varieties. Let $(X, \mathcal{O}_X)$ be an algebraic variety over a base field K. The K-algebra of *germs of regular functions* at the point x is the stalk at x of the sheaf $\mathcal{O}_X$:

$$\mathcal{O}_x := \mathcal{O}_{X,x} = \varinjlim_{V \ni x} \mathcal{O}_U(V)$$

where the inductive limit is taken over all open neighbourhoods V of x in U. Similarly, we can adapt Definition 1.2.69 for abstract varieties. The *K-algebra of rational functions* of X is given by

$$K(X) = \varinjlim_{\overline{U}=X} \mathcal{O}_X(U)$$

where U runs over the dense open sets in X. The following result then follows immediately from Proposition 1.2.77.

Proposition 1.3.21 *Let X be an algebraic variety over K. For any open dense set $U \subset X$ we set*

$$K(X) = \operatorname{Frac} \mathcal{O}_X(U) .$$

This result is particularly useful when U is an open affine set.

Rational Maps

Definition 1.3.22 If X and Y are algebraic varieties over a base field K a *rational map* $\varphi\colon X \dashrightarrow Y$ is an equivalence class of pairs (U, φ_U) where U is a dense open subset of X, φ_U is a map from U to Y and two pairs (U, φ_U) and (V, φ_V) are equivalent if and only if φ_U and φ_V are identical on the intersection $U \cap V$. The rational map φ is said to be *dominant* if for any representative (U, φ_U) of φ the image of φ_U is dense in Y. The map φ is *defined* at a point $x \in X$ if there is a representative (U, φ_U) of φ such that $x \in U$. There is an obvious order on the set of pairs representing φ and the largest open set on which φ is defined is called the *domain* of φ. We denote it by

$$\mathrm{dom}(\varphi) := \{x \in X \mid \varphi \text{ is defined at } x\}\ .$$

Remark 1.3.23 1. Any morphism is a rational map.
2. We will often use a dotted arrow $\dashrightarrow$ to emphasise that a rational map $\varphi\colon X \dashrightarrow Y$ is not necessarily defined at every point. Likewise, we will use the full arrow $\to$ to indicate that a rational map $\varphi\colon X \to Y$ is actually a morphism.
3. It is obvious that if $Y = K$ then a rational map $\varphi\colon X \dashrightarrow Y = K$ is simply a rational function as defined in 1.2.69.

Exercise 1.3.24 The given relation is an equivalence relation because we have required that the open sets should be dense. When X is irreducible it is enough to require that the open sets should be non empty.

Exercise 1.3.25 (See Exercise 1.2.56) Suppose that $K = \mathbb{R}$ or $\mathbb{C}$. Let $X \subset \mathbb{A}^n(K)$, and $Y \subset \mathbb{A}^N(K)$ be algebraic sets over K and let

$$\varphi\colon X \dashrightarrow Y$$

be a rational map. Prove that φ is the restriction of a rational map $\mathbb{A}^n(K) \dashrightarrow \mathbb{A}^N(K)$, or in other words that there are polynomials $P_1, \ldots, P_N \in K[X_1, \ldots, X_n]$ with coefficients in K and non zero polynomials $Q_1, \ldots, Q_N \in K[X_1, \ldots, X_n]$ with coefficients in K such that for any point $(x_1, \ldots, x_n) \in X$ at which φ is well-defined we have that

$$\varphi(x_1, \ldots, x_n) = \left(\frac{P_1(x_1, \ldots, x_n)}{Q_1(x_1, \ldots, x_n)}, \ldots, \frac{P_N(x_1, \ldots, x_n)}{Q_N(x_1, \ldots, x_n)}\right).$$

Smooth curves (see Definition 1.3.26) are exceptional in the sense that any rational map on a smooth curve can be extended to a morphism onto a projective target.

Proposition 1.3.26 *Let X be a non-singular curve, let Y be a projective variety and let $\varphi\colon X \dashrightarrow Y$ be a rational map. There is then a regular map $\Phi\colon X \to Y$ extending φ, by which we mean that at every point $P \in \mathrm{dom}(\varphi)$ we have $\Phi(P) = \varphi(P)$.*

Proof See [Har77, Proposition I.6.8]. □

It is not always possible to compose rational maps, but it is clearly possible to compose *dominant* rational maps. There is therefore a well-defined category of varieties and dominant rational maps. The "isomorphisms" in this category are called *birational* maps.

Definition 1.3.27 If X and Y are algebraic varieties over a base field K, a *birational map* $\varphi\colon X \dashrightarrow Y$ is a dominant rational map which has a dominant rational inverse, by which we mean a dominant rational map $\psi\colon Y \dashrightarrow X$ such that $\varphi \circ \psi = \mathrm{id}_Y$ and $\psi \circ \varphi = \mathrm{id}_X$ as rational maps. If there is a birational map $X \dashrightarrow Y$ we say that the varieties X and Y are *birationally equivalent* or simply *birational*. A *birational morphism* is a morphism $\varphi\colon X \to Y$ which has a dominant rational inverse.

Remark 1.3.28 In other words, a birational map $\varphi\colon X \dashrightarrow Y$ is a birational morphism if and only if $\mathrm{dom}(\varphi) \supset X$.

Example 1.3.29 The blow-up of a variety along a subvariety (see Appendix F for details of this construction) is a birational morphism.

Theorem 1.3.30 *Let X and Y be algebraic varieties over the same base field K. There is a map $\varphi \mapsto \varphi^*$ which associates a K-algebra morphism from $K(Y)$ to $K(X)$ to any dominant rational map from X to Y. Moreover, if X and Y are irreducible this map yields a bijection between the following two sets:*

1. *The set of dominant rational maps from X to Y*
2. *The set of K-algebra morphisms from $K(Y)$ to $K(X)$.*

Proof Let $\varphi\colon X \dashrightarrow Y$ be a dominant rational map and let (U, φ_U) be a pair representing φ. Let (V, f_V) be a pair representing an element of $K(Y)$. By hypothesis $\varphi_U(U)$ is dense in Y so $\varphi_U^{-1}(V)$ is a non-empty open set in X. It follows that $f \circ \varphi_U$ is regular on $\varphi_U^{-1}(V)$ and the pair $\left(\varphi_U^{-1}(V), f \circ \varphi_U\right)$ represents a rational function on X. This yields a K-algebra morphism $\varphi^*\colon K(Y) \to K(X)$.

Conversely, suppose that X and Y are irreducible and let $\theta\colon K(Y) \to K(X)$ be a morphism of K-algebras. We want to define a rational map φ from X to Y such that $\varphi^* = \theta$. By definition of a variety, Y can be covered by open affine subsets. Since Y is irreducible we can therefore assume that Y is a closed subset of $\mathbb{A}^N(K)$ and $\mathcal{A}(Y) = K[y_1, \dots, y_N]/\mathcal{I}(Y) = \mathcal{P}(Y)$. The functions $y_i|_Y \in \mathcal{P}(Y)$ are rational functions on Y and their images $\theta(y_i) \in K(X)$ are rational functions on X. For any $i = 1 \dots N$ let $U_i \subset X$ be the domain of $\theta(y_i)$. The open set $U := \cap_{i=1}^N U_i$ is non empty because X is irreducible. We define a map $\varphi_U : U \to \mathbb{A}^N(K)$ by $\varphi_U = (\theta(y_1), \dots \theta(y_N))$. Since $Y = \mathcal{Z}(\mathcal{I}(Y))$ and θ is a morphism of K-algebras it follows that $\varphi_U(x) \in Y$ for any $x \in U \subset X$. Since each component $\theta(y_i)$ of φ_U is regular on U it is easy to check that φ_U is a morphism. The pair (U, φ_U) represents a rational map $\varphi\colon X \dashrightarrow Y$ such that $\varphi^* = \theta$. □

Remark 1.3.31 If X is irreducible and there is a dominant rational map $\varphi\colon X \dashrightarrow Y$ then Y is also irreducible. To prove this, let $U \subset X$ be an open set on which φ is defined such that $\varphi(U)$ is dense in Y. Since X is irreducible, U is also irreducible and by Lemma 1.2.18, $\varphi(U)$ is irreducible.

Proposition 1.3.32 *Let K be a field and let X and Y be two algebraic varieties over K. If X and Y are irreducible then the following are equivalent.*

1. *The varieties X and Y are birationally equivalent,*
2. *There are non-empty open sets $U \subset X$ and $V \subset Y$ which are isomorphic as algebraic varieties,*
3. *The K-algebras* $\operatorname{Frac} \mathcal{R}(X)$ *and* $\operatorname{Frac} \mathcal{R}(Y)$ *are isomorphic,*
4. *The fields $K(X)$ and $K(Y)$ are isomorphic as K-algebras.*

Proof $3 \iff 4$ by Proposition 1.2.77.

$1 \implies 2$. Let $\varphi\colon X \dashrightarrow Y$ and $\psi\colon Y \dashrightarrow X$ be inverse rational maps and consider representatives (U, φ) of φ and (V, ψ) of ψ. The composition $\psi \circ \varphi$ is represented by $(\varphi^{-1}(V), \psi \circ \varphi)$ and by hypothesis $\psi \circ \varphi$ is the identity on $\varphi^{-1}(V)$. Similarly, $\varphi \circ \psi$ is the identity on $\psi^{-1}(U)$. By construction, the open sets $\varphi^{-1}(\psi^{-1}(U))$ in X and $\psi^{-1}(\varphi^{-1}(V))$ in Y are isomorphic.

$2 \implies 4$ by definition of a function field.

$4 \implies 1$ by Theorem 1.3.30. □

Corollary 1.3.33 *Let K be a field and let X and Y be algebraic varieties over K. Assume that X and Y are irreducible. If the fields $K(X)$ and $K(Y)$ are isomorphic as K-algebras then there are open sets $U \subset X$ and $V \subset Y$ such that U is isomorphic to V.*

Proof This result is an immediate corollary of Proposition 1.3.32. There is also a direct proof for algebraically closed K.[10] We can assume that X and Y are affine and in this case the finitely generated K-algebras $\mathcal{A}(X)$ and $\mathcal{A}(Y)$ are sub-algebras of $K(X)$. As $K(X) = \operatorname{Frac} \mathcal{A}(Y)$ and $\mathcal{A}(X)$ is finitely generated, the primitive element Theorem A.5.9 assures us there is a function $f \in K(X)$, $f \neq 0$ such that $\mathcal{A}(X) \subset \mathcal{A}(Y)[\frac{1}{f}]$. Similarly there is a $g \in K(X)$, $g \neq 0$ such that $\mathcal{A}(Y) \subset \mathcal{A}(X)[\frac{1}{g}]$. It follows that $\mathcal{A}(X)[\frac{1}{fg}] = \mathcal{A}(Y)[\frac{1}{fg}]$. There is an element $a \in \mathcal{A}(X)$ and an element $b \in \mathcal{A}(Y)$ such that $\mathcal{A}(X)[\frac{1}{fg}] = \mathcal{A}(X)[\frac{1}{a}]$ and $\mathcal{A}(Y)[\frac{1}{fg}] = \mathcal{A}(Y)[\frac{1}{b}]$. By Theorem 1.2.50, for any Zariski-closed subset F we can identify the algebras $\mathcal{A}(F)$ and $\mathcal{R}(F)$ and by Exercise 1.2.60 we have that $\mathcal{A}(\mathcal{D}(a)) = \mathcal{A}(X)[\frac{1}{a}]$ and $\mathcal{A}(\mathcal{D}(b)) = \mathcal{A}(Y)[\frac{1}{b}]$. The algebras $\mathcal{A}(\mathcal{D}(a))$ and $\mathcal{A}(\mathcal{D}(b))$ are therefore isomorphic and if K is algebraically closed this implies that $\mathcal{D}(a) \subset X$ and $\mathcal{D}(b) \subset Y$ are isomorphic algebraic varieties. □

Definition 1.3.34 Let X be an abstract algebraic variety. The sheaf of rational functions on X is the sheaf of K-algebras denoted $\mathcal{M} := \mathcal{M}_X$, defined on any open set of X by:

$$U \mapsto \operatorname{Frac} \mathcal{O}_X(U).$$

For any $x \in X$, $\mathcal{M}_x$ is canonically isomorphic to $\operatorname{Frac} \mathcal{O}_{X,x}$.

[10] This proof is taken from a lecture course by Antoine Chambert-Loir [CL98].

Proposition 1.3.35 *For any irreducible algebraic variety X, any non-empty open set $U \subset X$ and any $x \in U$ we have that*

$$\mathcal{M}_x = \operatorname{Frac} \mathcal{O}_{X,x} = \mathcal{M}_X(U) = \operatorname{Frac} \mathcal{R}(U) = K(X) .$$

In particular, the sheaf $\mathcal{M}_X$ is a sheaf of constant ring-valued functions.

Proof Since any sheaf of locally constant functions on an irreducible space is constant, it will be enough to prove the proposition for quasi-affine X. For quasi-affine X the statement follows directly from Proposition 1.2.77 (see also Proposition 1.3.21). □

Remark 1.3.36 Let $F \subset \mathbb{A}^n(K)$ be an irreducible algebraic set over a real closed or algebraically closed field. For any $x \in F$, the stalk $\mathcal{M}_x$ is isomorphic to the *field of rational functions* $K(F) = \operatorname{Frac} \mathcal{P}(F)$ of the irreducible variety F (see Corollary 1.2.79). The fact that $\operatorname{Frac} \mathcal{P}(F)$ is a field follows from the fact that $\mathcal{P}(F)$ is an integral domain because F is irreducible.

Rational Varieties

By Proposition 1.3.35, if an algebraic variety X over a base field K is irreducible its ring of rational functions is a field which we call the function field. By Noether's normalisation Lemma A.5.6, X (which is an integral variety and is hence reduced (implicit in Definition 1.3.1) and irreducible) is a finite degree extension of the field of rational fraction in n variable $K(X_1, \ldots, X_n)$ (here n is the dimension of X over K). Classically we say that the variety X is *rational* if and only if its field of functions is *isomorphic* to $K(X_1, \ldots, X_n)$. We now give another definition of rationality, equivalent to this one by Proposition 1.3.32 (exercise).

Definition 1.3.37 Let K be a field and let $\overline{K}$ be the algebraic closure of K.

1. An algebraic variety X of dimension n over K is *rational* (or *rational over K*) if and only if it is birationally equivalent to projective space $\mathbb{P}^n(K)$, or alternatively if there exist dense Zariski open sets $U \subset X$, $V \subset \mathbb{P}^n(K)$ and an isomorphism $U \stackrel{\simeq}{\to} V$ of K-algebraic varieties.
2. A quasi-projective variety X over K is *geometrically rational* (or *rational over $\overline{K}$*) if and only if the variety $X_{\overline{K}}$,[11] which is an algebraic variety over $\overline{K}$, is rational or alternatively if there exist dense Zariski open subsets $U \subset X_{\overline{K}}$, $V \subset \mathbb{P}^n(\overline{K})$ and an isomorphism $U \stackrel{\simeq}{\to} V$ of algebraic varieties over $\overline{K}$.

Proposition 1.3.38 *Any rational variety is geometrically rational.*

Remark 1.3.39 The converse of the above proposition is false, as we will see below.

[11] Definition of $X_{\overline{K}}$: if X is a subvariety of $\mathbb{P}^N(K)$ the subvariety $X_{\overline{K}} \subset \mathbb{P}^n(\overline{K})$ is defined by the same homogeneous ideal as X, see Definition 2.3.1.

Exercise 1.3.40 1. The algebraic varieties $\mathbb{P}^n(K)$ and $\mathbb{A}^n(K)$ are rational over K.
2. The complex surface $\mathbb{P}^1(\mathbb{C}) \times \mathbb{P}^1(\mathbb{C})$ is rational over $\mathbb{C}$.

Example 1.3.41 1. The complex Hirzebruch surfaces $\mathbb{F}_n$ (surfaces fibered over $\mathbb{P}^1(\mathbb{C})$ with fibre $\mathbb{P}^1(\mathbb{C})$ see Definition 4.2.1) are rational complex surfaces.
2. We will see examples in Section 4.4.10 of geometrically rational surfaces that are not rational. In particular, certain conic bundles—notably those of the form $x^2 + y^2 = P(z)$ for some $P \in \mathbb{R}[z]$—are geometrically rational but not rational.

1.4 Euclidean Topology

Let X be an algebraic variety over a field K. The set X is then also a topological space with the Zariski topology. If $K = \mathbb{C}$, X is a complex algebraic variety and if $K = \mathbb{R}$, X is a real algebraic variety. In both cases, X is equipped with a natural topology, more refined than the Zariski topology, called the *Euclidean topology*.[12]

Definition 1.4.1 Let X be a complex or real algebraic variety. The *Euclidean topology* on X is the topology generated by open sets of the form

$$V(U; f_1, \dots, f_r; \varepsilon) := \{x \in U \mid |f_l(x)| < \varepsilon, \text{ for } l = 1, \dots, r\}$$

where U is a Zariski open set in X, r is a natural number, $f_1, \dots, f_r$ are real or complex valued regular functions on U and $\varepsilon > 0$ is a real number.

Remark 1.4.2 Over $\mathbb{R}$ the open basis of semi-algebraic sets (Definition B.2.1) $V(U; f_1, \dots, f_r; \varepsilon)$ can be replaced by the basis of open sets of the form:

$$\{x \in U \mid f_1(x) > 0, \dots, f_r(x) > 0\} .$$

(which are also semi-algebraic).

Exercise 1.4.3 1. Prove that any Zariski-closed set is also closed in the Euclidean topology but the converse does not hold.
2. Prove that the above definition of the Euclidean topology using regular function is the same as the topology defined in a similar way using C^∞ functions.

Exercise 1.4.4 Prove that any morphism of real or complex varieties (which is continuous for the Zariski topology by definition) is also continuous for the Euclidean topology.

Theorem 1.4.5 *Let X be a complex algebraic variety. If X is irreducible then it is connected for the Euclidean topology.*

---

[12]Or alternatively, the *transcendental topology* or the *usual topology* or sometimes the *complex topology* if $K = \mathbb{C}$, and so on.

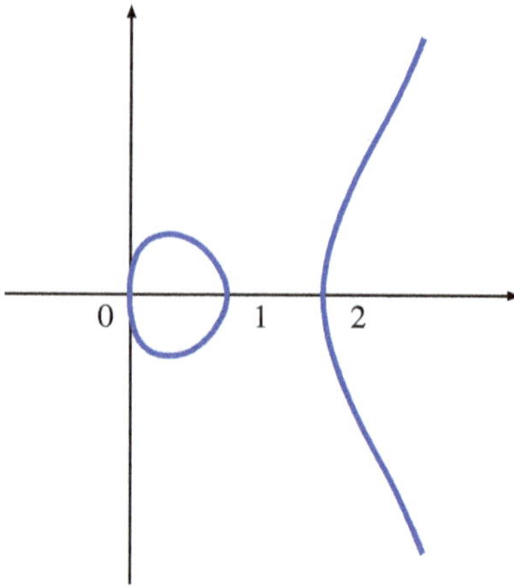

Fig. 1.1 $y^2 - x(x-1)(x-2) = 0$

Proof See [Sha94, VII.2, Theorem 1] for example. □

Remark 1.4.6 The converse is obviously false.

Remark 1.4.7 The statement of Theorem 1.4.5 is easily seen to be false over $\mathbb{R}$, even in dimension 1. Consider an irreducible plane cubic given by an equation $y^2 = (x - a)(x-b)(x-c)$ where a, b, c are pairwise distinct real numbers. The set of its real points $\mathcal{Z}\left(y^2 - (x-a)(x-b)(x-c)\right) \subset \mathbb{A}^2(\mathbb{R})$ has two connected components. This cubic is illustrated in Figure 1.1 for $a = 0, b = 1, c = 2$.

Exercise 1.4.8 (See [Ser56, Lemma 1 and Proposition 2].)

Let X be a complex quasi-projective algebraic variety and let $X \hookrightarrow \mathbb{P}^N(\mathbb{C})$ be a projective embedding. The topology induced on X by the Euclidean topology on $\mathbb{P}^N(\mathbb{C})$ is the Euclidean topology on X. Similarly, if X is a quasi-projective real algebraic variety and $X \hookrightarrow \mathbb{P}^N(\mathbb{R})$ is an embedding then the topology induced by the Euclidean topology on $\mathbb{P}^N(\mathbb{R})$ is the Euclidean topology on X.

Lemma 1.4.9 *Let X be a real or complex algebraic variety. If X is projective then X is compact with respect to the Euclidean topology*

Proof Both real and complex projective space are compact with respect to the Euclidean topology and X is a closed set in such a projective space. □

Remark 1.4.10 Recall that any quasi-projective real algebraic variety is affine by Proposition 1.3.11. Let X be a projective real algebraic variety. There is then an n such that X can be embedded as a Euclidean compact subset of $\mathbb{R}^n$.

This remark motivates the following definition.

Definition 1.4.11 A real or complex algebraic variety is said to be *complete* if it is compact with respect to the Euclidean topology.

Remark 1.4.12 It turns out to be possible to define completeness of varieties over any algebraically closed field—see [Har77, II.4] for example. The key fact about completeness is that any projective variety is complete.

Remark 1.4.13 In [BCR98, Definition 3.4.10], a real set is said to be complete if it is "closed and bounded" because unlike compactness this notion generalises to semi-algebraic sets over real closed fields other than $\mathbb{R}$. Of course, the two definitions coincide over $\mathbb{R}$.

Proposition 1.4.14 *Let X be a quasi-projective algebraic variety over $\mathbb{R}$ or $\mathbb{C}$.*

1. *If X is projective then it is complete.*
2. *If X is non-singular and complete then it is projective.*

Remark 1.4.15 Note that X is assumed quasi-projective. In particular, there exist complete non-singular complex algebraic varieties that are not projective. An example of such a variety due to Hironaka is given in [Har77, Appendix B.3.4.1].

Proof The first statement is simply Lemma 1.4.9. The second, whose proof is explained in Chapter 2, Theorem 2.3.7, is a corollary of Hironaka's theorem on resolution of singularities Theorem 1.5.54. □

1.5 Dimension and Smooth Points

When the base field is $\mathbb{R}$ or $\mathbb{C}$, algebraically "smooth" or "non-singular" points have certain similarities with points on topological (or differentiable) manifolds, notably because the only local analytic model is an open subset of the Euclidean topology on a finite dimensional vector space. The dimension of this space is determined by the local geometry of the variety. For more details, see [Mal67, Tou72] and Appendix B. In particular, a complex affine algebraic set F is non singular at a point x in F if and only if the space F with its Euclidean topology is an analytic variety in a neighbourhood of x (Definition 1.4.1).

We can analyse singularities locally without using the Euclidean topology via ring completions—see [Har77, Thm. 5.3, page 33] for more details. For example, $(0, 0)$ is an ordinary double point of the affine plane complex curve $C := \mathcal{Z}(xy)$ and every ordinary double point is locally analytically isomorphic to C but there is no Zariski open neighbourhood of C containing $(0, 0)$ which is isomorphic to some Zariski open neighbourhood of $(0, 0)$ in the curve $\mathcal{Z}(y^2 - x^2(x + 1))$. See [Har77, Example I.5.6.3] for more details.

As the following example taken from [BCR98, Example 3.3.11.b] shows, this analogy is of limited value over the real numbers. This example is a key illustration of the theory developed in this section and is completed by Exercise 1.5.31.

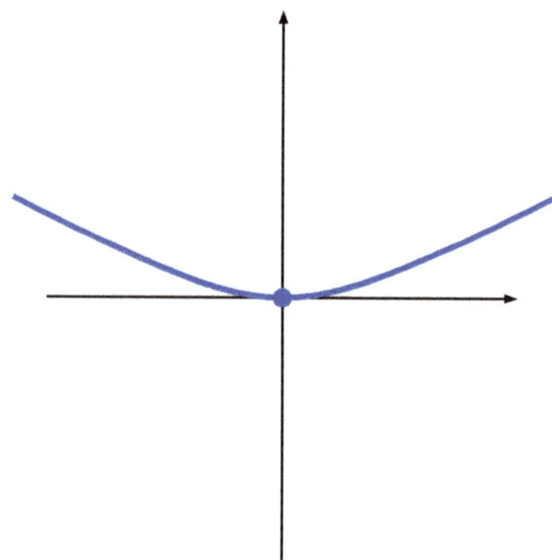

Fig. 1.2 $y^3 + 2x^2y - x^4 = 0$

Example 1.5.1 Consider the polynomial $P = y^3 + 2x^2y - x^4$ and set

$$C := \{(x, y) \in \mathbb{R}^2 \mid P(x, y) = 0\} .$$

If we consider C as a differentiable subset of $\mathbb{R}^2$ (see Figure 1.2) then it is a $\mathcal{C}^\infty$ submanifold, but $(0, 0)$ is a singular point (Definition 1.5.27) of the irreducible algebraic curve (Definition 1.5.9) $C = \mathcal{Z}(P)$.

Indeed, for any $(x, y) \in C$ we have that $x^2 = y(1 + \sqrt{1 + y})$. In some neighbourhood of $(0, 0)$, the function y is thus a smooth (and indeed analytic) function of x by the implicit function theorem. On the other hand, the partial derivatives of P all vanish at $(0, 0)$. The Zariski tangent space $T^{Zar}_{(0,0)}C$ (Definition 1.5.22) is therefore equal to $\mathbb{R}^2$ and the dimension of C is 1 by Definition 1.5.9. The point $(0, 0)$ is therefore a singular point of C by Definition 1.5.27.

Definition 1.5.2 For any set E a *chain* of *length* n of subsets of E is a sequence $E_0 \subsetneq E_1 \subsetneq \cdots \subsetneq E_n$ where the $E_i \subset E$ are all distinct.

Definition 1.5.3 The *dimension* (or *Krull dimension*) of a ring A is the supremum of the lengths of chains of prime ideals in A.

Remark 1.5.4 The dimension of a ring can be infinite even if the ring is Noetherian. See [Eis95, Exercise 9.6, page 229] for Nagata's example of an infinite dimensional Noetherian ring. On the other hand, any *local* (Definition A.3.7) Noetherian ring is of finite dimension (Exercise).

We recall a fundamental result from commutative algebra. In the theorem below, $\operatorname{trdeg}_K L$ denotes the transcendance degree of a field extension $L|K$, see Definition A.5.7.

Theorem 1.5.5 *Let K be a field and let A be a finitely generated integral K-algebra. We then have that*

$$\dim A = \operatorname{trdeg}_K \operatorname{Frac}(A) .$$

Proof By Noether's normalisation Lemma A.5.6, there is an integer $d \geqslant 0$ and an injective map $K[X_1, \ldots, X_d] \hookrightarrow A$ which makes A into an integral K-algebra over $K[X_1, \ldots, X_d]$, finitely generated as a $K[X_1, \ldots, X_d]$-module. It follows from Proposition A.5.3 that $\dim A = d$ and from the fact that Frac A is algebraic over $K(X_1, \ldots, X_d)$ that $\operatorname{trdeg}_K \operatorname{Frac}(A) = d$. □

Exercise 1.5.6 For any field K we have that $\dim K[X_1, \ldots, X_n] = n$.

Example 1.5.7 If $A := \mathbb{C}[X, Y]$ then the Krull dimension of A is 2. A chain realising this equality is given by $(0) \subsetneq (X) \subsetneq (X, Y)$. The transcendance degree of Frac A over $\mathbb{C}$ is 2 and the transcendance degree of Frac A over $\mathbb{R}$ is also 2.

Corollary 1.5.8 *Let A be an affine integral domain and let $\mathfrak{m} \subset A$ be a maximal ideal in A. We then have that* $\dim A_{\mathfrak{m}} = \dim A$.

Proof By Definition A.5.5 there is a field K over which A is a finitely generated integral K-algebra. By Theorem 1.5.5 the dimension of A is the common length of all maximal chains of prime ideals of A. If $\mathfrak{m}$ is a maximal ideal there is therefore a chain of prime ideals contained in $\mathfrak{m}$ of length $\dim A$. But the dimension of $A_{\mathfrak{m}}$ is the length of a maximal chain of prime ideals contained in $\mathfrak{m}$ by Proposition A.3.5. □

Definition 1.5.9 The *dimension* $\dim I$ of an ideal I in $K[X_1, \ldots, X_n]$ is the dimension of the quotient ring $K[X_1, \ldots, X_n]/I$. The *dimension* of an irreducible algebraic set $F \subset \mathbb{A}^n(K)$ over a field K is the dimension of its associated ideal $\mathcal{I}(F)$, or in other words the dimension of its ring of affine coordinates[13] $\mathcal{A}(F) = K[X_1, \ldots, X_n]/\mathcal{I}(F)$. We denote this quantity by $\dim F$ or $\dim_K F$.

Remark 1.5.10 We could have used the dimension of the ring $\mathcal{R}(F)$ rather than $\mathcal{A}(F)$ but if $K = \mathbb{R}$ (for example) the ring $\mathcal{R}(F)$ is not typically finitely generated (see Proposition 1.2.38). On the other hand, for any $x \in F$, the local ring $\mathcal{R}(F)_{\mathfrak{m}_x}$ is of the right dimension, see Proposition 1.5.41.

Remark 1.5.11 Any ideal in A has a natural A-module structure. The dimension of the A-module I is equal to the dimension of A whenever A is an integral domain for example. It is important not to confuse the dimension of the ideal $\dim I = \dim A/I$ with the dimension of I as an A-module. See [Eis95, Chapter 9] for more details.

Definition 1.5.12 In a ring A the *codimension* (or *height*) of a prime ideal I is the supremum of lengths of chains of prime ideals contained in I. It is denoted codim I.

Exercise 1.5.13 (Dimension and codimension of an ideal)

---

[13] See Proposition 1.2.27.

1. It follows from the correspondence theorem (see Proposition A.2.8) that if I is an ideal of A then $\dim I$ is the supremum of lengths of chains of prime ideals of A containing I.
2. For any prime ideal I in A we have that $\operatorname{codim} I = \dim A_I$.

Exercise 1.5.14 Let I be a prime ideal of dimension d in $\mathbb{C}[X_1, \ldots, X_n]$. If I is generated by polynomials with real coefficients then

$$\dim(I \cap \mathbb{R}[X_1, \ldots, X_n]) \leqslant d$$

as an ideal in $\mathbb{R}[X_1, \ldots, X_n]$.

It turns out we can do better: these two sets are of the same dimension.

Lemma 1.5.15 *Let I be a prime ideal of dimension d in $\mathbb{C}[X_1, \ldots, X_n]$. The ideal $I \cap \mathbb{R}[X_1, \ldots, X_n]$ is then a prime ideal of dimension d in $\mathbb{R}[X_1, \ldots, X_n]$.*

Proof Simply apply Proposition A.5.3 to the integral injective map $\mathbb{R}[X_1, \ldots, X_n]/(I \cap \mathbb{R}[X_1, \ldots, X_n]) \to \mathbb{C}[X_1, \ldots, X_n]/I$. □

Note that for any given prime ideal $I \subset K[X_1, \ldots, X_n]$ the dimension of the algebraic set $F = \mathcal{Z}(I)$ is equal to the dimension of the ideal $\mathcal{I}(F)$ which is not necessarily equal to the dimension of I—see Example 1.5.20.

Exercise 1.5.16 Let K be a field. Deduce from Exercise 1.5.6 that $\dim \mathbb{A}^n(K) = n$.

We now give a more direct definition of the dimension of an irreducible affine algebraic set (Proposition 1.5.19).

Definition 1.5.17 The *dimension* of a topological space X is the supremum of the lengths of chains of irreducible closed subsets of X. We denote this dimension by $\dim X$. If $X \neq \varnothing$ then $\dim X$ is a natural number or $+\infty$. By convention we set $\dim \varnothing = -\infty$.

Remark 1.5.18 This definition is well adapted to coarse topologies such as the Zariski topology which are almost combinatorical. As an exercise, the reader may check that any Hausdorff topological space has dimension 0.

For reasonable topological spaces such as topological manifolds (see Appendix B.5) it is better to define the dimension as being the maximal index of non-zero cohomological groups with compact support of the space (see Definition B.6.7).

Proposition 1.5.19 *Let K be a field. The dimension of an irreducible algebraic set $F \subset \mathbb{A}^n(K)$ over K is equal to its dimension as a topological space with the induced Zariski topology.*

When the base field is not algebraically closed this proposition should be applied to the ideal $\mathcal{I}(F)$, see Example 1.5.20.

Proof Left to the reader as an exercise. □

Example 1.5.20 The irreducible affine algebraic set

$$F := \mathcal{Z}(x^2 + y^2) \subset \mathbb{A}^2(\mathbb{R})$$

is a single point $(0, 0)$. The dimension of the quotient ring $\mathbb{R}[x, y]/(x^2 + y^2)$ is equal to 1 (see Example A.2.10). *A priori* this may seem counter-intuitive because the dimension of F as a topological space is 0—which is also the dimension of the quotient of the ring $\mathbb{R}[x, y]$ by the ideal $\mathcal{I}(F) = (x, y)$. We have that $(x, y) = \mathcal{I}(\mathcal{Z}(x^2 + y^2)) \neq (x^2 + y^2)$ and

$$\frac{\mathbb{R}[x, y]}{(x, y)} = \mathcal{P}(F) \neq \frac{\mathbb{R}[x, y]}{(x^2 + y^2)} .$$

The situation is better understood by considering the algebraic set $Z := \mathcal{Z}_{\mathbb{C}}(x^2 + y^2) \subset \mathbb{A}^2(\mathbb{C})$[14] which is a reducible complex curve. The point $(0, 0)$ is the intersection of the two irreducible components $\mathcal{Z}_{\mathbb{C}}(x - iy)$ and $\mathcal{Z}_{\mathbb{C}}(x + iy)$ and it is the only *real* point on the curve Z.

Example 1.5.21 Consider the affine algebraic set

$$F := \mathcal{Z}(y^2 - x^2(x - 2)) \subset \mathbb{A}^2(\mathbb{R})$$

shown in Figure 1.3. It is an irreducible algebraic set of dimension 1 and the points $(0, 0)$ and $(0, 2)$ are irreducible algebraic sets of dimension 0 contained in F. The fact that $(0, 0)$ appears isolated may seem at first glance to be contradictory with the fact that it is a point on a *curve*.

Non-singular Points

As the Example 1.5.21 shows, the meaning of the algebraic dimension of an algebraic set is not always obvious. In the following section, we will define "non-singular" or "smooth" points, where the dimension can be easily interpreted.

Definition 1.5.22 Let K be a field, let $F \subset \mathbb{A}^n(K)$ be an algebraic set and let $P_1, \ldots, P_l$ be a generating set for $\mathcal{I}(F)$. Consider a point $a \in F$. The *Zariski tangent space* to F at the point a, denoted $T_a^{Zar} F$, is the subspace of K^n given by

$$T_a^{Zar} F := \bigcap_{i=1}^{l} \left\{ x \in K^n \mid \sum_{j=1}^{n} \frac{\partial P_i}{\partial X_j}(a) x_j = 0 \right\} = \bigcap_{i=1}^{l} \ker d_a P_i \ .$$

[14] The set $\mathcal{Z}_{\mathbb{C}}(x^2 + y^2)$ is the set of complex zeros of $x^2 + y^2$, see Definition 1.2.12.

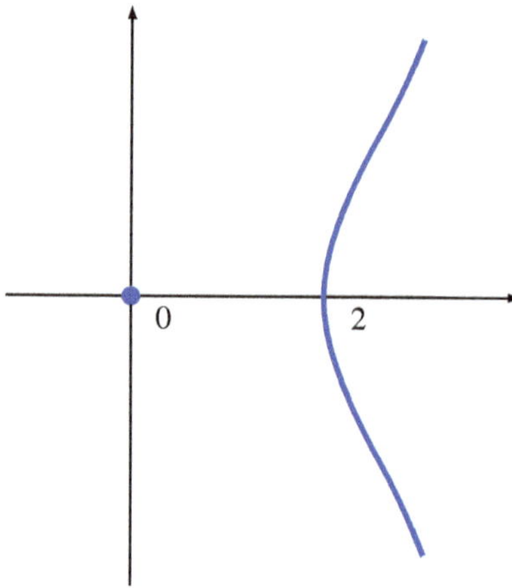

Fig. 1.3 $y^2 - x^2(x-2) = 0$

Remark 1.5.23 The partial derivative of a a polynomial with respect to one of its variables is well-defined over an arbitrary field, but its behaviour in nonzero characteristic p can be surprising. For example, if $P(X) = X^p$ then $\frac{\partial P}{\partial X =} pX^{p-1} = 0$.

Remark 1.5.24 (*Notation* $T_a^{Zar} F$) If $T_a F$ denotes the usual differential-geometric tangent space to $a \in F$ then in Example 1.5.1,

$$\mathbb{R}^2 = T_{(0,0)}^{Zar} C \neq T_{(0,0)} C = \mathcal{Z}(y) \simeq \mathbb{R} \ .$$

We will define non-singular points later on. In any such point, the Zariski tangent space and the usual tangent space coincide.

Proposition 1.5.25 *Let K be a field of characteristic zero and let I be a prime ideal in $K[X_1, \ldots, X_n]$. Set $A := K[X_1, \ldots, X_n]/I$. If d is the dimension of the ring A then for any set of generators (see Example A.3.14) $P_1, \ldots, P_l$ of I, the matrix $\left(\frac{\partial P_i}{\partial X_j}\right)_{\substack{i=1\ldots l \\ j=1\ldots n}}$ has rank $n-d$ over* $\mathrm{Frac}(A)$.

In particular, for any $a \in \mathcal{Z}(I)$, $\mathrm{rk}_K \left(\frac{\partial P_i}{\partial X_j}(a)\right)_{\substack{i=1\ldots l \\ j=1\ldots n}} \leqslant n-d$. Indeed, the proposition tells us that determinants of all the $(n-d+1) \times (n-d+1)$-sub-matrices of $\left(\frac{\partial P_i}{\partial X_j}\right)_{\substack{i=1\ldots l \\ j=1\ldots n}}$ vanish over $\mathrm{Frac}(A)$ so the determinants of all $(n-d+1) \times (n-d+1)$-submatrices of $\left(\frac{\partial P_i}{\partial X_j}(a)\right)_{\substack{i=1\ldots l \\ j=1\ldots n}}$ vanish.

Proof See [HP52, Chapter 10, Section 14, Theorem 1] or [Sam67, Chapter 2, Section 4.2, lemme 2]. □

Remark 1.5.26 By Proposition 1.5.25, if F is irreducible then

$$\dim_K T_a^{Zar} F \geqslant \dim_K F \ .$$

Definition 1.5.27 Let $F \subset \mathbb{A}^n(K)$ be an irreducible algebraic set and consider a point $a \in F$. We say that a is a *non singular point* of F if $\dim_K T_a^{Zar} F = \dim_K F$. Any point a which is not a non singular point of F will be said to be a *singular point*.

Remark 1.5.28 We set $d = \dim_K F$. If $P_1, \ldots, P_l$ is a generating set for $\mathcal{I}(F)$, a point a is a non singular point of F if and only if

$$\mathrm{rk}_K \left(\frac{\partial P_i}{\partial X_j}(a) \right)_{\substack{i=1\ldots l \\ j=1\ldots n}} = n - d \ .$$

In particular, if $K = \mathbb{R}$ (resp. $\mathbb{C}$), and a is a non singular point of F then some neighbourhood of a in the subset $F \subset K^n$ with its Euclidean topology (see Definition 1.4.1) is a differentiable submanifold of K^n whose real dimension is d (resp. $2d$).

Proposition 1.5.29 *Let $F \subset \mathbb{A}^n(\mathbb{R})$ be an irreducible algebraic set which in the Euclidean topology is a differentiable submanifold of $\mathbb{R}^n$ of dimension d in a neighbourhood of some point $a \in F$. We then have that* $\dim F = d$ *as a real algebraic set. In other words, the dimension of F as an algebraic set, or alternatively as a Zariski closed subset, is equal to its dimension as a differentiable manifold.*

Proof We follow the proof given in [BCR98, Proposition 2.8.14]. By hypothesis, the space $T_a F$ is a vector space of dimension d which is also an affine subspace of $\mathbb{R}^n$. The orthogonal projection[15] $p_a \colon F \to T_a F$ is a semi-algebraic map (see Definition B.2.2) which induces a bijection between some open semi-algebraic neighbourhood U of a in F and a semi-algebraic open set in $T_a F$. As the map p_a is semi-algebraic we have that the dimension associated to the Zariski topology (Definition 1.5.17) satisfies $\dim U = \dim p_a(U)$ by [BCR98, Theorem 2.8.8]. Since $p_a(U)$ is a non-empty Euclidean open set in $\mathbb{R}^d = T_a F$ its Zariski dimension is d. We now prove by induction on d that if $f \in \mathbb{R}[X_1, \ldots, X_d]$ vanishes on a non empty Euclidean open set V then f is the zero polynomial. If $d = 1$ then the result is immediate. Suppose now that $d > 1$ and our result holds for $d - 1$. Let $f \in \mathbb{R}[X_1, \ldots, X_d]$ be a polynomial function vanishing on V. We can write

$$f(X', X_d) = X_d^l f_l(X') + X_d^{l-1} f_{l-1}(X') + \cdots + f_0(X')$$

where $X' = (X_1, \ldots, X_{d-1})$, $l = \deg f$ and $\forall i = 0, \ldots, l$, $f_i \in \mathbb{R}[X_1, \ldots, X_{d-1}]$.

For any $X' \in V \cap \mathbb{R}^{d-1}$ the function $X_n \mapsto f(X', X_n)$ vanishes at every point of $V \cap \mathbb{R}$ and is therefore the zero polynomial. It follows that the polynomial functions f_i vanish at every point of $V \cap \mathbb{R}^{d-1}$ and therefore are identically zero by

[15] By which we mean the restriction to F of the orthogonal projection from $\mathbb{R}^n \to T_a F$ for some scalar product on $\mathbb{R}^n$.

the induction hypothesis. Any element $f \in \mathcal{I}(p_a(U))$ vanishes at every point of $p_a(U)$ so f is a zero function by the above. It follows that $\mathcal{I}(p_a(U)) = (0)$ and $\mathcal{Z}(\mathcal{I}(p_a(U))) = \mathbb{A}^d(\mathbb{R})$ which has dimension d by Exercise 1.5.16. □

Remark 1.5.30 Example 1.5.1 shows that a point of an algebraic set can be smooth differentiably but singular algebraically. The above result shows that despite this, the dimension of an algebraic set as a differentiable manifold and as an algebraic variety are the same.

Exercise 1.5.31 We go back to the polynomial

$$P = y^3 + 2x^2y - x^4$$

from Example 1.5.1 and consider the complex curve $F_{\mathbb{C}} := \mathcal{Z}_{\mathbb{C}}(P) \subset \mathbb{A}^2(\mathbb{C})$, i.e. the set of points $F_{\mathbb{C}} = \{(x, y) \in \mathbb{C}^2 \mid P(x, y) = 0\}$.

1. Prove that $F_{\mathbb{C}}$ is an irreducible algebraic set.
2. Prove that the dimension of $F_{\mathbb{C}}$ is 1.
3. Prove that the dimension of $T^{Zar}_{(x,y)} F_{\mathbb{C}}$ is 1 at every point in $(x, y) \in F_{\mathbb{C}}$ other than $(0, 0)$.
4. Deduce that $(0, 0)$ is a singular point of $F_{\mathbb{C}}$.
5. Prove that, unlike the real curve, the complex curve $F_{\mathbb{C}}$ is not a $\mathcal{C}^\infty$ submanifold of $\mathbb{C}^2$ in a neighbourhood of the point $(0, 0)$.

Whilst it is easy to check that the definitions of dimension and singular points of an affine algebraic set F do not depend on the precise equations used to define F or on the precise choice of embedding into affine space, it is still useful to have an intrinsic definition.

Definition 1.5.32 A Noetherian local ring A of maximal ideal $\mathfrak{m}$ and residue field $K = A/\mathfrak{m}$ is said to be *regular* if $\dim A = \dim_K \mathfrak{m}/\mathfrak{m}^2$, where we consider the natural K-vector space structure on $\mathfrak{m}/\mathfrak{m}^2$.

Proposition 1.5.33 *Let $F \subset \mathbb{A}^n(K)$ be an irreducible algebraic set over a field K. A point $a \in F$ is non-singular if and only if the local ring $\mathcal{O}_{F,a} = \mathcal{P}(F)_{\mathfrak{m}_a}$ of germs of regular functions at a is regular.*

Proof We assume that K is of characteristic zero. See [Liu02, Proposition IV.2.5 and Theorem IV.2.19] for a proof in arbitrary characteristic. Let a be the origin in K^n. The ideal $\mathfrak{m}_a$ is formed of polynomials that vanish at a. Consider the linear map $\theta_a \colon \mathfrak{m}_a \to (K^n)^\vee$ sending $P \in \mathfrak{m}_a$ to the linear form

$$x = (x_1, \dots, x_n) \mapsto \sum_{j=1}^{n} \frac{\partial P}{\partial X_j}(a) \cdot x_j \, .$$

The map θ_a induces an isomorphism $d_a \colon \mathfrak{m}_a/\mathfrak{m}_a^2 \to (K^n)^\vee$. The dual of $T_a^{Zar}(F) \subset K^n$ can be identified with a quotient of $(K^n)^\vee$ isomorphic via d_a to the quotient $\mathfrak{m}_a/(\mathfrak{m}_a^2 + \mathcal{I}(F))$. Let $\mathfrak{m}_{F,a}$ be the maximal ideal of the local ring $\mathcal{O}_{F,a}$. The K-vector space $\mathfrak{m}_a/(\mathfrak{m}_a^2 + \mathcal{I}(F))$ is then isomorphic to $\mathfrak{m}_{F,a}/\mathfrak{m}_{F,a}^2$. It follows from Definition 1.5.27 that a is a non-singular point of F if and only if $\dim \mathfrak{m}_{F,a}/\mathfrak{m}_{F,a}^2 = \dim F$. We now note that since F is irreducible, $\dim \mathcal{O}_{F,a} = \dim \mathcal{O}_F$ by Corollary 1.5.8. □

Proposition 1.5.34 *Let $F \subset \mathbb{A}^n(\mathbb{R})$ be a real algebraic set and let a be a point of F. The ring $\mathcal{O}_{F,a}$ is then a regular local ring of dimension d if and only if there exist $n-d$ polynomials $P_1, \ldots, P_{n-d} \in \mathcal{I}(F)$ and an open Euclidean set U in $\mathbb{R}^n$ containing a such that $F \cap U = \mathcal{Z}(P_1, \ldots, P_{n-d}) \cap U$ and*

$$\mathrm{rk}\left(\frac{\partial P_i}{\partial X_j}(a)\right)_{\substack{i=1\ldots n-d\\ j=1\ldots n}} = n-d\ .$$

Proof To prove that the condition is sufficient we note that the standard differential-geometric proof remains valid in the setting of semi-algebraic sets. We can assume that the determinant of the sub-matrix $\left(\frac{\partial P_i}{\partial X_j}(a)\right)_{\substack{i=1\ldots n-d\\ j=1\ldots n}}$ is non zero, so applying the implicit function theorem to $(X_1 - a_1, \ldots, X_d - a_d, P_1, \ldots, P_{n-d})$, we get a semi-algebraic diffeomorphism $\varphi\colon U \to V$ from U, a semi-algebraic open neighbourhood of 0 in $\mathbb{R}^n$ to V, a semi-algebraic open neighbourhood of a in $\mathbb{R}^n$ such that $\varphi((\mathbb{R}^d \times \{0\}) \cap U) = F \cap V$. The Zariski dimension of any irreducible component of F passing through a is therefore less than or equal to d by [BCR98, Theorem 2.8.8]. The dimension of the ring $\mathcal{O}_{F,a}$ is thus bounded below by d. Moreover, $\mathcal{O}_{F,a}$ is a quotient of $\mathcal{O}_{\mathbb{R}^n,a}/(P_1, \ldots, P_{n-d})$, which is a regular (and in particular integral) local ring of dimension d, and it follows that $\mathcal{O}_{F,a} = \mathcal{O}_{\mathbb{R}^n,a}/(P_1, \ldots, P_{n-d})$. This completes the proof of the proposition. □

Let I be a prime ideal of $\mathbb{R}[X_1, \ldots, X_n]$. We saw in Example 1.5.20 that it is possible to have $\dim \mathcal{I}(\mathcal{Z}(I)) < \dim I$. The previous proposition yields a characterisation of the case where these two dimensions are equal.

Corollary 1.5.35 *Let $I = (P_1, \ldots, P_l)$ be a prime ideal in $\mathbb{R}[X_1, \ldots, X_n]$. We then have that* $\dim(\mathcal{I}(\mathcal{Z}(I))) = \dim I$ *if and only if $\mathcal{Z}(I)$ contains a point a such that*

$$\mathrm{rk}_{\mathbb{R}}\left(\frac{\partial P_i}{\partial X_j}(a)\right)_{\substack{i=1\ldots l\\ j=1\ldots n}} = n - \dim I\ .$$

Proof We set $d = \dim I$. Let $a \in \mathcal{Z}(I)$ be such that the rank of the matrix $\left(\frac{\partial P_i}{\partial X_j}(a)\right)$ is $n-d$. We then have that $\mathcal{O}_{F,a}$ is a regular ring of dimension d by Proposition 1.5.34 and $\mathcal{Z}(I)$ is therefore an algebraic set of dimension d or in other words $\dim(\mathcal{I}(\mathcal{Z}(I))) = d$. □

Proposition 1.5.33 renders the notion of "non singular point" intrinsic and enables us to generalise it to abstract algebraic varieties.

Definition 1.5.36 Let X be an algebraic variety over a field K. A point $x \in X$ is said to be *non singular* (or *regular*) if the local ring $\mathcal{O}_{X,x}$ is a regular ring. The variety X is said to be *non singular* if all its points are non singular. The variety X is said to be *singular* if it has at least one singular point. We denote by $\operatorname{Sing} X$ the locus of singular points (or *singular locus*) of X and by $\operatorname{Reg} X := X \setminus \operatorname{Sing} X$ the locus of non singular points (or *regular locus*) of X.

We recall the definition of a *normal point* of a variety. For a curve, this means that there is only one *branch* (*i. e.* local irreducible component) of the curve passing through the point in question.

Definition 1.5.37 A quasi-projective algebraic variety X over a field K is said to be *normal at* $x \in X$ if the local ring $\mathcal{O}_{X,x}$ is integrally closed (Definition A.5.2) in $K(X)$. The variety X is said to be *normal* if it is normal at every point.

Example 1.5.38 Let X be a real irreducible quasi-projective algebraic variety. We know that X is affine and we can therefore assume that $X \subset \mathbb{R}^n$ as an algebraic set. Let A be the integral closure of $\mathcal{P}(X)$ in its fraction field. Since A is a finitely generated $\mathbb{R}$-algebra, we can assume that $A = \mathbb{R}[X_1, \ldots, X_p]/I$ for some ideal $I \subset \mathbb{R}[X_1, \ldots, X_p]$. We set $\widetilde{X} := \mathcal{Z}(I) \subset \mathbb{R}^p$. The variety $\widetilde{X}$ is then normal. We call it the *normalisation* of X and the birational map $\nu \colon \widetilde{X} \to X$ is called the *normalisation map*.

Exercise 1.5.39 Let $F \subset \mathbb{A}^2(\mathbb{R})$ be the affine cubic of equation $y^2 - x^2(x-2) = 0$ represented in Figure 1.3.

We then have that $\widetilde{F} = \mathbb{R}$ and the normalisation map is given as follows:

$$\begin{aligned} \mathbb{R} &\longrightarrow \mathbb{R}^2 \\ t &\longmapsto \left(t^2+2, t(t^2+2)\right) . \end{aligned}$$

Proposition 1.5.40 *Let X be a topological space and let $X = \bigcup_{i=1}^k X_i$ be a decomposition of X into not necessarily disjoint closed sets X_i. We then have that*

$$\dim X = \sup_{i \in \{1 \ldots k\}} \dim X_i .$$

Proposition 1.5.41 *Let X be an affine algebraic variety over an infinite base field K. If X is irreducible then for any $x \in X$ we have that* $\dim X = \dim \mathcal{O}_{X,x}$. *If X is not irreducible then the Krull dimension of $\mathcal{O}_{X,x}$ is the maximum of the dimensions of irreducible components of X containing x. We denote this number by* $\dim_x X$.

Exercise 1.5.42 Prove this lemma, starting with the case where K is algebraically closed (see [Per95, IV.2.3 et 2.9]) and the avoidance lemma A.3.12, using the fact that K is infinite.

Definition 1.5.43 Let X be an algebraic variety over an infinite field K and consider a point $x \in X$. The *dimension* of X at the point x is defined by

$$\dim_x X := \dim \mathcal{O}_{X,x} \ .$$

The *dimension* of X is the supremum of all these dimensions

$$\dim X := \sup_{x \in X} \dim_x X \ .$$

Remark 1.5.44 The dimension of an abstract variety at a point can be calculated in any open affine subset containing the point.

Definition 1.5.45 An algebraic variety of dimension 1 is called a *curve* and an algebraic variety of dimension 2 is called a *surface*.

Exercise 1.5.46 Let K be a field. Deduce from Exercise 1.5.16 that $\dim \mathbb{P}^n(K) = n$.

Definition 1.5.47 Let $x \in X$ be a non singular point in a variety of dimension n over a field K. A set of n elements $f_1, \dots, f_n \in \mathcal{O}_x$ is said to be a *local system of parameters* at x if every $f_i \in \mathfrak{m}_x$ and the classes $f_1, \dots, f_n$ form a basis of the K-vector space $\mathfrak{m}_x/\mathfrak{m}_x^2$.

Note that the analytic local inversion theorem fails in the algebraic setting. The parameters f_i only become local coordinates after refining the topology.

Exercise 1.5.48 If $K = \mathbb{C}$ there is an Euclidean open neighbourhood U of x (see Definition 1.4.1) on which the restrictions $f_1, \dots, f_n$, seen as functions on U, form a system of complex local analytic coordinates.

Proposition 1.5.49 *Let $x \in X$ be a non singular point on an algebraic variety of dimension n over a field K. Any system of local parameters at x generates the maximal ideal m_x in $\mathcal{O}_x$.*

Proof Let $f_1, \dots, f_n \in \mathcal{O}_x$ be a local system of parameters at x. We simply apply Nakayama's Lemma (A.2.11) to the finitely generated $\mathcal{O}_x$-module $M = \mathfrak{m}_x/\langle f_1, \dots, f_n\rangle$ and the ideal $\mathfrak{a} = \mathfrak{m}_x$. □

Exercise 1.5.50 Using Theorem 1.5.5, prove that if X is an irreducible algebraic variety over a field K then

$$\dim X = \operatorname{trdeg}_K K(X) \ .$$

The following theorem tells us that we can calculate the dimension of a variety X by calculating the dimension of each of its irreducible components at a non-singular point and taking the maximum of the numbers thus obtained.

Theorem 1.5.51 *Let X be an algebraic variety over a field of characteristic zero. If $X \neq \varnothing$ then the set* Sing X *of singular points of X is a strict closed subset of X. In other words, the set* Reg X *of non-singular points of X is a non-empty Zariski open subset.*

Proof As X is reduced by definition, we may without loss of generality assume that X is irreducible and affine. Assume that $X \subset \mathbb{A}^n(K)$ and set $d := \dim X$. As the field K is of characteristic zero, Definition 1.5.27 implies that Sing X is algebraic since it is defined as the set of zeros of the ideal generated by $\mathcal{I}(X) = (P_1, \ldots, P_l)$ and all the determinants of $(n-d) \times (n-d)$-submatrices of $\left(\frac{\partial P_i}{\partial X_j}\right)_{\substack{i=1\ldots l\\ j=1\ldots n}}$. In particular, if $\mathcal{I}(\operatorname{Sing} X) = \mathcal{I}(X)$ then the rank of the matrix

$$\left(\frac{\partial P_i}{\partial X_j}\right)_{\substack{i=1\ldots l\\ j=1\ldots n}}$$

over $\operatorname{Frac}(K[X_1, \ldots, X_n]/\mathcal{I}(X))$ is strictly less than $n-d$. By Proposition 1.5.25, the ideal $\mathcal{I}(\operatorname{Sing} X)$ is strictly larger than $\mathcal{I}(X)$ and it follows that $\operatorname{Sing} X = \mathcal{Z}(\mathcal{I}(\operatorname{Sing} X)) \subsetneq X = \mathcal{Z}(\mathcal{I}(X))$. □

Remark 1.5.52 1. There is a proof of the same result for algebraically closed K of arbitrary characteristic in [Har77, Chapitre I, Theorem 5.3].

2. The theorem holds for an arbitrary field K, see [Liu02, Proposition IV.2.24 and Corollary VIII.2.40(a)].

Definition 1.5.53 Let X be an algebraic variety over a field K. A *resolution of singularities* of X is a proper birational morphism $\pi : Y \to X$ which induces a biregular map $Y \setminus \pi^{-1}(\operatorname{Sing} X) \to X \setminus \operatorname{Sing} X$.

Recall that Hironaka's theorem on the resolution of singularities [Hir64] (see also [Kol07]) holds over any field of characteristic zero. We refer to [Wal35] for a proof for surfaces.

Theorem 1.5.54 (Hironaka 1964) *Let X be an algebraic variety over a field K of characteristic zero. There is then a non singular K-variety Y and a proper birational morphism $\pi : Y \to X$ which induces a biregular morphism $Y \setminus \pi^{-1}(\operatorname{Sing} X) \to X \setminus \operatorname{Sing} X$.*

Moreover if X is projective we can require the variety Y to be projective.

We end this section with the birational invariance of the number of Euclidean connected components.

Theorem 1.5.55 *Let X and Y be quasi-projective algebraic varieties over the same base field K which are both complete and non singular. If $K = \mathbb{R}$ or $\mathbb{C}$ and X and Y are birationally equivalent over K then they have the same number of connected components in the Euclidean topology.*

Remark 1.5.56 The normalisation map $\nu : \widetilde{C} \to C$ over a curve C such that $\widetilde{C}$ has two connected components whose images under ν meet in at least one point illustrates the fact that the "non singular" hypothesis is necessary in the above result. Consider for example the curve $\widetilde{C} = \mathcal{Z}((x_2, x_3) \cap (x_1, x_3 - x_0))$ in $\mathbb{P}^3_{x_0:x_1:x_2:x_3}$ and let ν be the restriction to $\widetilde{C}$ of the projection $\mathbb{P}^3 \to \mathbb{P}^2$, $(x_0 : x_1 : x_2 : x_3) \mapsto$

$(x_0 : x_1 : x_2)$. The (reducible) curve $\widetilde{C}$ is a disjoint union of two lines—C_1, whose equations are $x_2 = x_3 = 0$ and C_2, whose equations are $x_1 = 0, x_3 = x_0$. The curve $C := \nu(\widetilde{C})$ is connected and singular since $\nu(C_1) \cap \nu(C_2) = \{(1 : 0 : 0)\}$. The hypothesis "complete" is also necessary, because the ellipse and the hyperbola are birationally equivalent—if $E := \{x^2 + y^2 = 1\} \subset \mathbb{R}^2_{x,y}$, $H := \{u^2 - v^2 = 1\} \subset \mathbb{R}^2_{u,v}$ and $\varphi\colon E \dashrightarrow H$, $(x, y) \mapsto (\frac{1}{x}, \frac{y}{x})$ then φ is a birational map inducing an isomorphism between the dense open set $E \setminus \{x = 0\}$ in E and H—but $\#\pi_0(E) = 1$ whereas $\#\pi_0(H) = 2$.

Proof When $K = \mathbb{C}$ this statement is a corollary of Theorem 1.4.5. When $K = \mathbb{R}$ Proposition 1.3.26 implies that two non singular complete curves are birationally equivalent if and only if they are biregularly isomorphic and the theorem follows for curves. For real surfaces, the result follows from the factorisation of birational maps (Corollary 4.3.9) and the explicit description of the topology of a blow-up (Example 4.2.18). For the general case, see the proof of Theorem 2.3.12. □

Remark 1.5.57 An alternative proof is given in [DK81, Thm. 13.3] and there is a sketch proof in [BCR98, Theorem 3.4.12] which remains valid for a real closed field other than $\mathbb{R}$ if we replace "connected Euclidean" by "semi-algebraically connected".

1.6 Plane Curves

This section draws on [Che78] and [Ful89].

Definition 1.6.1 Let K be a field.

1. We say that two polynomials $P, Q \in K[X, Y]$ are *equivalent* if there is a non-zero $\lambda \in K^*$ such that $P = \lambda Q$.
2. An *affine plane curve* defined over K is then an equivalence class of non-constant polynomials for this relation.
3. Let $P \in K[X, Y]$ be a non-constant polynomial. We say that P *determines* (or is the *equation* of) the affine plane curve represented by P.
4. Similarly, let $P(X_0, X_1, X_2) \in K[X_0, X_1, X_2]$ be a non-constant homogeneous polynomial in three variables. We say that P determines a *projective plane curve*.
5. We say that an affine or projective plane curve is *irreducible* (resp. *reduced*) over K if P is irreducible over K (resp. has no multiple factors).

Remarks 1.6.2 (*Sets of points vs. equations*)

1. If P is an irreducible polynomial then the ideal generated by P is prime and its zero locus $\mathcal{Z}(P)$ is an irreducible topological space, see Definition 1.2.15. On the other hand, even though $\mathcal{Z}(P^2) = \mathcal{Z}(P)$, the polynomial P^2 is not irreducible. This illustrates the pitfalls that arise when mixing the two definitions of a "plane curve" used in this section- the algebraic set or the equivalence class of polynomials.

2. Consider a given plane curve. As any two equivalent polynomials have the same zero set C, the plane curve (in the sense of polynomials) determines C. The set C is an algebraic variety over K by definition and if $\dim C = 1$ it is an algebraic curve as defined in Definition 1.5.45. As mentioned in Exercise 1.2.68 on conics, this zero locus is often identified with the class of polynomials by abuse of notation, which can be risky.
   If $K = \mathbb{C}$ then there is a one-to-one correspondence between affine (resp. projective) *reduced* plane curves and 1-dimensional subvarieties of $\mathbb{A}^2(\mathbb{C})$ (resp. $\mathbb{P}^2(\mathbb{C})$). The definition given above generalises the notion of a dimension 1 subvariety of the plane by authorising multiple components—consider the double line of Exercise 1.2.68. Further on we will return to our original definition (1.3.1) which requires varieties to be reduced and our "multiple plane curves" will be thought of as special divisors of the plane. See Definition 2.6.1 for more details.
   If $K = \mathbb{R}$ the zero locus of an affine plane curve can be empty—consider $x^2 + y^2 + 1 = 0$ for example—or of dimension 0—consider $x^2 + y^2 = 0$. This problem will be resolved in Chapter 2 when we introduce $\mathbb{R}$-curves.

Let $\mathbb{A}^n(K)$ be affine space of dimension n over a field K. An affine change of coordinates is a bijective polynomial map $\Phi = (P_1, \ldots, P_n)\colon \mathbb{A}^n(K) \to \mathbb{A}^n(K)$ such that every P_i is a linear polynomial.

Exercise 1.6.3 Prove that any such morphism is the composition of a linear map and a translation.

Let $\Phi = (P, Q)\colon \mathbb{A}^2(K) \to \mathbb{A}^2(K)$ be an affine change of coordinates, *i. e.* $P(X, Y) = a_0 + a_1X + a_2Y$, $Q(X, Y) = b_0 + b_1X + b_2Y$ and

$$a_1b_2 - a_2b_1 \neq 0\,.$$

For any $f \in K[X, Y]$ we define $f^\Phi \in K[X, Y]$ by

$$f^\Phi(X, Y) = f(P(X, Y), Q(X, Y))\,.$$

The map $f \mapsto f^\Phi$ thus defined is an element of the automorphism group $\mathrm{Aut}(K[X, Y]|K)$. If C is a plane curve of equation f then we denote by C^Φ the plane curve of equation f^Φ.

Exercise 1.6.4 Describe the elements of $\mathrm{Aut}(K[X]|K)$. Describe the elements of $\mathrm{Aut}(K[X, Y]|K)$.

Definition 1.6.5 A property of a family of curves $f_1, \ldots, f_k$ and points $p_1, \ldots, p_l$ is said to be *invariant under affine changes of coordinates* if for any affine change of coordinates Φ the property also holds for the families $f_1^\Phi, \ldots, f_k^\Phi$ and $\Phi^{-1}(p_1), \ldots$ $\Phi^{-1}(p_l)$.

Exercise 1.6.6 The degree of a plane curve is invariant under affine change of coordinates.

Intersection Multiplicity

Let $P \in K[X, Y]$ be a polynomial whose constant term is 0. We can then write $P = P_\mu + P_{\mu+1} + \cdots + P_d$ where d is the degree of P and P_i is the degree i homogeneous part of P, $\mu \leqslant d$ and $P_\mu \neq 0$.

Definition 1.6.7 The integer $\mu > 0$ is called the *multiplicity* of P at $(0, 0)$. Let C be a plane curve and let a be a point of C. After affine change of coordinate we may assume that $a = (0, 0)$ and hence an equation for C is a polynomial whose constant term vanishes. The *multiplicity*, denoted $\mu_a(C)$, of the curve C at the point a is defined to be the multiplicity of P at $(0, 0)$. If $\mu_a(C) > 1$ we say that a is a *multiple point* of C with multiplicity $\mu_a(C)$. If $\mu_a(C) = 1$ then a is a *simple point*, if $\mu_a(C) = 2$ it is a *double point*, if $\mu_a(C) = 3$ it is a *triple point* and so on.

See [Ful89, Section 3.2] for a proof of the fact that the above definition is intrinsic and invariant under affine change of coordinates.

Applied to plane curves, Definition 1.5.27 yields the following.

Lemma 1.6.8 *A point $a \in C$ is a singular point of the curve C of equation P if and only if*

$$\frac{\partial P}{\partial x}(a) = 0 \quad et \quad \frac{\partial P}{\partial x}(a) = 0 \ .$$

Exercise 1.6.9 Deduce from Proposition 1.5.33 that the property of being a singular point of a plane curve is invariant under affine change of coordinates.

Exercise 1.6.10 Let C be a plane curve and let a be a point of C. The point a is a simple point of C if and only if it is a non singular point of C.

Recall that by Exercise 1.2.42, for any $a \in \mathbb{A}^n(K)$ the local ring $\mathcal{O}_{\mathbb{A}^n(K),a} \subset K(X_1, \ldots, X_n)$ at a is given by

$$\mathcal{O}_{\mathbb{A}^n(K),a} = \left\{ \frac{F}{G} \in K(X_1, \ldots, X_n) \mid G(a) \neq 0 \right\} \ .$$

Recall that $K[X_1, \ldots, X_n] \subset \mathcal{O}_{\mathbb{A}^n(K),a} \subset K(X_1, \ldots, X_n)$. Let $P_1, \ldots, P_l$ be elements of $K[X_1, \ldots, X_n]$. To simplify notation, we denote by $(P_1, \ldots, P_l)$ the ideal $(P_1, \ldots, P_l)\mathcal{O}_{\mathbb{A}^2(K),a}$ in $\mathcal{O}_{\mathbb{A}^2(K),a}$.

Definition 1.6.11 Let C_1 and C_2 be two affine plane curves over K of equations $P_1(x, y)$ and $P_2(x, y)$ which may be reducible or non-reduced. For any $a \in \mathbb{A}^2(K)$ we set

$$(C_1 \cdot C_2)_a := \dim_K \mathcal{O}_{\mathbb{A}^2(K),a}/(P_1, P_2) \ .$$

The number thus defined is called the *intersection multiplicity* of the curves C_1 and C_2 at a.

Remark 1.6.12 This number is invariant under affine changes of coordinates, see [Ful89, Section 3.3].

Bézout's Theorem

We say that the curves C_1 and C_2 intersect properly at a if C_1 and C_2 have no common component passing through a.

Theorem 1.6.13 (Characterisation of intersection multiplicity) *Let C_1 and C_2 be plane curves. Their intersection number at a has the following properties, and is moreover uniquely determined by them.*

1. $(C_1 \cdot C_2)_a$ *is an integer* $\geqslant 0$ *if* C_1 *and* C_2 *intersect properly at* a;
2. $(C_1 \cdot C_2)_a = 0$ *if and only if* $a \notin C_1 \cap C_2$;
3. *If* Φ *is an affine change of coordinates of* $\mathbb{A}^2(K)$ *and* $b = \Phi^{-1}(a)$ *then* $(C_1^\Phi \cdot C_2^\Phi)_b = (C_1 \cdot C_2)_a$;
4. $(C_1 \cdot C_2)_a = (C_2 \cdot C_1)_a$;
5. $(C_1 \cdot C_2)_a \geqslant \mu_a(C_1)\mu_a(C_2)$, *with equality if and only if the two curves do not have a common tangent line at* a;
6. *If* $C_1 = \sqcup_i C_{1,i}^{r_i}$ *and* $C_2 = \sqcup_j C_{2,j}^{s_j}$ *then*
$$(C_1 \cdot C_2)_a = \sum_{i,j} r_i s_j (C_{1,i} \cdot C_{2,j})_a \ ;$$
7. *If* P *and* Q *are equations of plane curves then*
$$(P \cdot Q)_a = (P \cdot (Q + AP))_a$$
for any polynomial $A \in K[X, Y]$.

Proof See [Ful89, Section 3.3]. □

Exercise 1.6.14 If P_i is a homogeneous degree d_i equation of the projective plane curve C_i then we have

$$(C_1 \cdot C_2)_a = \dim_K \mathcal{O}_{\mathbb{P}^2(K),a}/(\frac{P_1}{L^{d_1}}, \frac{P_2}{L^{d_2}})$$

where $L = 0$ is the equation of a line which does not pass through a. Prove that this number does not depend on the choice of L.

Definition 1.6.15 Let C_1 and C_2 be two projective plane curves which may be reducible or non-reduced. We set

$$(C_1 \cdot C_2) := \sum_{a \in \mathbb{P}^2(K)} (C_1 \cdot C_2)_a .$$

This is the *intersection number* of the curves C_1 and C_2.

Theorem 1.6.16 (Bézout's theorem: geometric statement) *Let C_1 and C_2 be two projective plane curves which may be reducible or non-reduced, of degrees d_1 and d_2 respectively. If C_1 and C_2 are defined over an algebraically closed field and have no common component then*

$$(C_1 \cdot C_2) = d_1 d_2$$

where the intersection points on the left hand side are counted with multiplicity.

Proof See [Per95, Chapitre VI]. □

Genus Formula

Theorem 1.6.17 *For any positive integer d we set*

$$g(d) := \frac{(d-1)(d-2)}{2} .$$

Let C be a non singular complex projective plane curve of degree d. If C is irreducible and of genus $g(C)$ then

$$g(C) = g(d) .$$

Proof (*See* [GH78, pages 219–220]) We project the curve C from a point p to a line L, where the point p should be chosen neither in C nor in L. After a linear change of coordinates we may assume that $p = [1 : 0 : 0]$ and $L = \{X = 0\}$ and we may also suppose that the line at infinity $\{Z = 0\}$ is not tangent to C. Let $F(X, Y, Z)$ be a homogeneous polynomial of degree d defining C. Taking coordinates $x = X/Z$, $y = Y/Z$ in the chart $Z \neq 0$ the affine equation of the curve will be denoted $f(x, y) = F(x, y, 1)$. Consider the projection $\pi_p \colon C \to \mathbb{P}^1$ whose expression in this open affine set is given by:

$$\pi_p \colon (x, y) \mapsto y .$$

The degree of the map $\pi_p \colon C \to \mathbb{P}^1$ is d. Close to a point $q \in C$ such that $(\partial f/\partial x)(q) \neq 0$ the function y is a local coordinate on C and π_p is not ramified. If $(\partial f/\partial x)(q) = 0$ then $(\partial f/\partial y)(q) \neq 0$ since C is non singular at q and the implicit function theorem implies that x is a local coordinate on C in a neighborhood of q. We then have a parameterisation $x \mapsto (x, y) = (x, y(x))$ of C in a neighbourhood of q, from which it follows that

$$f(x, y(x)) \equiv 0$$

and hence the chain rule implies that

$$\frac{\partial f}{\partial x} + \frac{\partial f}{\partial y} \cdot \frac{\partial y}{\partial x} \equiv 0 \quad \text{in a neighbourhood of } q \text{ in } C.$$

Table 1.1 Genuses of smooth plane curves of small degree

d	1	2	3	4	5	6
$g(d)$	0	0	1	3	6	10

It follows that the order of the zero of $\frac{\partial y}{\partial x}$ at q, which is equal to the branching order of $\pi_p : x \mapsto y(x)$ at q, is equal to the order of the zero of $\frac{\partial f}{\partial x}$ at q which is also equal to the intersection multiplicity of C with the curve of equation $\{\frac{\partial f}{\partial x} = 0\}$ at the point q. The equation $\{\frac{\partial f}{\partial x} = 0\}$ determines a curve of degree $d - 1$ in $\mathbb{P}^2$ so the total intersection multiplicity of this curve with C is $d(d-1)$ by Bézout's theorem. By hypothesis there are no points in $\{\frac{\partial f}{\partial x} = 0\} \cap C$ on the line at infinity $\{Z = 0\}$.

By the Riemann–Hurwitz theorem (E.2.18) it follows that

$$g(C) = -d + 1 + \frac{1}{2}\sum_{q \in C} b_{\pi_p}(q) = \frac{1}{2}(d-2)(d-1)\ .$$

□

Remark 1.6.18 An important consequence of the genus formula is the existence of projective curves that cannot be embedded in the projective plane. For example, an irreducible smooth curve of genus 2 cannot be embedded as a non-singular plane curve, since for all $d \in \mathbb{N}^*$, $\frac{1}{2}(d-2)(d-1)) \neq 2$ (Table 1.1).

1.7 Umbrellas

We end this chapter with a series of images showing some remarkable singular real algebraic varieties appearing in the article [FHMM16]. These surfaces are called *umbrellas* because their real locus consists of a two-dimensional surface attached to a one-dimensional *handle*. The first two umbrellas in this list are well known to experts in the field, the others are new.

Remark 1.7.1 Despite appearances, all the umbrellas in this list are irreducible in the Zariski topology

Whitney's Umbrella

This is the subvariety of $\mathbb{R}^3$ whose equation is

$$zx^2 = y^2 \quad \text{(Figure 1.4).}$$

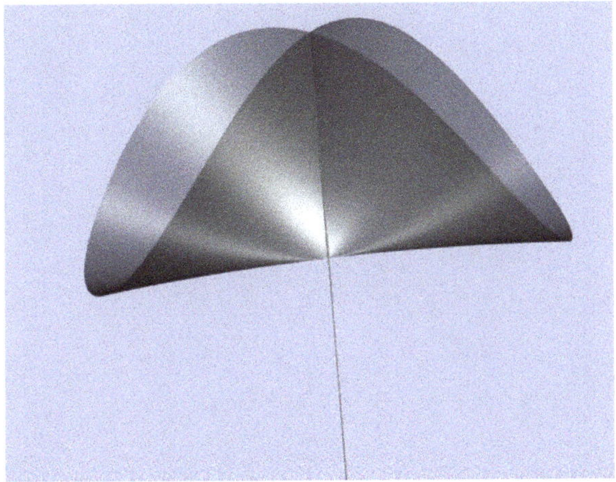

Fig. 1.4 Whitney's umbrella

The line $\{x = y = 0\}$ is the real singular locus of the surface. The half-line $\{x = y = 0, z \geqslant 0\}$ is contained in the Euclidean closure of the non singular locus which is simply the open surface $\{zx^2 = y^2, x \neq 0, y \neq 0\}$. The half line $\{x = y = 0, z < 0\}$ is contained in the Zariski closure of this surface, but not in its Euclidean closure (Figure 1.4).

Cartan's Umbrella

This is the subvariety of $\mathbb{R}^3$ whose equation is

$$z(x^2 + y^2) = x^3 \quad \text{(Figure 1.5).}$$

Once again, the line $\{x = y = 0\}$ is the real singular locus of this variety, but this time only the point $\{x = y = z = 0\}$ is contained in the Euclidean closure of the non singular locus $\{z(x^2 + y^2) = x^3, x \neq 0, y \neq 0\}$ (Figure 1.5).

Kollár's Umbrella

This is the subvariety of $\mathbb{R}^3$ whose equation is

$$x^2 + y^2z^2 - y^3 = 0 \quad \text{(Figure 1.6).}$$

see [KN15]).

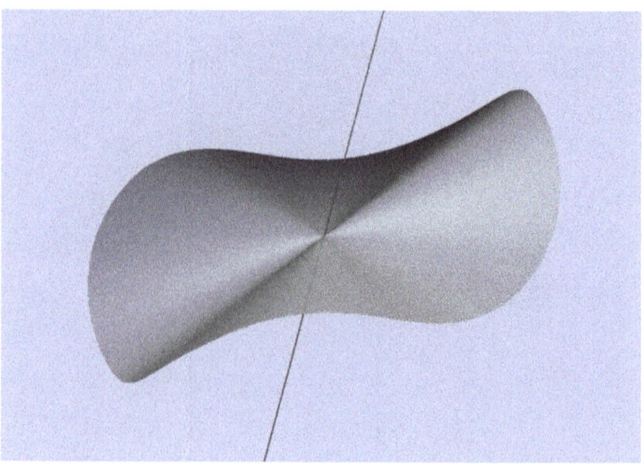

Fig. 1.5 Cartan's umbrella

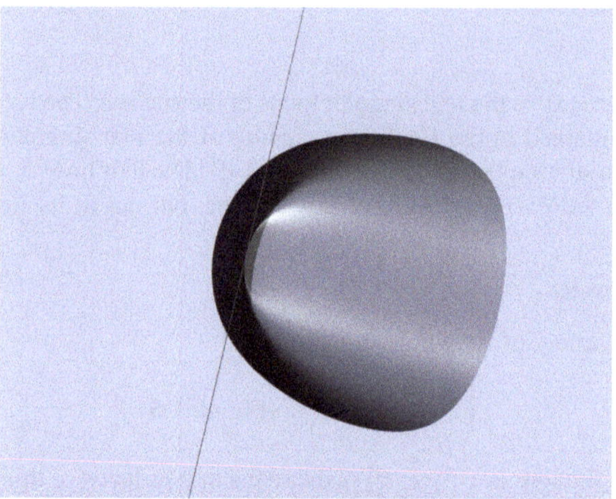

Fig. 1.6 Kollár's umbrella

As for the two next examples, the real singular locus of this umbrella is the line $\{x = y = 0\}$ and on this line only the point $\{x = y = z = 0\}$ is contained in the Euclidean closure of the non singular locus, which is given by $\text{Reg} = \{x^2 + y^2z^2 - y^3 = 0, x \neq 0, y \neq 0\}$ (Figure 1.6).

Cuspidal Umbrella

This is the subvariety of $\mathbb{R}^3$ whose equation is (Figure 1.7)

$$x^2 - y^2(y^3 - z^2) = 0 \quad \text{(Figure 1.7).}$$

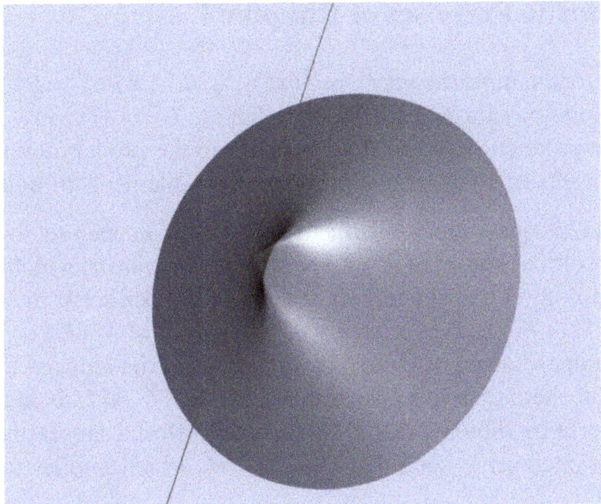

Fig. 1.7 Cuspidal umbrella

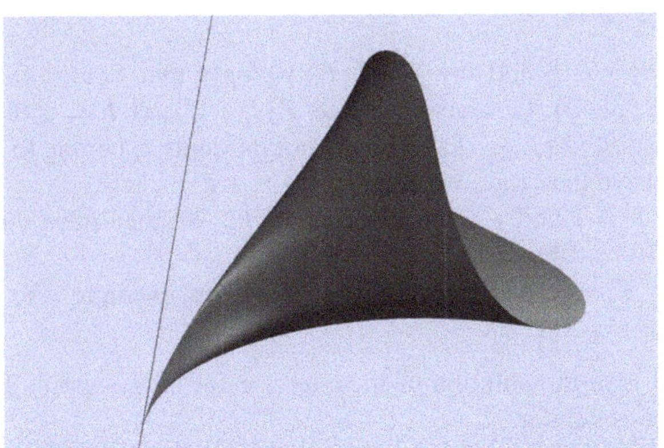

Fig. 1.8 Horned umbrella

Horned Umbrella

This is the subvariety of $\mathbb{R}^3$ whose equation is

$$x^2 + y^2\big((y - z^2)^2 + yz^3\big) = 0 \quad \text{(Figure 1.8)}$$

which expands to (Figure 1.8)

$$x^2 + y^4 + y^2z^4 + y^3z^3 - 2y^3z^2 = 0 \; .$$

1.8 Solutions to exercises of Chapter 1

1.2.2 Consider for example the set $F := \{(x, y) \in K^2 \mid x - y = 0\}$, which is closed in the Zariski topology on $\mathbb{A}^2(K)$. Its complement $U := \{(x, y) \in K^2 \mid x \neq y\}$ is not open in the product topology. Any open set in the product topology contains a product of open sets in $\mathbb{A}^1(K)$, i.e. a product of complements of finite subsets of K.

1.2.9 1. The set U is quasi-algebraic if and only if U is an open set in an algebraic set F (i.e. a Zariski closed subset of $\mathbb{A}^n(K)$ or $\mathbb{P}^n(K)$ for some n) which by definition of the induced topology is true if and only if there is an open set V in $\mathbb{A}^n(K)$ or $\mathbb{P}^n(K)$ such that $U = V \cap F$, i.e. U is a locally closed subset of $\mathbb{A}^n(K)$ or $\mathbb{P}^n(K)$.

2. If U is an open subset of $\overline{U}$ then by definition of the induced topology there is an open set V of the larger space such that $U = V \cap \overline{U}$ so U is locally closed. We prove the converse by showing that if U is a locally closed subset of the larger space then there is an open set V such that $U = V \cap \overline{U}$, which implies that U is open in $\overline{U}$.

By 1 we have $U = V \cap F$ for some open V and closed F and hence $U \subset V \cap \overline{U}$ since $U \subset V$ and $U \subset \overline{U}$. Conversely, $U \subset F$ and F is closed so $\overline{U} \subset F$, hence $V \cap \overline{U} \subset V \cap F = U$.

1.2.14 1. For any $x \in F$ and any $f \in \mathcal{I}(F)$ we have that $f(x) = 0$ by definition of $\mathcal{I}(F)$ so $x \in \mathcal{Z}(\mathcal{I}(F))$. Conversely if $x \in \mathcal{Z}(\mathcal{I}(F))$ and $F = \mathcal{Z}((f_1, \dots, f_l))$ for some family $f_i \in K[X_1, \dots, X_n]$ then the polynomials f_i belong to $\mathcal{I}(F)$ so for any $i = 1 \dots n$ we have that $f_i(x) = 0$ and hence $x \in F$.

2. Consider $n = 1$ and $I = (x^2)$, for example. We then have that $\mathcal{Z}(I) = \{x \in \mathbb{A}^1(K) \mid x^2 = 0\} = \{0\}$ and $\mathcal{I}(\mathcal{Z}(I)) = (x) \neq I$.

3. Consider $K = \mathbb{R}$ and $I = (x^2 + y^2 + 1)$, for example. We then have that $\mathcal{I}(\mathcal{Z}(I)) = \mathcal{I}(\varnothing) = \mathbb{R}[x, y]$.

1.2.21 We will give the solution to the exercise for affine space. The solution for projective space is similar.

1.a. Let F and G be two Zariski-closed subsets such that $\mathbb{A}^n(K) = F \cup G$. We will prove that at least one of these two subsets is the whole space. There are polynomials $f, g \in K[X_1, \dots, X_n]$ such that $F \subset \mathcal{Z}(f)$ and $G \subset \mathcal{Z}(g)$ and hence $\mathbb{A}^n(K) = \mathcal{Z}(f) \cup \mathcal{Z}(g) = \mathcal{Z}(fg)$. If both subspaces are not strict we can assume that f and g are non-zero. The field K is infinite so $\mathcal{Z}(fg) = \mathbb{A}^n(K)$ implies that $fg \equiv 0$ and hence either $f = 0$ or $g = 0$ from which it follows that either F or G is equal to $\mathbb{A}^n(K)$.

2. If $K = \{\alpha_1, \dots, \alpha_r\}$ then $f = \Pi_{\substack{1 \leqslant i \leqslant n \\ 1 \leqslant j \leqslant r}} (x_i - \alpha_j)$ satisfies $\mathbb{A}^n(K) = \mathcal{Z}(f)$ and $f \neq 0$. It follows that $\mathbb{A}^n(K) = \cup_{i,j} \mathcal{Z}(x_i - \alpha_j)$.

1.2.32 See the solution of Exercise 1.2.14(1) for a proof of the fact that the given condition is sufficient. (The argument given in this exercise remains valid for $\mathbb{P}^n(K)$). The condition is obviously necessary.

1.2.37 We will show that the inverse image of any closed set in $\mathbb{A}^1(K)$ is closed in U. A closed set in $\mathbb{A}^1(K)$ is either a finite set of points or the whole of $\mathbb{A}^1(K)$.

It is therefore enough to show that $f^{-1}(y) = \{x \in U \mid f(x) = y\}$ is closed[16] in U for any $y \in \mathbb{A}^1(K)$. This is a local condition so we may assume that $U \subset \mathbb{A}^n(K)$ is an open set and consider an open set V in U on which f can be represented as $\frac{g}{h}$ where $g, h \in K[X_1, \ldots, X_n]$ are such that for all $x \in V$, $h(x) \neq 0$. We then have that $f^{-1}(y) \cap V = \{x \in V \mid \frac{g(x)}{h(x)} = y\}$, but $\frac{g(x)}{h(x)} = y$ if and only if $(g(x) - yh(x)) = 0$. It follows that $f^{-1}(y) \cap V = \mathcal{Z}(g - yh) \cap V$ which is closed in V and hence $f^{-1}(y)$ is closed in U.

1.2.42 We note that the image of a polynomial function which does not vanish at x under the natural injection $K[X_1, \ldots, X_n] \to \mathcal{O}_{\mathbb{A}^n(K),x}$ which sends f to the class of the pair $(\mathbb{A}^n(K), f)$ is invertible in $\mathcal{O}_{\mathbb{A}^n(K),x}$ because if $f(x) \neq 0$ then $\frac{1}{f}$ is regular on the neighbourhood $\mathcal{D}(f)$ of x. It follows that there is a morphism

$$\varphi\colon K[X_1, \ldots, X_n]_{\mathfrak{m}_x} \to \mathcal{O}_{\mathbb{A}^n(K),x}$$

which sends the class of $\frac{g}{h}$ to the class of the pair $(\mathcal{D}(h), \frac{g}{h})$. This morphism is injective because $K[X_1, \ldots, X_n] \to \mathcal{O}_{\mathbb{A}^n(K),x}$ is injective.[17] We now consider an element of $\mathcal{O}_{\mathbb{A}^n(K),x}$ represented by a pair (U, f). By definition of a regular map, there is a neighbourhood $V \subset U$ of x and polynomials $g, h \in K[X_1, \ldots, X_n]$ such that h does not vanish on V and $\frac{g}{h} = f$ on V. The fraction $\frac{g}{h}$ represents an element of $K[X_1, \ldots, X_n]_{\mathfrak{m}_x}$ whose image under φ is equivalent to f.

1.2.51 1. The function f is regular on $K^2 \setminus \mathcal{Z}(x^2 + y^2 + 1)$ but is not polynomial—if there were a $p \in K[x, y]$ such that $\forall (x, y) \in K^2 \setminus \mathcal{Z}(x^2 + y^2 + 1)$ then we would have $p(x, y) = \frac{1}{x^2+y^2+1}$ and hence $p(x, y)(x^2 + y^2 + 1) = 1$ which is impossible since for any given y we would have $\deg_y p + 2 = 0$ (note that K is algebraically closed and hence infinite.)
2. See 1, noting that in this case $\mathcal{Z}(x^2 + y^2 + 1) = \varnothing$.

1.2.56 1. As $\varphi\colon F_1 \to F_2$ is a morphism for any $U \subset F_2$ and any regular function $f \in \mathcal{O}_{F_2}(U)$ we have that $f \circ \varphi \in \mathcal{O}_{F_1}(\varphi^{-1}(U))$. In particular for any global regular function $f \in \mathcal{O}_{F_2}(F_2)$, $f \circ \varphi \in \mathcal{O}_{F_1}(F_1)$—or in other words, by Theorem 1.2.50 for any $f \in \mathcal{P}(F_2)$ we have that $f \circ \varphi \in \mathcal{P}(F_1)$. Apply this to the functions $y_i|_{F_2}$ for $i = 1 \ldots m$. We then have that $y_i \circ \varphi \in \mathcal{P}(F_1)$ or in other words there are functions $f_i \in K[x_1, \ldots, x_n]$ such that for any $(x_1, \ldots, x_n) \in F_1$, $y_i \circ \varphi(x_1, \ldots, x_n) = f_i(x_1, \ldots, x_n)$. The result follows.
2. Similarly, it follows from Theorem 1.2.52 that for any $f \in \mathcal{S}_{F_2}^{-1}\mathcal{P}(F_2)$ we have that $f \circ \varphi \in \mathcal{S}_{F_1}^{-1}\mathcal{P}(F_1)$ (where $\mathcal{S}_{F_k} = \{h \in \mathcal{P}(F_k) \mid \forall x \in F_k, h(x) \neq 0\}$). Consider the functions $y_i|_{F_2}$ for any $i = 1 \ldots m$. We then have that $y_i \circ \varphi \in \mathcal{S}_{F_1}^{-1}\mathcal{P}(F_1)$ or in any words for any $i = 1 \ldots m$ there are polynomial functions $g_i \in \mathbb{R}[x_1, \ldots, x_n]$ and $h_i \in \mathbb{R}[x_1, \ldots, x_n]$ such that for any point $(x_1, \ldots, x_n) \in F_1, h_i(x_1, \ldots, x_n) \neq 0$ and $y_i \circ \varphi(x_1, \ldots, x_n) = \frac{g_i(x_1, \ldots, x_n)}{h_i(x_1, \ldots, x_n)}$. The result follows.

---

[16] Which would be obvious if the topology were Hausdorff!

[17] The localisation $S^{-1}A$ is a *flat* A-module by the universal property of localisation-, see Proposition A.3.2.

1.2.58 1 and 2 and 3. Denote by $H := \mathcal{Z}(xy-1) \subset \mathbb{A}^2(K)$. The map $K \setminus \{(0,0)\} \to H, x \mapsto (x, \frac{1}{x})$ is an isomorphism.

Similarly, if A and B are matrices in $\mathcal{M}_n$ then the coefficients of the matrix $AB - I_n$ are polynomials in the coefficients of A and B. We set $\mathcal{H} := \mathcal{Z}(AB - I_n) \subset \mathcal{M}_n \times \mathcal{M}_n \simeq K^{2n^2}$. The map $\mathbf{GL}_n(K) \to \mathcal{H}, A \mapsto (A, A^{-1})$ is an isomorphism.

Note that K^* and $\mathbf{GL}_n(K)$ are special cases of principal open sets—see Exercise 1.2.60 (1)—where the function f is given by $f\colon z \mapsto z$, respectively $f\colon A \mapsto \det A$.

1.2.59 1. Let d be the degree of H, i.e. $H = \mathcal{Z}(f)$ for some homogeneous polynomial f of degree d. Consider the degree d Veronese embedding, $\varphi_d\colon \mathbb{P}^n(K) \to \mathbb{P}^N(K)$ where $N = \binom{n+d}{n} - 1$ which sends $(x_0 : \cdots : x_n)$ to the $N+1$-tuplet of all degree d monomials in the $n+1$ variables $x_0, \ldots, x_n$. The image $\varphi_d(H)$ is then the intersection of a hyperplane H_0 in $\mathbb{P}^N(K)$ with the image of φ_d and via φ_d, the set $\mathbb{P}^n(K) \setminus H$ is a closed set in $\mathbb{P}^N(K) \setminus H_0$. The result follows because the complement of a hyperplane is affine. Indeed, consider coordinates on $\mathbb{P}^N(K)$ such that H_0 is the hyperplane of equation $x_0 = 0$. The map

$$\begin{array}{ccc} \mathbb{P}^N(K) \setminus H_0 & \longrightarrow & \mathbb{A}^N(K) \\ (x_0 : \cdots : x_N) & \longmapsto & (\frac{x_1}{x_0}, \ldots, \frac{x_N}{x_0}) \end{array}$$

is then an isomorphism.

2. To prove that $U := \mathbb{A}^2(K) \setminus \{(0,0)\}$ is not affine, we prove that $\mathcal{O}_{\mathbb{A}^2(K)}(U)$ is isomorphic to $K[x,y]$ or in other words that every regular function on U extends to a regular function on $\mathbb{A}^2(K)$. We cover U with the two open sets $U_1 := \mathcal{D}(x)$ and $U_2 := \mathcal{D}(y)$. Let $f\colon U \to K$ be a regular set. The restriction of f to U_1 is then of the form $\frac{g_1}{x^n}$ for some polynomial $g_1 \in K[x,y]$ and some natural number n. Moreover we can assume that x^n does not divide g_1. Similarly, $f = \frac{g_2}{y^m}$ on U_2. Since their restrictions coincide on $U_1 \cap U_2$ we have that $x^n g_2 = y^m g_1$. By the uniqueness of decompositions into irreducible elements in the factorial ring $K[x,y]$ we deduce that $n = m$ and $g_1 = g_2$, from which the theorem follows.

3. Theorem 1.2.53 tells us that the only regular functions on an irreducible (or indeed connected) projective algebraic variety are the constant functions and the Nullstellensatz A.5.12 (or its consequence Theorem 1.2.50) tells us that any affine set whose only regular functions are the constant functions is a point.

1.2.60 1. Consider the affine algebraic set

$$\begin{aligned} Z := \{&x \in K^{n+1} \mid \\ &\forall g \in \mathcal{I}(F), g(x_1, \ldots, x_n) = 0 \text{ and } x_{n+1} f(x_1, \ldots, x_n) - 1 = 0\}. \end{aligned}$$

By Exercise 1.2.56 it is clear that the restriction to Z of the projection $(x_1, \ldots, x_n, x_{n+1}) \mapsto (x_1, \ldots, x_n)$ is an isomorphism to $\mathcal{D}(f)$ inducing an isomorphism of K-algebras $\mathcal{R}(Z) \xrightarrow{\simeq} \mathcal{O}_F(\mathcal{D}(f))$.

Let $\beta\colon \mathcal{R}(F)[\frac{1}{f}] \to \mathcal{R}(Z)$ be the following morphism. For any $h = \sum_{i=0}^{d} h_i(\frac{1}{f})^i$ where each $h_i \in \mathcal{R}(F)$ the image of h under β is $(Z, x \mapsto \sum_{i=0}^{d} x_{n+1}^i h_i(x_1, \ldots, x_n))$. This morphism is injective, since if

$$\sum_{i=0}^{d} x_{n+1}^i h_i(x_1, \ldots, x_n)) \equiv 0$$

on Z then for any i we have that $h_i \equiv 0$ on F because K is infinite. Consider an element $a \in \mathcal{R}(Z)$ and a point $x \in Z$. There is a neighbourhood U of x in Z such that $a = \frac{b}{c}$ on U where $b, c \in \mathcal{P}(Z)$ and c does not vanish at any point of U. We can decompose $b(x_1, \ldots, x_n, x_{n+1}) = \sum_{i=0}^{d_1} x_{n+1}^i b_i(x_1, \ldots, x_n)$ and $c(x_1, \ldots, x_n, x_{n+1}) = \sum_{i=0}^{d_2} x_{n+1}^i c_i(x_1, \ldots, x_n)$. Since f is an invertible regular function on $\mathcal{D}(f)$ we have that $a = \frac{b'}{c'}$ on U where for any $x = (x_1, \ldots, x_n, x_{n+1}) \in Z$ we set $b'(x) = \sum_{i=0}^{d_1} f(x_1, \ldots, x_n)^{d_2-i} b_i(x_1, \ldots, x_n)$ and $c'(x) = \sum_{i=0}^{d_2} f(x_1, \ldots, x_n)^{d_2-i} c_i(x_1, \ldots, x_n)$. By construction c' is contained in $\mathcal{P}(F)$ and does not vanish at any point of U and $b' \in \mathcal{P}(F)[\frac{1}{f}]$. If we can show that it is possible to take $U = F$ then the proof is complete. We know that this is possible in two cases—if K is algebraically closed by Theorem 1.2.50 or if K is a real closed field (Definition A.5.18) by Theorem 1.2.52.

2. We have that $\mathcal{D}(f) \simeq \{(x, y, z) \in \mathbb{A}^3(\mathbb{R}) \mid z(x^2 + y^2) - 1 = 0\}$. In particular, $\mathbb{A}^2(\mathbb{R}) \setminus \{(0, 0)\}$ is affine, which is not the case for $\mathbb{A}^2(K) \setminus \{(0, 0)\}$ for any algebraically closed field K.

1.2.68 1. Let P be an irreducible polynomial of degree 2. If $\mathcal{Z}(P) \neq \varnothing$ then $\mathcal{I}(\mathcal{Z}(P))$ is a prime ideal and $\mathcal{Z}(P)$ is irreducible by Proposition 1.2.30.

2. Let $A := \mathcal{Z}(y - x^2)$ be a parabola and let $B := \mathcal{Z}(xy - 1)$ be a hyperbola. We have that $\mathcal{A}(A) = K[x, y]/(y - x^2) \simeq K[x, x^2] \simeq K[x]$ and $\mathcal{A}(B) = K[x, y]/(xy - 1) \simeq K[x, \frac{1}{x}]$ which is the localisation of $K[x]$ at x. Any morphism of K-algebras $h\colon K[x, \frac{1}{x}] \to K[x]$ must send x to a constant because x is invertible in $K[x, \frac{1}{x}]$. It follows that h is not surjective and cannot be an isomorphism. The two curves A and B are therefore not isomorphic.

3a. Since K is algebraically closed any symmetric matrix S with coefficients in K is congruent to a diagonal matrix $S = {}^tPDP$. Any degree 2 polynomial in three variables is therefore isomorphic to $x^2 + y^2 + z^2$ or $x^2 + y^2$ or x^2 after a linear change of variables. These last two polynomials are reducible so any projective conic defined by an irreducible polynomial is therefore isomorphic to $\mathcal{Z}(xz - y^2)$ which is the image of $\mathbb{P}^1(K)$ under the degree two embedding $\mathbb{P}^1(K) \to \mathbb{P}^2(K)$, $(u : v) \mapsto (u^2 : uv : v^2)$.

3b. Set $U := \mathbb{A}^1(K) \setminus \{0\}$. We have that $\Gamma(\mathbb{A}^1(K), \mathcal{O}_{\mathbb{A}^1(K)}) = K[x]$. Since U is an open set in $\mathbb{A}^1(K)$ we have the following equality of sheaves

$$\mathcal{O}_U = \mathcal{O}_{\mathbb{A}^1(K)}|_U .$$

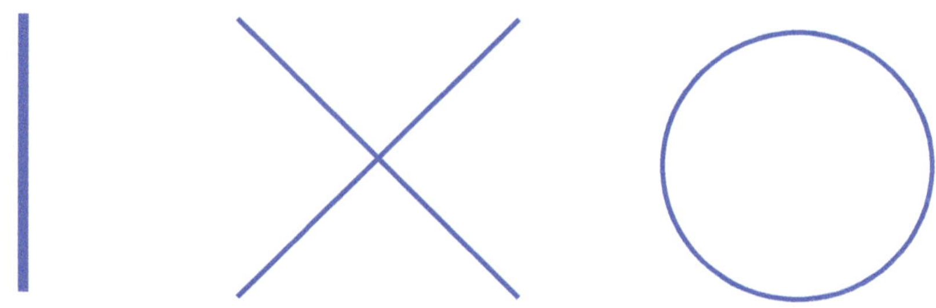

Fig. 1.9 Projective complex conics. By convention, a double curve is represented by a thickened line. From left to right: $r = 1, x^2 = 0$, $r = 2, x^2 + y^2 = 0$, $r = 3, x^2 + y^2 + z^2 = 0$

It follows that

$$\mathcal{O}_U(U) = \mathcal{O}_{\mathbb{A}^1(K)}(U)$$

which by Exercise 1.2.60 is equal to $K[x]_x = K[x, \frac{1}{x}]$ which is not isomorphic to $K[x]$. The result follows by Corollary 1.2.65.

3c. Let $P \in K[x, y]$ be a degree 2 irreducible polynomial and consider $C := \mathcal{Z}(P)$. Its projective closure is $\widehat{C} = \mathcal{Z}(Q) \subset \mathbb{P}^2(K)$ where $Q \in K[x, y, z]$ is the irreducible homogeneous polynomial $z^2 P(\frac{x}{z}, \frac{y}{z})$. The intersection of the projective conic $\widehat{C}$ with the line at infinity $L := \mathcal{Z}(z)$ is determined by the two variables homogeneous polynomial $Q(x, y, 0)$ which factors as the product of two polynomials of degree 1. If $Q(x, y, 0)$ has a double root then $\widehat{C}$ meets $\mathcal{Z}(z)$ in a unique point and $C = \widehat{C} \setminus \{\text{pt.}\} \simeq \mathbb{P}^1(K) \setminus \{\infty\} \simeq \mathbb{A}^1(K)$ by (3a). Likewise, if $Q(x, y, 0)$ has two distinct roots then C is isomorphic to $\mathbb{P}^1(K)$ minus two points, or in other words $\mathbb{A}^1(K)$ minus a point. Changing the coordinates so this point is at 0, we get the coordinate ring $\mathcal{A}(C) = K[x, \frac{1}{x}]$. See the second line of Figure 1.10.

4a. The classifying invariant is the rank r of the quadratic polynomial. This can be equal to 1, 2 or 3 and after linear change of coordinates the associated conics are given by equations $x^2 = 0$ (a double projective line), $x^2 + y^2 = 0$ (two projective lines meeting in a point) or $x^2 + y^2 + z^2 = 0$ (an irreducible projective conic). See Figure 1.9.

4b. Removing a point from a projective line leaves an affine line. Considering all possible intersections between the projective conics in the above classification and the line at infinity yields the following list: a double affine line, two affine lines meeting in a point, two parallel affine lines, an irreducible affine conic which may be an ellipse or a parabola. See Figure 1.10. (Linear changes of coordinates have been used to make these diagrams clearer.)

5a. Consider the projective real conics $\overline{C_1} := \mathcal{Z}(x^2 + y^2 + z^2)$ and $\overline{C_2} := \mathcal{Z}(x^2 + y^2 - z^2)$. The first is empty and the second is a circle so they are not isomorphic.

Any symmetric matrix with real coefficients is diagonalisable in an orthonormal basis so any irreducible homogeneous polynomial of degree 2 in three variables with non-empty zero locus can be written $x^2 \pm y^2 \pm z^2$ after linear change of basis.

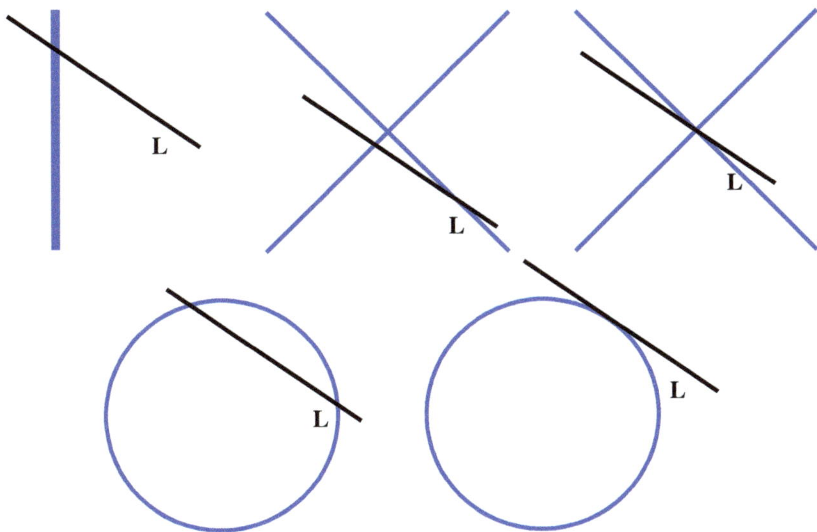

Fig. 1.10 Affine complex conics. L is the line at infinity—the actual affine conic is the complement of L. In order: a double affine line, two affine lines meeting in a point, two parallel affine lines, an ellipse and a parabola

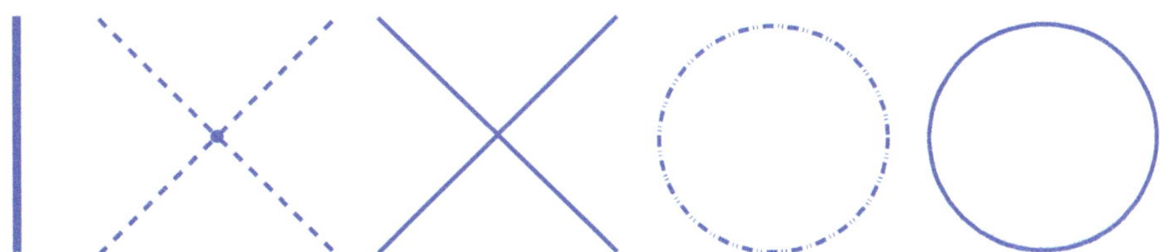

Fig. 1.11 Real projective conics. By convention, non-real curves are drawn as dotted lines. From left to right: $x^2 = 0, x^2 + y^2 = 0, x^2 - y^2 = 0, x^2 + y^2 + z^2 = 0, x^2 + y^2 - z^2 = 0$. Real locus: a real projective double line, an isolated real point, two non-parallel real projective lines, $\varnothing$, an ellipse

Permuting coordinates, any real projective irreducible conic is thus isomorphic to C_1 or C_2, cf. the two right hand conics in Figure 1.11.

5b. Consider the real affine conics $A := \mathcal{Z}(x^2 - y)$, $B := \mathcal{Z}(xy - 1)$, $C_1 := \mathcal{Z}(x^2 + y^2 + 1)$, $C_2 := \mathcal{Z}(x^2 + y^2 - 1)$ et $C_3 := \mathcal{Z}(x^2 - y^2 - 1)$. The set C_1 is empty. C_2 is compact and connected, C_3 is neither compact nor connected, A is not compact but is connected and C_3 is isomorphic to B via the change of variables $(x, y) \mapsto (x - y, x + y)$.

Let C be an affine conic defined by an irreducible polynomial and let $\widehat{C}$ be its projective completion. If $\widehat{C}$ is empty it is isomorphic to $\mathcal{Z}(x^2 + y^2 + z^2)$ and C is isomorphic to C_1. If $\widehat{C}$ is non empty then as in (c) we consider its intersection with the line at infinity $\mathcal{Z}(z)$. This intersection may be empty (two imaginary points): C then isomorphic to C_2. It may contain one real point, in which case C is isomorphic

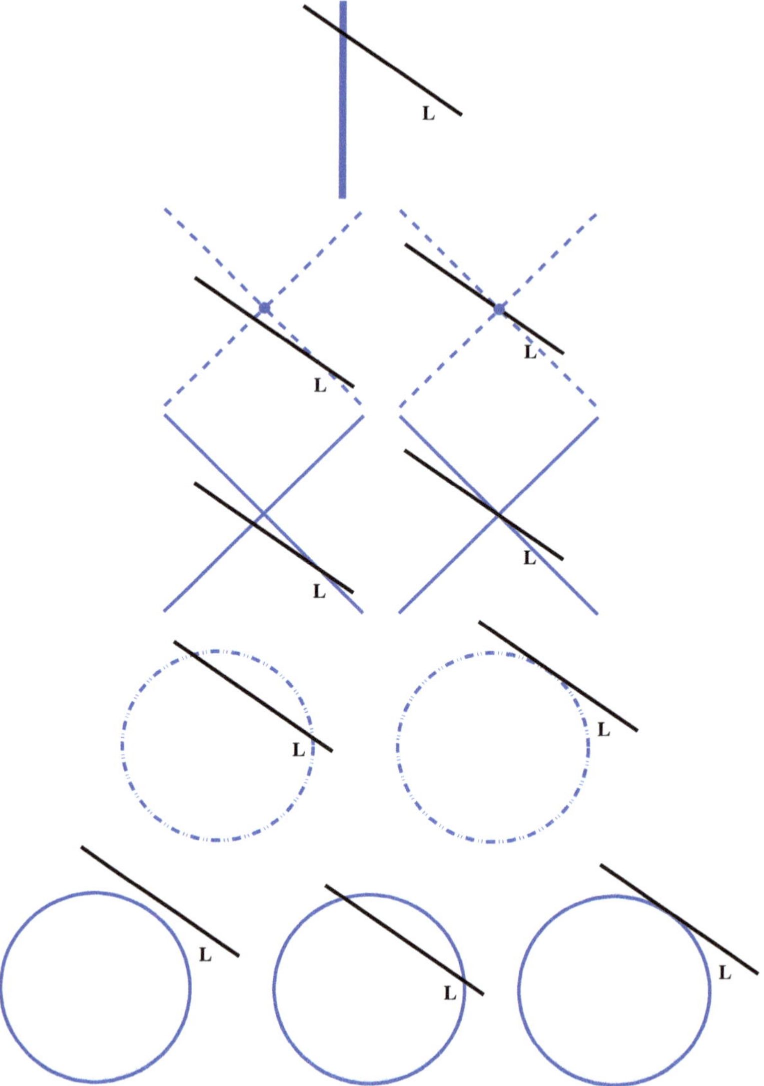

Fig. 1.12 Real affine conics. L is the line at infinity and the actual affine conic is the complement of L. Real locus in order: an affine double line, an isolated real point, $\varnothing$, two non-parallel affine lines, two affine parallel lines, $\varnothing$, $\varnothing$, an ellipse, a hyperbola, a parabola

to $\mathbb{P}^1(\mathbb{R})$ minus a point, which is isomorphic to A. Finally, the intersection may contain two real points, so C is isomorphic to $\mathbb{A}^1(\mathbb{R})$ minus a point, in which case C is isomorphic to $B \simeq C_3$. Compare with the last two lines of Figure 1.12.

5c. The real classifying invariant is the signature (s^+, s^-) of the quadratic polynomial. (The rank r is then given by $r = s^+ + s^-$). Permuting variables, there are five possibilities for the conic: $\{(1,0),(0,1)\}$, $\{(2,0),(0,2)\}$, $\{(1,1)\}$, $\{(3,0),(0,3)\}$, $\{(2,1),(1,2)\}$ corresponding respectively, after linear change of variables, to $x^2=0$ (a real projective double line), $x^2+y^2=0$ (two imaginary projective lines meeting in one real point), $x^2-y^2=0$ (two real projective lines meeting in one real point), $x^2+y^2+z^2=0$ (an irreducible projective conic without real points) and $x^2+y^2-z^2=0$ (an irreducible projective conic with real points). See Figure 1.11.

5d. We argue as in the complex case, see Figure 1.12.

1.2.80 1. This follows immediately from Corollary 1.2.79, $K(\mathbb{A}^n(K)) = \text{Frac}\ \mathcal{P}(\mathbb{A}^n(K)) = \text{Frac}\ K[X_1,\ldots,X_n]$.

2. Consider the inclusion map $i\colon \mathbb{A}^n(K) \hookrightarrow \mathbb{P}^n(K)$, $(x_1,\ldots,x_n) \mapsto (1 : x_1 : \cdots : x_n)$. The set $i(\mathbb{A}^n(K))$ is a dense open set in $\mathbb{P}^n(K)$, since it is a non-empty open set in $\mathbb{P}^n(K)$ which is irreducible since K is infinite by Exercise 1.2.21(1), so by Proposition 1.2.71, $K(\mathbb{P}^n(K)) = K(\mathbb{A}^n(K))$.

3a. It is immediate that $P(f_1, f_2) = 0$ in $K(f_1, f_2)$, so f_2 is algebraic over $K(f_1)$ and hence $K(C) = K(f_1, f_2)$ is a finite degree extension of $K(f_1)$. Similarly, $K(C)$ is a finite degree extension of $K(f_2)$.

If f_1 is transcendental over K then $K(f_1) \simeq K(x) \simeq K(X)$. If f_1 is algebraic over K then f_2 is transcendental over K by the hypothesis on the degree of P and $K(f_2) \simeq K(y) \simeq K(X)$.

3b. This follows from Corollary 1.2.79 and the Nullstellensatz. See Corollary A.5.13.

1.3.15 1. This is an immediate corollary of Proposition 1.2.61.

2. We use the same notation as in the definition: $X = \mathcal{Z}(J) \setminus \mathcal{Z}(I)$. Let f and g be two functions in $\mathcal{I}(\mathcal{Z}(I))$. We then have that $\mathcal{D}(f) \cap \mathcal{D}(g) \cap \mathcal{Z}(J) = X \setminus (\mathcal{Z}(f) \cup \mathcal{Z}(g)) = X \setminus (fg)$.

3. All open sets in X are of the form $X \setminus \mathcal{Z}(L)$ for some ideal L in $K[X_0,\ldots,X_n]$. Let $\{f_1,\ldots,f_l\}$ be a set of generators of L. We then have that $\mathcal{Z}(L) = \cap_{i=1}^l \mathcal{Z}(f_i)$ so $X \setminus \mathcal{Z}(L) = \cup_{i=1}^l \mathcal{D}(f_i)$. Note that if X is affine then $\mathcal{O}_X(X \setminus \mathcal{Z}(L)) = \cap_{i=1}^l \mathcal{O}_X(\mathcal{D}(f_i))$.

4. See [Ser55a, Proposition 1, page 234].

1.3.24 It is clear that the relationship $\sim$ is reflexive and symmetric. We now show that it is transitive: suppose that $(U,\varphi_U) \sim (V,\varphi_V)$ and $(V,\varphi_V) \sim (W,\varphi_W)$. We then have that $\varphi_U|_{U\cap V} = \varphi_V|_{U\cap V}$ and $\varphi_V|_{V\cap W} = \varphi_W|_{V\cap W}$ from which it follows that $\varphi_U|_{U\cap V\cap W} = \varphi_W|_{U\cap V\cap W}$. But V is dense in X so $U\cap V\cap W$ is dense in $U\cap W$ and hence $\varphi_U|_{U\cap W} = \varphi_W|_{U\cap W}$ or in other words $(U,\varphi_U) \sim (W,\varphi_W)$.

1.3.25 We use that same argument as in Exercise 1.2.56: pulling back the coordinate functions yields regular functions on some dense open set in X, namely the open set on which φ is defined. If this open set is written in the form $U = X \setminus \mathcal{Z}(I)$ for some ideal I generated by $f_1,\ldots,f_k$ then $U = \cup_{i=1}^k \mathcal{D}(f_i)$.

1.4.4 In fact the Euclidean topology is the coarsest topology for which all Zariski-continuous functions f are continuous with respect to the Euclidean norm on the target space.

1.5.14 By Lemma A.2.9, $I_{\mathbb{R}} := I \cap \mathbb{R}[X_1, \ldots, X_n]$ is a prime ideal in $\mathbb{R}[X_1, \ldots, X_n]$ and $I = I_{\mathbb{R}} \otimes \mathbb{C}[X_1, \ldots, X_n]$ by hypothesis. For any chain of prime ideals of length l, $I_{\mathbb{R}} = J_0 \subsetneq J_1 \subsetneq \cdots \subsetneq J_l \subsetneq \mathbb{R}[X_1, \ldots, X_n]$ we have that $I = J_0 \otimes \mathbb{C}[X_1, \ldots, X_n] \subsetneq J_1 \otimes \mathbb{C}[X_1, \ldots, X_n] \subsetneq \cdots \subsetneq J_l \otimes \mathbb{C}[X_1, \ldots, X_n] \subsetneq \mathbb{C}[X_1, \ldots, X_n]$ is a chain of prime ideals of length l. It follows that $\dim I_{\mathbb{R}} \leqslant \dim I$ by Exercise 1.5.13(1).

1.5.50 Let $U \subset X$ be a non-empty affine open set. As X is irreducible, U is dense and hence $K(X) = K(U)$ by Proposition 1.2.71. As $\dim X = \dim U$ by Remark 1.5.44, we simply apply Theorem 1.5.5 to the ring of affine coordinates of U.

Chapter 2
$\mathbb{R}$-Varieties

In the introduction to Chapter 1 we warned the reader that our category of *real algebraic varieties* was insufficient for certain purposes. In this chapter we introduce complex varieties with a conjugation map, which Atiyah [Ati66] calls "real spaces".

In this introduction we will assume for simplicity that our varieties are projective. Let $X \subset \mathbb{P}^n(\mathbb{C})$ be a complex algebraic set defined by *real* homogeneous equations. The set $V \subset \mathbb{P}^n(\mathbb{R})$ of real solutions to these equations, which is simply $X \cap \mathbb{P}^n(\mathbb{R})$, is then a real algebraic set. Both X and V are sometimes called *real* varieties in the literature, depending on the type of problem being studied. It is tempting to distinguish the objects V and X by calling V a real algebraic variety (as in Chapter 1) and X an algebraic variety defined over $\mathbb{R}$. Some authors make this distinction—see [BK99, Hui95] for example—but not all—see [Sil89, DIK00] for example. It is fairly common to consider that a "real algebraic variety" and an "algebraic variety defined over $\mathbb{R}$" are the same thing, namely a complex algebraic variety which has a set of real defining equations, or alternatively, a complex variety stable under conjugaison.

In practice we can mostly specify which point of view we are using on a case by case basis, since many problems require just one point of view or the other. Occasionally, however, we will need to jump between definitions in the course of a single argument. We have chosen to call a pair of a complex algebraic variety and a conjugation map an *algebraic* $\mathbb{R}$*-variety* (see Definition 2.1.10) and reserve the expression *real algebraic variety* for algebraic subsets of $\mathbb{P}^n(\mathbb{R})$. Note that the "real varieties" defined in [Sil89, I.2] and [DIK00] are our $\mathbb{R}$-varieties.

This chapter deals with $\mathbb{R}$-varieties and their relationship with the real algebraic varieties defined in the previous chapter. After defining $\mathbb{R}$-varieties and studying their main properties in Section 2.1, we explain to what extent an $\mathbb{R}$-variety determines a real algebraic variety in Section 2.2. In the subsequent section we will consider the following question: given a real algebraic variety, does it determine an $\mathbb{R}$-variety? We end Section 2.2 with a summary of the logical relations between real algebraic varieties, $\mathbb{R}$-varieties and schemes over $\mathbb{R}$, achieving thereby one of the goals stated in the Introduction. The final part of this chapter deals with refinements and consequences of this theory. Section 2.5, which is technically difficult and can be skipped

F. Mangolte, *Real Algebraic Varieties*, Springer Monographs in Mathematics,
https://doi.org/10.1007/978-3-030-43104-4_2

on first reading, deals with sheaves and bundles, Section 2.6 deals with divisors and Section 2.7 deals with ℝ-plane curves.

2.1 Real Structures on Complex Varieties

In this section we introduce complex varieties to the study of real varieties. The following example illustrates their usefulness: further on, Example 2.1.29 illustrates the usefulness of abstract real structures on complex varieties.

Example 2.1.1 (*Continuation of Example* 1.5.20) Let us return to the real irreducible algebraic variety $F := \mathcal{Z}(x^2 + y^2) \subset \mathbb{A}^2(\mathbb{R})$ which is an isolated point $(0, 0)$. Consider the algebraic set $X := \mathcal{Z}_{\mathbb{C}}(x^2 + y^2) \subset \mathbb{A}^2(\mathbb{C})$ which is a reducible complex curve. The restriction of $\sigma : (x, y) \mapsto (\overline{x}, \overline{y})$ to X is an involution sending X to itself: its set of fixed points is $F = X^\sigma = \{(0, 0)\}$. The complex algebraic *curve* X has a unique *real* point. The point $(0, 0)$ is the intersection of the two irreducible components $\mathcal{Z}_{\mathbb{C}}(x - iy)$ and $\mathcal{Z}_{\mathbb{C}}(x + iy)$ and it is the only real point of X. We have $\dim X = 1$ and $\dim F = 0$.

Going further, consider the variety $V := \mathcal{Z}(x^2 + y^2 - z) \subset \mathbb{A}^3(\mathbb{R})$ and the morphism $\pi\colon V \to \mathbb{A}^1(\mathbb{R})$, $(x, y, z) \mapsto z$. For any $z_0 \in \mathbb{A}^1(\mathbb{R})$ the fibre $\pi^{-1}(z_0)$ is an algebraic subset of the affine plane $\mathcal{Z}((z - z_0)) \simeq \mathbb{A}^2(\mathbb{R})$. If $z_0 > 0$, $\pi^{-1}(z_0)$ is a non singular real curve; $\pi^{-1}(0) \simeq F$ on the other hand is a point and for all $z_0 < 0$, $\pi^{-1}(z_0)$ is empty. Consider $Y := \mathcal{Z}_{\mathbb{C}}(x^2 + y^2 - z) \subset \mathbb{A}^3(\mathbb{C})$ and $\pi_{\mathbb{C}}\colon Y \to \mathbb{A}^1(\mathbb{C})$, $(x, y, z) \mapsto z$. For any z_0 the preimage $\pi_{\mathbb{C}}^{-1}(z_0)$ is an algebraic subset of the affine plane $\mathcal{Z}((z - z_0)) \simeq \mathbb{A}^2(\mathbb{C})$. Consider a point $z_0 \in \mathbb{A}^1(\mathbb{R}) \subset \mathbb{A}^1(\mathbb{C})$. If $z_0 > 0$ then $\pi_{\mathbb{C}}^{-1}(z_0)$ is a non singular complex curve whose real locus is a non singular real curve. If $z_0 = 0$ then $\pi_{\mathbb{C}}^{-1}(0) \simeq X$ is a singular complex curve whose real locus is a point. If $z_0 < 0$ then $\pi_{\mathbb{C}}^{-1}(z_0)$ is a non singular complex curve whose real locus is empty.

The complex variety Y provides a deeper understanding of this example. The real variety V can be recovered as the set of fixed points of the involution defined by complex conjugation on $\mathbb{C}^3$. More generally, we will seek to imitate the standard conjugation map. On $\mathbb{A}^n(\mathbb{C}) = \mathbb{C}^n$ we denote by $\sigma_{\mathbb{A}} := \sigma_{\mathbb{A}^n}$ the involution

$$\sigma_{\mathbb{A}}\colon \begin{cases} \mathbb{A}^n(\mathbb{C}) & \longrightarrow & \mathbb{A}^n(\mathbb{C}) \\ (z_1, \ldots, z_n) & \longmapsto & (\overline{z_1}, \ldots, \overline{z_n}) \end{cases} .$$

In particular, for any $z \in \mathbb{C}$, $\sigma_{\mathbb{A}^1}(z) = \overline{z}$. Similarly, on $\mathbb{P}^n(\mathbb{C})$ we denote by $\sigma_{\mathbb{P}} := \sigma_{\mathbb{P}^n}$ the standard conjugation map

$$\sigma_{\mathbb{P}}\colon \begin{cases} \mathbb{P}^n(\mathbb{C}) & \longrightarrow & \mathbb{P}^n(\mathbb{C}) \\ (x_0 : x_1 : \cdots : x_n) & \longmapsto & (\overline{x_0} : \overline{x_1} : \cdots : \overline{x_n}) \end{cases} .$$

We can recover $\mathbb{R}^n \subset \mathbb{C}^n$ as the set of fixed points of $\sigma_{\mathbb{A}^n}$ and the real projective plane $\mathbb{P}^n(\mathbb{R}) \subset \mathbb{P}^n(\mathbb{C})$ as the set of fixed points of $\sigma_{\mathbb{P}^n}$. We will generalise this situation to an arbitrary (algebraic or analytic) complex variety. In other words, we will introduce *real structures* (analogues of $\sigma_{\mathbb{A}}$ and $\sigma_{\mathbb{P}}$) on complex varieties: see Definition 2.1.10 for more details. We note immediately that for general $X \subset \mathbb{C}^n$ it is not enough to consider the restriction of $\sigma_{\mathbb{A}}$ to X for two reasons. Firstly, we have to require that this restriction induces a morphism from X to X (i.e. $\sigma_{\mathbb{A}}(X) \subset X$). Secondly, a given complex variety X can have several different *real forms* (see Definition 2.1.13) corresponding to different real structures. In other words, there are pairs of complex varieties X_1 and X_2 defined by real polynomials which are isomorphic as complex varieties but do not have an isomorphism defined over $\mathbb{R}$: see Example 2.1.29 for an example.

Let f be a holomorphic function (such as a polynomial) defined in a neighbourhood of $z_0 = (z_{0,1}, \dots, z_{0,n}) \in \mathbb{C}^n$ by

$$f(z) = \sum_{k \in \mathbb{N}^n} a_k (z_1 - z_{0,1})^{k_1} \dots (z_n - z_{0,n})^{k_n} .$$

There is then a *conjugate* holomorphic function of f, denoted ${}^\sigma f$, defined in a neighbourhood of $\overline{z_0} = (\overline{z_{0,1}}, \dots, \overline{z_{0,n}}) \in \mathbb{C}^n$ by

$${}^\sigma f(z) = \sum_{k \in \mathbb{N}^n} \overline{a_k} (z_1 - \overline{z_{0,1}})^{k_1} \dots (z_n - \overline{z_{0,n}})^{k_n}$$

or in other words ${}^\sigma f = \overline{f} \circ \sigma_{\mathbb{A}^n} = \sigma_{\mathbb{A}^1} \circ f \circ \sigma_{\mathbb{A}^n}$. If F is a subset of $\mathbb{C}^n$ defined by the vanishing of the functions $f_1, \dots, f_k$ then

$$\overline{F} := \{z \in \mathbb{C}^n | \sigma_{\mathbb{A}^n}(z) \in F\}$$

is the set of common zeros of the functions ${}^\sigma f_1, \dots, {}^\sigma f_k$. It follows that if $F \subset \mathbb{A}^n(\mathbb{C})$ is a complex algebraic affine set then $\overline{F} \subset \mathbb{A}^n(\mathbb{C})$ is also a complex algebraic affine set.

Remark 2.1.2 Note that ${}^\sigma f$ and $\overline{f}$ are not the same thing. If f is a holomorphic function then ${}^\sigma f$ is also holomorphic whereas $\overline{f} = \sigma_{\mathbb{A}} \circ f$ anti-holomorphic. Passing from f to ${}^\sigma f$ simply involves conjugating coefficients. If f is a polynomial then ${}^\sigma f$ is also a polynomial, unlike $\overline{f}$. The coefficients of the polynomial f are real if and only if ${}^\sigma f = f$.

Exercise 2.1.3 (Sheaf on a conjugate algebraic set)

1. *Let $\mathcal{O}$ be the sheaf of regular functions on $\mathbb{A}^n(\mathbb{C})$ (resp. $\mathbb{P}^n(\mathbb{C})$). We define a sheaf ${}^\sigma\mathcal{O}$ on $\mathbb{A}^n(\mathbb{C})$ (resp. $\mathbb{P}^n(\mathbb{C})$) by setting*

$${}^\sigma\mathcal{O}(U) := \left\{{}^\sigma f \mid f \in \mathcal{O}(\overline{U})\right\} .$$

for every open set U in $\mathbb{A}^n(\mathbb{C})$ (resp. $\mathbb{P}^n(\mathbb{C})$)
Prove that ${}^\sigma\mathcal{O} = \mathcal{O}$.

2. *Let $F \subset \mathbb{A}^n(\mathbb{C})$ be an affine algebraic set. The sheaf of regular functions on F is denoted $\mathcal{O}_F$ and the sheaf of regular functions on $\overline{F}$ is denoted $\mathcal{O}_{\overline{F}}$ (These are sheaves deduced from $\mathcal{O}$: equipped with these sets, F and $\overline{F}$ are sub-varieties of $\mathbb{A}^n(\mathbb{C})$—see Definition 1.3.7 and Example 1.3.8).*
Prove that if $\overline{F} = F$ then ${}^{\sigma}\mathcal{O}_F := ({}^{\sigma}\mathcal{O})_F$ is a sheaf on F which is equal to $\mathcal{O}_F$ by the above. We then say that $\mathcal{O}_F$ is an ℝ-sheaf*: see Definition 2.2.1 for more details.*

Proposition 2.1.4 *Let $X \subset \mathbb{A}^n(\mathbb{C})$ be an algebraic set. The restriction of $\sigma_{\mathbb{A}^n}$ to X is an involution of X if and only if X can be defined by real polynomials.*

Let $X \subset \mathbb{P}^n(\mathbb{C})$ be an algebraic set. The restriction of $\sigma_{\mathbb{P}^n}$ to X is an involution of X if and only if X can be defined by real homogeneous polynomials.

Proof If $X = \mathcal{Z}(P_1, \ldots, P_l)$ then by definition we have that $\overline{X} = \mathcal{Z}({}^{\sigma}P_1, \ldots, {}^{\sigma}P_l)$ and the restriction $\sigma_{\mathbb{A}}|_X$ is an endomorphism of X if and only if $\overline{X} = X$. Suppose that $\overline{X} = X$. We then have that $\mathcal{Z}(P_1, \ldots, P_l) = \mathcal{Z}({}^{\sigma}P_1, \ldots, {}^{\sigma}P_l) = \mathcal{Z}(\frac{1}{2}(P_1 + {}^{\sigma}P_1), \ldots, \frac{1}{2}(P_l + {}^{\sigma}P_l), \frac{1}{2i}(P_1 - {}^{\sigma}P_1), \ldots, \frac{1}{2i}(P_l - {}^{\sigma}P_l))$. The proposition follows on noting that for any polynomial P with complex coefficients the polynomials $\frac{1}{2}(P + {}^{\sigma}P)$ and $\frac{1}{2i}(P - {}^{\sigma}P)$ have real coefficients. The converse is immediate.

Similarly, if X is a projective algebraic variety defined in $\mathbb{P}^n(\mathbb{C})$ by homogeneous polynomial equations

$$P_1(z_0, \ldots, z_n) = \cdots = P_l(z_0, \ldots, z_n) = 0\,,$$

then the variety $\overline{X}$ defined by ${}^{\sigma_{\mathbb{A}^{n+1}}}P_1(z_0, \ldots, z_n) = \cdots = {}^{\sigma_{\mathbb{A}^{n+1}}}P_l(z_0, \ldots, z_n) = 0$ is an algebraic subvariety of $\mathbb{P}^n(\mathbb{C})$. It is easy to check that if P is a homogeneous polynomial then $\frac{1}{2}(P + {}^{\sigma}P)$ and $\frac{1}{2i}(P - {}^{\sigma}P)$ are homogenous polynomials The restriction of $\sigma_{\mathbb{P}}$ to X is therefore an endomorphism of X if and only if X can be defined by real homogeneous polynomials. □

Before generalising the above to abstract varieties we need the following definition.

Definition 2.1.5 Let $\mathcal{L}$ be a sheaf of complex functions over a topological space X. The *anti-sheaf* $\overline{\mathcal{L}}$ of $\mathcal{L}$ is defined over any open set U in X by

$$\overline{\mathcal{L}}(U) := \{\overline{f} := \sigma_{\mathbb{A}} \circ f \mid f \in \mathcal{L}(U)\}\,.$$

More generally, let X be a topological space and let $\mathcal{L}$ be a sheaf of maps to $\mathbb{C}^n$.[1] We define the sheaf $\overline{\mathcal{L}}$ over any open set U of X by

$$\overline{\mathcal{L}}(U) := \{\overline{f} := \sigma_{\mathbb{A}^n} \circ f \mid f \in \mathcal{L}(U)\}\,.$$

[1]Note that $\mathcal{L}$ is no longer a sheaf of rings, but a sheaf of vector spaces.

Definition 2.1.6 Let $(X, \mathcal{O}_X)$ be a complex algebraic variety (resp. a complex analytic space[2]). The *conjugate* variety (resp. the *conjugate* analytic space) of X is defined to be the topological space X equipped with the anti-sheaf of $\mathcal{O}_X$

$$\overline{X} := (X, \overline{\mathcal{O}_X}) .$$

Exercise 2.1.7 *If F is the subset of $\mathbb{C}^n$ defined by the vanishing of functions $f_1, \ldots, f_k$ then $\overline{F} := \{z \in \mathbb{C}^n | \sigma_{\mathbb{A}^n}(z) \in F\}$ is the vanishing locus of the functions ${}^\sigma f_1, \ldots, {}^\sigma f_k$. If $F \subset \mathbb{A}^n(\mathbb{C})$ is a complex affine algebraic set then $\overline{F}$ is a complex affine algebraic set and $\sigma_{\mathbb{A}}$ induces an isomorphism of varieties from $(\overline{F}, \mathcal{O}_{\overline{F}})$ to the conjugate variety $(F, \overline{\mathcal{O}_F})$.*

Let $(X, \mathcal{O}_X)$ and $(Y, \mathcal{O}_Y)$ be complex algebraic varieties (resp. complex analytic spaces). In particular, $\mathcal{O}_X$ and $\mathcal{O}_Y$ are sheaves of complex valued functions. Recall that a map $\varphi\colon X \to Y$ is *regular* (resp. *holomorphic*) if and only if it is continuous and for any function $f \in \mathcal{O}_Y(V)$ the function $f \circ \varphi$ belongs to $\mathcal{O}_X(\varphi^{-1}(V))$. (See Definition 1.3.4).

Definition 2.1.8 A map $\varphi\colon (X, \mathcal{O}_X) \to (Y, \mathcal{O}_Y)$ is *anti-regular* (resp. *anti-holomorphic*) if and only if it is continuous and for every open set V in Y and every function $f \in \mathcal{O}_Y(V)$ the function $\overline{f} \circ \varphi$ belongs to $\mathcal{O}_X(\varphi^{-1}(V))$.

Remark 2.1.9 If X is a complex algebraic variety (resp. complex analytic space) and $\mathcal{O}_X$ is its sheaf of regular functions (resp. holomorphic functions) the anti-sheaf $\overline{\mathcal{O}_X}$ is the sheaf of anti-regular (resp. anti-holomorphic) functions. A continuous map $\varphi\colon X \to Y$ is anti-regular (or anti-holomorphic) from $(X, \mathcal{O}_X)$ to $(Y, \mathcal{O}_Y)$ if and only if it is regular (or holomorphic) when considered as a map from $(X, \mathcal{O}_X)$ to the conjugate variety $(Y, \overline{\mathcal{O}_Y})$—see Exercise 2.1.7.

As promised in the introduction, we now generalise the involutions $\sigma_{\mathbb{A}}$ and $\sigma_{\mathbb{P}}$ to complex varieties. (We invite the reader to compare this definition with Atiyah's "real structures on a bundle" in [Ati66].)

Definition 2.1.10 (*Real structure*) A *real structure* on a complex algebraic variety (resp. complex analytic space) X is an anti-regular (resp. anti-holomorphic) global involution σ on X.

Examples 2.1.11 (*Basic examples*)

1. $\sigma_{\mathbb{A}}$ on $\mathbb{A}^n(\mathbb{C})$;
2. $\sigma_{\mathbb{P}}$ on $\mathbb{P}^n(\mathbb{C})$;
3. $(x : y) \mapsto (-\overline{y} : \overline{x})$ on $\mathbb{P}^1(\mathbb{C})$.

---

[2]In complex analytic geometry the term *variety* is usually only used for non singular complex analytic spaces see Appendix D.

Definition 2.1.12 (ℝ-*variety*) In short, we will say that a pair (X, σ) is an ℝ-*variety* if X is a complex variety and σ is a real structure on X. If necessary we will specify whether (X, σ) is an *algebraic* ℝ-*variety* or *analytic* ℝ-*variety*. On occasion we will wish to authorise our analytic varieties to be singular: we will then call them *analytic* ℝ-*spaces*.

Definition 2.1.13 An ℝ-variety (X, σ) is also called a *real form* of the complex variety X.

Example 2.1.14 Real forms of Lie groups provide a rich family of examples. See [MT86] for more details.

Remark 2.1.15 Generalising ℝ-varieties to complex analytic varieties is particularly useful when studying real K3 surfaces (Definition 4.5.3), 2-dimensional complex ℝ-toruses (Definition 4.5.22), real elliptic surfaces (Definition 4.6.1) and real Moishezon varieties (Definition 6.1.4).

Remark 2.1.16 Let (X, σ) be an ℝ-variety and let $U \subset X$ be an open affine set. The set $\sigma(U)$ is then also an open affine set, since σ is a homeomorphism. Moreover, if $\varphi\colon U \to \mathbb{A}^n(\mathbb{C})$ is an embedding of U as an affine algebraic variety of ideal $I = (P_1, \ldots, P_l) \subset \mathbb{C}[X_1, \ldots, X_n]$ then $\sigma_{\mathbb{A}} \circ \varphi \circ \sigma : \sigma(U) \to \mathbb{A}^n(\mathbb{C})$ is an embedding of $\sigma(U)$ as an affine variety of ideal ${}^{\sigma}I = ({}^{\sigma}P_1, \ldots, {}^{\sigma}P_l) \subset \mathbb{C}[X_1, \ldots, X_n]$.

Definition 2.1.17 A pair (Y, τ) is an ℝ-*subvariety* of (X, σ) if and only if $Y \subset X$ is a complex subvariety of X and $\tau = \sigma|_Y$.

By definition, an ℝ-variety (X, σ) is *quasi-affine* (resp. *affine*, resp. *quasi-projective*, resp. *projective*) if the *complex* variety X has a regular embedding $\varphi\colon X \hookrightarrow \mathbb{A}^n(\mathbb{C})$ (resp. $\varphi\colon X \hookrightarrow \mathbb{A}^n(\mathbb{C})$ with closed image, resp. $\psi : X \hookrightarrow \mathbb{P}^n(\mathbb{C})$, resp. $\psi : X \hookrightarrow \mathbb{P}^n(\mathbb{C})$ with closed image). The central question is whether there is always a regular embedding such that $\varphi \circ \sigma = \sigma_{\mathbb{A}} \circ \varphi$ (resp. $\psi \circ \sigma = \sigma_{\mathbb{P}} \circ \psi$). In other words, is (X, σ) isomorphic as a ℝ-variety to a ℝ-subvariety of $(\mathbb{A}^n(\mathbb{C}), \sigma_{\mathbb{A}})$ (resp. $(\mathbb{P}^n(\mathbb{C}), \sigma_{\mathbb{P}})$)? The answer to this question is yes: this is one of the main results of the theory. Any quasi-projective ℝ-variety can be defined by equations with real coefficients: see Theorem 2.1.33.

The well known identification (see [Ser56] for more details) of a complex projective algebraic variety with an analytic variety is compatible with its real structure.

Proposition 2.1.18 *Let X be a complex projective algebraic variety. The variety X then has a real structure if and only if there is an anti-holomorphic involution on the analytic space underlying X.*

Proof Let X^h be the underlying analytic space of X, by which we mean that X^h is the set X with its Euclidean topology and the sheaf $\mathcal{O}^h_X$ of holomorphic functions associated to the sheaf $\mathcal{O}_X$. If X is projective then the conjugate variety $\overline{X}$ is also projective. Let $\sigma : X^h \to X^h$ be an anti-holomorphic involution and let $\psi : X^h \to$

$\overline{X^h}$ be the canonical map induced by the identity on topological spaces. The map $\sigma \circ \psi \colon X^h \to X^h$ is holomorphic and X is projective so by Serre's **GAGA** theorems [Ser56] it is regular for the Zariski topology. In other words, $\sigma : X \to X$ is an anti-regular involution. □

Consider $X \subset \mathbb{P}^n(\mathbb{C})$ and let $\psi \colon X \to \mathbb{P}^N(\mathbb{C})$ be a morphism of complex varieties. We denote by ${}^{\sigma}\psi := \sigma_{\mathbb{P}} \circ \psi \circ \sigma_{\mathbb{P}}$.

Proposition 2.1.19 (Conditions for the existence of a real structure) *If a complex quasi-projective variety $X \subset \mathbb{P}^n(\mathbb{C})$ has a real structure then there is an isomorphism $\psi \colon X \to \overline{X}$ satisfying ${}^{\sigma}\psi \circ \psi = \mathrm{id}_X$.*

Proof If σ is a real structure on X then we simply set $\psi := \overline{\sigma} = \sigma_{\mathbb{P}} \circ \sigma$. We then have that $\psi^{-1} = \sigma^{-1} \circ \sigma_{\mathbb{P}}{}^{-1} = \sigma \circ \sigma_{\mathbb{P}}$. Moreover, ${}^{\sigma}\psi = \sigma_{\mathbb{P}} \circ \psi \circ \sigma_{\mathbb{P}} = \sigma_{\mathbb{P}} \circ (\sigma_{\mathbb{P}} \circ \sigma) \circ \sigma_{\mathbb{P}} = \sigma \circ \sigma_{\mathbb{P}}$. □

Remark 2.1.20 We insist on the fact that a real structure σ is an involution (i.e. $\sigma \circ \sigma = \mathrm{id}$). The following example by Shimura [Shi72a, p. 177] (see also [Sil92, p. 152]) shows that a complex variety can be isomorphic to its conjugate without having a real structure! (The variety in question has an anti-isomorphism or order 4 but no anti-isomorphism of order 2.)

Exercise 2.1.21 (Curves without real structures) *Let m be an odd number, let $a_0 \in \mathbb{R}$ be a real number and let $a_k \in \mathbb{C} \setminus \mathbb{R}$, $k = 1, \ldots, m$ be non real complex numbers. Consider the curve $C_{m,a_0,\ldots,a_m}$ which is the projective completion (i.e. the Zariski closure of the image of the affine curve under the inclusion $j \colon \mathbb{A}^2(\mathbb{C}) \hookrightarrow \mathbb{P}^2(\mathbb{C})$—see Lemma 1.2.43 and Exercise 1.2.44) of the affine plane curve of equation*

$$y^2 = a_0 x^m + \sum_{k=1}^{m} \left(a_k x^{m+k} + (-1)^k \overline{a_k} x^{m-k} \right) .$$

1. *Prove that the curve $C_{m,a_0,\ldots,a_m}$ is isomorphic to its conjugate via the map $\varphi \colon (x, y) \mapsto (-\frac{1}{x}, \frac{i}{x^m} y)$ for $(x, y) \neq (0, 0)$ and $\varphi(0, 0) = (0, 0)$.*
2. *Prove that φ induces an anti-isomorphism of $C_{m,a_0,\ldots,a_m}$ of order 4.*
3. *Assume that the number a_0, the numbers a_k and the numbers $\overline{a_k}$ are all algebraically independent over $\mathbb{Q}$.*
   (a) Prove that the only automorphisms of $C_{m,a_0,\ldots,a_m}$ are the identity and $\rho \colon (x, y) \mapsto (x, -y)$. (Use Exercise 1.2.80(3a).)
   (b) Deduce that $C_{m,a_0,\ldots,a_m}$ has no real structure.

See Section 5.5 and [KK02, Theorem 5.1] for examples of complex surfaces with no real structure, or even with no anti-automorphism.

Definition 2.1.22 The *real locus*, or *real part* of an $\mathbb{R}$-variety (X, σ) is the set of fixed points $X^{\sigma} := \{x \in X \mid \sigma(x) = x\}$ of the real structure. By analogy with the set of real points of a scheme defined over $\mathbb{R}$ the set of fixed points of σ is often denoted

$$X(\mathbb{R}) := X^\sigma$$

when there is no possible confusion.

Remark 2.1.23 Obviously, if (Y, τ) is an $\mathbb{R}$-subvariety of (X, σ) then $Y(\mathbb{R}) \subset X(\mathbb{R})$.

Examples 2.1.24 (*Real loci of Examples* 2.1.11)

1. $\mathbb{A}^n(\mathbb{R})$;
2. $\mathbb{P}^n(\mathbb{R})$;
3. $\varnothing$.

Definition 2.1.25 Let (X, σ) and (Y, τ) be $\mathbb{R}$-varieties. A *morphism of* $\mathbb{R}$*-varieties* (or *regular map of* $\mathbb{R}$*-varieties*) $(X, \sigma) \to (Y, \tau)$ is a morphism of complex varieties $\varphi \colon X \to Y$ which commutes with the real structures

$$\forall x \in X, \quad \varphi(\sigma(x)) = \tau(\varphi(x)) \ .$$

Remark 2.1.26 $\mathbb{R}$-varieties (X, σ) and (Y, τ) are therefore *isomorphic* if and only if there is an isomorphism $X \overset{\varphi}{\simeq} Y$ of complex varieties commuting with the real structures. Indeed, if $\varphi \colon X \to Y$ commutes with the real structures i.e. $\varphi \circ \sigma = \tau \circ \varphi$ then $\varphi^{-1} \colon Y \to X$ is a morphism of $\mathbb{R}$-varieties; for any $y \in Y$ and $x = \varphi^{-1}(y)$ we have that $\varphi(\sigma(\varphi^{-1}(y))) = \varphi(\sigma(x)) = \tau(\varphi(x)) = \tau(y)$ and hence $\sigma(\varphi^{-1}(y)) = \varphi^{-1}(\tau(y))$.

Definition 2.1.27 Let (X, σ) and (Y, τ) be $\mathbb{R}$-varieties. A *rational map of* $\mathbb{R}$*-varieties* $(X, \sigma) \dashrightarrow (Y, \tau)$ is a rational map of complex varieties

$$\varphi \colon X \dashrightarrow Y$$

which commutes with the real structures

$$\forall x \in \operatorname{dom}(\varphi) \subset X, \quad \varphi(\sigma(x)) = \tau(\varphi(x)) \ .$$

Remark 2.1.28 Denoting the Galois group by $G := \operatorname{Gal}(\mathbb{C}|\mathbb{R})$, the involution σ (resp. τ) equips X (resp. Y) with a G-action. A regular map of $\mathbb{R}$-varieties $(X, \sigma) \to (Y, \tau)$ is then by definition a G-equivariant regular map of complex varieties. Similarly, a rational map of $\mathbb{R}$-varieties is a G-equivariant rational map of complex varieties.

If X is a projective algebraic variety defined in some $\mathbb{P}^n(\mathbb{C})$ by homogeneous polynomial equations

$$P_1(z_0, \ldots, z_n) = \cdots = P_l(z_0, \ldots, z_n) = 0 \ ,$$

then, as we have seen above, the variety X has a real structure induced by $\sigma_{\mathbb{P}}\colon \mathbb{P}^n(\mathbb{C}) \to \mathbb{P}^n(\mathbb{C})$ if and only if we can assume the polynomials P_i have *real* coefficients, or in other words if the homogeneous ideal generated by the P_is has a system of generators with real coefficients. If this is the case then the real locus of the $\mathbb{R}$-variety $(X, \sigma_{\mathbb{P}}|_X)$ is simply $X(\mathbb{R}) = X \cap \mathbb{P}^n(\mathbb{R})$. Similarly, if X is an affine algebraic variety defined in $\mathbb{A}^n(\mathbb{C})$ by polynomial equations

$$P_1(z_1, \dots, z_n) = \cdots = P_l(z_1, \dots, z_n) = 0 ,$$

then $\sigma_{\mathbb{A}}\colon \mathbb{A}^n(\mathbb{C}) \to \mathbb{A}^n(\mathbb{C})$ induces a real structure on the complex variety X if and only if we can assume the polynomials P_i have real coefficients and in this case the real locus of the $\mathbb{R}$-variety $(X, \sigma_{\mathbb{A}}|_X)$ is given by

$$X(\mathbb{R}) = X \cap \mathbb{A}^n(\mathbb{R}) .$$

Note that the variety X may however have other real structures than the restriction of $\sigma_{\mathbb{P}}$ or $\sigma_{\mathbb{A}}$.

Example 2.1.29 (*Two distinct real structures on the same complex variety*) Consider the affine algebraic plane curve $C \subset \mathbb{A}^2(\mathbb{C})$ determined by the equation $y^2 = x^3 - x$. As this equation has real coefficients, the conjugation $\sigma_{\mathbb{A}}$ restricted to C yields a real structure. If we set $\sigma_1 := \sigma_{\mathbb{A}}|_C$ then (C, σ_1) is an $\mathbb{R}$-variety whose set of real points $C(\mathbb{R}) = \mathcal{Z}\left(y^2 - x(x-1)(x+1)\right) \cap \mathbb{A}^2(\mathbb{R})$ has two connected components in the Euclidean topology—see Figure 2.1.

Now let us consider σ_2, the restriction to C of the anti-regular involution $\mathbb{A}^2(\mathbb{C}) \to \mathbb{A}^2(\mathbb{C})$, $(x, y) \mapsto (-\overline{x}, i\overline{y})$. We check that $\sigma_2(C) \subset C$ so the pair (C, σ_2) is an $\mathbb{R}$-variety whose real structure is not induced by $\sigma_{\mathbb{A}}$. Let C' be the curve of equation $y^2 = x^3 + x$ in $\mathbb{A}^2(\mathbb{C})$ end let ζ be a square root of $-i$, $\zeta^2 = -i$. The morphism $\varphi\colon \mathbb{A}^2(\mathbb{C}) \to \mathbb{A}^2(\mathbb{C})$, $(x, y) \mapsto (ix, \zeta y)$ is an automorphism of $\mathbb{A}^2(\mathbb{C})$ whose

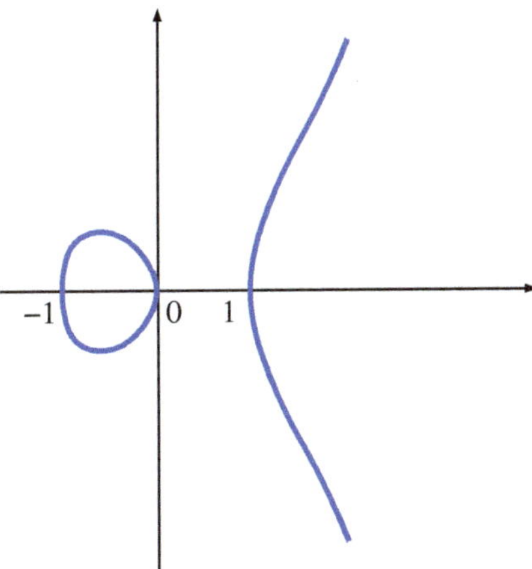

Fig. 2.1 $C : y^2 = x(x-1)(x+1)$

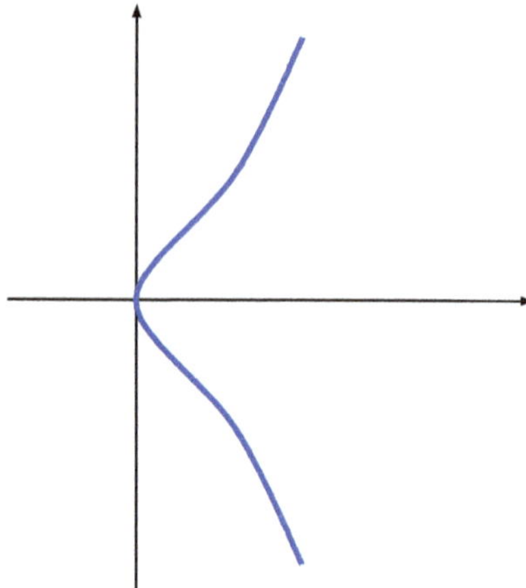

Fig. 2.2 $C' : y^2 = x(x - i)(x + i)$

restriction $\varphi|_C : C \to C'$ is an isomorphism of complex varieties. Set $\sigma' := \varphi|_C \circ \sigma_2 \circ \varphi^{-1}|_{C'}$: the $\mathbb{R}$-curves (C, σ_2) and (C', σ') are then isomorphic. It is easy to check that $\sigma' = \sigma_{\mathbb{A}}|_{C'}$. The set of real points $C'(\mathbb{R}) = \mathcal{Z}\left(y^2 - x(x-i)(x+i)\right) \cap \mathbb{A}^2(\mathbb{R})$ has only one connected components—see Figure 2.2. The $\mathbb{R}$-varieties (C, σ_1) and (C, σ_2) are therefore not isomorphic by Proposition 2.1.38 below.

In the above example, the *abstract* $\mathbb{R}$-variety (C, σ_2) is isomorphic to the $\mathbb{R}$-variety (C', σ') whose real structure is induced by the real structure on the surrounding space. The fact that there is always an $\mathbb{R}$-subvariety of some $\mathbb{A}^n$ isomorphic to a given affine abstract $\mathbb{R}$-variety is guaranteed by the fundamental Theorem 2.1.30 below. We insist on the fact that the isomorphism of complex varieties $C \to C'$ is not always induced by an automorphism of the surrounding space.

Theorem 2.1.30 (Real embedding of an affine $\mathbb{R}$-variety) *Let (X, σ) be an algebraic $\mathbb{R}$-variety. If the complex variety X is affine, $X \hookrightarrow \mathbb{A}^m(\mathbb{C})$ then there is an affine algebraic set $F \subset \mathbb{A}^n(\mathbb{C})$ such that $\sigma_{\mathbb{A}}(F) \subset F$ and there is an isomorphism of $\mathbb{R}$-varieties*

$$(F, \sigma_{\mathbb{A}}|_F) \simeq (X, \sigma) .$$

In particular, the ideal $\mathcal{I}(F)$ is generated by real polynomials or in other words there is an ideal $I \subset \mathbb{R}[X_1, \ldots, X_n]$ such that $\mathcal{I}(F) = I_{\mathbb{C}}$ and $\mathcal{A}(X)$ is isomorphic to $\mathcal{A}(F) = (\mathbb{R}[X_1, \ldots, X_n]/I) \otimes_{\mathbb{R}} \mathbb{C}$.

Remark 2.1.31 Note that $n \neq m$ in general.

This theorem is a reformulation—modulo Lemma A.7.3—of the following result.

Lemma 2.1.32 *Let (X, σ) be an affine algebraic $\mathbb{R}$-variety. There is then a real affine algebraic set $V \subset \mathbb{A}^n(\mathbb{R})$ with defining ideal $I = \mathcal{I}(V) \subset \mathbb{R}[X_1, \ldots, X_n]$ such that the $\mathbb{R}$-algebra $\mathcal{A}(V) = \mathbb{R}[X_1, \ldots, X_n]/I$ is isomorphic to the $\mathbb{R}$-algebra of affine invariant coordinates $\mathcal{A}(X)^\sigma = \{f \in \mathcal{A}(X) \mid {}^\sigma f = f\}$ of X.*

Proof The above result is a special case of the scheme-theoretic result stating that there is an equivalence between the data of an affine scheme X over $\mathbb{C}$ with a real structure σ and the data of a real scheme X_0, namely that if $X = \operatorname{Spec} A$ then $X_0 = \operatorname{Spec} A^\sigma$. See Section 2.4 for more details. □

Theorem 2.1.33 (Real embedding of a quasi-projective $\mathbb{R}$-variety) *Let (X, σ) be an algebraic $\mathbb{R}$-variety. If the complex algebraic variety X is projective (resp. quasi-projective), $X \hookrightarrow \mathbb{P}^m(\mathbb{C})$ then there is a projective (resp. quasi-projective) algebraic set $F \subset \mathbb{P}^n(\mathbb{C})$ such that $\sigma_\mathbb{P}(F) \subset F$ and there is an isomorphism of $\mathbb{R}$-varieties*

$$(F, \sigma_\mathbb{P}|_F) \simeq (X, \sigma)\ .$$

Remark 2.1.34 We insist on the fact that, as in the affine case, $n \neq m$ in general.

Proof The above statement is a special case of the scheme-theoretic statement that there is an equivalence between the data of a quasi-projective scheme X over $\mathbb{C}$ with a real structure σ and the data of a real scheme X_0 such that $X_0 = X/\langle\sigma\rangle$. See Section 2.4 for more details. □

Like many other authors, Silhol [Sil89] states the above result as a special case of a general result of the Galois descent theory developed first by Weil [Wei56, Theorem 7] then Grothendieck [Gro95, Théorème 3]. See also Borel–Serre [BS64, Proposition 2.6, p. 129]. We give an alternative Proof of Theorem 2.1.33 in Section 2.6, namely Theorem 2.6.44.

In Example 2.1.29, $\sigma_\mathbb{A}$ and $\sigma_\mathbb{A}'\colon (x, y) \mapsto (-\overline{x}, i\overline{y})$ are distinct real structures on $\mathbb{A}^2(\mathbb{C})$. The $\mathbb{R}$-varieties $\big(\mathbb{A}^2(\mathbb{C}), \sigma_\mathbb{A}\big)$ and $\big(\mathbb{A}^2(\mathbb{C}), \sigma_\mathbb{A}'\big)$, however, are isomorphic via the map $\varphi\colon (x, y) \mapsto (ix, \zeta y)$. In this situation we say that the real structures are *equivalent*.

Definition 2.1.35 Two real structures σ and τ on a complex variety X are *equivalent* if they are conjugate under an automorphism of the complex variety X or in other words if there is an automorphism φ of X such that

$$\sigma = \varphi^{-1} \circ \tau \circ \varphi$$

In other words, σ and τ are equivalent if there is an isomorphism of $\mathbb{R}$-varieties, $\varphi\colon (X, \sigma) \to (X, \tau)$.

Remark 2.1.36 Two real forms (see Definition 2.1.13), (X, σ) and (X, τ) of a complex variety X are isomorphic if and only if the real structures σ and τ are equivalent.

Example 2.1.37 It is proved in [Kam75] that all real structures on the affine complex plane are equivalent.

We recall that for any $\mathbb{R}$-variety (X, σ) we define $\#\pi_0(X^\sigma) = \#\pi_0(X(\mathbb{R}))$ to be the number of connected components of the real locus in the Euclidean topology.

Proposition 2.1.38 (Real locus and isomorphism) *An isomorphism of $\mathbb{R}$-varieties $\varphi\colon (X,\sigma) \to (Y,\tau)$ induces a homeomorphism between X^σ and Y^τ in the Euclidean topology. In particular*

$$\#\pi_0(X^\sigma) = \#\pi_0(Y^\tau) \quad \textit{or in other words} \quad \#\pi_0(X(\mathbb{R})) = \#\pi_0(Y(\mathbb{R}))\ .$$

Proof Start by noting that for a any given real structure the Euclidean topology on the real locus is simply the topology induced by the Euclidean topology on the complex variety. As φ is a homeomorphism for the Euclidean topology (see Exercise 1.4.4) and commutes with the real structures, it induces a bijection $X^\sigma \to Y^\tau$ between the fixed loci which is a homeomorphism. □

Corollary 2.1.39 (Real locus and equivalence) *Let σ and τ be real structures on a complex variety X. If σ and τ are equivalent then X^σ and X^τ are homeomorphic for the Euclidean topology and in particular*

$$\#\pi_0(X^\sigma) = \#\pi_0(X^\tau)\ .$$

Proof The real structures σ and τ are equivalent so there is an isomorphism of $\mathbb{R}$-varieties $\varphi\colon (X,\sigma) \to (X,\tau)$. □

Example 2.1.40 (*Two real forms on the same complex variety*) We return to the two complex algebraic curves C and C' studied in Example 2.1.29 whose equations in $\mathbb{A}^2(\mathbb{C})$ are $y^2 = x^3 - x$ and $y^2 = x^3 + x$ respectively. It is easy to check that the set of real points of $C(\mathbb{R}) \subset \mathbb{A}^2(\mathbb{R})$ has two connected components, see Figure 2.1, and that the set of real points of $C'(\mathbb{R}) \subset \mathbb{A}^2(\mathbb{R})$ has only one connected component, see Figure 2.2. In particular, by Proposition 2.1.38, the $\mathbb{R}$-curves (C,σ_1) and (C,σ_2) are not isomorphic.

The complex variety C therefore has two non-equivalent real structures $\sigma_1 = \sigma_{\mathbb{A}}|_C\colon (x,y) \mapsto (\overline{x},\overline{y})$ and $\sigma_2 = \varphi^{-1}|_{C'} \circ \sigma_{\mathbb{A}}|_{C'} \circ \varphi|_C\colon (x,y) \mapsto (-\overline{x}, i\overline{y})$. It is interesting to note that these non equivalent real structures are restrictions of real structures $\sigma_{\mathbb{A}}$ and $\varphi^{-1} \circ \sigma_{\mathbb{A}} \circ \varphi$ on $\mathbb{A}^2(\mathbb{C})$ which are equivalent by definition.

Remark 2.1.41 (*Non-standard real structure on the projective line*) We have already met the antipodal map on the Riemann sphere:

$$\sigma_{\mathbb{P}}'\colon \mathbb{P}^1(\mathbb{C}) \to \mathbb{P}^1(\mathbb{C}), \quad (x_0 : x_1) \mapsto (-\overline{x_1} : \overline{x_0})$$

which is a real structure on $\mathbb{P}^1(\mathbb{C})$ whose set of fixed points is empty and which is therefore not equivalent to $\sigma_{\mathbb{P}}$.

Exercise 2.1.42 (Real structures on a complex torus) *Find four pairwise non-equivalent real structures on $\mathbb{P}^1(\mathbb{C}) \times \mathbb{P}^1(\mathbb{C})$. (There are in fact exactly four classes of real structures on $\mathbb{P}^1(\mathbb{C}) \times \mathbb{P}^1(\mathbb{C})$.)*

Remark 2.1.43 Until recently it was not known whether the number of equivalence classes of real structures on a given complex variety was finite. See [DIK00, Appendix D] for a review of this question.

In [Les18], John Lesieutre constructs a variety of dimension 6 with a discrete automorphism group which cannot be generated by a finite number of generators and which has infinitely many non-isomorphic real forms. In [DO19], Dinh and Oguiso use different methods to construct examples of projective varieties of any dimension greater than one with non-finitely automorphism generated group. Their work also provides examples of real varieties of any dimension greater than one with infinitely many non-isomorphic real forms. In [DFM18], Dubouloz, Freudenburg and Moser–Jauslin construct affine rational varieties with infinitely many pairwise non-isomorphic real forms in every dimension ≥ 4.

Surprisingly, this finiteness question is still open for rational surfaces. See Benzerga's work [Ben16a, Ben16b, Ben17] for the most recent results on this question.

2.2 $\mathbb{R}$-Varieties and Real Algebraic Varieties

For a given quasi-projective $\mathbb{R}$-variety (X, σ) we seek to define a sheaf of regular functions on $X(\mathbb{R})$ with which $X(\mathbb{R})$ becomes a real algebraic variety as in Definition 1.3.9. By Theorem 2.1.33 and Exercise 2.1.3 the structural sheaf satisfies ${}^{\sigma}\mathcal{O}_X = \mathcal{O}_X$, which justifies the following definition. Recall that a real structure is a Zariski homeomorphism and in particular if U is open in X then so is $\sigma(U)$. Let $\mathcal{L}$ be a sheaf of $\mathbb{C}^n$-valued functions. For any open set U in X and any map $f \in \mathcal{L}(U)$ we denote by ${}^{\sigma}f : \sigma(U) \to \mathbb{C}^n$ the map $\overline{f} \circ \sigma = \sigma_{\mathbb{A}} \circ f \circ \sigma$. We then have that ${}^{\sigma}f \in \mathcal{L}(\sigma(U))$ which generalises the notion of conjugate function introduced at the beginning of Section 2.1.

Definition 2.2.1 Let (X, σ) be an $\mathbb{R}$-variety and let $\mathcal{L}$ be a sheaf of $\mathbb{C}^n$-valued functions. The sheaf ${}^{\sigma}\mathcal{L}$ defined on any open set U of X by

$$
{}^{\sigma}\mathcal{L}(U) := \{{}^{\sigma}f \mid f \in \mathcal{L}(\sigma(U))\} \ .
$$

is a sheaf on X called the *conjugate sheaf*. We say that $\mathcal{L}$ is an $\mathbb{R}$-*sheaf* if and only if ${}^{\sigma}\mathcal{L} = \mathcal{L}$. Note that this is required to be an equality, not an isomorphism.

From a cohomological point of view, the sheaves $\mathcal{L}$ and ${}^{\sigma}\mathcal{L}$ are similar. (See [Liu02, Section 5.2] for an introduction to sheaf cohomology.) In particular, we have the following proposition.

Proposition 2.2.2 *Let (X, σ) be an $\mathbb{R}$-variety and let $\mathcal{L}$ be a coherent sheaf (Definition C.6.7) of $\mathbb{C}^n$-valued functions. We then have that*

$$
\dim_{\mathbb{C}} H^k(X, {}^{\sigma}\mathcal{L}) = \dim_{\mathbb{C}} H^k(X, \mathcal{L}) \ .
$$

Proof See [Sil89, I.(1.9)]. □

Let (X, σ) be a quasi-projective $\mathbb{R}$-variety. We saw above that the sheaf of $\mathbb{C}$-algebras $\mathcal{O}_X$ is an $\mathbb{R}$-sheaf: ${}^\sigma\mathcal{O}_X = \mathcal{O}_X$. In particular, for any open set U in X, the morphism

$$\begin{array}{ccc} \mathcal{O}_X(U) & \longrightarrow & \mathcal{O}_X(\sigma(U)) \\ f & \longmapsto & {}^\sigma f \end{array}$$

is a ring isomorphism.

Remark 2.2.3 We can prove more: this map is an anti-isomorphism of $\mathbb{C}$-algebras. Let us prove anti-linearity: for any $\lambda \in \mathbb{C}$ and for any regular function f on U we have that ${}^\sigma(\lambda f) = \overline{\lambda f} \circ \sigma = \overline{\lambda}(\overline{f} \circ \sigma) = \overline{\lambda}({}^\sigma f)$.

If A is an $\mathbb{R}$-algebra equipped with a G-action, where $G := \mathrm{Gal}(\mathbb{C}|\mathbb{R})$, and σ is the corresponding involution of A then we denote by $A^G := A^\sigma = \{a \in A \mid \sigma(a) = a\}$ the sub-algebra of invariants of A (see Definition A.7.2).

Let (X, σ) be an $\mathbb{R}$-variety. A subset $U \subset X$ is said to be *invariant* if and only if $\sigma(U) = U$. Any such subset inherits a G-action: since σ is a homeomorphism, for any open set U in X the intersection $U \cap \sigma(U)$ is an invariant open set in X. For any invariant open set U we say that a local section f over U is *invariant* if ${}^\sigma f = f$. Let $\mathcal{F}$ be an $\mathbb{R}$-sheaf of functions on X. We denote by $\mathcal{F}_{X(\mathbb{R})}$ the sheaf of its restrictions to $X(\mathbb{R})$, see Definition C.1.6 and by $\mathcal{F}^G_{X(\mathbb{R})}$ its invariant subset. We apply this definition to $\mathcal{O}_X$, which is an $\mathbb{R}$-sheaf of functions on X, and obtain a sheaf

$$(\mathcal{O}_X)^G_{X(\mathbb{R})} := \left((\mathcal{O}_X)_{X(\mathbb{R})}\right)^G$$

of real-valued functions on $X(\mathbb{R})$. It takes some work to prove that these functions are $\mathbb{R}$-valued, since a priori they are $\mathbb{C}$-valued—see below for the proof.

Let us describe the local sections of this new sheaf. Let $\Omega \subset X(\mathbb{R})$ be an open subset in the induced topology. We check first that any $f \in (\mathcal{O}_X)^G_{X(\mathbb{R})}(\Omega)$ is $\mathbb{R}$-valued. As f is invariant, for any $x \in \Omega$ we have that $f(x) = ({}^\sigma f)(x) = \overline{f(\sigma(x))}$ and since x is a point in $X(\mathbb{R})$ we have that $\sigma(x) = x$ so $f(x) \in \mathbb{R}$. By definition of $(\mathcal{O}_X)_{X(\mathbb{R})}$ there is an open neighbourhood $U \subset X$ of x and an element $g \in \mathcal{O}_X(U)$ such that $g|_{U\cap\Omega} = f|_{U\cap\Omega}$. Replacing U by $U \cap \sigma(U)$ and g by $\frac{1}{2}(g + {}^\sigma g)$ we get an element $g \in (\mathcal{O}_X(U))^G$ such that $g|_{U\cap\Omega} = f|_{U\cap\Omega}$. In other words, the local sections of $(\mathcal{O}_X)^G_{X(\mathbb{R})}$ over an open set Ω in $X(\mathbb{R})$ are as follows.

$$\begin{aligned}(\mathcal{O}_X)^G_{X(\mathbb{R})}(\Omega) = \big\{f\colon \Omega \to \mathbb{R} \mid \forall x \in \Omega,\\ \exists U \text{ open invariant neighbourhood of } x \text{ in } X \text{ and}\\ \exists g \in (\mathcal{O}_X(U))^G \mid g|_{U\cap\Omega} = f|_{U\cap\Omega}\big\}\,.\end{aligned}$$

We invite the reader to compare the following theorem with Theorem 2.1.30.

Theorem 2.2.4 *Let $F \subset \mathbb{A}^n(\mathbb{C})$ be a complex affine algebraic set such that $\mathcal{I}(F)$ is generated by polynomials with real coefficients. In particular, $F(\mathbb{R}) := F \cap \mathbb{A}^n(\mathbb{R})$ is a real algebraic affine set.*

If $F(\mathbb{R})$ is dense in F with respect to the Zariski topology then

$$\mathcal{O}_{F(\mathbb{R})} \simeq (\mathcal{O}_F)^G_{F(\mathbb{R})} .$$

Proof Let $I \subset \mathbb{R}[X_1, \dots, X_n]$ be an ideal and let $F = \mathcal{Z}_{\mathbb{C}}(I) \subset \mathbb{A}^n(\mathbb{C})$ be the complex algebraic set whose ideal is $\mathcal{I}(F) = I_{\mathbb{C}}$ and whose sheaf of regular functions is $\mathcal{O}_F$. The set $F(\mathbb{R}) = F \cap \mathbb{A}^n(\mathbb{R}) = \mathcal{Z}_{\mathbb{R}}(I) \subset \mathbb{A}^n(\mathbb{R})$ is then a real algebraic set whose sheaf of regular functions will be denoted by $\mathcal{O}_{F(\mathbb{R})}$. By hypothesis F is stable under $\sigma_{\mathbb{A}}$. By Proposition C.3.12 these sheaves are isomorphic if and only if their stalks are isomorphic.

Let $\Omega \subset F(\mathbb{R})$ be a Zariski open subset in $\mathbb{A}^n(\mathbb{R})$ and let f be an element of $\mathcal{O}_{F(\mathbb{R})}(\Omega)$. Passing to a smaller open set if necessary, we can assume that on Ω $f = \frac{p}{q}$ where p, q are polynomials with real coefficients and q does not vanish at any point of Ω. There is then an open set U of F in $\mathbb{A}^n(\mathbb{C})$ on which q does not vanish and hence $f \in \mathcal{O}_{F(\mathbb{R})}(\Omega)$ can be extended to a regular function $f_{\mathbb{C}} \in \mathcal{O}_F(U)$ such that ${}^{\sigma} f_{\mathbb{C}} = f_{\mathbb{C}}$. As $F(\mathbb{R})$ is dense in F, the germ of the extension $f_{\mathbb{C}}$ of f is uniquely determined by the germ of f. It follows that $\mathcal{O}_{F(\mathbb{R})} \simeq (\mathcal{O}_F)^G_{F(\mathbb{R})}$. □

Theorem 2.2.4 motivates our next definition.

Definition 2.2.5 Let (X, σ) be an $\mathbb{R}$-variety. We say that (X, σ) has *enough real points* if and only if $X(\mathbb{R})$ is Zariski-dense in X.

Exercise 2.2.6 *Let $I \subset \mathbb{R}[X_1, \dots, X_n]$ be a radical ideal and let $F = \mathcal{Z}_{\mathbb{C}}(I) \subset \mathbb{A}^n(\mathbb{C})$ be the associated complex algebraic set as in Definition 1.2.12. Let $(F, \sigma_{\mathbb{A}}|_F)$ be the associated affine $\mathbb{R}$-variety.*

1. *Prove that the $\mathbb{R}$-variety $(F, \sigma_{\mathbb{A}}|_F)$ has enough real points if and only if $\mathcal{I}(\mathcal{Z}(I)) \subset I$ in $\mathbb{R}[X_1, \dots, X_n]$.*
2. *Prove that the $\mathbb{R}$-variety $(F, \sigma_{\mathbb{A}}|_F)$ has enough real points if and only if I is a real ideal, see Definition A.5.14.*

Exercise 2.2.7 *Prove that the $\mathbb{R}$-variety $(F = \mathcal{Z}_{\mathbb{C}}(x^2 + y^2), \sigma_{\mathbb{A}}|_F)$—which has a non-empty real locus—does not have enough real points. (See Example 2.1.1). Further prove that $\mathcal{O}_{F(\mathbb{R})} \neq (\mathcal{O}_F)^G_{F(\mathbb{R})}$.*

Theorem 2.2.9 below characterises those $\mathbb{R}$-varieties that have enough real points. In particular, any irreducible non singular $\mathbb{R}$-variety with non-empty real locus has enough real points.

Lemma 2.2.8 *Let (X, σ) be an algebraic $\mathbb{R}$-variety, let $a \in X(\mathbb{R})$ be a real point and let $\mathfrak{m}_a$ be the maximal ideal of the local ring $\mathcal{O}_{X,a}$. We then have that*

$$\dim_{\mathbb{C}} \mathfrak{m}_a/\mathfrak{m}_a^2 = \dim_{\mathbb{R}}((\mathfrak{m}_a/\mathfrak{m}_a^2)^G) .$$

Proof As a is real σ induces an anti-linear involution on $\mathcal{O}_{X,a}$ and by Lemma A.7.3 there is a basis of $\mathfrak{m}_a/\mathfrak{m}_a^2$ whose elements are all σ-invariant. □

Theorem 2.2.9 (Density of the real locus in the complex variety)

1. *The space $\mathbb{A}^n(\mathbb{R})$ is dense in $\mathbb{A}^n(\mathbb{C})$ for the Zariski topology.*
2. *Let $V \subset \mathbb{A}^n(\mathbb{C})$ be an irreducible affine algebraic set whose ideal $I = \mathcal{I}(V)$ is generated by polynomials with real coefficients. The real locus $V(\mathbb{R}) = V \cap \mathbb{A}^n(\mathbb{R})$ is Zariski dense in V if and only if it contains at least one non singular point of V.*
3. *Let (X, σ) be an algebraic $\mathbb{R}$-variety. The real locus $X(\mathbb{R})$ is Zariski dense in every irreducible component Z of X containing a non singular real point if and only if $X(\mathbb{R})$ is not contained in the singular locus of X. In other words, $\overline{X(\mathbb{R})}^{Zar} \cap Z = Z$ if and only if* (Reg Z) $\cap X(\mathbb{R})$ *is non empty.*

Corollary 2.2.10 *Let (X, σ) be an algebraic $\mathbb{R}$-variety. If the complex variety X is irreducible and non singular and if $X(\mathbb{R}) \neq \varnothing$ then (X, σ) has enough real points, or in other words $\overline{X(\mathbb{R})}^{Zar} = X$.*

The behaviour of the Euclidean topology is very different.

Proposition 2.2.11 *The real locus $X(\mathbb{R})$ of an algebraic $\mathbb{R}$-variety (X, σ) is closed in X for the Euclidean topology.*

Proof The real structure σ is continuous for the Euclidean topology and the real locus $X(\mathbb{R}) = \{x \in X \mid x = \sigma(x)\}$ is therefore closed in X because the Euclidean topology is Hausdorff. □

Proof of Theorem 2.2.9 1. We reuse the argument of Proposition 1.5.29. Assume for the moment that we have proved that if a polynomial function vanishes on all real affine points then it is identically zero—this will be proved below by induction on the dimension. If $\mathcal{Z}(I)$ is a closed subset of $\mathbb{A}^n(\mathbb{C})$ containing $\mathbb{A}^n(\mathbb{R})$ then for any $f \in I$ the function f vanishes on every point of $\mathbb{A}^n(\mathbb{R})$ and by assumption f is the zero polynomial. It follows that $I = (0)$ and $\mathcal{Z}(I) = \mathbb{A}^n(\mathbb{C})$.

Let us now prove that for any n, any polynomial vanishing on all real affine points is identically zero. For $n = 1$, the result is immediate. Suppose that $n > 1$ and the induction hypothesis holds for $n - 1$. Let $f \in \mathbb{C}[X_1, \dots, X_n]$ be a polynomial function vanishing on $\mathbb{R}^n$. We can write

$$f(X', X_n) = X_n^d f_d(X') + X_n^{d-1} f_{d-1}(X') + \dots + f_0(X')$$

where $X' = (X_1, \dots, X_{n-1})$, $d = \deg f$ and $\forall i = 0, \dots, d$, $f_i \in \mathbb{C}[X_1, \dots, X_{n-1}]$.

For any $X' \in \mathbb{R}^{n-1}$ the function $X_n \mapsto f(X', X_n)$ vanishes at every real point so $f(X', X_n)$ is the zero polynomial for any fixed X'. It follows that the polynomial functions f_i vanish for every real $X' \in \mathbb{R}^{n-1}$ and are therefore identically zero by the induction hypothesis.

2. As V is irreducible in $\mathbb{A}^n(\mathbb{C})$, $I = \mathcal{I}(V)$ is a prime ideal in $\mathbb{C}[X_1, \ldots, X_n]$ and $I_{\mathbb{R}} := I \cap \mathbb{R}[X_1, \ldots, X_n]$ is a prime ideal in $\mathbb{R}[X_1, \ldots, X_n]$ (Lemma A.2.9). We then have that $V = \mathcal{Z}_{\mathbb{C}}(I_{\mathbb{R}})$ and $V(\mathbb{R}) = \mathcal{Z}(I_{\mathbb{R}})$. Set $d = \dim_{\mathbb{C}} V$: by the Nullstellensatz (Corollary A.5.13), we have that $\dim I = d$ (see Definition 1.5.9) and $\dim I_{\mathbb{R}} = d$ by Lemma 1.5.15. Note that *a priori* $\dim_{\mathbb{R}} V(\mathbb{R})$ is not necessarily equal to d: see Example 1.5.20 or 2.2.15.

We now use the fact that there is a non singular point $a = (a_1, \ldots, a_n) \in (\operatorname{Reg} V) \cap \mathbb{A}^n(\mathbb{R})$. By Remark 1.5.28, V is a differentiable submanifold of dimension $2d \leqslant 2n$ at a or in other words there is a Euclidean neighbourhood W of a in $\mathbb{C}^n$ such that $W \cap V$ is a Euclidean neighbourhood of a in V of real dimension $2d$ and $W \cap V(\mathbb{R})$ is a Euclidean neighbourhhood of a in $V(\mathbb{R})$ of real dimension d (take an open chart (W, φ) where $W = \sigma(W)$ and justify that φ can be chosen G-equivariant using Lemma A.7.3 if necessary). The subvariety $V(\mathbb{R})$ is therefore a submanifold of real dimension d at a. The real algebraic set $V(\mathbb{R})$ then has Zariski dimension d by Proposition 1.5.29, or in other words the dimension of the ideal $\mathcal{I}(V(\mathbb{R}))$ is equal to d. There is therefore a length d chain of prime ideals in $\mathbb{R}[X_1, \ldots, X_n]$ containing $\mathcal{I}(V(\mathbb{R}))$. As $\mathcal{I}(V(\mathbb{R})) \supset I_{\mathbb{R}}$ by definition if $\mathcal{I}(V(\mathbb{R}))$ were different from $I_{\mathbb{R}}$ we would get a chain of length $d + 1$ of prime ideals containing $I_{\mathbb{R}}$, contradicting the fact that $\dim I_{\mathbb{R}} = d$. It follows that $\mathcal{I}(\mathcal{Z}(I_{\mathbb{R}})) = I_{\mathbb{R}}$ and hence $\overline{V(\mathbb{R})} = V$ by 2.2.6(1).

3. We can assume that X is irreducible. By definition of a algebraic variety, X can be covered by open affine subsets. By hypothesis we can therefore chose an open affine subset U in X such that $U \cap X(\mathbb{R})$ is not contained in the singular locus of X (and in particular, U is not empty and since X is irreducible, U is Zariski-dense). Replacing U by $U \cap \sigma(U)$ if necessary we can assume that U is stable under σ. As U is affine (see Exercise 1.3.15(4)) the $\mathbb{R}$-variety $(U, \sigma|_U)$ is isomorphic to an affine $\mathbb{R}$-variety $(V, \sigma_{\mathbb{A}}|_V) \subset (\mathbb{A}^n(\mathbb{C}), \sigma_{\mathbb{A}})$ by Theorem 2.1.30 so we now simply apply (2) to this affine $\mathbb{R}$-variety. $V \cap \mathbb{A}^n(\mathbb{R})$ is dense in $V \cap \mathbb{A}^n(\mathbb{C}) = V$ and we note that $U \cap X(\mathbb{R}) = \varphi^{-1}(V \cap \mathbb{A}^n(\mathbb{R}))$ for any $\mathbb{R}$-isomorphism $\varphi \colon U \to V$. $\square$

Example 2.2.12 (*Reducible, singular, non empty and non dense*) We return to Example 1.5.20. Consider the reducible affine algebraic $\mathbb{R}$-variety

$$(V, \sigma) := (\mathcal{Z}_{\mathbb{C}}(x^2 + y^2), \sigma_{\mathbb{A}}|_V)$$

whose real locus is the isolated point $a = (0, 0)$. By definition we have that $\mathcal{O}_{V,a} = \left(\frac{\mathbb{C}[x,y]}{(x^2+y^2)}\right)_{(0,0)}$ whence $\dim \mathcal{O}_{V,a} = \dim \mathcal{O}_{V,a}^G = 1$ et $\dim_{\mathbb{C}} \mathfrak{m}_{V,a}/\mathfrak{m}_{V,a}^2 = \dim_{\mathbb{R}}((\mathfrak{m}_{V,a}/\mathfrak{m}_{V,a}^2)^G) = 2$, illustrating the fact that a is a real singular point of the 1-dimensional complex variety. *A contrario* we have that $\dim \mathcal{O}_{V(\mathbb{R}),a} = \dim_{\mathbb{R}} \mathfrak{m}_{V(\mathbb{R}),a}/\mathfrak{m}_{V(\mathbb{R}),a}^2 = 0$ illustrating the fact that the real algebraic variety $\{a\}$ is a zero-dimensional non singular variety.

Example 2.2.13 (*Irreducible, singular, dense*) We return to Example 1.5.21. Consider the affine algebraic $\mathbb{R}$-curve

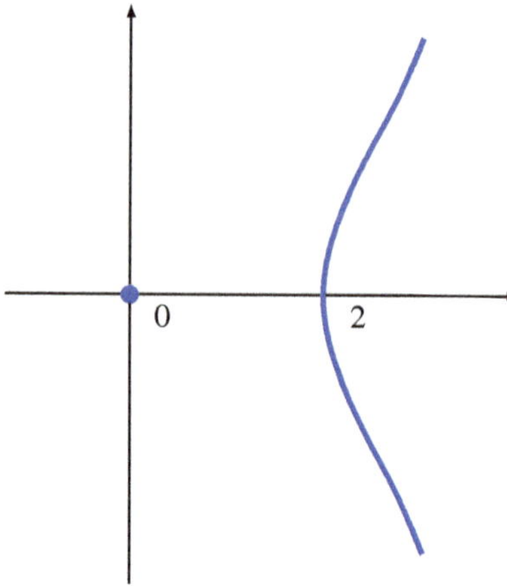

Fig. 2.3 $V(\mathbb{R}) = \{y^2 - x^2(x-2) = 0\} \subset \mathbb{A}^2(\mathbb{R})$

$$(V, \sigma) := (\mathcal{Z}_{\mathbb{C}}(y^2 - x^2(x-2)), \sigma_{\mathbb{A}}|_V)$$

whose real locus is shown in Figure 2.3. The Zariski closure in $\mathbb{A}^2(\mathbb{C})$ of the "branch" $(\operatorname{Reg} V) \cap V(\mathbb{R}) = V(\mathbb{R}) \cap \{x > 1\}$ is V.

Remark 2.2.14 The point $(0, 0)$ is not, however, contained in the Euclidean closure of the branch $V(\mathbb{R}) \cap \{x > 1\}$.

Example 2.2.15 (*Irreducible, singular, non empty and non dense*) This is an example of an irreducible singular algebraic set V whose real locus is neither empty nor Zariski dense in V. Consider

$$P(x, y) = ((x+i)^2 + y^2 - 1)((x-i)^2 + y^2 - 1) + x^2 = \\ x^4 + 2x^2y^2 + y^4 - 4y^2 + 4 + x^2$$

which is a polynomial in $\mathbb{R}[x, y]$. The set $V := \mathcal{Z}_{\mathbb{C}}(P) \subset \mathbb{A}^2(\mathbb{C})$ is an irreducible algebraic set and its real locus contains exactly two points. Indeed, let $P_1(x, y) = P(x, y) - x^2$ and set $V_1 := \mathcal{Z}(P_1)$. If (x, y) is a real point of V_1 then $y^2 = 1 - (x+i)^2$ or $y^2 = 1 - (x-i)^2$. As x and y are real, x must be identically zero so $y = \pm\sqrt{2}$ and $V_1(\mathbb{R}) = \{(0, \sqrt{2}), (0, -\sqrt{2})\}$. We will now prove that we also have that $V(\mathbb{R}) = \{(0, \sqrt{2}), (0, -\sqrt{2})\}$. Note that if $P(x, y) = 0$ for some real x then this implies that $P_1(x, y) = x^4 + 2x^2y^2 + y^4 - 4y^2 + 4$ is a negative or zero real number. Considering P_1 as a degree 2 polynomial in the variable $Y = y^2$ with coefficients in $\mathbb{R}[x]$ we see that its discriminant is equal to $-4x^2$. If x is non zero then this discriminant is strictly negative so for real x and y, $P(x, y) = 0$ if and only if $P_1(x, y) = 0$. We leave it as an exercise for the reader to show that P is irreducible, a long but unsurprising calculation. (We constructed the polynomial P by starting from the polynomial P_1 and looking for a perturbation of P_1 preserving the two real points in V_1, whose existence is guaranteed by Brusotti's Theorem 2.7.10.).

Exercise 2.2.16 *Construct a similar example from the example given in Remark 1.2.31(2).*

Theorem 2.2.17 *Let (X, σ) be a quasi-projective algebraic ℝ-variety. If the variety (X, σ) has* enough real points, *or in other words if $X(\mathbb{R})$ is Zariski dense in X, then the real locus equipped with the restriction of the structural sheaf, $\left(X(\mathbb{R}), (\mathcal{O}_X)^G_{X(\mathbb{R})}\right)$, is a real algebraic variety as in Definition 1.3.9.*

Proof This follows easily from Theorem 2.1.33 and the projective analogue of Theorem 2.2.4. □

Corollary 2.2.18 *Let (X, σ) be a quasi-projective algebraic ℝ-variety. If the complex variety X is irreducible and non singular and $X(\mathbb{R}) \neq \varnothing$ then $\left(X(\mathbb{R}), (\mathcal{O}_X)^G_{X(\mathbb{R})}\right)$ is a real algebraic variety.*

Proof See Corollary 2.2.10. □

The following proposition justifies the introduction of a third type of morphism between ℝ-varieties, somewhere between regular maps Definition 2.1.25 and rational maps Definition 2.1.27.

Proposition 2.2.19 *Let (X, σ) and (Y, τ) be ℝ-varieties with enough real points and let*

$$\psi : (X, \sigma) \dashrightarrow (Y, \tau)$$

be a rational map of ℝ-varieties. If the domain of ψ contains the real locus $X(\mathbb{R})$, then ψ induces by restriction a regular map of real algebraic varieties $\left(X(\mathbb{R}), (\mathcal{O}_X)^G_{X(\mathbb{R})}\right) \to \left(Y(\mathbb{R}), (\mathcal{O}_Y)^G_{Y(\mathbb{R})}\right)$.

Proof See Exercise 2.2.26(2). □

Definition 2.2.20 Let (X, σ) and (Y, τ) be ℝ-varieties.
A *rational ℝ-regular map* or *real morphism*

$$\psi : (X, \sigma) \dashrightarrow (Y, \tau)$$

is a rational map of ℝ-varieties such that $X(\mathbb{R}) \subset \mathrm{dom}(\psi)$.

Remark 2.2.21 A morphism of ℝ-varieties is of course always a rational ℝ-regular map but the converse is false.

Proposition 2.2.22 *Let (X, σ) and (Y, τ) be quasi-projective ℝ-varieties. Suppose that these varieties have enough real points. The following then hold.*

1. *A rational ℝ-regular map of ℝ-varieties $(X, \sigma) \dashrightarrow (Y, \tau)$ induces a regular map of real algebraic varieties*

$$\left(X(\mathbb{R}), (\mathcal{O}_X)^G_{X(\mathbb{R})}\right) \to \left(Y(\mathbb{R}), (\mathcal{O}_Y)^G_{Y(\mathbb{R})}\right) .$$

2. *Conversely, any regular map of real algebraic varieties*

$$\left(X(\mathbb{R}), (\mathcal{O}_X)^G_{X(\mathbb{R})}\right) \to \left(Y(\mathbb{R}), (\mathcal{O}_Y)^G_{Y(\mathbb{R})}\right)$$

is the restriction of an $\mathbb{R}$-regular rational map $\psi\colon (X, \sigma) \dashrightarrow (Y, \tau)$.

3. *Any rational map of $\mathbb{R}$-varieties $(X, \sigma) \dashrightarrow (Y, \tau)$ induces a rational map of real algebraic varieties*

$$\left(X(\mathbb{R}), (\mathcal{O}_X)^G_{X(\mathbb{R})}\right) \dashrightarrow \left(Y(\mathbb{R}), (\mathcal{O}_Y)^G_{Y(\mathbb{R})}\right) .$$

4. *Conversely, any rational map*

$$\left(X(\mathbb{R}), (\mathcal{O}_X)^G_{X(\mathbb{R})}\right) \dashrightarrow \left(Y(\mathbb{R}), (\mathcal{O}_Y)^G_{Y(\mathbb{R})}\right)$$

is the restriction of a rational map $(X, \sigma) \dashrightarrow (Y, \tau)$.

Proof Left for the reader as an exercise. □

Remark 2.2.23 We insist on (2) in the above proposition: the complex extension of a real regular map is not generally regular. The map $(x, y) \mapsto \frac{1}{x^2+y^2+1}$ from $\mathbb{A}^2(\mathbb{R})$ to $\mathbb{A}^1(\mathbb{R})$ is a regular map of real algebraic varieties but does not extend to a morphism of $\mathbb{R}$-varieties.

Remark 2.2.24 The "isomorphisms" corresponding to $\mathbb{R}$-regular rational maps are the *$\mathbb{R}$-biregular birational maps*. Note that it is important the map be both birational and $\mathbb{R}$-biregular: blowing up a real point (or in other words, contracting a (-1)-real curve) on an $\mathbb{R}$-surface (see Definition 4.1.26 for more details) is an $\mathbb{R}$-regular birational map but it is not $\mathbb{R}$-biregular.

Definition 2.2.25 Let (X, σ) and (Y, τ) be $\mathbb{R}$-varieties.
A *$\mathbb{R}$-biregular birational map* or *real isomorphism*

$$\psi\colon (X, \sigma) \dashrightarrow (Y, \tau)$$

is a birational map of $\mathbb{R}$-varieties inducing a biregular map of real algebraic varieties

$$\left(X(\mathbb{R}), (\mathcal{O}_X)^G_{X(\mathbb{R})}\right) \overset{\cong}{\to} \left(Y(\mathbb{R}), (\mathcal{O}_Y)^G_{Y(\mathbb{R})}\right) .$$

Exercise 2.2.26 (Use Exercises 1.2.56 and 1.3.25) *Let $F_1 \subset \mathbb{A}^n(\mathbb{C})$ and $F_2 \subset \mathbb{A}^m(\mathbb{C})$ be affine algebraic sets stable under $\sigma_{\mathbb{A}}$ so that $(F_1, \sigma_{\mathbb{A}}|_{F_1})$ and $(F_2, \sigma_{\mathbb{A}}|_{F_2})$ are affine $\mathbb{R}$-varieties and let $\varphi\colon F_1 \dashrightarrow F_2$ be a rational map of complex varieties.*

1. *Prove that φ is a morphism of $\mathbb{R}$-varieties if and only if there are polynomial functions $f_1, \ldots, f_m \in \mathbb{R}[x_1, \ldots, x_n]$ such that for any point $(x_1, \ldots, x_n) \in F_1$,*

$$\varphi(x_1, \ldots, x_n) = (f_1(x_1, \ldots, x_n), \ldots, f_m(x_1, \ldots, x_n)) .$$

In this case, $F_1 \subset \mathrm{dom}(\varphi)$ and $\varphi\colon F_1 \to F_2$ is a morphism of complex varieties.

2. *Prove that φ is an $\mathbb{R}$-regular birational map if and only if there are polynomial functions $g_1, \ldots, g_m \in \mathbb{R}[x_1, \ldots, x_n]$ and $h_1, \ldots, h_m \in \mathbb{R}[x_1, \ldots, x_n]$ such that for every point $(x_1, \ldots, x_n) \in F_1(\mathbb{R})$, $h_1(x_1, \ldots, x_n) \neq 0, \ldots, h_m(x_1, \ldots, x_n) \neq 0$ and*

$$\varphi(x_1, \ldots, x_n) = \left(\frac{g_1(x_1, \ldots, x_n)}{h_1(x_1, \ldots, x_n)}, \ldots, \frac{g_m(x_1, \ldots, x_n)}{h_m(x_1, \ldots, x_n)} \right) .$$

In this case $F_1(\mathbb{R}) \subset \mathrm{dom}(\varphi)$ and if F_1 and F_2 have enough real points then $\varphi|_{F_1(\mathbb{R})}\colon F_1(\mathbb{R}) \to F_2(\mathbb{R})$ is a regular map of real algebraic varieties with the induced structure.

2.2.1 Non Singular $\mathbb{R}$-Varieties

A non singular complex variety of complex dimension n is naturally a real differential manifold of dimension $2n$ with the Euclidean topology. For example, for any non singular projective algebraic variety $X \subset \mathbb{P}^N(\mathbb{C})$ we have that X inherits a differential submanifold structure from $\mathbb{P}^N(\mathbb{C})$. If X is stable under $\sigma_{\mathbb{P}}$ and $X(\mathbb{R}) \neq \varnothing$ then $X(\mathbb{R})$ is a real algebraic variety by Corollary 2.2.18. The variety $X(\mathbb{R})$ inherits a Euclidean topology from $\mathbb{P}^N(\mathbb{R})$ (the same as in Definition 1.4.1) and can be thought of as a differential submanifold of $\mathbb{P}^N(\mathbb{R})$.

Proposition 2.2.27 *Let (X, σ) be an $\mathbb{R}$-variety. If the complex variety X is non singular and has complex dimension n then the set X with its Euclidean topology is a differential manifold of real dimension $2n$. If moreover $X(\mathbb{R}) \neq \varnothing$ then the set $X(\mathbb{R})$ with its euclidean topology is a differentiable manifold of real dimension n.*

We invite the reader to compare this result with Remark 1.5.28. We recall that under the hypotheses of the above proposition, $X(\mathbb{R})$ is Euclidean closed but Zariski dense in X. See Corollary 2.2.10 and Proposition 2.2.11 for more details.

Proof As we have seen above, as $\mathcal{O}_X$ is an $\mathbb{R}$-sheaf, the morphism

$$\begin{array}{ccc} \mathcal{O}_X(U) & \longrightarrow & \mathcal{O}_X(\sigma(U)) \\ f & \longmapsto & {}^{\sigma}f \end{array}$$

is a ring isomorphism for any open set U in X. As the variety X is non singular and of dimension n we can find a local system of parameters $\{\varphi_x\}_{x\in X}$—see Definition 1.5.47. Exercise 1.5.48 tells us that in terms of local coordinates we get a set of systems (U_x, φ_x) where $\varphi_x : U_x \to \mathbb{C}^n$ is analytic and on refining this open cover using Euclidean open sets we can assume that $\forall x \in X$, $U_{\sigma(x)} = \sigma(U_x)$ and

$$\forall x \in X, \quad {}^{\sigma}(\varphi_x) = \varphi_{\sigma(x)} \,. \tag{2.1}$$

where ${}^{\sigma}(\varphi_x) = \sigma_{\mathbb{A}} \circ \varphi_x \circ \sigma$.

It follows that if $(z_1, \ldots, z_n)_x$ is a system of local coordinates satisfying (2.1), then the system $(\Re(z_1), \Im(z_1), \ldots, \Re(z_n), \Im(z_n))_x$ is a system of real local coordinates for the manifold structure, equivalent to the complex local system of coordinates $(z_1, \overline{z_1}, \ldots, z_n, \overline{z_n})$.

The real structure σ then transforms $(z_1, \overline{z_1}, \ldots, z_n, \overline{z_n})_x$ into

$$(\overline{z_1}, z_1, \ldots, \overline{z_n}, z_n)_{\sigma(x)} \, .$$

In particular, if $x \in X(\mathbb{R})$ is a non singular point of X then by (2.1), ${}^{\sigma}(\varphi_x) = \varphi_{\sigma(x)} = \varphi_x$ from which it follows that $\sigma_{\mathbb{A}} \circ \varphi_x = \varphi_x \circ \sigma$ and if $y \in U_x \cap X(\mathbb{R})$ then $\overline{\varphi_x(y)} = \varphi_x(y)$. The local coordinates of a real point are therefore real and the restriction of φ_x to $X(\mathbb{R})$ induces a system of real smooth (and in fact analytic) local coordinates $(\Re(z_1), \ldots, \Re(z_n))$ on $X(\mathbb{R})$ in a neighbourhood of x.

Alternatively, we can bypass the first part of this argument by using Lemma 2.2.8. Let x be a real point of X: there is then a system of local parameters which is invariant under σ. By Exercise 1.5.48 we can derive from this an invariant system of local coordinates. □

The underlying $2n$-dimensional manifold structure on the non singular complex variety X is not only *orientable* (since a holomorphic change of coordinate map has a positive determinant), but also *oriented*. Any isomorphism $\mathbb{R}^{2n} \simeq \mathbb{C}^n$ yields an orientation on $\mathbb{R}^{2n}$ by pull back and the complex structure on X yields such an isomorphism. (See Exercise B.5.11 for more details).

Proposition 2.2.28 *Let (X, σ) be a non-singular ℝ-variety. The real structure σ is a diffeomorphism of the $2n$-dimensional oriented manifold X which preserves the orientation if n is even and reverses it otherwise.*

Proof This follows immediately from the previous proof. The map σ takes $(z_1, \overline{z_1}, \ldots, z_n, \overline{z_n})_x$ to $(\overline{z_1}, z_1, \ldots, \overline{z_n}, z_n)_{\sigma(x)}$, so the determinant of its differential is $(-1)^n$. □

2.2.2 Compatible Atlas

Exercise 2.2.29 *If X is a non singular complex analytic variety of dimension n we can reframe the definition of the conjugate variety using a maximal atlas $(U_i, \varphi_i)_i$ determining the complex structure on X: the complex structure of the conjugate variety $(X, \overline{\mathcal{O}_X})$ is given by the atlas $(U_i, \sigma_{\mathbb{A}^n} \circ \varphi_i)_i$.*

Definition 2.2.30 A *compatible atlas* on a smooth analytic ℝ-variety (X, σ) of dimension n is an atlas $\mathcal{A} = \{(U_i, \varphi_i \colon U_i \to \mathbb{C}^n)\}_i$ on the complex analytic variety X satisfying the following conditions. (Recall that ${}^{\sigma}\varphi_i = \sigma_{\mathbb{A}} \circ \varphi_i \circ \sigma$.)

1. The atlas is globally stable for the real structure, or in other words

$$(U_i, \varphi_i) \in \mathcal{A} \implies (\sigma(U_i), {}^{\sigma}\varphi_i) \in \mathcal{A} \,;$$

2. If $U_i \cap X(\mathbb{R}) \neq \varnothing$ then $U_i = \sigma(U_i)$ and ${}^{\sigma}\varphi_i = \varphi_i$;
3. If $U_i \cap X(\mathbb{R}) = \varnothing$ then $U_i \cap \sigma(U_i) = \varnothing$.

Exercise 2.2.31 *Give a compatible atlas for* $(\mathbb{P}^1(\mathbb{C}), \sigma_{\mathbb{P}})$.

Proposition 2.2.32 *Every smooth analytic* $\mathbb{R}$*-variety has a compatible atlas.*

Proof This follows from the existence of local systems of parameters satisfying (2.1). □

2.3 Complexification of a Real Variety

We have seen that the real locus of an $\mathbb{R}$-variety is a real algebraic variety whenever it is Zariski dense. In this section we will study the converse: given a real algebraic variety V, is there an $\mathbb{R}$-variety whose real locus is isomorphic to V?

Let K be a field and let $L \supset K$ be an extension of K. The set $\mathbb{A}^n(K)$ is then a subspace of $\mathbb{A}^n(L)$ and $\mathbb{P}^n(K)$ is a subset of $\mathbb{P}^n(L)$.

Definition 2.3.1 (*Revisions of Definition* 1.2.12) Let $F \subset \mathbb{A}^n(K)$ be an algebraic set over K of ideal $I = \mathcal{I}(F) \subset K[X_1, \dots X_n]$. We define the algebraic set F_L over L to be the set $\mathcal{Z}_L(I)$ of zeros of I in $\mathbb{A}^n(L)$:

$$F_L := \mathcal{Z}_L(I) \subset \mathbb{A}^n(L) \,.$$

Similarly, if $F \subset \mathbb{P}^n(K)$ is a projective algebraic set of homogeneous ideal $I = \mathcal{I}(F) \subset K[X_0, \dots X_n]$ then we define an algebraic set

$$F_L := \mathcal{Z}_L(I) \subset \mathbb{P}^n(L) \,.$$

More generally, if $U = F \setminus F' \subset \mathbb{A}^n(K)$ is a quasi-affine set and $I = \mathcal{I}(F) \subset K[X_1, \dots, X_n]$ and $I' = \mathcal{I}(F') \subset K[X_1, \dots, X_n]$ are the associated ideals then we can define a quasi-affine set

$$U_L := F_L \setminus F'_L = \mathcal{Z}_L(I) \setminus \mathcal{Z}_L(I') \subset \mathbb{A}^n(L) \,.$$

And finally if $U = F \setminus F' \subset \mathbb{P}^n(K)$ is a quasi-projective algebraic set and $I = \mathcal{I}(F) \subset K[X_0, \dots, X_n]$ and $I' = \mathcal{I}(F') \subset K[X_0, \dots, X_n]$ are the associated homogeneous ideals then we define a set

$$U_L := F_L \setminus F'_L = \mathcal{Z}_L(I) \setminus \mathcal{Z}_L(I') \subset \mathbb{P}^n(L) \,.$$

Any real algebraic set (which here will be assumed affine to simplify the notation) $F \subset \mathbb{R}^n$ with vanishing ideal $I := \mathcal{I}(F) \subset \mathbb{R}[X_1, \ldots, X_n]$ is therefore naturally associated to a *complexification* $F_\mathbb{C} := \mathcal{Z}_\mathbb{C}(\mathcal{I}(F)) = \mathcal{Z}_\mathbb{C}(I) \subset \mathbb{C}^n$ which is just the set of *complex* common zeros of the real polynomials vanishing on F. Note that the ideal I is made up of polynomials with real coefficients whereas $F_\mathbb{C} \subset \mathbb{C}^n$ is a set of complex points. As $F_\mathbb{C}$ is defined by polynomials with real coefficients, $\sigma_\mathbb{A}(F_\mathbb{C}) \subset F_\mathbb{C}$ and the restriction σ of the standard real structure $\sigma_\mathbb{A} \colon (x_1, \ldots, x_n) \mapsto (\overline{x_1}, \ldots, \overline{x_n})$ to $F_\mathbb{C}$ is a real structure with which $(F_\mathbb{C}, \sigma)$ is an $\mathbb{R}$-variety. Our initial real algebraic variety can be recovered as the set of fixed points of $F = (F_\mathbb{C})^\sigma$.

The above construction depends heavily on the equations defining F. The following definition enables us to consider abstract complexifications, by which we mean complexifications which are independent of a particular embedding into affine or projective space, or alternatively independent of a choice of equations.

Definition 2.3.2 Let $(V, \mathcal{O}_V)$ be a real algebraic variety. A pair $((X, \sigma), j)$ is a *complexification* of V if (X, σ) is an $\mathbb{R}$-variety with enough real points and $j \colon V \to X$ is an injective map inducing an isomorphism of real algebraic varieties

$$(V, \mathcal{O}_V) \xrightarrow{\simeq} (X(\mathbb{R}), (\mathcal{O}_X)^G_{X(\mathbb{R})}) \,.$$

A complexification $((X, \sigma), j)$ of a real algebraic variety V is *quasi-projective* (resp. *non singular*) if X is a quasi-projective (resp. non singular) complex variety.

Let $((X, \sigma), j)$ be a complexification of a real algebraic variety V and let $\psi \colon (X, \sigma) \dashrightarrow (Y, \tau)$ be an $\mathbb{R}$-biregular birational map. It is easy to check that $((Y, \tau), \psi \circ j)$ is then a complexification of V. Indeed, since $X(\mathbb{R})$ is dense in X and ψ is birational the set $Y(\mathbb{R}) = \psi(X(\mathbb{R}))$ is dense in Y. The following proposition establishes the converse.

Proposition 2.3.3 *Let V be a real algebraic variety and let $((X, \sigma), j)$ be a complexification of V. Then for any complexification $((X', \sigma'), j')$ of V, there is a unique $\mathbb{R}$-biregular birational map $\psi \colon (X, \sigma) \dashrightarrow (X', \sigma')$, $X(\mathbb{R}) \subset \mathrm{dom}(\psi)$ such that the following diagram commutes.*

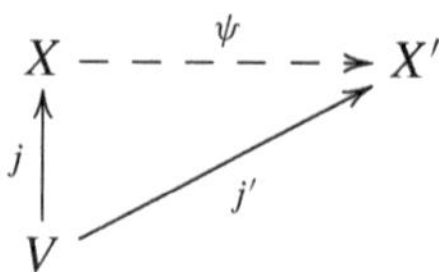

Proof We start by proving the proposition in the case where V, X and X' are affine. The uniqueness of the map for affine varieties will then enable us to glue complexifications and $\mathbb{R}$-biregular birational maps on open affine subsets of V to prove the general result. By hypothesis the morphism $h = j' \circ j^{-1} \colon X(\mathbb{R}) \to X'(\mathbb{R})$ is an isomorphism of real algebraic varieties. By the solution to Exercise 1.2.56(2), there is a morphism defined on an open neighbourhood of $X(\mathbb{R})$ in X extending $j' \circ j^{-1}$.

As $X(\mathbb{R})$ is dense in X, the rational map $\psi\colon (X,\sigma) \dashrightarrow (X',\sigma')$ induced by this extension is an $\mathbb{R}$-biregular birational map uniquely determined by $j' \circ j^{-1}$. □

Proposition 2.3.4 *Any real affine algebraic set has an affine complexification. Any real projective algebraic set has a projective complexification.*

Proof Let $X \subset \mathbb{A}^n(\mathbb{R})$ be a real affine algebraic set and let $I = \mathcal{I}(X) \subset \mathbb{R}[X_1, \dots, X_n]$ be its ideal. The set X is then the set of real zeros of $\mathcal{Z}(I) \subset \mathbb{A}^n(\mathbb{R})$ and the Zariski closure, $X_{\mathbb{C}}$ of X in $\mathbb{A}^n(\mathbb{C})$ is the set of complex zeros $\mathcal{Z}_{\mathbb{C}}(I) \subset \mathbb{A}^n(\mathbb{C})$ by Remark 1.2.13. By construction the $\mathbb{R}$-variety $(X_{\mathbb{C}}, \sigma_{\mathbb{A}}|_{X_{\mathbb{C}}})$ has enough real points; denoting by $j\colon X \hookrightarrow X_{\mathbb{C}}$ the inclusion map, the pair $\big((X_{\mathbb{C}}, \sigma_{\mathbb{A}}|_{X_{\mathbb{C}}}), j\big)$ is then an affine complexification of X. Similarly, if $X \subset \mathbb{P}^n(\mathbb{R})$ is a real projective algebraic set and $I = \mathcal{I}(X) \subset \mathbb{R}[X_0, \dots, X_n]$ is its homogeneous ideal then we take $\mathcal{Z}_{\mathbb{C}}(I) \subset \mathbb{P}^n(\mathbb{C})$, the set of complex zeros of I. □

Remark 2.3.5 We have seen that any real projective variety is also affine, and therefore has an affine complexification.

A complex projective algebraic variety is not generally affine, so a projective $\mathbb{R}$-variety is not typically affine, and neither is a projective complexification.

Certain real affine algebraic varieties also have projective complexifications, and these will be studied in Theorem 2.3.7 below.

Remark 2.3.6 Let X be a quasi-projective real algebraic variety with $X = V \setminus W \subset \mathbb{P}^n(\mathbb{R})$. Let $I_V \subset \mathbb{R}[X_0, \dots, X_n]$ be the homogeneous ideal of V and let $I_W \subset \mathbb{R}[X_0, \dots, X_n]$ be the homogeneous ideal of W. The set $V_{\mathbb{C}} = \mathcal{Z}_{\mathbb{C}}(I_V)$ is a projective complexification of V by the above and $W_{\mathbb{C}} = \mathcal{Z}_{\mathbb{C}}(I_W)$ is a projective complexification of W. The variety $X_{\mathbb{C}} = V_{\mathbb{C}} \setminus W_{\mathbb{C}}$ is therefore a quasi-projective complexification of X.

We recall Definition 1.4.11 which states that a real algebraic variety is complete if and only if it is compact for the Euclidean topology.

Theorem 2.3.7 *Any non singular complete real affine algebraic variety has a non singular projective complexification.*

Before proving this theorem we state some very useful lemmas concerning birational morphisms of $\mathbb{R}$-varieties. Let (X,σ) be an $\mathbb{R}$-variety and let $x \in X(\mathbb{R})$ be a real point. We denote by C_x the *connected component* of $X(\mathbb{R})$ containing x. Throughout this section, connected means connected in the Euclidean topology.

Lemma 2.3.8 *Let (X,σ) be an $\mathbb{R}$-variety and let $x \in X(\mathbb{R}) \cap \operatorname{Reg} X$ be a regular real point. The Euclidean connected component $C_x \subset X(\mathbb{R})$ is not then contained in any strict Zariski closed subset of X.*

Proof By Proposition 1.5.29, x has a connected Euclidean open neighbourhood $U \subset X(\mathbb{R})$ homeomorphic to a non empty subset of $\mathbb{R}^n$ where n is the Zariski dimension of X at x. As $U \subset C_x$ and any strict Zariski closed subset of X is of strictly positive codimension the result follows. □

Lemma 2.3.9 *Let $\varphi\colon (Y,\tau) \to (X,\sigma)$ be a birational morphism of $\mathbb{R}$-varieties and let $Z \subset Y$ be the smallest Zariski closed subset such that $\varphi|_{Y\setminus Z}$ is an isomorphism onto its image. Consider a point $y \in Y(\mathbb{R}) \cap \operatorname{Reg} Y$: the connected Euclidean component $C_{\varphi(y)}$ is not then contained in $\varphi(Z)$.*

Proof As $\operatorname{codim} Z > 0$, $C_y \cap (Y \setminus Z) \neq \varnothing$ by Lemma 2.3.8. It follows that $\varphi(C_y) \cap (X \setminus \varphi(Z)) \neq \varnothing$ and as the image of a connected subset under a continuous map is still connected, $\varphi(C_y) \subset C_{\varphi(y)}$ and hence $C_{\varphi(y)} \cap (X \setminus \varphi(Z)) \neq \varnothing$. □

Proposition 2.3.10 *Let (X,σ) be an $\mathbb{R}$-variety and let $\varphi\colon (Y,\tau) \to (X,\sigma)$ be a resolution of singularities of X. Suppose that the connected component of a real singular point $x \in X(\mathbb{R})$ is contained in the singular locus $C_x \subset \operatorname{Sing} X$. We then have that $\varphi^{-1}(x) \cap Y(\mathbb{R}) = \varnothing$.*

Proof By Theorem 1.5.51, $\operatorname{Sing} X$ is a strict Zariski closed subset of X. The result then follows from Lemma 2.3.9 applied to $Z = \pi^{-1}(\operatorname{Sing} X)$ using Definition 1.5.53. □

Proof of Theorem 2.3.7 Let V be a non singular real affine algebraic variety which is compact for the Euclidean topology. By Proposition 2.3.4, V has an affine complexification $((X,\sigma), j)$. By Theorem 2.2.9, $X(\mathbb{R}) \simeq V$ does not meet $\operatorname{Sing} X$. We consider a projective completion (X',σ') of (X,σ): in particular, X is a subvariety of X' and $\sigma = \sigma'|_X$. By Hironaka's resolution of singularities Theorem 1.5.54 there is a non singular projective $\mathbb{R}$-variety (Y,τ) and a birational morphism $\pi : (Y,\tau) \to (X',\sigma')$ of $\mathbb{R}$-varieties which is an isomorphism on $\pi^{-1}(\operatorname{Reg} X') \to \operatorname{Reg} X'$. As $X(\mathbb{R}) \subset \operatorname{Reg} X$, the restriction of the composition $(Y,\tau) \to (X,\sigma)$ to $X(\mathbb{R})$ is an isomorphism.

As V is compact, $X(\mathbb{R})$ is also compact, so it is closed in $X'(\mathbb{R})$ for the Euclidean topology. It follows that for every $x \in X'(\mathbb{R}) \setminus X(\mathbb{R})$ there is an inclusion $C_x \subset X'(\mathbb{R}) \setminus X(\mathbb{R})$ and Proposition 2.3.10 tells us that $\pi^{-1}(X'(\mathbb{R}) \setminus X(\mathbb{R})) \cap Y(\mathbb{R}) = \varnothing$. We can therefore conclude that $((Y,\tau), (\pi|_{Y(\mathbb{R})})^{-1} \circ j)$ is a non singular projective complexification of V. □

Remark 2.3.11 In the above proof, $X'(\mathbb{R}) \setminus X(\mathbb{R})$ may be non empty. In Example 2.6.38, examined in detail below, we consider the set

$$W := \mathcal{Z}(16(x_1^2 + x_2^2) - (x_1^2 + x_2^2 + x_3^2 + 3)^2) \subset \mathbb{A}^3(\mathbb{R})\,.$$

and the projective complexification given by

$$\widehat{W}_{\mathbb{C}} := \mathcal{Z}\left(16(x_1^2 + x_2^2) - (x_1^2 + x_2^2 + x_3^2 + 3x_0^2)^2\right) \subset \mathbb{P}^3(\mathbb{C})\,.$$

The $\mathbb{R}$-variety $(\widehat{W}_{\mathbb{C}}, \sigma_{\mathbb{P}}|_{\widehat{W}_{\mathbb{C}}})$ contains real points that do not belong to the torus of revolution $W_{\mathbb{C}}(\mathbb{R}) = W$. Indeed, if $x_1^2 + x_2^2 \leqslant 16$ then the point $\left(0 : x_1 : x_2 : \sqrt{4\sqrt{(x_1^2 + x_2^2)} - (x_1^2 + x_2^2)}\right)$ belongs to $\widehat{W}_{\mathbb{C}}(\mathbb{R}) \setminus W_{\mathbb{C}}(\mathbb{R})$. The $\mathbb{R}$-morphism $\psi : \mathbb{P}^1(\mathbb{C}) \times \mathbb{P}^1(\mathbb{C}) \to \widehat{W}_{\mathbb{C}}$ is a resolution of singularities of $\widehat{W}_{\mathbb{C}}$.

We use the above results to prove Theorem 1.5.55 for $\mathbb{R}$-varieties.

Theorem 2.3.12 *Let $\varphi\colon (Y,\tau) \to (X,\sigma)$ be a birational morphism of non singular $\mathbb{R}$-varieties. If the real loci $X(\mathbb{R})$ et $Y(\mathbb{R})$ are compact for the Euclidean topology then they have the same number of connected components.*

$$\#\pi_0(Y(\mathbb{R})) = \#\pi_0(X(\mathbb{R}))\ .$$

Proof Let $Z \subset Y$ be the smallest Zariski closed subset such that $\varphi|_{Y\setminus Z}$ is an isomorphism onto its image. The map φ is continuous for the Euclidean topology so $\#\pi_0(Y(\mathbb{R})) \geqslant \#\pi_0(X(\mathbb{R}))$. To prove the opposite inequality, assume there are two distinct connected components Y_1 and Y_2 in $Y(\mathbb{R})$ such that $\varphi(Y_1) \cap \varphi(Y_2)$ is non empty. Let U be an open Euclidean neighbourhood of $x \in \varphi(Y_1) \cap \varphi(Y_2)$ in $X(\mathbb{R})$. We then have that $U \cap \varphi(Y_1) \neq \varnothing$ and $U \cap \varphi(Y_2) \neq \varnothing$. Indeed for $i = 1, 2$, $\varphi^{-1}(U) \cap Y_i$ is a non empty open space in $Y(\mathbb{R})$ and as Y is non singular $\varphi^{-1}(U) \cap Y_i \setminus Z$ is non empty by Lemma 2.3.8. As X is non singular we can assume that U is homeomorphic to a non empty open set in $\mathbb{R}^n$, where n is the dimension of X, which by the above is cut into two disjoint parts by the algebraic subset $\varphi(Z)$. The codimension of $\varphi(Z)$ is at least two because φ is a birational morphism (see [Sha94, II.4.4, Theorem 2] for example) which contradicts the fact that $\varphi(Z)$ disconnects the open set U. This yields a contradiction. □

The behaviour of an $\mathbb{R}$-variety away from its real points is often irrelevant for the study of the real locus $X(\mathbb{R})$—but not always. We saw in Remark 2.3.11 an example where we needed to consider the non real points of the complex variety.

Definition 2.3.13 A quasi-algebraic affine or projective set U over K is said to be *geometrically irreducible* if the set $U_{\overline{K}}$ (see Definition 2.3.1) defined over the algebraic closure $\overline{K}$ of K is irreducible.

A quasi-projective algebraic set V over K, is said to be *geometrically irreducible* if the image U of V under embedding into a projective space over K is geometrically irreducible. Under these circumstances the image under any projective embedding of V is geometrically irreducible by Exercise 2.3.14.

An $\mathbb{R}$-variety (X,σ) is said to be *irreducible* if and only if X is irreducible as a complex variety.

Exercise 2.3.14 *Check that if $\varphi\colon V \to \mathbb{P}^N(K)$ and $\varphi'\colon V \to \mathbb{P}^{N'}(K)$ are two projective embeddings of V then $\varphi(V)_{\overline{K}}$ is irreducible if and only if $\varphi'(V)_{\overline{K}}$ is irreducible.*

Proposition 2.3.15 *Let K be a field.*

1. *An algebraic set over K which is geometrically irreducible is irreducible.*
2. *An algebraic variety over K which is geometrically irreducible is irreducible.*
3. *A real algebraic variety V is geometrically irreducible if and only if it has an irreducible complexification.*

4. *Let (X, σ) be a quasi-projective algebraic $\mathbb{R}$-variety with enough real points. We then have that (X, σ) is irreducible if and only if the real algebraic variety $\left(X(\mathbb{R}), (\mathcal{O}_X)^G_{X(\mathbb{R})}\right)$ is geometrically irreducible.*

Proof Left as an exercise for the reader. □

Remark 2.3.16 Recall that by Corollary 2.2.10 the real locus of a non singular irreducible algebraic $\mathbb{R}$-variety is Zariski dense whenever it is non empty.

Exercise 2.3.17 (Review of Example *2.1.1*)

1. *The real algebraic set $F := \mathcal{Z}(x^2 + y^2) \subset \mathbb{A}^2(\mathbb{R})$ is geometrically irreducible.*
2. *On the other hand, the $\mathbb{R}$-variety (V, σ), where $V := \mathcal{Z}_{\mathbb{C}}(x^2 + y^2) \subset \mathbb{A}^2(\mathbb{C})$ and $\sigma = \sigma_{\mathbb{A}}|_V$, is not irreducible.*
3. *This appears to contradict the fact that $V^\sigma = F$—what is happening?*

2.3.1 *Rational Varieties*

Definition 2.3.18 (Rational $\mathbb{R}$-varieties)

1. An $\mathbb{R}$-variety (X, σ) of dimension n is *rational* (or $\mathbb{R}$-*rational*) if it is birationally equivalent to the $\mathbb{R}$-variety $(\mathbb{P}^n(\mathbb{C}), \sigma_{\mathbb{P}})$, or in other words if there is a birational map of $\mathbb{R}$-varieties $(X, \sigma) \dashrightarrow (\mathbb{P}^n(\mathbb{C}), \sigma_{\mathbb{P}})$.
2. An $\mathbb{R}$-variety (X, σ) of dimension n is *geometrically rational* (or $\mathbb{C}$-rational) if and only if the complex variety X is rational, or in other words if there is a birational map of complex varieties $X \dashrightarrow \mathbb{P}^n(\mathbb{C})$.

Remark 2.3.19 We invite the reader to compare this definition with Definition 1.3.37 in the first chapter. Note that "geometric" irreducibility and rationality behave differently: a geometrically irreducible variety is irreducible, whereas a rational variety is geometrically rational.

Proposition 2.3.20 *Any $\mathbb{R}$-rational $\mathbb{R}$-variety is $\mathbb{C}$-rational*

Remark 2.3.21 The converse of the above proposition is false, an example being given by $\mathbb{P}^1(\mathbb{C})$ with its anti-holomorphic involution $z \mapsto -\frac{1}{\bar{z}}$. See Remark 2.1.41 for more details. Chapter 4 contains many 2-dimensional examples.

Proposition 2.3.22 *Let (X, σ) be a quasi-projective non singular $\mathbb{R}$-variety. If (X, σ) is $\mathbb{R}$-rational and has non zero dimension then $X(\mathbb{R})$ is connected and non empty.*

Proof This follows from Theorem 2.3.12 since $\mathbb{P}^n(\mathbb{R})$ is connected and non empty for all $n > 0$. □

2.4 $\mathbb{R}$-Varieties, Real Algebraic Varieties and Schemes Over $\mathbb{R}$—a Comparison

This section reviews the various types of $\mathbb{R}$-varieties met so far and the logical relationships between them. We have identified two different types of real variety: real algebraic varieties and $\mathbb{R}$-varieties. In total, there are five different incarnations of real algebraic varieties:

1. The real locus of a set of real equations.

2a. A complex variety defined by equations with real coefficients.

2b. A complex variety with an anti-regular involution.
These last two cases of special complex varieties are equivalent if we make the extra assumption that the variety is *quasi-projective*.

3a. A scheme defined over $\mathbb{R}$.

3b. A scheme defined over $\mathbb{C}$ with a real structure.
Once again, these last two cases are equivalent if we make the assumption that the scheme is *quasi-projectif.*

At the end of the day, the last four definitions are all equivalent for quasi-projective varieties and only the first is different. A variety of type (1) can be thought of as the *germ* of a variety of type (2a) in a neighbourhood of the real locus.

Moreover, any such variety has two topologies and two associated structures

- Zariski topology and algebraic variety structure.
- Euclidean topology and analytic variety structure.

There is a dictionary translating algebraic structures into underlying analytic structures. For example, the (anti)-regular maps become (anti)-holomorphic. This "translation" is not however an equivalence unless the variety is *projective*. See Appendix D.5 for more details.

Let us examine these structures in more detail.

1. (Section 1.3) A *real algebraic variety* (resp. *complex algebraic variety*) is a topological space X with a subsheaf $\mathcal{O}_X$ of the sheaf of functions with a finite covering of *affine* open sets U, by which we mean that $(U, \mathcal{O}_X|_U)$ is isomorphic to the zero set $\mathcal{Z}(I) \subset \mathbb{A}^n(\mathbb{R})$ of an ideal $I \subset \mathbb{R}[X_1, \dots, X_n]$ with the sheaf of functions which are locally rational fractions without *real* poles (resp. the set of zeros $\mathcal{Z}(I) \subset \mathbb{A}^n(\mathbb{C})$ of an ideal $I \subset \mathbb{C}[X_1, \dots, X_n]$ with the sheaf of functions which are locally rational functions without poles). Varieties X and Y are isomorphic if and only if there exists a *biregular* map $X \to Y$.
2. (Section 2.1) An $\mathbb{R}$-*variety* (X, σ) is a complex variety X with an anti-regular involution (or in other words a *real structure*) σ. The $\mathbb{R}$-varieties (X, σ) and (Y, τ) are isomorphic if there is a *biregular* isomorphism of complex varieties that commutes with the real structure. The varieties (X, σ) and (Y, τ) are *birationally $\mathbb{R}$-biregularly isomorphic* if there is a birational map $\varphi\colon X \dashrightarrow Y$ commuting with real structure such that $X(\mathbb{R}) \subset \operatorname{dom}(\varphi)$ and $Y(\mathbb{R}) \subset \operatorname{dom}(\varphi^{-1})$.
(Section 2.3) A *complexification* of a real algebraic variety V is an $\mathbb{R}$-variety

(X, σ) with enough real points whose real locus $X(\mathbb{R})$ is isomorphic to V as a real algebraic variety.

(a) (Section 2.1) Any quasi-projective $\mathbb{R}$-variety can be realised as a variety defined by real coefficients (as can its principal sheaves, see Section 2.5).
(b) (Section 2.2) A quasi-projective $\mathbb{R}$-variety with enough real points induces by restriction a real algebraic variety structure on its real locus. A morphism of quasi-projective $\mathbb{R}$-varieties with enough real points induces a regular map of real algebraic varieties.
(c) (Section 2.3) Conversely, any quasi-projective real algebraic variety has a complexification which is an $\mathbb{R}$-variety with enough real points. Any morphism of quasi-projective real algebraic varieties can be extended to a rational $\mathbb{R}$-regular map of $\mathbb{R}$-varieties.
(d) (Section 2.3) Two $\mathbb{R}$-varieties which are complexifications of isomorphic real algebraic varieties are birationally $\mathbb{R}$-isomorphic but not generally isomorphic.

3. This paragraph requires some knowledge of schemes—see [Duc14] or [Liu02] for more details. See also [Ben16b, Section 3.1] for a more specific discussion of realisations of schemes over $\mathbb{R}$. We leave it is an exercise for the reader to check the claims made below.

A scheme over a field K (or a K-schema) is a scheme X with a scheme morphism (called the *structural map*) $X \to \operatorname{Spec} K$. Throughout this paragraph, we assume X is of finite type over K (or in other words that X is covered by a finite number of spectra of finitely generated K-algebras). Two $\mathbb{R}$-schemes X and Y are *birationally $\mathbb{R}$-biregularly isomorphic* if there is a birational map $\varphi\colon X \dashrightarrow Y$ of $\mathbb{R}$-schemes such that φ is regular at every $\mathbb{R}$-rational point of X and φ^{-1} is regular at every $\mathbb{R}$-rational point of Y. Let X be a scheme over $\mathbb{C}$ equipped with an involution σ lifting complex conjugation $\sigma_{\mathbb{A}}^* = \operatorname{Spec}(z \mapsto \bar{z})\colon \operatorname{Spec}\mathbb{C} \to \operatorname{Spec}\mathbb{C}$: we call such an involution a *real structure on* X. If X is quasi-projective then by [BS64, Proposition 2.6] there is a scheme $Z = X/\langle\sigma\rangle$ over $\mathbb{R}$ and an isomorphism of $\mathbb{C}$-schemes $\varphi\colon X \to Z \times_{\operatorname{Spec}\mathbb{R}} \operatorname{Spec}\mathbb{C}$ such that $\sigma = \varphi^{-1} \circ (\operatorname{id} \times \sigma_{\mathbb{A}}^*) \circ \varphi$. Moreover, the pair (Z, φ) is uniquely determined by the pair (X, σ) up to $\mathbb{R}$-isomorphism. For example if $X = \operatorname{Spec} A$ is affine then $Z = \operatorname{Spec} A^{\sigma}$.
Implicitly, most types of algebraic varieties used in this book are different manifestations of $\mathbb{R}$-schemes of finite type.

(a) The set $X(\mathbb{R})$ of $\mathbb{R}$-rational points of a scheme X over $\mathbb{R}$ with the restriction of the structural sheaf is a real algebraic variety. A morphism of $\mathbb{R}$-schemes induces a morphism of real algebraic varieties.
(b) Conversely, any quasi-projective real algebraic variety can be obtained as the set of $\mathbb{R}$-rational points of a scheme X over $\mathbb{R}$. Any morphism of quasi-projective real algebraic varieties can be extended to an $\mathbb{R}$-regular map of schemes over $\mathbb{R}$.

(c) Any two schemes over $\mathbb{R}$ whose real loci are isomorphic as real algebraic varieties are birationally $\mathbb{R}$-biregularly isomorphic.
(d) Let Z be a scheme of finite type over $\mathbb{R}$. We can associate to it the following $\mathbb{R}$-variety: X is the topological space of $\mathbb{C}$-rational points of the $\mathbb{C}$-scheme $Z \times_{\operatorname{Spec}\mathbb{R}} \operatorname{Spec}\mathbb{C}$, The pair (X, σ) is the $\mathbb{R}$-variety obtained on equipping X with the real structure $\sigma := \mathrm{id} \times \operatorname{Spec}(z \mapsto \bar{z})$. We denote by $X(\mathbb{R})$ the set of closed points fixed by σ. If $Z(\mathbb{R})$ is the set of $\mathbb{R}$-rational points of the $\mathbb{R}$-scheme Z then $X(\mathbb{R}) = Z(\mathbb{R})$. A morphism of schemes over $\mathbb{R}$ induces a morphism of $\mathbb{R}$-varieties.
(e) Conversely, if (X, σ) is an $\mathbb{R}$-variety then there is a $\mathbb{C}$-scheme Z such that $Z(\mathbb{C}) = X$, [Har77, II.2.6] and there is an involutive morphism $\sigma_Z\colon Z \to Z$ lifting $\sigma_{\mathbb{A}}^*\colon \operatorname{Spec}\mathbb{C} \to \operatorname{Spec}\mathbb{C}$ such that $\sigma_Z|_{Z(\mathbb{C})} = \sigma$. As we have seen above, if X is quasi-projective then (Z, σ_Z) corresponds to an $\mathbb{R}$-scheme. A morphism of $\mathbb{R}$-varieties induces a morphism of schemes over $\mathbb{R}$.

2.4.1 Real Forms of a ℂ-Scheme

By the above, Definition 2.1.13 can be reformulated scheme theoretically as follows.

Definition 2.4.1 A *real form* of a scheme X over $\mathbb{C}$ is a scheme X_0 over $\mathbb{R}$ whose complexification $X_0 \times_{\operatorname{Spec}\mathbb{R}} \operatorname{Spec}\mathbb{C}$ is isomorphic to X.

2.4.2 Notations X, $X(\mathbb{R})$, $X(\mathbb{C})$, $X_{\mathbb{C}}$, $X_{\mathbb{R}}$

We now briefly discuss the various notations the reader may meet in the literature.

As in scheme theory, where by abuse of notation the structural morphism $Z \to \operatorname{Spec}\mathbb{R}$ is often omitted, the abbreviation X for the $\mathbb{R}$-variety (X, σ) is often used. Consequently, the notation $X_{\mathbb{C}}$ for the variety X is often used to emphasise the fact that we are concentrating on the complex variety and "forgetting" σ. Some authors, particularly of the Russian school, use the notation $X_{\mathbb{C}}$ or $\mathbb{C}X$ for the complex locus and $X_{\mathbb{R}}$ or $\mathbb{R}X$ for the real locus of $\mathbb{R}$-varieties.

Remark 2.4.2 In case that wasn't confusing enough, there is another object called $X_{\mathbb{R}}$ in the literature, constructed using extension of scalars. In the embedded case, it simply means separating the real and imaginary parts of the equations of a complex variety. From the scheme point of view this corresponds to taking the scheme morphism $\operatorname{Spec}\mathbb{C} \to \operatorname{Spec}\mathbb{R}$ associated to the inclusion $\mathbb{R} \hookrightarrow \mathbb{C}$ and compose maps $X \to \operatorname{Spec}\mathbb{C} \to \operatorname{Spec}\mathbb{R}$ to see that a scheme over $\mathbb{C}$ is *necessarily* a scheme over $\mathbb{R}$. For example, if $X \subset \mathbb{A}^n(\mathbb{C})$ is defined by r equations

$$\left\{P_i(z_1, \ldots, z_n) = 0\right\}_{i=1,\ldots,r}$$

then $X_{\mathbb{R}} \subset \mathbb{A}^{2n}(\mathbb{R})$ is defined by the $2r$ equations

$$\big\{\Re(P_i(x_1 + iy_1, \ldots, x_n + iy_n) = 0), \\ \Im(P_i(x_1 + iy_1, \ldots, x_n + iy_n) = 0\big\}_{i=1,\ldots,r} .$$

Let X be an algebraic variety defined over $\mathbb{C}$ which for simplicity we will assume to be non singular. Consider the product variety $Z := X \times \overline{X}$ with the anti-regular involution $\sigma_Z\colon (x, y) \mapsto (\overline{y}, \overline{x})$. The set of real points of the $\mathbb{R}$-variety (Z, σ_Z) is then a real algebraic variety as in Definition 1.3.9, homeomorphic in the Euclidean topology to the topological manifold underlying the complex variety X. Some authors use $X_{\mathbb{R}} = Z(\mathbb{R})$ to denote this *underlying* real algebraic variety.

2.5 Coherent Sheaves and Algebraic Bundles

We will now generalise the above constructions to certain sheaves and vector bundles needed in the development of the theory.

2.5.1 Coherent $\mathbb{R}$-Sheaves

Let (X, σ) be an $\mathbb{R}$-variety, let $\mathcal{L}$ be a quasi-coherent sheaf of $\mathcal{O}_X$-modules (see Theorem C.7.3) and let U be an open affine set in X. The space of sections $M := \mathcal{L}(\sigma(U))$ is then an $\mathcal{O}_X(\sigma(U))$-module. We define an $\mathcal{O}_X(U)$-module ${}^{\sigma}M$ by equipping the group M with the following $\mathcal{O}_X(U)$-twisted action.

$$(f, m) \mapsto {}^{\sigma}f \cdot m \tag{2.2}$$

where

$$(f, m) \mapsto f \cdot m$$

denotes the $\mathcal{O}_X(\sigma(U))$-action on M.

Definition 2.5.1 Let (X, σ) be an $\mathbb{R}$-variety and let $\mathcal{L}$ be a quasi-coherent sheaf of $\mathcal{O}_X$-modules. The *conjugate sheaf* ${}^{\sigma}\mathcal{L}$ is the sheaf of $\mathcal{O}_X$-modules defined over U by declaring ${}^{\sigma}\mathcal{L}(U)$ to be the twisted $\mathcal{O}_X(U)$-module ${}^{\sigma}M$. We say that $\mathcal{L}$ is an $\mathbb{R}$-*sheaf* if and only if $\mathcal{L} = {}^{\sigma}\mathcal{L}$. This is required to be an equality, not simply an isomorphism.

Remark 2.5.2 These definitions generalise Definition 2.2.1. Indeed, for any open set U in X, there is an equality of $\mathcal{O}_X(U)$-modules ${}^{\sigma}\mathcal{L}(U) = \mathcal{L}(\sigma(U))$ provided the right hand side is equipped with the twisted action (2.2). In particular, if $\mathcal{L}$ is a sheaf of $\mathbb{C}^n$-valued functions then ${}^{\sigma}\mathcal{L}(U) = \{{}^{\sigma}f \mid f \in \mathcal{L}(\sigma(U))\}$. Moreover, $\mathcal{L}$ is an $\mathbb{R}$-sheaf if and only if ${}^{\sigma}\mathcal{L}(U) = \mathcal{L}(U)$ for any open set U in X.

Our definition of an $\mathbb{R}-$sheaf is motivated by the following result which explicits the relationship between $\mathbb{R}$-sheaves on an $\mathbb{R}$-variety (X, σ) and sheaves of invariant functions. *A priori* an $\mathbb{R}$-sheaf is only a sheaf which is globally fixed by σ.

Lemma 2.5.3 *Let (X, σ) be a quasi-projective $\mathbb{R}$-variety and let $\mathcal{L}$ be a quasi-coherent sheaf of $\mathcal{O}_X$-modules. If $\mathcal{L}$ is an $\mathbb{R}$-sheaf then there is a quasi-coherent sheaf of $\mathcal{O}_X$-modules $\mathcal{L}_0$ such that for any open affine subset $U \subset X$,*

$$\mathcal{L}(U \cap \sigma(U)) \simeq \mathcal{L}_0(U \cap \sigma(U)) \otimes_{\mathbb{R}} \mathbb{C}$$

and $\forall f \in \mathcal{L}_0(U \cap \sigma(U))$, ${}^{\sigma} f = f$. When this is the case we will say that f has real coefficients.

Proof Recall that by definition σ is a homeomorphism for the Zariski topology on X and in particular if U is a Zariski open set in X then the intersection $U \cap \sigma(U)$ is also Zariski open. Moreover, by Exercise 1.3.15(4), the open set $U \cap \sigma(U)$ is affine. It will therefore be enough to prove the result for an affine $\mathbb{R}$-variety so by Theorem 2.1.33 we may assume we are in the case where $X \subset \mathbb{A}^n(\mathbb{C})$ and $\mathcal{I}(X) \subset \mathbb{R}[X_1, \ldots, X_n]$. Under these hypotheses we have that $\sigma = \sigma_{\mathbb{A}}|_X$ and

$$\mathcal{O}_X(X) = \mathcal{A}(X) = (\mathbb{R}[X_1, \ldots, X_n]/\mathcal{I}(X)) \otimes_{\mathbb{R}} \mathbb{C}\ .$$

Let M be the $\mathcal{A}(X)$-module of global sections of the $\mathcal{O}_X$-module $\mathcal{L}(X)$. By hypothesis, σ induces a Galois action on M for which, on equipping the subgroup of fixed points M^G with its natural $\mathcal{A}(X(\mathbb{R}))$-module structure, we have that

$$M = M^G \otimes_{\mathcal{A}(X(\mathbb{R}))} (\mathcal{A}(X(\mathbb{R})) \otimes_{\mathbb{R}} \mathbb{C})\ .$$

We then simply define $\mathcal{L}_0$ to be the sheaf associated to the $\mathcal{A}(X(\mathbb{R}))$-module M^G. See Definition C.7.2 for more details, □

We will make intensive use of coherent $\mathbb{R}$-sheaves, particularly invertible sheaves, see Definition C.5.8. These are in bijective correspondence with line bundles, see Corollary 2.5.13.

Let $(X, \mathcal{O}_X)$ be an *affine* real or complex algebraic variety and let $\mathcal{F}$ be a quasi-coherent sheaf. The set of global sections $\Gamma(X, \mathcal{F})$ is then a $\Gamma(X, \mathcal{O}_X)$-module. If $\mathcal{F}$ is locally free then this module is *projective*, by which we mean that it is a direct summand of a free $\Gamma(X, \mathcal{O}_X)$-module, see Definition A.4.6.

The next lemma requires us to generalise Definition C.7.2. Let M be a $\Gamma(X, \mathcal{O}_X)$-module and let $\mathcal{O}_X \otimes_{\Gamma(X,\mathcal{O}_X)} M$ be the sheaf of $\mathcal{O}_X$-modules associated to the presheaf $U \mapsto \mathcal{O}_X(U) \otimes_{\Gamma(X,\mathcal{O}_X)} M$. If $(X, \mathcal{O}_X)$ is a *complex* variety then $\mathcal{O}_X(U) = \Gamma(X, \mathcal{O}_X)_f$ for any principal open set $U = \mathcal{D}(f)$ and $\mathcal{O}_X \otimes_{\Gamma(X,\mathcal{O}_X)} M$ can be identified with the sheaf $\widetilde{M}$ of Definition C.7.2. In particular, $\big(\mathcal{O}_X \otimes_{\Gamma(X,\mathcal{O}_X)} M\big)(U) = \widetilde{M}(U) = M_f$ for any principal open set $U = \mathcal{D}(f)$. If $(X, \mathcal{O}_X)$ is a *real* variety then for any open set U in X, $\mathcal{O}_X(U)$ can be identified with the inductive limit $\varinjlim_{\mathcal{D}(f)\supset U} \Gamma(X, \mathcal{O}_X)_f$ of the localisations $\Gamma(X, \mathcal{O}_X)_f$ where f runs over the set of regular functions which do not vanish on any point of U and $\left(\mathcal{O}_X \otimes_{\Gamma(X,\mathcal{O}_X)} M\right)(U) \simeq \varinjlim_{\mathcal{D}(f)\supset U} M_f$.

The special case of locally free finitely generated sheaves leads us directly to vector bundles.

Lemma 2.5.4 *Let $(X, \mathcal{O}_X)$ be a real or complex affine algebraic variety. Let $\mathcal{F}$ be a sheaf of finitely generated locally free $\mathcal{O}_X$-modules. The $\Gamma(X, \mathcal{O}_X)$-module $\Gamma(X, \mathcal{F})$ of global sections of $\mathcal{F}$ is then projective and finitely generated. Conversely, let M be a projective finitely generated $\Gamma(X, \mathcal{O}_X)$-module. The associated $\mathcal{O}_X$-module $\mathcal{O}_X \otimes_{\Gamma(X,\mathcal{O}_X)} M$ is then finitely generated and locally free.*

Proof Left as an exercise for the reader. □

If $(X, \mathcal{O}_X)$ is a *complex* variety then every locally free finitely generated $\mathcal{O}_X$-module $\mathcal{F}$ is equal to the sheaf $\widetilde{\Gamma(X, \mathcal{F})}$ associated to its $\Gamma(X, \mathcal{O}_X)$-module $\Gamma(X, \mathcal{F})$ of global sections.

Proposition 2.5.5 *If $(X, \mathcal{O}_X)$ is a* complex *affine algebraic variety then the map $M \mapsto \widetilde{M}$ yields a bijective correspondence between finitely generated projective $\Gamma(X, \mathcal{O}_X)$-modules and finitely generated locally free $\mathcal{O}_X$-modules.*

Proof See [Har77, Corollary II.5.5]. □

On the other hand, as the following example shows, if $(X, \mathcal{O}_X)$ is a *real* affine variety then there are finitely generated locally free sheaves which are not associated to $\Gamma(X, \mathcal{O}_X)$-modules.

Example 2.5.6 Based on [BCR98, Example 12.1.5], see also [FHMM16, Example 5.35].

Let $P \in \mathbb{R}[x, y]$ be the polynomial defined by

$$P(x, y) = x^2(x-1)^2 + y^2$$

which has exactly two real zeros, $a_0 = (0, 0)$ and $a_1 = (1, 0)$. Set $U_i = \mathbb{R}^2 \setminus \{a_i\}$ for $i = 0, 1$. The Zariski open subsets U_0 and U_1 form an open covering of $\mathbb{A}^2(\mathbb{R})$. We define a locally free coherent rank 1 sheaf $\mathcal{F}$ by gluing the sheaves $\mathcal{O}_{\mathbb{A}^2(\mathbb{R})}|_{U_0}$ and $\mathcal{O}_{\mathbb{A}^2(\mathbb{R})}|_{U_1}$ over $U_0 \cap U_1$ using the transition function $\psi_{01} = P$ on $U_0 \cap U_1$. In other words, two sections $s_0 \in \mathcal{O}_{\mathbb{A}^2(\mathbb{R})}|_{U_0}(V_0)$ and $s_1 \in \mathcal{O}_{\mathbb{A}^2(\mathbb{R})}|_{U_1}(V_1)$ on the Zariski open sets V_0 and V_1 are glued together if and only if $\psi_{01}s_1 = s_0$ over $V_0 \cap V_1$.

The $\mathcal{O}_{\mathbb{A}^2(\mathbb{R})}$-module $\mathcal{F}$ is not generated by its global sections because any global section s of $\mathcal{F}$ vanishes at a_1. Indeed, the restriction s_i of s to U_i is a regular function on U_i for $i = 0, 1$. The gluing condition is $\psi_{01}s_1 = s_0$ on $U_0 \cap U_1$. Set $s_i = g_i/h_i$ where $g_i, h_i \in \mathbb{R}[x, y]$, with $h_i \neq 0$ at every point on U_i and g_i, h_i coprime for $i = 0, 1$. The gluing condition implies that $Ph_0g_1 = g_0h_1$ on $\mathbb{R}^2$. As P is irreducible and $h_1(a_0) \neq 0$ the polynomial P divides g_0 or in other words there is an $\lambda \in \mathbb{R}^*$ such that $g_0 = \lambda P g_1$ and $h_1 = \lambda^{-1}h_0$. In particular $g_0(a_1) = 0$ and hence $s(a_1) = 0$. It follows that the quasi-coherent sheaf $\mathcal{F}$ on $\mathbb{A}^2(\mathbb{R})$ is not generated by global sections. *A fortiori*, there is no $\Gamma(\mathbb{A}^2(\mathbb{R}), \mathcal{O}_{\mathbb{A}^2(\mathbb{R})})$-module whose associated sheaf is $\mathcal{F}$.

Note that the module of global sections $\Gamma(\mathbb{A}^2(\mathbb{R}), \mathcal{F})$ is isomorphic to $\Gamma(\mathbb{A}^2(\mathbb{R}), \mathcal{O}_{\mathbb{A}^2(\mathbb{R})}) = \mathcal{R}(\mathbb{R}^2)$ *via* the map $(s_0, s_1) \mapsto s_1 = \frac{g_1}{h_1}$ since $h_1 = \lambda^{-1} h_0$ does not vanish at any point of $\mathbb{R}^2$.

2.5.2 *Algebraic Vector Bundles*

Definition 2.5.7 Let $(X, \mathcal{O}_X)$ be an algebraic variety over a field K. A *rank r pre-algebraic vector bundle* over X is a K-vector bundle (E, π), see Definition C.3.5, where E is an algebraic variety over K, $\pi : E \to X$ is a regular map and the homeomorphisms $\psi_i : \pi^{-1}(U_i) \overset{\simeq}{\to} U_i \times K^r$ are biregular maps. More generally, a *pre-algebraic vector bundle* has constant rank on every connected component of X.

Remark 2.5.8 On an affine real algebraic variety the vector bundles defined above are called *pre-algebraic* in [BCR98] but *algebraic* in the previous version [BCR87].

Consider a pre-algebraic (resp. rank r) vector bundle on X. Its sheaf of algebraic local sections is then naturally equipped with a $\mathcal{O}_X$-module structure which is locally free (resp. of rank r).

Proposition 2.5.9 *Let $(X, \mathcal{O}_X)$ be an algebraic variety over a base field K. There is a bijective correspondence between the class of finitely generated locally free (resp. of rank r) coherent sheaves on X and isomorphism classes of pre-algebraic (resp. rank r) vector bundles on X.*

Proof See [BCR98, Proposition 12.1.3]. □

If $(X, \mathcal{O}_X)$ is a *complex* variety, pre-algebraic bundles are well behaved, as we saw in Proposition 2.5.5. If $(X, \mathcal{O}_X)$ is a *real* variety, the pre-algebraic line bundle associated to the sheaf $\mathcal{F}$ of Example 2.5.6 is not generated by its global sections, illustrating the fact that on a real variety the notion of pre-algebraic vector bundles is not particularly useful and motivating thereby the following definition.

Definition 2.5.10 A pre-algebraic vector bundle (E, π) on an affine real algebraic variety X is said to be *algebraic* if it is isomorphic to a pre-algebraic subbundle of a direct sum of structural sheaves. Similarly, a finitely generated locally free sheaf is said to be *algebraic* if its associated vector bundle is algebraic.

Remark 2.5.11 (*Real and complex bundles*)

1. Proposition 2.5.5 implies that any pre-algebraic vector bundle on an affine complex algebraic variety is algebraic.
2. On a real affine algebraic variety the vector bundles defined above were said to be *algebraic* in [BCR98, Definition 12.1.6] but were *strongly algebraic* in [BCR87].

Definition 2.5.12 A rank one algebraic vector bundle is called a *line bundle.*

Corollary 2.5.13 *Let $(X, \mathcal{O}_X)$ be a real or complex algebraic variety. There is a bijective correspondence between isomorphism classes of invertible algebraic sheaves on X and (algebraic) line bundles on X.*

Proof This follows immediately from Proposition 2.5.9. □

Theorem 2.5.14 *Let $(X, \mathcal{O}_X)$ be a real affine algebraic variety and let (E, π) be a pre-algebraic vector bundle on X. The bundle E is then algebraic if and only if there is a finitely generated projective $\Gamma(X, \mathcal{O}_X)$-module M such that the $\Gamma(X, \mathcal{O}_X)$-module of algebraic sections of (E, π) is isomorphic to the $\Gamma(X, \mathcal{O}_X)$-module $\mathcal{O}_X \otimes_{\Gamma(X,\mathcal{O}_X)} M$.*

Proof See [BCR98, Theorem 12.1.7]. □

As in [Hui95], we see that Definition 2.5.10 of "nice" vector bundles on a real algebraic variety V, which may initially seem unnatural, simply says that "nice" vector bundles are precisely those that can be obtained by restricting an ℝ-vector bundle on some complexification (X, σ) of V.

Let (X, σ) be a quasi-projective algebraic ℝ-variety with enough real points (see Definition 2.2.5 and Theorem 2.2.9) and let $\mathcal{L}$ be a finitely generated locally free ℝ-sheaf. It is immediate that the restriction $\mathcal{L}_0|_{X(\mathbb{R})}$ of the sheaf $\mathcal{L}_0$ defined in Lemma 2.5.3 is a finitely generated locally free sheaf on the real algebraic variety $\left(X(\mathbb{R}), (\mathcal{O}_X)^G_{X(\mathbb{R})}\right)$.

Theorem 2.5.15 *Let (X, σ) be a quasi-projective algebraic ℝ-variety with enough real points and let $\mathcal{L}$ be a finitely generated locally free ℝ-sheaf. The finitely generated locally free sheaf $\mathcal{L}_0|_{X(\mathbb{R})}$ on the real algebraic variety $\left(X(\mathbb{R}), (\mathcal{O}_X)^G_{X(\mathbb{R})}\right)$ is then algebraic.*

Corollary 2.5.16 *Let (X, σ) be a quasi-projective algebraic ℝ-variety with enough real points and let (E, π) be a* topological *vector bundle on the real algebraic variety $\left(X(\mathbb{R}), (\mathcal{O}_X)^G_{X(\mathbb{R})}\right)$.*

The vector bundle (E, π) is then algebraic if and only if there is a pre-algebraic ℝ-vector bundle $(\mathcal{E}, \eta)$ on (X, σ) whose restriction $(\mathcal{E}|_{X(\mathbb{R})}, \eta|_{X(\mathbb{R})})$ is isomorphic to $(E \otimes \mathbb{C}, \pi \otimes \mathbb{C})$.

Remark 2.5.17 In other words, a *topological* ℝ-vector space E on a real affine algebraic variety V is *algebraic* if and only if tensoring with ℂ yields the restriction to V of an algebraic ℂ-vector bundle $\mathcal{E}$ equipped with a real structure on some complexification $V_{\mathbb{C}}$ of V.

2.6 Divisors on a Projective $\mathbb{R}$-Variety

This section draws on [Liu02, Chapter 7], where the interested reader will find all the proofs left out below. A handful of statements and proofs in this section require some knowledge of sheaf cohomology, for which we also refer to [Liu02, Section 5.2].

2.6.1 Weil Divisors

Definition 2.6.1 Let X be a quasi-projective irreducible normal complex algebraic variety (Definition 1.5.37). This is not the weakest possible hypothesis we could make: everything that follows holds on any variety that is *non singular in codimension* 1.

- A *prime divisor* on X is an irreducible closed subvariety of X of codimension 1.
- A *Weil divisor* on X is a *codimension* 1 *cycle*, i.e. a finite formal sum of prime divisors with integer coefficients[3]

$$D = \sum_{\substack{A \text{ prime Weil} \\ \text{divisor on } X}} a_A A\,, \quad a_A \in \mathbb{Z} \text{ almost all zero.}$$

- Let $D = \sum a_A A$ be a divisor. For any prime divisor A in X, the integer a_A is called the *multiplicity*, denoted $\operatorname{mult}_A(D)$, of D along A.
- The *support* of a divisor is the subvariety

$$\operatorname{Supp} D = \bigsqcup_{a_A \neq 0} A\,.$$

- If all the coefficients vanish, i.e. $\operatorname{Supp} D = \varnothing$, we write $D = 0$.
- If all the coefficients are positive or zero D is said to be *effective* and we write $D \geqslant 0$.

We denote by $Z^1(X)$ we set of all Weil divisors on X. By definition, $Z^1(X)$ is the free abelian group generated by prime divisors.

Example 2.6.2 1. If X is a curve then the prime divisors on X are the points of X. We define the *degree* of a Weil divisor $\sum_{i=1}^{s} a_i D_i$ to be the sum of the coefficients

$$\deg D = \sum_{i=1}^{s} a_i\,.$$

---

[3]Or in other words—zero except for a finite number of them.

2. If X is a projective surface then the prime divisors on X are the irreducible curves in X. There is then no intrinsic definition of the degree of a divisor but we can define the degree with respect to a choice of very ample divisor or projective embedding.
3. If $X = \mathbb{P}^n$ then prime divisors are irreducible hypersurfaces. The degree of a hypersurface D_i is then well-defined (it is the degree of a polynomial generating the principal ideal $\mathcal{I}(D_i)$, see [Har77, Chapitre I]) and the *degree* of a Weil divisor $\sum_{i=1}^{s} a_i D_i \in Z^1(\mathbb{P}^n)$ is defined by

$$\deg D = \sum_{i=1}^{s} a_i \deg D_i \ .$$

If $f \in K(X)^* = \mathbb{C}(X)^*$ is a rational function not identically zero (see Definition 1.2.69 and Remark1.2.74) and A is a prime divisor we define the *multiplicity* $\mathrm{mult}_A(f)$ of f along A as follows:

- $\mathrm{mult}_A(f) = k > 0$ if f vanishes along A to order k;
- $\mathrm{mult}_A(f) = -k$ if f has a pole of order k along A (i.e. if $\frac{1}{f}$ vanishes along A to order k;
- $\mathrm{mult}_A(f) = 0$ in all other cases.

We can associate to any rational function $f \in K(X)^*$ a divisor $\mathrm{div}(f) \in Z^1(X)$ defined by

$$\mathrm{div}(f) := \sum_{\substack{A \text{ prime Weil} \\ \text{divisor in } X}} \mathrm{mult}_A(f) A \ .$$

Note that $\mathrm{div}(f) \in Z^1(X)$ since $\mathrm{mult}_A(f)$ vanishes for almost all prime divisors A. Such divisors are called *principal divisors*. Since $\mathrm{div}(fg) = \mathrm{div}(f) + \mathrm{div}(g)$ the set of such divisors is a subgroup $\mathcal{P}(X)$ in $Z^1(X)$.

Exercise 2.6.3 *Prove that for any rational function f on $\mathbb{P}^n$ we have that*

$$\deg(\mathrm{div}(f)) = 0 \ .$$

Definition 2.6.4 Two divisors D, D' on a variety X are said to be *linearly equivalent* if $D - D'$ is a principal divisor. We denote by $D \sim D'$ the equivalence relation thus defined and by

$$\mathrm{Cl}(X) := Z^1(X)/\mathcal{P}(X) = Z^1(X)/\sim$$

the group of divisors modulo linear equivalence.

Exercise 2.6.5 *Prove that the group $\mathrm{Cl}(\mathbb{P}^n)$ is isomorphic to $\mathbb{Z}$ and it is generated by the linear class of the divisor $1H$ associated to a hyperplane $H \subset \mathbb{P}^n$.*

Example 2.6.6 Let C be a projective plane curve of degree d—see Definition 1.6.1—and let L be a line in $\mathbb{P}^2(\mathbb{C})$. The curve C is then linearly equivalent to d times the line L. In particular, any projective conic (see Exercise 1.2.68) is linearly equivalent to the double line $2L$.

2.6.2 *Cartier Divisors*

Let X be an algebraic variety, let $U \subset X$ be an open subset and let $f \in K(U)^*$ be a rational function which is not identically zero on U. By definition there is then a dense open subset $V \subset U$ such that $\forall p \in V$, $f(p) = \frac{g(p)}{h(p)}$ for some $g, h \in \mathcal{O}_X(V)$.

Definition 2.6.7 A *Cartier divisor* (or locally principal divisor) on an algebraic variety X is a global section of the quotient sheaf arising from the following exact sequence of multiplicative sheaves

$$1 \longrightarrow \mathcal{O}_X^* \longrightarrow \mathcal{M}_X^* \longrightarrow \mathcal{M}_X^*/\mathcal{O}_X^* \longrightarrow 1 \tag{2.3}$$

where $\mathcal{O}_X^*$ is the sheaf of regular functions that do not vanish at any point and $\mathcal{M}_X^*$ is the sheaf of rational functions that are not identically zero[4]. We denote by

$$\mathrm{Div}(X) := \Gamma(X, \mathcal{M}_X^*/\mathcal{O}_X^*)$$

the group of Cartier divisors. The group law on $\mathrm{Div}(X)$ is abelian and is written additively.

Definition 2.6.8 A Cartier divisor is said to be *principal* if it is associated to a global rational function. We say that two divisors D_1 and D_2 are *linearly equivalent* if $D_1 - D_2$ is principal. We then write $D_1 \sim D_2$ as for Weil divisors. The subgroup of $\mathrm{Div}(X)$ of principal divisors is isomorphic to $\mathcal{P}(X)$ and we denote by

$$\mathrm{CaCl}(X) := \mathrm{Div}(X)/\mathcal{P}(X) = \mathrm{Div}(X)/\sim$$

the group of Cartier divisors modulo linear equivalence.

Proposition 2.6.9 *Let X be an algebraic variety. The group* $\mathrm{CaCl}(X)$ *is a subgroup of the cohomology group* $H^1(X, \mathcal{O}^*)$.

Proof We consider the long exact sequence associated to the short exact sequence (2.3). Part of this long exact sequence is given by $H^0(X, \mathcal{M}_X^*) \xrightarrow{f} H^0(X, \mathcal{M}_X^*/\mathcal{O}_X^*) \xrightarrow{g} H^1(X, \mathcal{O}_X^*)$. By definition, the image of $H^0(X, \mathcal{M}_X^*)$ under f is the group of principal divisors so g induces an inclusion

$$\mathrm{CaCl}(X) \hookrightarrow H^1(X, \mathcal{O}^*) .$$

□

[4]Of course, $\mathcal{M}_X^*(X) = K(X)^*$. The notation $\mathcal{M}_X$, chosen to emphasise the fact that the corresponding analytic sheaf is the sheaf of meromorphic functions, is used to avoid confusion with the canonical sheaf $\mathcal{K}_X$. See Definition 2.6.26 for more details.

Let $D = (U_i, f_i)_i \in \mathrm{Div}(X)$ be a Cartier divisor described with respect to an open covering $\{U_i\}_i$ of X. There are therefore germs of regular functions $g_i, h_i \in \mathcal{O}_X(U_i)$ such that

$$f_i = \frac{g_i}{h_i} \quad \text{and} \quad \frac{g_i}{h_i} \cdot \left(\frac{g_j}{h_j}\right)^{-1} \in \mathcal{O}_X^*(U_i \cap U_j).$$

Let D be a Cartier divisor on X. For any prime divisor A on X we define the *multiplicity* $\mathrm{mult}_A(D)$ of D on A as follows. If D is represented by $(U_i, f_i)_{i \in I}$ then we set $\mathrm{mult}_A(D) = \mathrm{mult}_A(f_i)$: since by hypothesis $\frac{f_i}{f_j}$ is nowhere vanishing, the value $\mathrm{mult}_A(D)$ does not depend on i. If a Cartier divisor D is represented by data $(U_i, f_i)_{i \in I}$ then we associate to it a Weil divisor

$$[D] := \sum_{\substack{A \text{ prime divisor} \\ \text{on } X}} \mathrm{mult}_A(D) A.$$

The map $\mathrm{Div}(X) \to Z^1(X)$, $D \mapsto [D]$ thus defined is a group morphism.

Proposition 2.6.10 *Let X be an irreducible complex variety.*

1. *If X is normal then the map* $\mathrm{Div}(X) \to Z^1(X)$, $D \mapsto [D]$ *is injective and induces an injective morphism*

$$\mathrm{CaCl}(X) \to \mathrm{Cl}(X)\ .$$

2. *If X is non singular then $D \mapsto [D]$ is an isomorphism*

$$\mathrm{Div}(X) \simeq Z^1(X)$$

and the induced morphism

$$\mathrm{CaCl}(X) \simeq \mathrm{Cl}(X)$$

is an isomorphism.

Proof See [Har77, II.6]. □

2.6.3 Line Bundles

We recall that an (algebraic) complex line bundle is an algebraic vector bundle of fiber ℂ as in Definition 2.5.7. We further remark that over ℂ, any pre-algebraic vector bundle is algebraic, as in Remark 2.5.11(1). The sheaf of sections of such a bundle is an invertible sheaf, see Definition C.5.8, and the correspondence thus induced between isomorphism classes of line bundles and invertible sheaves is one-to-one, see Proposition 2.5.9.

To any Cartier divisor D represented by $(U_i, f_i)_i$ we can associate the sub-sheaf $\mathcal{O}_X(D) \subset \mathcal{M}_X$ defined by $\mathcal{O}_X(D)|_{U_i} = f_i^{-1}\mathcal{O}_X|_{U_i}$. The sheaf $\mathcal{O}_X(D)$ is an invertible sheaf over X. By abuse of notation we will also denote by $\mathcal{O}_X(D)$ the associated line bundle. More explicitly, the line bundle $\mathcal{O}_X(D)$ is given by the data of the open cover $\{U_i\}_{i \in I}$ of X and the transition functions $f_{ij} \colon U_i \cap U_j \to \mathbb{C}^*$ where $f_{ij} = f_j|_{U_i \cap U_j} \circ f_i^{-1}|_{U_i \cap U_j}$. The total space of the bundle is the quotient of the disjoint union $\sqcup_i (U_i \times \mathbb{C})$ by the equivalence relation $(x, z) \sim (x, f_{jk}(x)z)$ for any pair of open sets U_j, U_k containing x. This quotient is well defined because these functions satisfy the *cocycle condition*:

$$f_{ik} = f_{ij} f_{jk} \quad \text{sur} \quad U_i \cap U_j \cap U_k \quad \forall i, j, k \ .$$

By construction, D is effective if and only if $\mathcal{O}_X(-D) \subset \mathcal{O}_X$. If U is an open subset of X then $\mathcal{O}_X(D)|_U = \mathcal{O}_U(D|_U)$.

Definition 2.6.11 The line bundle $\mathcal{O}_X(D)$ is the line bundle *associated* to D.

We denote by $\mathrm{Pic}(X)$ the *Picard group* of line bundles modulo isomorphism with group operation given by tensor product and by $\rho \colon \mathrm{Div}(X) \to \mathrm{Pic}(X)$ the map associating to a divisor D the isomorphism class of the line bundle $\mathcal{O}_X(D)$.

Proposition 2.6.12 *Let X be a complex algebraic variety. The Picard group* $\mathrm{Pic}(X)$ *is isomorphic to the cohomology group* $H^1(X, \mathcal{O}^*)$.

Proof See [Har77, III, Exercise 4.5] or [GH78, Section 1.1] for an analytic version of this theorem. □

Example 2.6.13 Consider $X = \mathbb{P}^n$. By Exercise 2.6.5, the group $\mathrm{Cl}(\mathbb{P}^n)$ is isomorphic to $\mathbb{Z}$ and it is generated by the class of a hyperplane $H \subset \mathbb{P}^n$. The Picard group $\mathrm{Pic}(\mathbb{P}^n)$ is therefore isomorphic to $\mathbb{Z}$ and has a natural generator, namely the line bundle associated to H. By convention, we denote this line bundle by $\mathcal{O}_{\mathbb{P}^n}(1) := \mathcal{O}_{\mathbb{P}^n}(H)$. The other generator of $\mathrm{Pic}(\mathbb{P}^n)$ is its dual bundle, denoted $\mathcal{O}_{\mathbb{P}^n}(-1) := \mathcal{O}_{\mathbb{P}^n}(1)^\vee$.

By extension, we write $\mathcal{O}_{\mathbb{P}^n}(k) := \mathcal{O}_{\mathbb{P}^n}(1)^{\otimes k}$ and $\mathcal{O}_{\mathbb{P}^n}(-k) := \mathcal{O}_{\mathbb{P}^n}(-1)^{\otimes k}$ for any positive integer k. In particular, $\mathcal{O}_{\mathbb{P}^n}(0) = \mathcal{O}_{\mathbb{P}^n}$. It follows that the line bundle associated to the divisor kH is $\mathcal{O}_{\mathbb{P}^n}(k)$ for any $k \in \mathbb{Z}$. See [Ser55a, Chapitre III, Section 2] for the original construction of the sheaves $\mathcal{O}(k)$.

Definition 2.6.14 The line bundle $\mathcal{O}_{\mathbb{P}^n}(1)$ is called *Serre's twisting sheaf* and the line bundle $\mathcal{O}_{\mathbb{P}^n}(-1)$ is called the *tautological bundle*. See Section F.1 for a direct construction of this bundle.

Exercise 2.6.15 *Consider an integer* $d>1$. *Prove that the vector space* $\Gamma(\mathbb{P}^n, \mathcal{O}^n_{\mathbb{P}}(dH))$ *of global sections of the line bundle* $\mathcal{O}_{\mathbb{P}^n}(d)$ *is exactly the space of degree* d *homogeneous polynomials in* $n+1$ *variables. Deduce that* $\dim H^0\left(\mathbb{P}^n, \mathcal{O}^n_{\mathbb{P}}(d)\right) = \binom{n+d}{d}$.

Proposition 2.6.16 *Let X be an irreducible quasi-projective complex algebraic variety.*

1. *For any* $D_1, D_2 \in \mathrm{Div}(X)$ *we have that*

$$\rho(D_1 + D_2) = \mathcal{O}_X(D_1) \otimes \mathcal{O}_X(D_2) .$$

2. *The map* $\rho\colon \mathrm{Div}(X) \to \mathrm{Pic}(X)$ *induces an isomorphism*

$$\mathrm{CaCl}(X) \simeq \mathrm{Pic}(X) .$$

Proof See [Har77, II.6]. □

By abuse of notation we will often write $D \in \mathrm{Pic}(X)$ for the linear class of a divisor $D \in \mathrm{Div}(X)$.

Corollary 2.6.17 *Let X be a non singular irreducible quasi-projective complex algebraic variety. There are isomorphisms*

$$\mathrm{Cl}(X) \simeq \mathrm{CaCl}(X) \simeq \mathrm{Pic}(X) \simeq \mathrm{Div}(X)/\mathcal{P}(X) .$$

Definition 2.6.18 Let D be a divisor on an algebraic variety X. The *linear system* $|D|$ is the set of effective divisors which are linearly equivalent to D. We identify this set with the projectivisation of the complex vector space $H^0(X, \mathcal{O}_X(D))$ of global sections of $\mathcal{O}_X(D)$.

We have that $H^0(X, \mathcal{O}_X(D)) = \{f \in K(X)^* \mid D + (f) \geqslant 0\} \cup \{0\}$. If this complex vector space is of finite dimension then any basis $\{s_0, \dots, s_N\}$ of $H^0(X, \mathcal{O}_X(D))$ is a set of global rational functions on X which enables us to defined a rational map

$$\varphi_D\colon \begin{cases} X \dashrightarrow \mathbb{P}(H^0(X, \mathcal{O}_X(D))) = \mathbb{P}^N(\mathbb{C}) \\ x \mapsto (s_0(x) : \cdots : s_N(x)) . \end{cases}$$

Remark 2.6.19 The map φ_D depends on a choice of basis for $H^0(X, \mathcal{O}_X(D))$ and is only determined by D up to automorphism of $\mathbb{P}(H^0(X, \mathcal{O}_X(D)))$.

Definition 2.6.20 A divisor D on a variety X is *very ample* if the rational map φ_D is a morphism embedding X in $\mathbb{P}(H^0(X, \mathcal{O}_X(D)))$. A divisor D is *ample* if one of its multiples mD, $m \geqslant 1$, is very ample.

Likewise, an invertible sheaf $\mathcal{L}$ is *very ample* if it is associated to a very ample divisor $\mathcal{L} = \mathcal{O}_X(D)$, and it is *ample* if $\mathcal{L}^{\otimes m}$ is very ample for some $m \geqslant 1$.

Proposition 2.6.21 *An abstract algebraic variety (constructed by "gluing together" affine algebraic varieties as in Definition 1.3.1) is projective if and only if it has an ample divisor.*

Proof Suppose that D is an ample divisor on X. There is then a multiple mD, $m \geqslant 1$, which is very ample and the associated morphism φ_{mD} embeds X as a closed subvariety of projective space. Conversely, let X be a projective algebraic variety and let $\varphi\colon X \to \mathbb{P}^N$ be an embedding. For any hyperplane H in $\mathbb{P}^N$ the divisor $\varphi^*(H)$ is a very ample divisor on X (or in terms of line bundles, $\varphi^*(\mathcal{O}_{\mathbb{P}^N}(1))$ is very ample on X). The divisor $\varphi^*(H)$ is the divisor of the *hyperplane section* of X relative to the embedding φ. □

Definition 2.6.22 A divisor D on an algebraic variety X (which we will assume *complete* in order to be sure that the maps φ_{mD} exist) is *big* if there exists an $m > 0$ for which the dimension of the image of the rational map $\varphi_{mD}\colon X \dashrightarrow \mathbb{P}(H^0(X, \mathcal{O}_X(mD)))$ is maximal, or in other words, if

$$\dim \varphi_{mD}(X) = \dim X \ .$$

Likewise, a line bundle $\mathcal{L}$ is *big* if for some $m > 0$ we have that

$$\varphi_{\mathcal{L}^{\otimes m}}(X) = \dim X \ .$$

Example 2.6.23 1. Any ample line bundle is of course big.
2. The pull back of an ample line bundle along a generically finite map is a big line bundle. See [Laz04, Section 2.2] for more details.

Theorem 2.6.24 *If X is a normal variety (which is the case in particular, for any non singular variety) then a line bundle $\mathcal{L}$ is big if and only if there is some $m > 0$ for which the rational map $\varphi_{\mathcal{L}^{\otimes m}} : X \dashrightarrow \mathbb{P}(H^0(X, \mathcal{O}_X(mD)))$ is birational onto its image.*

Proof This result follows from the existence of the Iitaka fibration. See [Laz04, Section 2.2] for more details. □

Remark 2.6.25 The bigness of a line bundle is invariant under birational transformations.

If X is a non singular quasi-projective complex algebraic variety then the sheaf of regular differential forms (see [Liu02, Chapter 6] or [Har77, II.8] for regular differential forms and Definition D.3.2 for holomorphic differential forms) of degree 1 on X, denoted $\Omega_X := \Omega^1_X$, is a locally free finitely generated sheaf. The associated vector bundle, also denoted Ω_X, has rank equal to the dimension of X and its determinant bundle $\det \Omega_X$ is a line bundle.

Definition 2.6.26 Let X be a non singular quasi-projective complex algebraic variety. The *canonical bundle* on X is the complex line bundle defined by

$$\mathcal{K}_X := \det \Omega_X = \bigwedge^n \Omega_X \ .$$

The *canonical divisor* of X denotes any divisor associated to the canonical bundle

$$\mathcal{O}_X(K_X) = \mathcal{K}_X \ .$$

It is customary to talk about "the" canonical divisor, even though such divisors are only defined up to linear equivalence.

Exercise 2.6.27 *Prove that $\mathcal{K}_{\mathbb{P}^n}$ is isomorphic to the line bundle $\mathcal{O}_{\mathbb{P}^n}(-n-1)$.*

Exercise 2.6.28 (See [CM09, Theorem 4.3]) *Let X be a non singular projective variety. Prove that if $H^0(X, \mathcal{O}_X(-K_X)) \neq 0$ and $H^0(X, \Omega^1_X) = 0$ then $H^0(X, \Omega^1_X(K_X)) = 0$.*

Using Serre duality (Theorem D.2.5) deduce that

$$H^2(X, \Theta_X) = 0$$

where Θ_X is the tangent bundle.

Definition 2.6.29 A non singular projective variety X is said to be of *general type* if its canonical bundle $\mathcal{K}_X$ is big.

2.6.4 Galois Group Action on the Picard Group

Let (X, σ) be an $\mathbb{R}$-surface: we denote by σ the involution induced on the divisor group of X. If $D = \sum n_i D_i$ is a Weil divisor on X then $\sigma D := \sum n_i \sigma(D_i)$. If $D = (U_i, f_i)_i$ is a Cartier divisor on X then $\sigma D = (\sigma(U_i), {}^\sigma f_i)_i$. If $\mathcal{L}$ is a line bundle on X with cocycle (U_{ij}, g_{ij}) then the conjugate sheaf (Definition 2.5.1) ${}^\sigma\mathcal{L}$ is the line bundle on X of cocycle $(\sigma(U_{ij}), {}^\sigma g_{ij})$.

Proposition 2.6.30 *Let X be projective. If D is a Cartier divisor and $\mathcal{O}_X(D)$ is the associated invertible sheaf then*

$$\mathcal{O}_X(\sigma D) = {}^\sigma(\mathcal{O}_X(D)).$$

Conversely, if $\mathcal{L}$ is an invertible sheaf on X, D is a divisor associated to $\mathcal{L}$ and D' is a divisor associated to ${}^\sigma\mathcal{L}$ then $D' \sim \sigma D$.

Proof Let $D = (U_i, f_i)_i$ be a Cartier divisor. The sheaf $\mathcal{O}_X(D)$ is determined by the cocycle $(g_{ij})_{ij} = (\frac{f_i}{f_j})_{ij}$. Indeed, $\Gamma(U, \mathcal{O}_X(D)) = \{f \in \mathcal{O}_X(U) \mid (f) + D \geqslant 0\}$. Let $(s_i)_i$ be a family of local sections of $\mathcal{O}_X(D)$. We then have that

$$\forall i, j, s_i = g_{ij} s_j \ . \tag{2.4}$$

By definition of the conjugate sheaf, $({}^\sigma s_i)_i$ is a family of local sections of the sheaf ${}^\sigma(\mathcal{O}_X(D))$ and by (2.4) we have that

$$\forall i, j, {}^\sigma s_i = {}^\sigma g_{ij} {}^\sigma s_j \ . \tag{2.5}$$

The proof follows on noting that $\mathcal{O}_X(\sigma D)$ is determined by the cocycle $({}^\sigma g_{ij})_{ij} = (\frac{{}^\sigma f_i}{{}^\sigma f_j})_{ij}$. □

Proposition 2.6.31 *Let D be a divisor invariant under (X, σ). There is then a basis $\{s_0, \ldots, s_N\}$ of the complex vector space $H^0(X, \mathcal{O}_X(D)) = \{f \in K(X)^* \mid D + (f) \geqslant 0\} \cup \{0\}$ consisting of invariant functions ${}^\sigma s_i = s_i$, $i = 0, \ldots, N$.*

Proof Follows immediately from Lemma A.7.3. □

Theorem 2.6.32 *Let (X, σ) be an irreducible non singular complex projective algebraic $\mathbb{R}$-variety. If $X(\mathbb{R}) \neq \varnothing$ then for any divisor D linearly equivalent to $\sigma(D)$ there is a divisor D' linearly equivalent to D such that $D' = \sigma(D')$. In other words,*[5]

$$\mathrm{Div}(X)^G/\mathcal{P}(X)^G = \mathrm{Pic}(X)^G \ .$$

Proof See [Sil89, pp. 19–20]. □

Example 2.6.33 ($\mathrm{Div}(X)^G/\mathcal{P}(X)^G \neq \mathrm{Pic}(X)^G$) The example of the conic X in $\mathbb{P}^2$ of equation $x_0^2 + x_1^2 + x_3^2 = 0$ shows that when $X(\mathbb{R}) = \varnothing$, $\mathrm{Pic}(X)^G$ can be larger than $\mathrm{Div}(X)^G/\mathcal{P}(X)^G$. In this example, $\mathrm{Pic}(X)^G = \mathrm{Pic}(X) = \mathbb{Z}$ which is generated by a point, but all the invariant divisors are of even degree and there is an exact sequence

$$0 \to \mathrm{Div}(X)^G/\mathcal{P}(X)^G \longrightarrow \mathrm{Pic}(X)^G \longrightarrow \mathbb{Z}/2\mathbb{Z} \to 0 \ .$$

Up till now we have studied the Picard group of linear divisor classes. We now present another group of divisor classes, the *Néron–Severi* group.

Definition 2.6.34 Let X be a non singular complex projective variety and let $\mathrm{Pic}^0(X)$ be the connected component of $\mathrm{Pic}(X)$ containing the identity ($\mathrm{Pic}^0(X)$ is the *Picard variety* of X, see Definition D.6.6). The *Néron–Severi group* $\mathrm{NS}(X)$ is the group of components of $\mathrm{Pic}(X)$:

$$0 \to \mathrm{Pic}^0(X) \longrightarrow \mathrm{Pic}(X) \longrightarrow \mathrm{NS}(X) \to 0 \ .$$

Two divisors in the same class in the Néron Severi group are said to be *algebraically equivalent.*[6]

Theorem 2.6.35 (Néron–Severi theorem) *Let X be a non singular complex projective variety. The group $\mathrm{NS}(X)$ is then finitely generated.*

Proof See [GH78, IV.6, pp. 461–462]. □

---

[5]Scheme-theoretically, if X is a scheme defined over $\mathbb{R}$ satisfying the hypotheses of the theorem then $\mathrm{Pic}(X) = \mathrm{Pic}(X_\mathbb{C})^G$.

[6]See [GH78, III.5] for an explanation of this term. The term "numerically equivalent" is also common in the literature: see [Ful98, Section 19.3] for more details.

Definition 2.6.36 Let X be a non singular complex projective variety. The rank of the Neron–Severi group $\rho(X) := \operatorname{rk} \operatorname{NS}(X) = \operatorname{rk}(\operatorname{Pic}(X)/\operatorname{Pic}^0(X))$ is called the *Picard number* of X. Let (X, σ) be a non singular projective $\mathbb{R}$-variety. If $X(\mathbb{R})$ is non empty then the *real Picard number* of (X, σ) is the rank of the *real* Néron–Severi group $\rho_{\mathbb{R}}(X) := \operatorname{rk}(\operatorname{Pic}(X)^G/\operatorname{Pic}^0(X)^G)$.

Proposition 2.6.37 *Let X be a non singular complex projective variety such that $q(X) = \dim H^1(X, \mathcal{O}_X) = 0$. We then have that*

$$\operatorname{NS}(X) \simeq \operatorname{Pic}(X) .$$

Proof It follows from the exact sequence (D.3) following Proposition D.6.7 that if $q(X) = 0$ then the group $\operatorname{Pic}^0(X)$ is trivial. □

2.6.5 *Projective Embeddings*

We have seen that any compact real affine algebraic variety has a projective complexification. The aim of this section is to study these projective models using ample divisors.

Example 2.6.38 ($\mathbb{R}$-*embedding of the product torus*) This example draws on [BCR98, Example 3.2.8]. Let V be the product torus $V := \mathcal{Z}\left(t^2 + u^2 - 1\right) \times \mathcal{Z}\left(v^2 + w^2 - 1\right) \subset \mathbb{A}^2(\mathbb{R}) \times \mathbb{A}^2(\mathbb{R})$ and let W be the quartic torus in $\mathbb{R}^3_{x_1,x_2,x_3}$ obtained by rotating the circle of centre $(2, 0)$ and radius 1 in the (x_1, x_3) plan around the x_3 axis

$$W := \mathcal{Z}(16(x_1^2 + x_2^2) - (x_1^2 + x_2^2 + x_3^2 + 3)^2) \subset \mathbb{A}^3(\mathbb{R}) .$$

Both of these real algebraic sets are diffeomorphic to the torus with the Euclidean topology $V \approx W \approx \mathbb{S}^1 \times \mathbb{S}^1$.

Consider W as a subset of $\mathbb{P}^3(\mathbb{R})$ via the inclusion $\mathbb{R}^3_{x_1,x_2,x_3} \subset \mathbb{P}^3(\mathbb{R})_{x_0:x_1:x_2:x_3}$. The polynomial map

$$\begin{array}{rccc} \varphi\colon & V & \longrightarrow & W \\ & (t, u, v, w) & \longmapsto & (1 : t(2 + v) : u(2 + v) : w) \end{array}$$

is bijective and its inverse $\varphi^{-1}\colon W \to V$,

$$\varphi^{-1}(x_0 : x_1 : x_2 : x_3) = \left(x_1x_0/\rho, x_2x_0/\rho, (\rho - 2x_0^2)/x_0^2, x_3/x_0\right)$$

where $\rho = (x_1^2 + x_2^2 + x_3^2 + 3x_0^2)/4$, is a regular map of real algebraic varieties since $W \cap \{x_0 = 0\} = \varnothing$.

The map φ is therefore an isomorphism of real algebraic varieties and the algebras $\mathcal{R}(V)$ and $\mathcal{R}(W)$ are isomorphic by Corollary 1.3.20: the algebras $\mathcal{P}(V)$ and $\mathcal{P}(W)$, however, are different, since the first is regular, unlike the second. Consider the projective complexifications of the toruses V and W: $\overline{V}_{\mathbb{C}} \simeq \mathbb{P}^1(\mathbb{C}) \times \mathbb{P}^1(\mathbb{C})$ for the first and the singular quartic hypersurface

$$\widehat{W}_{\mathbb{C}} := \mathcal{Z}(16(x_1^2 + x_2^2) - (x_1^2 + x_2^2 + x_3^2 + 3x_0^2)^2) \subset \mathbb{P}^3(\mathbb{C}) .$$

for the second. The map φ is then the restriction of a birational map of $\mathbb{R}$-varieties

$$\psi : \mathbb{P}^1(\mathbb{C}) \times \mathbb{P}^1(\mathbb{C}) \to \widehat{W}_{\mathbb{C}}$$

which is a resolution of singularities of $\widehat{W}_{\mathbb{C}}$.

Note that ψ is a morphism of $\mathbb{R}$-varieties but ψ^{-1} is only a rational map. Note also that as $\widehat{W}_{\mathbb{C}}$ is a quartic in $\mathbb{P}^3(\mathbb{C})$ which is birational to $\mathbb{P}^1(\mathbb{C}) \times \mathbb{P}^1(\mathbb{C})$ it must be singular. Indeed, $\mathbb{P}^1(\mathbb{C}) \times \mathbb{P}^1(\mathbb{C})$ is a rational surface whereas a non singular quartic in $\mathbb{P}^3$ is a non rational surface (called a *K3 surface*, see Definition 4.5.3). The $\mathbb{R}$-surfaces $(\mathbb{P}^1(\mathbb{C}) \times \mathbb{P}^1(\mathbb{C}), \sigma_{\mathbb{P}} \times \sigma_{\mathbb{P}})$ and $(\widehat{W}_{\mathbb{C}}, \sigma_{\mathbb{P}}|_{\widehat{W}_{\mathbb{C}}})$ are birationally equivalent but not isomorphic.

2.6.6 Review of Theorem 2.1.33

We have seen that a variety X embedded in $\mathbb{P}^n(\mathbb{C})$ and stable by the conjugation $\sigma_{\mathbb{P}}$ has a natural real structure σ induced by $\sigma_{\mathbb{P}}$. Note that if X is a projective complex variety with a real structure σ then its image under an arbitrary projective embedding is not always stable under $\sigma_{\mathbb{P}}$, but we can always find a real embedding by Theorem 2.6.44 below. We will give a proof of this theorem based on the Nakai–Moishezon criterion. Of course, Theorem 2.6.44 implies Theorem 2.1.33 for which we have only provided a reference for the proof. In what follows, up to and including the proof of Theorem 2.6.44, we will not use Theorem 2.1.33.

The key fact to remember is that if X is a complex projective variety then for any real structure σ on X the $\mathbb{R}$-variety (X, σ) has an equivariant embedding in projective space.

2.6.7 Nakai–Moishezon Criterion

See [Har77, Appendix A, p. 424] for the definition and main properties of intersection theory on varieties of arbitrary dimension. If the global variety has a real structure then this intersection theory is compatible with the real structure. If r is the dimension of a non singular variety Y and $D_1, D_2, \dots, D_r$ are divisors on Y then their intersection product $(D_1 \cdot D_2 \cdots D_r)$ belongs to $\mathbb{Z}$ and only depends on the linear class of the divisors D_i. In particular, if the D_is are hypersurfaces meeting transversally then $(D_1 \cdot D_2 \cdots D_r)$ is equal to the number of points in the intersection of the D_is.

Theorem 2.6.39 (Nakai–Moishezon criterion) *Let D be a Cartier divisor on a complex projective algebraic variety X. The divisor D is then ample on X if and only if for any irreducible subvariety $Y \subset X$ of dimension r we have that*

$$(D|_Y)^r > 0 .$$

Proof See [Har77, Appendix A, Theorem 5.1, p. 434], for example. The above statement also holds for singular X, but requires a modified intersection theory. See [Kle66, Ful98] for more details. □

Corollary 2.6.40 (Nakai–Moishezon criterion for surfaces) *A divisor D on a non singular irreducible complex projective algebraic surface X is ample if and only if $(D)^2 > 0$ and $D \cdot C > 0$ for any irreducible curve C in X.*

Proof Simply set $Y = X$ in the general criterion to obtain $(D)^2 > 0$ and for any irreducible curve $C \subset X$, $D \cdot C > 0$. □

Definition 2.6.41 A divisor D on a variety X is *nef* (for *numerically eventually free*[7]) if for any irreducible subvariety $Y \subset X$ of dimension r we have that

$$(D|_Y)^r \geqslant 0 .$$

Similarly, a line bundle $\mathcal{L}$ is *nef* if and only if it is associated to a nef divisor $\mathcal{L} = \mathcal{O}_X(D)$.

Remark 2.6.42 Any ample bundle is of course nef.

Proposition 2.6.43 *Let X be a complex projective variety with a real structure σ. There is then an ample divisor D such that $D = \sigma D$.*

Proof Let H be an ample divisor on X. For any irreducible subvariety $Y \subset X$ of dimension r the conjugate subvariety σY is irreducible and of dimension r and by the Nakai–Moishezon criterion (Theorem 2.6.39) we have that $(H|_{\sigma Y})^r > 0$. Since the real structure is involutive, $(\sigma H)|_Y = \sigma(H|_{\sigma Y})$ and since the real structure is compatible with the intersection product we get that $((\sigma H)|_Y)^r = (H|_{\sigma Y})^r > 0$. By the Nakai–Moishezon criterion, σH is ample, as is

$$D := H + \sigma H .$$

□

Theorem 2.6.44 *Let (X, σ) be an algebraic $\mathbb{R}$-variety. If the complex algebraic variety X is quasi-projective then there is an $\mathbb{R}$-embedding*

[7] If the linear system $|mD|$ is free for some $m > 0$ (eventually free), then D is nef. The incorrect interpretation *numerically effective* often appears in the literature, but considering (-1)-curves—see Definition 4.3.2—we see that a divisor can be effective without being either nef or linearly equivalent to a nef divisor.

$$\varphi\colon (X, \sigma) \hookrightarrow (\mathbb{P}^N(\mathbb{C}), \sigma_{\mathbb{P}}) .$$

Proof We start by assuming X is projective, so by Proposition 2.6.43, there is an ample divisor D_0 and a positive integer m such that $D = mD_0$ is very ample on X and satisfies $\sigma D = D$. By Proposition 2.6.31, there is a basis $\{s_0, \dots, s_N\}$ of $H^0(X, \mathcal{O}_X(D))$ such that ${}^\sigma s_i = s_i$, $i = 0, \dots, N$. As the divisor D is very ample, the map

$$\varphi_D\colon \begin{cases} X \dashrightarrow & \mathbb{P}^N(\mathbb{C}) \\ x \mapsto & (s_0(x) : \cdots : s_N(x)) \end{cases}$$

is a morphism which induces an isomorphism of $\mathbb{R}$-varieties

$$(X, \sigma) \simeq (\varphi_D(X), \sigma_{\mathbb{P}}|_{\varphi_D(X)}) .$$

Now consider a quasi-projective variety $U = X \setminus Y$, where X is a projective $\mathbb{R}$-variety and $Y \subset X$ is a closed $\mathbb{R}$-subvariety of X. We have just proved the existence of an $\mathbb{R}$-embedding; $\varphi\colon (X, \sigma) \hookrightarrow (\mathbb{P}^N(\mathbb{C}), \sigma_{\mathbb{P}})$: in particular, φ is a homeomorphism onto its image $\varphi(X \setminus Y) = \varphi(X) \setminus \varphi(Y)$ and φ therefore induces an embedding of U as a quasi-projective algebraic set

$$(U, \sigma|_U) \simeq (\varphi(X) \setminus \varphi(Y), \sigma_{\mathbb{P}}|_{\varphi(X)\setminus\varphi(Y)}) .$$

□

2.6.8 *Degree of a Subvariety of Projective Space*

Classically, we define the degree of a subvariety of $\mathbb{P}^N$ using its Hilbert polynomial [Har77, Section I.7] and only subsequently prove that this definition is equivalent to the definition given below.

Definition 2.6.45 (*Degree of a subvariety of projective space*) The *degree* of an n dimensional subvariety X of $\mathbb{P}^N$ is the degree of the 0-cycle $D := (H \cdot X)$ obtained on intersecting X with a general codimension n projective subspace H in $\mathbb{P}^N$.

There is a hidden difficulty in the above definition, namely finding the coefficients of the 0-cycle $D := (H \cdot X)$ for an arbitrary X. See the section preceding [Har77, Theorem 7.7, p. 53] for more details. If X is complex and non singular then by Bertini's Theorem D.9.1 if we choose a sufficiently general H then the 0-cycle D is the sum of all points in $H \cap X$.

Definition 2.6.46 (*Complex degree*) The *complex degree* of a complex projective algebraic variety is the smallest degree of any of its embeddings in a complex projective space $\mathbb{P}^N(\mathbb{C})$.

Definition 2.6.47 (*Real degree*) Let (X, σ) be a projective $\mathbb{R}$-variety. The *real degree* of (X, σ) is the smallest degree of a real embedding in projective space $(\mathbb{P}^N(\mathbb{C}), \sigma_{\mathbb{P}})$.

The real degree exists by Proposition 2.6.43. As any real embedding is also a complex embedding, the real degree is not smaller than the complex degree. The minimal degree of a complex projective embedding is frequently strictly smaller than the minimal degree of a real projective embedding. The simplest example is that of conic without real points, whose complex degree is 1 but whose real degree is 2. Let X be the projective plane curve defined by the equation $x^2 + y^2 + z^2 = 0$ with the restriction of $\sigma_{\mathbb{A}}$. The curve X is isomorphic as an abstract complex curve to the curve $\mathbb{P}^1(\mathbb{C})$ and has degree 1 embeddings—namely lines—in every $\mathbb{P}^n(\mathbb{C})$. None of these embeddings can be real because any embedding as an $\mathbb{R}$-line has real points. The following proposition generalises this principle.

Proposition 2.6.48 *Let $X \subset \mathbb{P}^n(\mathbb{C})$ be a algebraic subvariety, stable under $\sigma_{\mathbb{P}}$. If the degree of X is odd then $X(\mathbb{R}) \neq \varnothing$.*

Proof We can assume that $r := n - \dim X > 0$. Let H be a projective subspace of dimension r in $\mathbb{P}^n$ which is not contained in X. By hypothesis, the degree of the 0-cycle $D := (H \cdot X)$ is odd. In particular, the real part of D has odd degree and its support consists of an odd number of points so it is non empty. □

2.7 ℝ-Plane Curves

We end this chapter by applying the above theory to plane curves. We refer to Section 1.6 of the first chapter for the general definitions. Bézout's theorem on plane curves, given in Chapter 1, is here applied to $\mathbb{R}$-curves. It will be generalised to curves on other surfaces in Chapter 4.

Theorem 2.7.1 (Bézout's theorem for $\mathbb{R}$-plane curves) *Let C_1 and C_2 be projective plane $\mathbb{R}$-curves of degrees d_1 and d_2 respectively*

1. *If C_1 and C_2 have no common component then*

$$(C_1 \cdot C_2) = d_1 d_2 \ .$$

2. *If the intersection $C_1(\mathbb{R}) \cap C_2(\mathbb{R})$ is finite then*

$$(C_1(\mathbb{R}) \cdot C_2(\mathbb{R})) \leqslant d_1 d_2 \ .$$

3. *If moreover the branches of C_1 and C_2 are transverse at every point then the number of intersection points $\#(C_1(\mathbb{R}) \cap C_2(\mathbb{R}))$ is congruent modulo 2 to the product $d_1 d_2$.*

Proof We simply defined the intersection multiplicity modulo 2 at a point $a \in \mathbb{A}^2(\mathbb{R})$ of two affine plane ℝ-curves C_1 and C_2 of equations $P_1(x, y)$ and $P_2(x, y)$ to be

$$(C_1 \cdot C_2)_a^{\mathbb{R}} := \dim_{\mathbb{R}} \mathcal{O}_{\mathbb{A}^2(\mathbb{R}),a}/(P_1, P_2) \mod 2 ;$$

and the intersection number modulo 2 to be

$$(C_1 \cdot C_2)^{\mathbb{R}} := \sum_{a \in C_1(\mathbb{R}) \cap C_2(\mathbb{R})} (C_1 \cdot C_2)_a^{\mathbb{R}} \mod 2 .$$

We then apply Theorem 1.6.16 to the complex curves C_1 and C_2. □

We recall the genus formula proved in Chapter 1, Theorem 1.6.17. If C is a non singular irreducible projective plane curve of genus $g = g(C)$ then

$$g = \frac{(d-1)(d-2)}{2} .$$

The real locus of a non singular projective ℝ-curve is a compact differentiable variety of dimension 1. It is therefore homeomorphic to a finite union of disjoint embedded circles.

Theorem 2.7.2 (Harnack 1876) *Let (C, σ) be a non singular projective plane ℝ-curve of degree d. Let s be the number of connected components of $C(\mathbb{R})$. We then have that*

$$s \leqslant \frac{(d-1)(d-2)}{2} + 1 = g(C) + 1 . \tag{2.6}$$

Remark 2.7.3 Further on we will give an elementary proof of this inequality based on Bézout's theorem. It is useful to note that the number of connected components of a plane curve of degree d is bounded above by $\frac{(d-1)(d-2)}{2} + 1$ even when C is singular. First of all, it is enough to prove the result when C is irreducible. If not, C is defined by a product of polynomials of degrees d_1 and d_2, so that $d = d_1 + d_1$ and

$$\frac{(d_1-1)(d_1-2)}{2} + 1 + \frac{(d_2-1)(d_2-2)}{2} + 1 \leqslant \frac{(d-1)(d-2)}{2} + 1 .$$

We then show that we can assume that $C(\mathbb{R})$ contains at least one component of dimension 1 using Brusotti's Theorem 2.7.10 as in Corollary 3.3.20. The proof then follows the proof for the smooth case given below, see [BR90, Second proof of 5.3.2].

Remark 2.7.4 More generally, for any non singular projective ℝ-curve(C, σ) (note that C is not assumed to be plane), we have that $s \leqslant g(C) + 1$, where $g(C)$ is the genus of the topological surface C. We will give two proofs of this in Chapter 3 and Corollary 3.3.7. We will also see in Chapter 3 that this inequality can be generalised to higher dimension using Smith theory.

Lemma 2.7.5 *There is a real projective curve of degree d which passes through any given set of* $\binom{d+2}{2} - 1 = \frac{1}{2}(d+2)(d+1) - 1$ *points in* $\mathbb{P}^2(\mathbb{R})$.

Proof The number of degree d monomials in three variables is $\binom{d+2}{2}$. We deduce from this a bijection between the set of degree d curves in the real projective plane and a real projective space of dimension $\frac{1}{2}(d+2)(d+1) - 1$. □

Proposition 2.7.6 *For any point* $p \in \mathbb{RP}^2$,

$$\pi_1(\mathbb{RP}^2, p) \simeq \mathbb{Z}_2 .$$

Proof Consider $\mathbb{RP}^2$ as the quotient of $\mathbb{S}^2$ by the antipodal map. □

Definition 2.7.7 A simple closed curve in the real projective space is an *oval* if it is homotopic to 0 and a *pseudo-line* if it is not homotopically trivial.

Lemma 2.7.8 (Ovals and pseudo-lines) *Let* (C, σ) *be a non singular projective plane* ℝ*-curve of degree* d.

1. *If* d *is even all the connected components of* $C(\mathbb{R})$ *are ovals.*
2. *If* d *is odd then one connected component of* $C(\mathbb{R})$ *is a pseudo-line and all the others are ovals.*
3. *Any curve meets any oval in an even number of intersection points, counted with multiplicity.*

Proof The proof is left as an exercise. Use Bézout's theorem. □

Proof of Theorem 2.7.2 Suppose that $d > 2$. We argue by contradiction: suppose that Γ is a non singular irreducible plane ℝ-curve of degree d whose real locus has at least $g(d) + 1$ connected components. Let $h = g(d) + 1$ and $\Omega_1, \dots, \Omega_h$ be ovals in $\Gamma(\mathbb{R})$: there is at least one other component in $\Gamma(\mathbb{R})$. Choose $\frac{1}{2}d(d-1) - 1$ points on $\Gamma(\mathbb{R})$. Since $\frac{1}{2}d(d-1) - 1 \geqslant g(d) + 1$ for any $d > 2$ we can choose one point on each of the ovals $\Omega_1, \dots, \Omega_h$ and the other points on some other connected component of $\Gamma(\mathbb{R})$. Consider an ℝ-curve Δ of degree $d - 2$ passing through these $\frac{1}{2}d(d-1) - 1$ points. The curves Γ and Δ have no common components because Γ is irreducible and the degree of Δ is $d - 2$. By Bézout's theorem, the number of intersection points of Γ with Δ counted with multiplicity is less than or equal to $d(d-2)$. If Δ meets an oval Ω_i with multiplicity 1 then Δ meets Ω_i at some other point, so that $\Gamma \cdot \Delta \geqslant \frac{1}{2}d(d-1) - 1 + g(d) + 1 = (d-1)^2$ which is larger than $d(d-2)$. The theorem follows. □

The bound (2.6) is optimal: Harnack's bound is realised for any degree d:

Proposition 2.7.9 *For any* $d \in \mathbb{N}^*$ *there is a non singular projective plane* ℝ*-curve* (C, σ) *of degree* d *whose real locus* $C(\mathbb{R})$ *contains* $s = \frac{(d-1)(d-2)}{2} + 1$ *connected components.*

Proof See [BCR98, pp. 287–288] or [BR90, 5.3.11] for Harnack's construction. □

The constructions of the curves described above often use explicit deformations of reducible curves. We can often prove the existence of configurations of ovals of given degree without explicit constructions using Brusotti's useful theorem.

Theorem 2.7.10 (Brusotti's theorem) *Let $C \subset \mathbb{P}^2(\mathbb{R})$ be a degree d real plane curve whose singularities are ordinary double points. Suppose given a local deformation of each of the ordinary double points. There is then a deformation of the curve C in the space of real curves of degree d which realises each of the local deformations.*

Proof See [BR90, Section 5.5]. □

As well as (2.6) which gives a bound on the number of connected components, we have restrictions on the positions of ovals of plane $\mathbb{R}$-curves.

Definition 2.7.11 The complement $\mathbb{RP}^2 \setminus \Omega$ of a oval in the real projective plane has two connected components. One of these is diffeomorphic to the disc and is called the *interior* of the oval, and the other is diffeomorphic to a Moebius band. We say that another oval is *contained* in Ω if it is contained in its interior. An oval component of a real curve is said to be *empty* if it does not contain any other oval component. A family E is said to be a *nest* of ovals if and only if it is totally ordered by inclusion.

Definition 2.7.12 An oval is said to be *positive* (or *even*) if it is contained in an even number of ovals and *negative* (or *odd*) otherwise.[8]

Theorem 2.7.13 (Petrovskii's inequalities) *Let (C, σ) be a non-singular projective plane $\mathbb{R}$-curve of even degree $d = 2k$. Let p be the number of even ovals of $C(\mathbb{R})$ and let n be the number of negative ovals. We then have that*

$$p - n \leqslant \frac{3}{8}d(d-2) + 1 = \frac{3}{2}k(k-1) + 1\ ;$$
$$n - p \leqslant \frac{3}{8}d(d-2) = \frac{3}{2}k(k-1)\ .$$

See [Pet33, Pet38] or [Arn71]. In Chapter 3, Theorem 3.3.14 we prove these inequalities using double covers.

Corollary 2.7.14 *Let (C, σ) be a non singular projective plane $\mathbb{R}$-curve of even degree $d = 2k$. Let p be the number of positive ovals of $C(\mathbb{R})$ and n be the number of negative ovals. Then we have that*

$$p \leqslant \frac{7}{4}k^2 - \frac{9}{4}k + \frac{3}{2}\quad ;\qquad n \leqslant \frac{7}{4}k^2 - \frac{9}{4}k + 1\ .$$

Proof For any curve of even degree $d = 2k$, Harnack's inequality (2.6) gives $p + n \leqslant 2k^2 - 3k + 2$. Adding with the Petrovskii inequalities yields the desired result. □

[8] See [Pet38, p. 190] for a justification of this terminology.

Remark 2.7.15 (*Ragsdale's conjecture*) A famous, but incorrect, conjecture by Ragsdale [Rag06] states that p and n actually satisfy the inequalities $p \leqslant \frac{3}{2}k(k-1)+1$, et $n \leqslant \frac{3}{2}k(k-1)$. We will come back to this conjecture in Chapter 3, at the end of Section 3.5.

When the curve does not have any nest of ovals, all ovals are positive and Petrovskii's first inequality gives us the following.

Corollary 2.7.16 *Let C be a non singular projective plane $\mathbb{R}$-curve of even degree $d = 2k$ without a nest of ovals. The number of ovals $s := \#\pi_0(C(\mathbb{R}))$ is then bounded by*

$$s \leqslant \frac{3}{2}k(k-1)+1 .$$

Corollary 2.7.17 *The maximal even degree d curves, by which we mean the curves with the maximal number of connected components in their real locus, namely $\frac{(d-1)(d-2)}{2}+1$, (see Definition 3.3.10) have at least one nesting from degree 6 onwards.*

2.8 Solutions to exercises of Chapter 2

2.1.3 1. Let U be an open set in $\mathbb{A}^n(\mathbb{C})$ and consider $f \in {}^\sigma\mathcal{O}(U)$. By definition there is a function $g \in \mathcal{O}(\sigma_{\mathbb{A}}(U))$ such that $f = {}^\sigma g$ so $f = \overline{g} \circ \sigma_{\mathbb{A}} \colon U \to \mathbb{C}$ is regular and hence $f \in \mathcal{O}(U)$. The opposite inclusion $\mathcal{O}(U) \subset {}^\sigma\mathcal{O}(U)$ is proved by a similar argument.

2. Apply Definition 1.3.7 to the sheaf ${}^\sigma\mathcal{O}$ and the subspace F to get the sheaf ${}^\sigma\mathcal{O}_F$. If U is an open subset of F then $\overline{U}$ is an open set of $\overline{F}$ and hence of F by hypothesis. A function $f \colon U \to \mathbb{C}$ belongs to ${}^\sigma\mathcal{O}_F(U)$ if and only if for any point x in U there is a neighbourhood V of x in $\mathbb{A}^n(\mathbb{C})$ and a function $g \in {}^\sigma\mathcal{O}(V)$ such that $g(y) = f(y)$ for any $y \in V \cap U$. By the previous question $g \in \mathcal{O}(V)$ and hence ${}^\sigma\mathcal{O}_F = \mathcal{O}_F$.

2.1.7 The sets F and $\overline{F}$ are subsets of $\mathbb{A}^n(\mathbb{C})$ and $\mathcal{O}_{\overline{F}} = (\mathcal{O}_{\mathbb{A}^n})_{\overline{F}}$ (see Definition 1.3.7). The restriction $\sigma_{\mathbb{A}} : \overline{F} \to F$ is clearly bijective. Moreover, $\sigma_{\mathbb{A}}$ is continuous since if $Z = \mathcal{Z}(I)$ is a Zariski closed subset of F defined by an ideal I in $\mathbb{C}[X_1, \ldots, X_n]$ then ${\sigma_{\mathbb{A}}}^{-1}(Z) = \sigma_{\mathbb{A}}(Z) = \overline{Z} = \mathcal{Z}({}^\sigma I)$ where ${}^\sigma I := \{{}^\sigma f \mid f \in I\}$. Finally, $\sigma_{\mathbb{A}}|_{\overline{F}}$ induces an isomorphism of ringed spaces (see Exercise C.5.3) $(\overline{F}, \mathcal{O}_{\overline{F}}) \to (F, \overline{\mathcal{O}_F})$ because if U is an open subset of F then $\sigma_{\mathbb{A}}(U)$ is an open subset of $\overline{F}$ and if $f \in \overline{\mathcal{O}_F}(U)$ then $f \circ \sigma_{\mathbb{A}} \colon \sigma_{\mathbb{A}}(U) \to \mathbb{C}$ is regular or in other words $f \circ \sigma_{\mathbb{A}} \in \mathcal{O}_{\overline{F}}(\sigma_{\mathbb{A}}(U))$. Indeed, as $f \in \overline{\mathcal{O}_F(U)}$ there is a function $f_0 \in \mathcal{O}_F(U)$ such that $f = \overline{f_0}$ and it follows that $f \circ \sigma_{\mathbb{A}} = \overline{f_0} \circ \sigma_{\mathbb{A}} = {}^\sigma f_0$. As f_0 is regular on U, ${}^\sigma f_0$ is regular on $\sigma_{\mathbb{A}}(U)$.

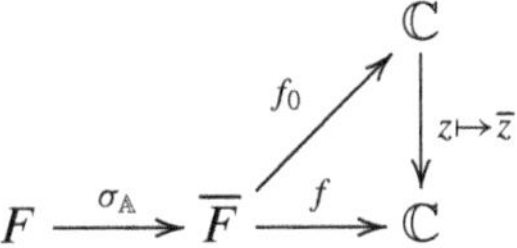

2.1.21 1. Recall that if C is the zero locus of a polynomial P then $\overline{C}$ is the zero locus of ${}^{\sigma}P$. A straightforward calculation shows that $(\varphi \circ \varphi)(x, y) = (x, y)$ so φ is an involutive automorphism of $\mathbb{A}^2(\mathbb{C})$ and in particular $\varphi^{-1} = \varphi$. Now consider $P(x, y) = y^2 - a_0 x^m - \sum_{k=1}^{m} \left(a_k x^{m+k} + (-1)^k \overline{a_k} x^{m-k}\right)$. On substituting $P(\varphi(x, y))$ we obtain $-\frac{y^2}{x^{2m}} + a_0 \frac{1}{x^m} + \sum_{k=1}^{m} \left(a_k \frac{1}{x^{m-k}} + (-1)^k \overline{a_k} \frac{1}{x^{m+k}}\right)$ and hence $-x^{2m} P(\varphi(x, y)) = {}^{\sigma}P(x, y)$.

2. Set $\tau = \sigma_{\mathbb{A}} \circ \varphi$. We then have that $\tau(x, y) = (-\frac{1}{\overline{x}}, -\frac{i\overline{y}}{\overline{x}^m})$ et $(\tau \circ \tau)(x, y) = (x, -y)$.

3a. Restricting the projection $(x, y) \mapsto x$ we exhibit the curve $C := C_{m,a_0,\dots,a_m}$ as a degree 2 covering of $\mathbb{P}^1(\mathbb{C})$. Its function field $\mathbb{C}(C)$ is therefore a degree two extension of $\mathbb{C}(x) = \mathbb{C}(\mathbb{P}^1(\mathbb{C}))$. Moreover, there is a one-to-one correspondence between automorphisms of C and automorphisms of the field $\mathbb{C}(C)$.[9] The two elements of the automorphism group of the extension $\mathbb{C}(C)|\mathbb{C}(x)$ are represented by id_C and ρ. Any automorphism of $\mathbb{C}(C)$ therefore induces an automorphism of Frac $(\mathbb{C}[x, y]/(P))$. If the coefficients of the one-variable polynomial $P(x, y) - y^2$ are independent over $\mathbb{Q}$ then the only non trivial automorphism is represented by ρ.

3b. By Proposition 2.1.19, if C has a real structure then there is an isomorphism between C and $\overline{C}$ satisfying ${}^{\sigma_{\mathbb{A}}}\psi \circ \psi = \mathrm{id}_C$.

Moreover, it follows from 3a that the only isomorphisms between $C_{m,a_0,\dots,a_m}$ and its conjugate are φ and $\varphi'\colon (x, y) \mapsto (-\frac{1}{x}, -\frac{i}{x^m} y)$, but $\varphi \circ {}^{\sigma_{\mathbb{A}}}\varphi = (\varphi') \circ ({}^{\sigma_{\mathbb{A}}}(\varphi')) = \rho \neq \mathrm{id}_{C_{m,a_0,\dots,a_m}}$. It follows that if a_0, a_k, $\overline{a_k}$ are independent over $\mathbb{Q}$ then the curve $C_{m,a_0,\dots,a_m}$ has no real structure.

2.1.42 We have two non-equivalent real structures on $\mathbb{P}^1(\mathbb{C})$:

$$\sigma_{\mathbb{P}}\colon (x_0 : x_1) \mapsto (\overline{x_0} : \overline{x_1})$$

et

$$\sigma_{\mathbb{P}}'\colon (x_0 : x_1) \mapsto (-\overline{x_1} : \overline{x_0})$$

which give rise to three non-equivalent structures on $\mathbb{P}^1(\mathbb{C}) \times \mathbb{P}^1(\mathbb{C})$: the involution $\sigma_{\mathbb{P}} \times \sigma_{\mathbb{P}}$ whose fixed locus is the torus $\mathbb{T}^2 = \mathbb{S}^1 \times \mathbb{S}^1$ and the involutions $\sigma_{\mathbb{P}} \times \sigma_{\mathbb{P}}'$ and $\sigma_{\mathbb{P}}' \times \sigma_{\mathbb{P}}'$ whose fixed loci are empty.

The fourth structure is $((x : y), (z : t)) \mapsto \left((\overline{z} : \overline{t}), (\overline{x} : \overline{y})\right)$ whose fixed locus is the sphere $\mathbb{S}^2$.

---

[9] As an automorphism of C is also a birational transformation of C we simply apply Theorem 1.3.30 which states there is a one-to-one correspondence between automorphisms of $\mathbb{C}(C)$ and birational transformations of C. The stronger correspondence used in this proof relies on the fact that C is a smooth projective curve.

2.2.6 1. We have that $F(\mathbb{R}) = \mathcal{Z}(I)$ and $\overline{F(\mathbb{R})} = \mathcal{Z}_{\mathbb{C}}(\mathcal{I}(F(\mathbb{R})))$.

If $\mathcal{I}(\mathcal{Z}(I)) \subseteq I$ then $\mathcal{Z}_{\mathbb{C}}(\mathcal{I}(F(\mathbb{R}))) \supseteq \mathcal{Z}_{\mathbb{C}}(I)$ or in other words $\overline{F(\mathbb{R})} \supseteq F$ so $F(\mathbb{R})$ is dense in F.

If $F(\mathbb{R})$ is dense in F then $\mathcal{Z}_{\mathbb{C}}(\mathcal{I}(F(\mathbb{R}))) = F = \mathcal{Z}_{\mathbb{C}}(I)$. As the ideal I is radical the ideal $I_{\mathbb{C}} = I \otimes_{\mathbb{R}[X_1,\ldots,X_n]} \mathbb{C}[X_1, \ldots, X_n]$ is also radical. It follows by the Nullstellensatz that $\mathcal{I}_{\mathbb{C}}(F(\mathbb{R})) \subseteq I_{\mathbb{C}}$ and hence $\mathcal{I}(F(\mathbb{R})) \subseteq I$.

2. This follows immediately from (1) using Theorem A.5.15.

2.2.7 Set $I = (x^2 + y^2)$: we then have that $F = \mathcal{Z}_{\mathbb{C}}(I) = \{x \pm iy = 0\}$ and the real locus is $F(\mathbb{R}) = \mathcal{Z}(I) = \{(0, 0)\}$ and $\mathcal{I}(\mathcal{Z}(I)) = (x, y) \subsetneq I$ in $\mathbb{R}[X_1, \ldots, X_n]$.

We set $a = (0, 0)$. On the one hand, $\mathcal{O}_{F(\mathbb{R}),a} = \left(\frac{\mathbb{R}[x,y]}{(x,y)}\right)_{\mathfrak{m}_{F(\mathbb{R}),a}} = \mathbb{R}$ and on the other hand $\left(\mathcal{O}_F^G|_{F(\mathbb{R})}\right)_a = \mathcal{O}_{F,a}^G = \left(\left(\frac{\mathbb{C}[x,y]}{(x^2+y^2)}\right)_{\mathfrak{m}_{F,a}}\right)^G \supsetneq \mathbb{R}$ since the class of the polynomial x modulo $(x^2 + y^2)$ belongs to $\mathcal{O}_{F,a}^G$ since its coefficients are real.

2.2.26 1. φ is a morphism of $\mathbb{R}$-varieties if and only if

- φ is an morphism of complex varieties and
- $\varphi \circ \sigma_{\mathbb{A}}|_{F_1} = \sigma_{\mathbb{A}}|_{F_2} \circ \varphi$.

By Exercise 1.2.56 the first condition is equivalent to the existence of polynomial functions $f_1, \ldots, f_m \in \mathbb{C}[x_1, \ldots, x_n]$ such that for every $(x_1, \ldots, x_n) \in F_1$, $\varphi(x_1, \ldots, x_n) = (f_1(x_1, \ldots, x_n), \ldots, f_m(x_1, \ldots, x_n))$. The second condition is equivalent to

$$\varphi\left(\overline{x_1}, \ldots, \overline{x_n}\right) = \overline{\varphi(x_1, \ldots, x_n)}\ ,$$

which simply means that for every $(x_1, \ldots, x_n) \in F_1$ and every $i = 1 \ldots m$,

$$f_i\left(\overline{x_1}, \ldots, \overline{x_n}\right) = \overline{f_i(x_1, \ldots, x_n)}\ .$$

i.e. for every $i = 1 \ldots m$, ${}^{\sigma} f_i = f_i$ or in other words f_i has real coefficients.

2. φ is an $\mathbb{R}$-regular rational map if and only if

- φ is a rational map of $\mathbb{R}$-varieties;
- $F_1(\mathbb{R}) \subset \mathrm{dom}(\varphi)$.

In other words, φ is an $\mathbb{R}$-regular rational map if and only if

- φ is a rational map of complex varieties
- $\varphi \circ \sigma_{\mathbb{A}}|_{F_1} = \sigma_{\mathbb{A}}|_{F_2} \circ \varphi$;
- $F_1(\mathbb{R}) \subset \mathrm{dom}(\varphi)$.

By Exercise 1.3.25, the first condition is equivalent to the existence of polynomial functions $g_1, \ldots, g_m \in \mathbb{C}[x_1, \ldots, x_n]$ and $h_1, \ldots, h_m \in \mathbb{C}[x_1, \ldots, x_n]$ such that for any $(x_1, \ldots, x_n) \in \mathrm{dom}(\varphi)$,

$$\varphi(x_1, \ldots, x_n) = \left(\frac{g_1(x_1, \ldots, x_n)}{h_1(x_1, \ldots, x_n)}, \ldots, \frac{g_m(x_1, \ldots, x_n)}{h_m(x_1, \ldots, x_n)}\right)\ .$$

The map φ is therefore an $\mathbb{R}$-regular rational map if and only if g_i and h_i have real coefficients and the functions h_i do not vanish at any point of $F_1(\mathbb{R})$.

2.2.31 The usual atlas is a compatible atlas because the functions defining the open sets have real coefficients. We set

$$U_0 := \{(x_0 : x_1) \in \mathbb{P}^1(\mathbb{C}) \mid x_0 \neq 0\}$$

and

$$\varphi_0 \colon \begin{cases} U_0 & \longrightarrow \mathbb{C} \\ (x_0 : x_1) & \longmapsto \frac{x_1}{x_0} \end{cases}.$$

Similarly, set $U_1 := \{(x_0 : x_1) \in \mathbb{P}^1(\mathbb{C}) \mid x_1 \neq 0\}$ and

$$\varphi_1 \colon \begin{cases} U_1 & \longrightarrow \mathbb{C} \\ (x_0 : x_1) & \longmapsto \frac{x_0}{x_1} \end{cases}.$$

We then have that

$$^{\sigma}\varphi_0 \colon \begin{cases} \sigma(U_0) & \stackrel{\sigma_{\mathbb{P}}}{\to} & U_0 & \stackrel{\varphi_0}{\to} & \mathbb{C} & \stackrel{\sigma_{\mathbb{A}}}{\to} & \mathbb{C} \\ (x_0 : x_1) & \longmapsto & (\overline{x_0} : \overline{x_1}) & \longmapsto & \frac{\overline{x_1}}{\overline{x_0}} & \longmapsto & \frac{x_1}{x_0} \end{cases}$$

and

$$^{\sigma}\varphi_1 \colon \begin{cases} U_1 & \longrightarrow \mathbb{C} \\ (x_0 : x_1) & \longmapsto \frac{x_0}{x_1} \end{cases}.$$

2.3.14 Use Exercise 1.2.56(3) to write the isomorphism

$$\varphi' \circ \varphi^{-1} \colon \varphi'(V) \to \varphi(V)$$

in homogeneous coordinates then check that $\varphi' \circ \varphi^{-1}$ extends to an isomorphism $\varphi(V)_{\overline{K}} \to \varphi'(V)_{\overline{K}}$.

2.3.17 1. $\mathcal{I}(F) = (x, y)$ so $F_{\mathbb{C}} = \{(0, 0\}$ is a complexification of F which is irreducible so F is geometrically irreducible.

2. $V = \mathcal{Z}_{\mathbb{C}}(x + iy) \cup \mathcal{Z}_{\mathbb{C}}(x - iy)$.

3. The $\mathbb{R}$-variety (V, σ) does not have enough real points so it is not a complexification of F.

2.6.15 See [Ser55a, Chapitre III, Section 2] if necessary.

2.6.27 To simplify notation we will prove this result only for $n = 2$. Take a system of linear homogeneous coordinates $(x_0 : x_1 : x_2)$ and let $U_k := \mathbb{P}^2 \setminus \mathcal{Z}(x_k)$ be the standard open affine set defined by $x_k \neq 0$. Consider U_0 with its coordinates u_1, u_2. Sections of $\mathcal{K}_{\mathbb{P}^2}$ on U_0 are all of the form $p(u_1, u_2)\, du_1 \wedge du_2$. We will calculate the poles and zeros of the section $du_1 \wedge du_2$ outside of U_0. There is only one divisor outside of U_0, namely $x_0 = 0$, so it is enough to check the multiplicity along this divisor. We will calculate in U_1 with coordinates v_0, v_2 such that $(1 : u_1 : u_2) = (v_0 : 1 : v_2)$. In other words, $u_1 = \frac{1}{v_0}$ and $u_2 = \frac{v_2}{v_0}$, from which we get that

$$du_1 \wedge du_2 = \left(-\frac{1}{v_0^2}\,dv_0\right) \wedge \left(\frac{v_0\,dv_2 - v_2\,dv_0}{v_0^2}\right) = -\frac{1}{v_0^3}\,dv_0 \wedge dv_2\ .$$

This form therefore has a pole of order 3 along $v_0 = 0$ as claimed.

2.6.28 Since $H^0(X, \mathcal{O}_X(-K_X)) \neq 0$, there is an *effective* divisor C linearly equivalent to $-K_X$.

There is an exact sequence

$$0 \to \mathcal{O}_X(-C) \to \mathcal{O}_X \to \mathcal{O}_C \to 0$$

which on tensorising with Ω^1_X gives us

$$0 \to \Omega^1_X(K_X) \to \Omega^1_X \to \Omega^1_X|_C \to 0$$

whose initial terms in the long exact sequence are

$$0 \to H^0(X, \Omega^1_X(K_X)) \to H^0(X, \Omega^1_X) \to \cdots$$

and the conclusion follows because $H^0(X, \Omega^1_X) = 0$.

For the second question simply note that Θ_X is the dual of Ω^1_X and apply Theorem D.2.5.

Chapter 3
Topology of Varieties with an Involution

Equipped with the Euclidean topology, an $\mathbb{R}$-variety (X, σ) is a topological space with a continuous involution. In this chapter we study the action of this involution on the homology of the topological space X.

We start with preliminary results on involutive modules, Poincaré duality and characteristic classes and then present Smith theory and its applications to $\mathbb{R}$-varieties. The main consequences of this theory are constraints on the topology of the real locus depending on the topology of the complex variety. Most of the time these constraints take the form of upper and lower bounds on various topological invariants, such as the Smith-Thom (3.8), Harnack (3.9), Petrovskii (3.11) and (3.12), Comessatti (3.14) inequalities.

If the variety X is non singular then X and $X(\mathbb{R})$ are topological and differentiable manifolds. We recall that- as in Definition 1.4.1 and Proposition 2.2.27—any non singular n-dimensional complex variety is also a differentiable manifold of dimension $\dim_{\mathbb{R}} X = 2n$ and its real locus is a differentiable manifold of dimension $\dim_{\mathbb{R}} X(\mathbb{R}) = n$ if $X(\mathbb{R}) \neq \varnothing$. This enables us to apply topological tools such as Poincaré duality and characteristic classes which yield various constraints on the topological invariants of these spaces expressed in the form of congruences, such as the results due to Rokhlin (3.15) and Gudkov–Kharlamov–Krakhnov (3.16).

We then turn our attention on $\mathbb{R}$-curves, especially plane curves, which leads to a discussion of the first part of Hilbert's famous sixteenth problem.

In the following section we consider the Galois cohomology of X's homology and define the various different forms of *Galois-maximality*. In many cases this method will enable us to calculate the homology of the real locus using the Galois group action on the homology of the complex variety. This method yields preciser bounds than those obtained in previous sections.

We end this chapter with a discussion of *algebraic cycles*, by which we mean homology classes represented by algebraic subvarieties.

F. Mangolte, *Real Algebraic Varieties*, Springer Monographs in Mathematics,
https://doi.org/10.1007/978-3-030-43104-4_3

3.1 Homology and Cohomology of $\mathbb{R}$-Varieties

Unless otherwise stated, the homology and cohomology used here will always be *singular* homology. Of course when dealing with the underlying topological space of a differentiable manifold or a real or complex quasi-projective variety with its Euclidean topology, we determine singular homology by calculating simplicial homology—see Remark B.3.3 and [Hat02, Section 2.1]. The homology groups of a *compact* topological or differentiable manifold are finitely generated, as are those of a *projective* real or complex algebraic variety—see [Hat02, Corollary A.8] for more details. We denote by $H_k(X, L; A)$ the kth homology group and by $H^k(X, L; A)$ the kth cohomology group of the pair (X, L) with values in an abelian group A: typically A will be a ring or a field such as $A = \mathbb{Z}_2, \mathbb{Z}, \mathbb{Q}, \mathbb{R}$ or $\mathbb{C}$. If $L = \varnothing$ then we will write "X" rather than "$X, \varnothing$". See the appendix, Section B.3 for basic homology and cohomology theory.

As is standard practice, we will denote by

$$\mathbb{Z}_m := \mathbb{Z}/m\mathbb{Z}$$

the cyclic group of order $m > 1$.(Be careful not to confuse this notation with the profinite group $\mathbb{Z}_p = \varprojlim \mathbb{Z}/p^n\mathbb{Z}$, where p is a fixed prime number and n runs over all natural numbers.)

3.1.1 Involutive Modules

An *involutive module* is a pair (M, σ) where M is a $\mathbb{Z}$-module with a linear involution σ. Any involutive module is a G*-group* where G is the group $\{1, \sigma\} \simeq \mathbb{Z}_2$. Indeed, any $\mathbb{Z}$-module is an abelian group and any abelian group is equipped with a unique $\mathbb{Z}$-module structure. We denote by M^G or M^σ the submodule of M of elements which are invariant under σ and by $M^{-\sigma}$ the submodule of anti-invariant elements.

Lemma 3.1.1 *Let M be a free $\mathbb{Z}$-module of finite rank n equipped with a linear involution σ. There is then a basis of M*

$$(a_1, \dots, a_r, b_1, \dots, b_\lambda, c_{\lambda+1}, \dots, c_{n-r})$$

such that

1. *for any $1 \leqslant i \leqslant r$, $\sigma(a_i) = a_i$;*
2. *for any $1 \leqslant i \leqslant \lambda$, $\sigma(b_i) = a_i - b_i$;*
3. *for any $\lambda \leqslant i \leqslant n - r$, $\sigma(c_i) = -c_i$.*

In other words, M can be decomposed as a direct sum

$$M_1 \oplus M_2 \oplus B_1 \oplus \cdots \oplus B_\lambda$$

where $\sigma|_{M_1} = \mathrm{id}_{M_1}$, $\sigma|_{M_2} = -\,\mathrm{id}_{M_2}$ *and* $\sigma|_{B_i}$ *has a matrix of the form* $\begin{pmatrix} 0 & 1 \\ 1 & 0 \end{pmatrix}$.

Proof By convention[1] we denote by $1 = \sigma \circ \sigma$ the identity map on M. The invariant submodule $\ker(1-\sigma)$ is a direct factor of M and the morphism induced by $1+\sigma$ onto $M/\ker(1-\sigma)$ is identically zero because $(1-\sigma)\circ(1+\sigma) \equiv 0$. In other words, the map induced by σ on $M/\ker(1-\sigma)$ is $-\,\mathrm{id}$. In any basis for M extending a basis for $\ker(1-\sigma)$ the matrix of σ is therefore of the form

$$\begin{pmatrix} \mathrm{I}_r & N \\ 0 & -\mathrm{I}_{n-r} \end{pmatrix}.$$

Using matrices of the form $\begin{pmatrix} \mathrm{I}_r & B \\ 0 & C \end{pmatrix}$ to make base changes we see that we can reduce N modulo 2 and replace N by any matrix of the form NC for some invertible matrix C. This completes the proof of the lemma.

To prove the second part of the lemma, simply consider the basis given by

$$(a_1 - b_1, \ldots, a_\lambda - b_\lambda, b_1, \ldots, b_\lambda)$$

of the submodule generated by $(a_1, \ldots, a_\lambda, b_1, \ldots, b_\lambda)$ and reorganise terms. □

Note that the integer λ appearing in the above lemma corresponds to the dimension $\lambda := \lambda_\sigma$ of the $\mathbb{Z}_2$-vector space $(1+\sigma)(M \otimes_{\mathbb{Z}} \mathbb{Z}_2)$ (See Appendix A.4 for the definition of the tensor product $\otimes$). It is therefore an invariant of the involutive module (M, σ). Similarly, the rank $r := r_\sigma = \mathrm{rk}\, M^\sigma$ (which is also equal to $\mathrm{rk}\, M_1 + \lambda_\sigma$) of the invariant submodule M^σ is independant of the choice of basis of M. We therefore have the following proposition.

Proposition 3.1.2 *Let (M, σ) and (N, τ) be free finitely generated involutive $\mathbb{Z}$-modules and let $(M, \sigma) \to (N, \tau)$ be a G-equivariant isomorphism (which is another way of saying that $(M, \sigma) \to (N, \tau)$ is an isomorphism of involutive modules). We then have that $\lambda_\sigma = \lambda_\tau$ and $r_\sigma = r_\tau$.*

Definition 3.1.3 Let M be a free $\mathbb{Z}$-module of finite rank n with a linear involution σ. We define the *Comessatti characteristic*[2] of (M, σ) to be the dimension $\lambda := \lambda_\sigma$ of the $\mathbb{Z}_2$-vector space $(1+\sigma)(M \otimes_{\mathbb{Z}} \mathbb{Z}_2)$.

As M is an *abelian* G-group the Galois cohomology sets $H^k(G, M)$ are abelian groups (indeed, they turn out to be $\mathbb{Z}_2$ vector spaces). See [Ser94, Section I.5] for a general definition of the cohomology groups $H^k(G, M)$.

---

[1] This convention derives from the fact that the identity is the multiplicative unit in the group algebra $\mathbb{Z}[G]$.

[2] Terminology due to Silhol [Sil89, I.(3.5.1), page 15].

Proposition 3.1.4 *Let (M, σ) be an involutive module and set $G = \{1, \sigma\}$. For any $k > 0$ we have that $H^k(G, M) \simeq H^{k+2}(G, M)$ and*

$$\begin{aligned} H^0(G, M) &= \ker(1-\sigma) &&= M^\sigma \ ; \\ H^1(G, M) &= \ker(1+\sigma)/\operatorname{Im}(1-\sigma) &&= M^{-\sigma}/\operatorname{Im}(1-\sigma) \ ; \\ H^2(G, M) &= \ker(1-\sigma)/\operatorname{Im}(1+\sigma) &&= M^\sigma/\operatorname{Im}(1+\sigma) \ . \end{aligned}$$

Proof See [Wei94, Theorem 6.2.2], noting that the *norm* on $\mathbb{Z}[G]$ is the element $1+\sigma$. □

Lemma 3.1.5 *Let E be a $\mathbb{Z}_2$-vector space equipped with a linear involution σ.*[3] *We then have that $E^\sigma = E^{-\sigma}$ and $(1+\sigma)E = (1-\sigma)E$ from which it follows that*

$$H^1(G, E) = H^2(G, E)$$

and setting $\lambda_\sigma = \dim_{\mathbb{Z}_2}(1+\sigma)E$ we have that

$$\begin{aligned} \dim_{\mathbb{Z}_2} E^G &= \dim_{\mathbb{Z}_2} E - \lambda_\sigma \ ; \\ \dim_{\mathbb{Z}_2} H^1(G, E) &= \dim_{\mathbb{Z}_2} E - 2\lambda_\sigma \ . \end{aligned}$$

Proof Simply note that $E^G = \ker(1+\sigma)$. □

Proposition 3.1.6 *Let M be a* free *$\mathbb{Z}$-module of rank n with a G-action. The group G also acts on the $\mathbb{Z}_2$-vector space $M_2 = M \otimes_{\mathbb{Z}} \mathbb{Z}_2 = M/2M$ and on setting $r = \operatorname{rk} M^\sigma$ and $\lambda = \dim_{\mathbb{Z}_2}(1+\sigma)M_2$ we get that*

$$\begin{aligned} \dim_{\mathbb{Z}_2} H^1(G, M) &= \operatorname{rk} M^{-\sigma} - \lambda = n - r - \lambda \ ; \\ \dim_{\mathbb{Z}_2} H^2(G, M) &= r - \lambda \ ; \\ \dim_{\mathbb{Z}_2} M_2^\sigma &= n - \lambda \ ; \\ \dim_{\mathbb{Z}_2} H^1(G, M_2) &= n - 2\lambda \ . \end{aligned}$$

Proof Simply apply Lemmas 3.1.1 and 3.1.5. □

We conclude this subsection with a useful result on involutive integral lattices, by which we mean free $\mathbb{Z}$-modules of finite rank n equipped with an integer-valued symmetric bilinear form (see Definition A.6.5) and an involution.

Proposition 3.1.7 *Let (M, Q) be an integral quadratic lattice with an involutive isometry σ. The discriminant of the restriction of Q to the invariant (resp. anti-invariant) part of M satisfies*

$$|\det(Q|_{M^\sigma})| = |\det(Q|_{M^{-\sigma}})| = 2^\lambda \ .$$

[3]The space E has a unique $\mathbb{Z}$-module structure obtained by composing the $\mathbb{Z}_2$ action with the unique ring morphism $\mathbb{Z} \to \mathbb{Z}_2$.

Proof Applying Lemma A.6.9 to the invariant submodule gives us $|\det(Q|_{M^\sigma})| = |\det(Q|_{M^{-\sigma}})| = [M : M^\sigma \oplus M^{-\sigma}]$. Lemma 3.1.1 then completes the proof of the proposition. □

3.1.2 Poincaré Duality on $\mathbb{R}$-Varieties

Poincaré duality holds for all topological manifolds—see Definition B.5.1—and therefore for all non singular $\mathbb{R}$-varieties in particular. Here we will deal with varieties whose underlying Euclidean topological space is compact: see Theorem B.7.1 for the non-compact case.

Proposition 3.1.8 *Let (M, σ) be an oriented compact topological manifold of dimension n equipped with an orientation-preserving (resp. reversing) involution. Consider an integer $k \in \{0, .., n\}$. The Poincaré duality isomorphism (Corollary B.7.2)*

$$\begin{aligned} D_M \colon H^k(M; \mathbb{Z}) &\xrightarrow{\cong} H_{n-k}(M; \mathbb{Z}) \\ \phi &\longmapsto [M] \frown \phi \end{aligned}$$

is then equivariant (resp. anti-equivariant) for the $G = \mathbb{Z}_2$ action determined by σ.

Proof Simply apply the fact that the cap-product is natural (Proposition B.7.5) to the continuous map $\sigma : M \to M$ for $l = n$ and $\alpha = [M]$:

$$\sigma_*([M]) \frown \phi = \sigma_*\left([M] \frown \sigma^*(\phi)\right) .$$

As the linear map σ_* is involutive it follows that

$$\sigma_*(D_M(\phi)) = D_M(\sigma^*(\phi)) \quad (\text{resp.} \quad \sigma_*(D_M(\phi)) = -D_M(\sigma^*(\phi)))$$

if $\sigma_*([M]) = [M]$ (resp. $\sigma_*([M]) = -[M]$). □

Corollary 3.1.9 *Let (X, σ) be a non singular projective $\mathbb{R}$-variety (or compact analytic $\mathbb{R}$-variety) of dimension n. The Poincaré duality isomorphism,*

$$\begin{aligned} D_X \colon H^k(X; \mathbb{Z}) &\xrightarrow{\cong} H_{2n-k}(X; \mathbb{Z}) \\ \phi &\longmapsto [X] \frown \phi \end{aligned}$$

is then equivariant if n is even and anti-equivariant if n is odd.

Proof By Proposition 2.2.27 the complex variety X with its Euclidean topology has an oriented real differentiable manifold structure of dimension $2n$. By Proposition 2.2.28 the real structure σ is orientation preserving if n is even and orientation reversing if n is odd. □

3.1.3 Orientability and Characteristic Classes

A complex variety is always orientable and oriented—see Exercise B.5.11 and Remark E.2.2(4)- which is not always the case for a real variety. In this subsection we will prove some results concerning the orientability of the real locus of an $\mathbb{R}$-variety. The first non-trivial case is that of surfaces because any non-singular $\mathbb{R}$-curve has real dimension 1 and is therefore orientable (see Remark B.5.5).

We start our investigations with non singular hypersurfaces in $\mathbb{P}^3$.

Proposition 3.1.10 *Let $X_d \subset \mathbb{P}^3(\mathbb{C})$ be a non singular real algebraic surface defined by a polynomial of degree d with real coefficients. Assume that $X_d(\mathbb{R}) = X_d \cap \mathbb{P}^3(\mathbb{R})$ is non empty. The compact topological surface $X_d(\mathbb{R})$ is then orientable if and only if d is even.*

Proof By Poincaré duality (see Proposition B.7.17), a topological surface $V \subset \mathbb{P}^3(\mathbb{R})$ is orientable if and only if its homology class $[V] \in H_2(\mathbb{P}^3(\mathbb{R}); \mathbb{Z}_2)$ vanishes. The group $H_2(\mathbb{P}^3(\mathbb{R}); \mathbb{Z}_2)$ is generated by the class of a real hyperplane $H \subset \mathbb{P}^3(\mathbb{R})$. The class $[X_d(\mathbb{R})] = d[H]$ therefore vanishes in $H_2(\mathbb{P}^3(\mathbb{R}); \mathbb{Z}_2)$ if and only if d is even. □

Remark 3.1.11 The group $H_4(\mathbb{P}^3(\mathbb{C}); \mathbb{Z})$ is generated by the class of a complex hyperplane $H \subset \mathbb{P}^3(\mathbb{C})$ so we have that $[X_d] = d[H]$ in $H_4(\mathbb{P}^3(\mathbb{C}); \mathbb{Z})$. This makes it tempting to reason using the complex variety: this idea is illustrated in Example 3.1.17. See also Section 3.7.

When the degree d is odd we have a stronger result.

Proposition 3.1.12 *Let $X_d \subset \mathbb{P}^3(\mathbb{C})$ be a non singular algebraic surface defined by a polynomial of degree d with real coefficients. If d is odd then $X_d(\mathbb{R}) \neq \varnothing$ and has a unique non-orientable connected component: there may be other components which are orientable. Moreover, this unique non orientable connected component has odd Euler characteristic.*

Proof 1. By Proposition 2.6.48, we have that $X_d(\mathbb{R}) \neq \varnothing$.

2. By Proposition 3.1.10 the real locus $X_d(\mathbb{R})$ has at least one non-orientable connected component since a line in $\mathbb{P}^3(\mathbb{R})$ transverse to $X_d(\mathbb{R})$ meets it in a odd number of points, namely d.
3. There cannot be any other non orientable component because in $\mathbb{P}^3(\mathbb{R})$ any two non orientable surfaces must meet. Let H be a plane which is transverse to $X_d(\mathbb{R})$ (such a plane exists by Bertini's theorem D.9.1)- the curve cut out on H by each non orientable component contains a pseudo-line, but any two pseudo-lines in $H \simeq \mathbb{RP}^2$ always meet. See [BR90, 5.1.6] for more details if necessary.
4. The statement about the Euler characteristic follows from the fact that any non-orientable surface in $\mathbb{RP}^3$ is cobordant with $\mathbb{RP}^2$ (see [BW69]) so is necessarily diffeomorphic to the connected sum of $\mathbb{RP}^2$ with a finite number of Klein bottles $\mathbb{K}^2$. We recall that the connected sum $\mathbb{V}\#\mathbb{K}^2$ of a non orientable surface $\mathbb{V}$ with a Klein bottle is homeomorphic to the connected sum $\mathbb{V}\#\mathbb{T}^2$ of $\mathbb{V}$ with a torus). □

Example 3.1.13 A non singular cubic surface $X_3(\mathbb{R})$ in $\mathbb{RP}^3$ is homeomorphic to one of the following surfaces. (In this list, $\mathbb{V}_g$ denotes the non-orientable surface whose topological Euler characteristic is $2-g$ and $\sqcup$ denotes disjoint union as in notation 4.2.15)

$$\mathbb{V}_1 = \mathbb{RP}^2, \quad \mathbb{V}_3, \quad \mathbb{V}_5, \quad \mathbb{V}_7, \quad \mathbb{RP}^2 \sqcup \mathbb{S}^2 .$$

See [BR90, Proposition 5.6.4] for an elementary proof of this fact. Anticipating Chapter 4, we can use the fact that a non singular cubic in $\mathbb{P}^3(\mathbb{C})$ is a Del Pezzo surface of degree 3, which implies that the complex surface X_3 is isomorphic to the blow up of $\mathbb{P}^2(\mathbb{C})$ in 6 points in general position, by which we mean that they do not all lie on one conic (they are not *coconic*) and no three of them are on the same line. Consider the case where this set of 6 points is globally fixed by $\sigma_{\mathbb{P}}$. The number of these points in non real conjugate pairs can be $0, 2, 4$ or 6, giving the first 4 possibilities. (See Example 4.2.18 for the calculation of the topology of a blow-up at a point). The last topological type can be realised by blowing up a real point on a conic bundle over $\mathbb{P}^1$ which is $\mathbb{R}$-minimal and has four singular fibres as in Example 4.2.8. The real locus of such a bundle is a disjoint union of two spheres which after blow up gives us the desired topological form. To be sure that this complex surface really is the blow up of $\mathbb{P}^2$ in 6 points we need the minimal model, which is a Hirzebruch surface by Exercise 4.2.11, to be of index 1. To prove this, recall that every singular fibre consists of two non real conjugate (-1)-curves, which gives us four contractions, to which we add our extra blow-up and the contraction of the exceptional section which is a (-1)-curve by hypothesis. See also [Sil89, Section VI.5].

Returning to abstract surfaces, the differentiable manifold structure inherited from the non singular real or complex algebraic structure means we can use the characteristic classes of the tangent bundle. See [MS74, Sections 4 et 14] for the construction and main properties of these classes.

Definition 3.1.14 Let (X, σ) be a non singular $\mathbb{R}$-variety of dimension n. We denote by T_X the differential tangent bundle of the $2n$-dimensional manifold underlying X and if $X(\mathbb{R}) \neq \varnothing$ we denote by $T_{X(\mathbb{R})}$ the differential tangent bundle of the n dimensional manifold underlying $X(\mathbb{R})$. The bundle $T_{X(\mathbb{R})}$ is a real vector bundle of rank n and its kth Stiefel–Whitney class

$$\mathrm{w}_k(X(\mathbb{R})) := \mathrm{w}_k(T_{X(\mathbb{R})}) \in H^k(X(\mathbb{R}); \mathbb{Z}_2)$$

is called the kth *Stiefel–Whitney class of* $X(\mathbb{R})$.

The bundle T_X is a real vector bundle of rank $2n$ and its kth Stiefel–Whitney class

$$\mathrm{w}_k(X) := \mathrm{w}_k(T_X) \in H^k(X; \mathbb{Z}_2)$$

is called the kth *Stiefel–Whitney class of* X. The bundle T_X has a natural rank n complex vector bundle structure and its kth Chern class

$$c_k(X) := c_k(T_X) \in H^{2k}(X; \mathbb{Z})$$

is called the kth *Chern class of X*.

The first Stiefel–Whitney class $\mathrm{w}_1(V)$ of a differentiable compact manifold V vanishes if and only if V is orientable. See [MS74, Problem 12.A] for more details. The vanishing of $\mathrm{w}_1(X(\mathbb{R}))$ therefore detects the orientability of the real locus $X(\mathbb{R})$: on the other hand, the first Stiefel–Whitney class of X always vanishes because X is orientable. The key result is therefore the following.

Proposition 3.1.15 *Let (X, σ) be a non singular $\mathbb{R}$-variety whose real locus is non empty. The variety $X(\mathbb{R})$ is then orientable if and only if*

$$w_1(X(\mathbb{R})) = 0 \quad in \quad H^1(X(\mathbb{R}); \mathbb{Z}_2) .$$

It is not always easy to calculate the characteristic classes of the real locus but we can sometimes use the characteristic classes of the complex variety.

Proposition 3.1.16 *Let X be a non singular complex projective variety of dimension n. The Stiefel–Whitney and Chern classes of X satisfy the following relationships.*

1. $\mathrm{w}_{2k+1}(X) = 0$;
2. $\mathrm{w}_{2k}(X) \equiv c_k(X) \mod 2$.

Moreover, if σ is a real structure on X then

3. $\sigma^* c_k(X) = (-1)^k c_k(X)$;
4. $\sigma^* \mathrm{w}_{2k}(X) = \mathrm{w}_{2k}(X)$.

Proof The first two equations are proved in [MS74, Problem 14.B]. For the third, first note that the image under σ of the tangent bundle T_X is isomorphic to the conjugate bundle $\overline{T_X}$ and then apply [MS74, Lemma 14.9]. The final equation is a consequence of the first three. □

Example 3.1.17 Let $X_d \subset \mathbb{P}^3(\mathbb{C})$ be a non singular algebraic surface of degree d. We then have that $\mathrm{w}_2(X_d) = 0$ if and only if d is even. Indeed, $\mathrm{w}_2(X) \equiv c_1(X)$ mod 2 and $c_1(X) = c_1(dH) = dc_1(H)$ for some hyperplane $H \subset \mathbb{P}^3(\mathbb{C})$.

Theorem 3.1.18 *Let (X, σ) be a non singular $\mathbb{R}$-variety of even dimension $n = 2m$ with non empty real locus such that $b_1(X; \mathbb{Z}_2) = 0$.*

If $\mathrm{w}_2(X) \in H^2(X; \mathbb{Z}_2)$, the second Stiefel–Whitney class of X, vanishes then $X(\mathbb{R})$ is orientable.

Remark 3.1.19 This theorem is particularly useful when the variety X is simply connected.

Proof of Theorem 3.1.18 The proof below draws on [DK00, 2.9.1 Remark, page 753]. The basic idea is to apply a result due to Edmonds.

Lemma 3.1.20 ([Edm81, Theorem 3]) *Let V be an oriented manifold with a* spin structure *(see Section B.5.2) equipped with a $\mathcal{C}^\infty$ involution σ fixing both the orientation and a spin structure. If the stable locus V^σ is non empty then it is orientable.*

The complex variety X with the Euclidean topology is an oriented manifold—see Proposition 2.2.27. An oriented differentiable manifold V has a spin structure if and only if $\mathrm{w}_2(T_V) = 0$, as in Proposition B.5.20. By [LM89, Chapter II, Theorem 2.1], as $b_1(X; \mathbb{Z}_2) = 0$, this variety has only one spin structure which must therefore be fixed by σ. As n is even, σ is orientation preserving by Proposition 2.2.28. The proposition follows on applying Edmond's theorem.

Example 3.1.21 The real locus of a real K3 surface (see Definition 4.5.3) is either empty or orientable. See the proof of Proposition 4.5.6 for more details.

3.2 Smith Theory

As in [Bre72, Chapitre III] we present a version of Smith theory based on simplicial homology which in principle is only valid for triangulable topological spaces—see Definition B.3.2. This class contains all differentiable manifolds and all complex or real quasi-projective varieties with their Euclidean topology—see Remark B.3.3.

Let (X, σ) be a projective $\mathbb{R}$-variety of dimension n. We equip the complex variety X and the real locus $X(\mathbb{R})$ with their Euclidean topology so that they become compact triangulable topological spaces as in Remark B.3.3. We do not assume that X is non singular: we do however assume it is projective, which guarantees it is compact for the Euclidean topology and that all its homology groups are finitely generated.

Remark 3.2.1 Most of the results given in this section remain valid for a compact complex analytic space with an anti-holomorphic involution. In particular, the following results hold for any compact Kähler variety X—see Definition D.3.4—equipped with an anti-holomorphic involution σ. This generalisation will be useful in the study of real K3 surfaces in Chapter 4.

The involution σ equips X with a Galois group action where $G = \mathrm{Gal}(\mathbb{C}|\mathbb{R}) \simeq \mathbb{Z}_2$. Let $L \subset X$ be a subvariety that is stable under σ (or in other words an $\mathbb{R}$-subvariety). A triangulation of the pair (X, L) is a simplicial pair $(\widetilde{X}, \widetilde{L})$—see Definition B.3.1—equipped with a G-action. In particular, σ acts simplicially on $\widetilde{X}$. As X is compact, $\widetilde{X}$ is a finite simplicial complex. Passing to a barycentric subdivision if necessary, we may assume that if σ fixes a simplex s of $\widetilde{X}$ then it fixes all the vertices of s as in [Bre72, Proposition III.1.1]. We denote by $\widetilde{X}^G$ the subcomplex fixed by σ and set $\widetilde{L}^G = \widetilde{L} \cap \widetilde{X}^G$. The complex $\widetilde{X}^G$ (resp. $\widetilde{L}^G$) is then a triangulation of $X(\mathbb{R})$ (resp. $L(\mathbb{R})$). We denote by $C(\widetilde{X}, \widetilde{L}; \mathbb{Z}_2)$ the chain group of the pair $(\widetilde{X}, \widetilde{L})$ with coefficients in $\mathbb{Z}_2$.

Theorem 3.2.2 *The sequence*

$$0 \to \rho C(\widetilde{X}, \widetilde{L}; \mathbb{Z}_2) \oplus C(\widetilde{X}^G, \widetilde{L}^G; \mathbb{Z}_2) \xrightarrow{i} \qquad (3.1)$$
$$C(\widetilde{X}, \widetilde{L}; \mathbb{Z}_2) \xrightarrow{\rho} \rho C(\widetilde{X}, \widetilde{L}; \mathbb{Z}_2) \to 0$$

where i is the sum of canonical injections and $\rho = 1 + \sigma$, is an exact sequence of chain complexes (we recall that $1 = \sigma \circ \sigma$ is the identity map).

Proof The second to last arrow is obviously surjective and if s is a simplex fixed by σ then $\rho(s) = 2s = 0$ which implies that i is injective. It remains to prove that the sequence is exact at $C(\widetilde{X}, \widetilde{L}; \mathbb{Z}_2)$. For every n we simply consider the set of n-chains in the orbit of s for every n-simplex $s \subset \widetilde{X} \setminus \widetilde{L}$. It is clear that $\rho \circ \rho = 2(1+\sigma) = 0$ so $\operatorname{Im} i \subset \ker \rho$. Consider an element $s \in \ker \rho$. If s is invariant then $\rho(s) = 2s = 0$ et $s = i(s) \in \operatorname{Im} i$. If $s \not\subset \widetilde{X}^G$ then any n-chain in the orbit of s can be written as $k_1 s + k_2 \sigma(s)$ where $k_i \in \{0, 1\}$ for $i = 1, 2$ and corresponds to the unique element $k_1 + k_2\sigma$ in the group algebra $\Lambda := \mathbb{Z}_2[G]$. The sequence (3.1) reduces to a sequence of vector spaces

$$0 \to \rho\Lambda \xrightarrow{i} \Lambda \xrightarrow{\rho} \rho\Lambda \to 0\,.$$

The sequence is immediately exact: Λ is a 2-dimensional vector space over $\mathbb{Z}_2$ and $\ker(\sigma : \Lambda \to \Lambda) = \langle \sigma \rangle$ is a vector subspace of dimension 1. □

Remark 3.2.3 More generally, if G is a group of prime order p then there is an exact sequence analogous to (3.1) for chains with coefficients in $\mathbb{Z}_p$. See [Bre72, Chapitre III, Theorem 3.1] for more details.

The singular homology groups of the pairs (X, L) and $(X(\mathbb{R}), L(\mathbb{R}))$ are isomorphic to the homology groups associated to the simplicial complexes $C(\widetilde{X}, \widetilde{L}; \mathbb{Z}_2)$ and $C(\widetilde{X}^G, \widetilde{L}^G; \mathbb{Z}_2)$ (see [Hat02, Section 2.1] for more details). The exact sequence (3.1) therefore induces a long exact sequence of homology groups:

$$\cdots \xrightarrow{\rho_{k+1}} H_{k+1}(\rho C(\widetilde{X}, \widetilde{L}; \mathbb{Z}_2)) \xrightarrow{\gamma_{k+1}} \qquad (3.2)$$
$$H_k(\rho C(\widetilde{X}, \widetilde{L}; \mathbb{Z}_2)) \oplus H_k(X(\mathbb{R}), L(\mathbb{R}); \mathbb{Z}_2) \xrightarrow{i_k}$$
$$H_k(X, L; \mathbb{Z}_2) \xrightarrow{\rho_k} H_k(\rho C(\widetilde{X}, \widetilde{L}; \mathbb{Z}_2)) \xrightarrow{\gamma_k} \cdots$$

where ρ_k is the map induced by ρ on each homology group.

Remark 3.2.4 Let σ_* be the action of σ on $H_k(X, L; \mathbb{Z}_2)$. The restriction of σ_* to $\operatorname{Im} i_k$ is then the identity.

For any given k we can deduce from the exact sequence (3.2) the following exact sequence

$$0 \to H_{k+1}(\rho C(\widetilde{X}, \widetilde{L}; \mathbb{Z}_2))/\operatorname{Im} \rho_{k+1} \to \qquad (3.3)$$
$$H_k(\rho C(\widetilde{X}, \widetilde{L}; \mathbb{Z}_2)) \oplus H_k(X(\mathbb{R}), L(\mathbb{R}); \mathbb{Z}_2) \to H_k(X, L; \mathbb{Z}_2) \to \operatorname{Im} \rho_k \to 0\,.$$

The homology groups appearing in the exact sequence above are $\mathbb{Z}_2$-vector spaces. For $k = 0, \ldots, 2n$ we set $a_k := \dim_{\mathbb{Z}_2} \operatorname{Im} \rho_k$ and $c_k := \dim_{\mathbb{Z}_2} H_k(\rho C(\widetilde{X}, \widetilde{L}; \mathbb{Z}_2))$ and we get

$$\begin{array}{llll}
0 & = \dim H_{2n}(X, L) & -c_{2n} & -a_{2n} \\
0 & = \dim H_{2n-1}(X, L) & -c_{2n-1} + c_{2n} & -a_{2n} - a_{2n-1} \\
\vdots & \vdots & \vdots & \vdots \\
0 & = \dim H_{n+1}(X, L) & -c_{n+1} + c_{n+2} & -a_{n+2} - a_{n+1} \\
\dim H_n(X(\mathbb{R}), L(\mathbb{R})) & = \dim H_n(X, L) & -c_n + c_{n+1} & -a_{n+1} - a_n \\
\vdots & \vdots & \vdots & \vdots \\
\dim H_k(X(\mathbb{R}), L(\mathbb{R})) & = \dim H_k(X, L) & -c_k + c_{k+1} & -a_{k+1} - a_k \\
\vdots & \vdots & \vdots & \vdots \\
\dim H_1(X(\mathbb{R}), L(\mathbb{R})) & = \dim H_1(X, L) & -c_1 + c_2 & -a_2 - a_1 \\
\dim H_0(X(\mathbb{R}), L(\mathbb{R})) & = \dim H_0(X, L) & -c_0 + c_1 & -a_1 - a_0 \,.
\end{array}$$

Summing these equalities and noting that $c_0 = a_0 = 0$ we get that

$$\sum_{l=0}^{n} \dim H_l(X(\mathbb{R}), L(\mathbb{R}); \mathbb{Z}_2) = \sum_{k=0}^{2n} \bigl(\dim H_k(X, L; \mathbb{Z}_2) - 2a_k\bigr) \,. \tag{3.4}$$

For ease of notation we will assume that $L = \varnothing$. We now interpret the groups $H_r(\rho C(\widetilde{X}; \mathbb{Z}_2))$ geometrically using the projection $p \colon X \to Y$ from X to the orbit space $Y := X/G$ (which is triangulable—see [Bre72, III.Section 1, page 117] for more details). The topological space

$$Y = X/G$$

is the *topological quotient* (or *topological quotient space*) of X by G: in other words, we equip Y with the finest topology rendering p continuous. We consider $X(\mathbb{R})$ as a subspace of both X and Y, we means essentially that we identify the ramification locus $X(\mathbb{R})$ and the branching locus $p(X(\mathbb{R}))$. The projection p is a double cover ramified along $X(\mathbb{R})$ and the restriction $X \setminus X(\mathbb{R}) \to Y \setminus X(\mathbb{R})$ of p is a non ramified double cover (see [Hat02, Section 1.3]).

Proposition 3.2.5 *The groups $H_k(\rho C(\widetilde{X}; \mathbb{Z}_2))$ appearing in the exact sequence* (3.2) *are isomorphic to the homology groups of the pair $(Y, X(\mathbb{R}))$:*

$$\forall k, \quad H_k(\rho C(\widetilde{X}; \mathbb{Z}_2)) \simeq H_k(Y, X(\mathbb{R}); \mathbb{Z}_2) \,.$$

Proof We start by proving that

$$H_k(\rho C(\widetilde{X}, \widetilde{L}; \mathbb{Z}_2)) \simeq H_k(\widetilde{X}/G, \widetilde{X}^G \cup \widetilde{L}/G; \mathbb{Z}_2)$$

as in [Bre72, III.(3.4)]. Let s be a simplex in $\widetilde{X} \setminus \widetilde{L}$. Then we have that $\rho((k_1 + k_2\sigma)(s)) = (k_1 + k_2)\rho(s)$ and $\rho((k_1 + k_2\sigma)(s)) = 0$ if and only if $k_1 + k_2 = 0$ where $s \in \widetilde{X}^G$. The map $\rho\colon C(\widetilde{X}, \widetilde{L}; \mathbb{Z}_2) \to C(\widetilde{X}, \widetilde{L}; \mathbb{Z}_2)$ therefore has the same kernel as the composition

$$C(\widetilde{X}, \widetilde{L}; \mathbb{Z}_2) \xrightarrow{j} C(\widetilde{X}, \widetilde{X}^G \cup \widetilde{L}; \mathbb{Z}_2) \xrightarrow{\pi} C(\widetilde{X}/G, \widetilde{X}^G \cup \widetilde{L}/G; \mathbb{Z}_2)\ .$$

The images of these morphisms are therefore isomorphic via the map given by $\rho(c) \mapsto (\pi \circ j)(c)$ for any chain c. Passing to homology groups for $L = \varnothing$ we get that

$$H_k(\rho C(\widetilde{X}; \mathbb{Z}_2)) \simeq H_k(\widetilde{X}/G, \widetilde{X}^G; \mathbb{Z}_2)$$

and

$$H_k(\widetilde{X}/G, \widetilde{X}^G; \mathbb{Z}_2) \simeq H_k(Y, X(\mathbb{R}); \mathbb{Z}_2)\ .$$

□

We note for later use that the following diagram is commutative

$$\begin{array}{ccccc}
0 \to \rho C(\widetilde{X}) \oplus C(\widetilde{X}^G) & \xrightarrow{i} & C(\widetilde{X}) & \xrightarrow{\rho} & \rho C(\widetilde{X}) \to 0 \\
\downarrow{\scriptstyle 0+\mathrm{id}} & & \downarrow & & \downarrow{\scriptstyle \simeq} \\
0 \to C(\widetilde{X}^G) & \longrightarrow & C(\widetilde{X}/G) & \longrightarrow & C(\widetilde{X}/G, \widetilde{X}^G) \to 0
\end{array}$$

and therefore induces a commutative diagram of homology groups

$$\begin{array}{ccccc}
\to H_k(\rho C(\widetilde{X})) \oplus H_k(X(\mathbb{R})) & \longrightarrow & H_k(X) & \longrightarrow & H_k(\rho C(\widetilde{X})) \to \\
\downarrow & & \downarrow & & \downarrow{\scriptstyle \simeq} \\
\to H_k(X(\mathbb{R})) & \longrightarrow & H_k(Y) & \longrightarrow & H_k(Y, X(\mathbb{R})) \to
\end{array} \tag{3.5}$$

The exact sequence (3.2) implies the following theorem.

Theorem 3.2.6 *Let (X, σ) be a projective $\mathbb{R}$-surface. There is a homology long exact sequence:*

$$\begin{aligned}
\cdots \to H_k(Y, X(\mathbb{R}); \mathbb{Z}_2) \oplus H_k(X(\mathbb{R}); \mathbb{Z}_2) \to H_k(X; \mathbb{Z}_2) \to \qquad (3.6)\\
H_k(Y, X(\mathbb{R}); \mathbb{Z}_2) \xrightarrow{\Delta_k} H_{k-1}(Y, X(\mathbb{R}); \mathbb{Z}_2) \oplus H_{k-1}(X(\mathbb{R}); \mathbb{Z}_2) \to \cdots
\end{aligned}$$

where $G = \{1, \sigma\} \simeq \mathbb{Z}_2$, $X(\mathbb{R}) = X^G$ and $Y = X/G$ is the quotient topological space of X by G.

Moreover, it follows from diagram (3.5) that the second component of Δ_k is the boundary map δ_k of the homology sequence associated to the pair $(Y, X(\mathbb{R}))$

$$H_k(Y, X(\mathbb{R}); \mathbb{Z}_2) \xrightarrow{\delta_k} H_{k-1}(X(\mathbb{R}); \mathbb{Z}_2) \to H_{k-1}(Y; \mathbb{Z}_2)\ . \tag{3.7}$$

3.3 Upper Bounds on Betti Numbers

Definition 3.3.1 Let (X, L) be a pair of topological spaces (if $L = \varnothing$ we simply write X) with $L \subset X$ such that for any k, $\dim_{\mathbb{Q}} H_k(X, L; \mathbb{Q}) < \infty$. The kth *Betti number* $b_k(X, L)$ of (X, L) is the dimension of the kth homology group of (X, L) with coefficients in $\mathbb{Q}$

$$b_k(X, L) := \dim_{\mathbb{Q}} H_k(X, L; \mathbb{Q}) \ .$$

We denote by $b_k(X, L; \mathbb{Z}_2)$ the kth *Betti number with coefficients in* $\mathbb{Z}_2$, or in other words

$$b_k(X, L; \mathbb{Z}_2) := \dim_{\mathbb{Z}_2} H_k(X, L; \mathbb{Z}_2) \ .$$

We will denote by

$$b_*(X) = \sum b_k(X) \quad \text{and} \quad b_*(X; \mathbb{Z}_2) = \sum b_k(X; \mathbb{Z}_2)$$

the total Betti number of X and the total Betti number of X with coefficients in $\mathbb{Z}_2$.

Remark 3.3.2 Of course, as

$$H_k(X, L; \mathbb{C}) = H_k(X, L; \mathbb{Q}) \otimes_{\mathbb{Q}} \mathbb{C}$$

(see Section B.4), we have that

$$b_k(X, L) = \dim_{\mathbb{C}} H_k(X, L; \mathbb{C}) \ .$$

Remark 3.3.3 The Betti numbers with coefficients in $\mathbb{Z}_2$ are not always equal to the Betti numbers. See Section B.4 for the general theory and the exercise below for some examples.

Exercise 3.3.4 Prove the following statements.

1. If $X = \mathbb{RP}^2$ then $b_1(X) = 0$ but $b_1(X; \mathbb{Z}_2) = 1$.
2. If $X = \mathbb{S}^1 \times \mathbb{S}^1$ then $b_1(X) = b_1(X; \mathbb{Z}_2) = 2$.
3. If $X = \mathbb{K}^2$ is a Klein bottle then $b_1(X) = 1$ but $b_1(X; \mathbb{Z}_2) = 2$.

[Hint: use universal coefficients, see Corollary B.4.5.]

Proposition 3.3.5 *Let (X, σ) be a projective $\mathbb{R}$-variety of dimension n and let $L \subset X$ be an $\mathbb{R}$-subvariety. The Betti numbers with coefficients in $\mathbb{Z}_2$ satisfy the following equation:*

$$\sum_{k=0}^{2n} b_k(X, L; \mathbb{Z}_2) - \sum_{l=0}^{n} b_l(X(\mathbb{R}), L(\mathbb{R}); \mathbb{Z}_2) \equiv 0 \mod 2$$

and in particular

$$\sum_{k=0}^{2n} b_k(X;\mathbb{Z}_2) - \sum_{l=0}^{n} b_l(X(\mathbb{R});\mathbb{Z}_2) \equiv 0 \mod 2.$$

Proof As X is projective it is compact for the Euclidean topology and its homology groups are therefore finitely generated. We then simply apply (3.4). □

Theorem 3.3.6 (Smith-Thom inequality) *Let (X,σ) be a projective $\mathbb{R}$-variety of dimension n and let $L \subset X$ be an $\mathbb{R}$-subvariety. The Betti numbers with coefficients in $\mathbb{Z}_2$ satisfy the following inequalities*

$$\sum_{l=0}^{n} b_l(X(\mathbb{R}), L(\mathbb{R});\mathbb{Z}_2) \leqslant \sum_{k=0}^{2n} b_k(X, L;\mathbb{Z}_2)$$

and in particular

$$\sum_{l=0}^{n} b_l(X(\mathbb{R});\mathbb{Z}_2) \leqslant \sum_{k=0}^{2n} b_k(X;\mathbb{Z}_2). \tag{3.8}$$

Proof This follows immediately from (3.4) as in the previous proof. □

For $\mathbb{R}$-varieties defined by explicit equations, Thom [Tho65] and Milnor [Mil64] independantly established an upper bound on $\sum_{l=0}^{n} b_l(X(\mathbb{R});\mathbb{Z}_2)$ depending on the degrees of the defining equations. This result is proved using Morse theory for manifolds with boundary. The bound given depends only on the complex variety so this inequality is of the same type as (3.8).

For curves, Theorem 3.3.6 has the following corollary, often called *Harnack's theorem* or *Harnack's inequality* despite that fact that the general statement was due to Klein, Harnack having proved it only for plane curves.

Corollary 3.3.7 (Harnack's inequality) *Let (C,σ) be a non singular projective irreducible $\mathbb{R}$-curve, let $g(C)$ be the genus of the topological surface underlying the complex curve C and let $s(C(\mathbb{R}))$ be the number of connected components of the real locus $C(\mathbb{R})$. We then have that*

$$s(C(\mathbb{R})) \leqslant g(C) + 1. \tag{3.9}$$

Proof We will give two proofs of (3.9), the first being an application of (3.8) and the second being Klein's original proof, see [Kle82, page 72].

1. As the topological surface C is compact, connected, boundary free and orientable of genus $g := g(C)$ its Betti numbers are $b_0(C) = b_2(C) = 1$ and $b_1(C) = 2g$. As $C(\mathbb{R})$ is a smooth compact curve by Proposition 2.2.27 (or by linearisation as below), it is homeomorphic to a disjoint union of $s := s(C(\mathbb{R}))$ circles and it follows that $b_0(C(\mathbb{R})) = b_1(C(\mathbb{R})) = s$.

2. The involution σ reverses the orientation on the topological surface C, so in a neighbourhood of any fixed point it can be linearised as a symmetry. Indeed, the topological surface C has a 2 dimensional $\mathcal{C}^\infty$ manifold structure and if P is a fixed point of σ then σ induces a linear involution on the tangent space $T_{X,p}$. It follows that the quotient $Y = C/\langle\sigma\rangle$ is a connected surface with boundary whose boundary, which may be empty, can be identified with $C(\mathbb{R})$. The base of the two-to-one covering map $C \setminus C(\mathbb{R}) \to Y \setminus \partial Y$ is connected because a connected surface minus its boundary remains connected. It follows that $C \setminus C(\mathbb{R})$ has at most two connected components and if it has two components they are exchanged by σ. The topological surface C minus all its invariant circles except one is connected. If we had $s(C(\mathbb{R})) > g(C) + 1$ then we could cut C along $g + 1$ circles without disconnecting it, which contradicts Riemann's definition of the genus, see Definition E.1.2. □

We recover Harnack's theorem 2.7.2 from Chapter 2:

Corollary 3.3.8 *Let (C, σ) be an irreducible non singular projective* plane *$\mathbb{R}$-curve of degree d. We then have that*

$$s(C(\mathbb{R})) \leqslant \frac{(d-1)(d-2)}{2} + 1 . \tag{3.10}$$

Proof Simply apply the genus formula (Theorem 1.6.17) to (3.9). □

Remark 3.3.9 Harnack's inequality (3.9) is optimal in the strongest possible sense: for any integer $g \geqslant 0$ and any integer s such that $0 \leqslant s \leqslant g + 1$ there is a projective non singular $\mathbb{R}$-curve (C, σ) such that $g(C) = g$ and $s(C(\mathbb{R})) = s$, see Section 3.5.

Definition 3.3.10 An irreducible non singular projective $\mathbb{R}$-curve (C, σ) is said to be *maximal* (we will also say that (C, σ) is an *M-curve*) if and only if Harnack's bound (3.9) is attained, or in other words if

$$s = g + 1 .$$

More generally, a non singular $\mathbb{R}$-variety (X, σ) of dimension n is said to be *maximal* (we will also say that (X, σ) is an *M-variety*) if the inequality (3.8) is an equality, or in other words if

$$b_*(X(\mathbb{R}); \mathbb{Z}_2) = b_*(X; \mathbb{Z}_2).$$

By (3.4) $b_*(X; \mathbb{Z}_2) - b_*(X(\mathbb{R}); \mathbb{Z}_2) = \sum_{k=0}^{2n} 2a_k$ so we can "measure" the non-maximality of an $\mathbb{R}$-variety using the quantity $a = \sum a_k$.

Definition 3.3.11 Consider an integer $a \in \mathbb{N}$. An $\mathbb{R}$-variety (X, σ) of dimension n is said to be an *$(M - a)$-variety* whenever

$$b_*(X; \mathbb{Z}_2) - b_*(X(\mathbb{R}); \mathbb{Z}_2) = 2a .$$

This terminology comes from the curve case, as illustrated by the exercise below.

Exercise 3.3.12 Prove that a non singular irreducible projective $\mathbb{R}$-curve (C, σ) of genus g with s connected components is an *$(M-a)$-curve* if and only if

$$s = g + 1 - a \ .$$

Example 3.3.13 1. (See Example 2.1.29.) A non singular plane projective cubic with two connected components (such as the cubic $zy^2 - x(x-z)(x+z) = 0$) is an M-curve; a non singular projective plane cubic with one connected component (such as the cubic of equation $zy^2 = x^3 + z^2x = 0$) is an $(M-1)$-curve. The original complex curve is of genus 1.
2. (See Example 4.2.19.) The quadric sphere of equation $x^2 + y^2 + z^2 - w^2 = 0$ in $\mathbb{P}^3$ is an $(M-1)$-surface: the torus $x^2 + y^2 - z^2 - w^2 = 0$ is an M-surface. In both cases the complex variety X is isomorphic to $\mathbb{P}^1(\mathbb{C}) \times \mathbb{P}^1(\mathbb{C})$ whence $\sum_{r=0}^{4} b_r(X; \mathbb{Z}_2) = 4$.

Maximal plane $\mathbb{R}$-curves have other constraints on their topology than those given by (3.10). For example, any plane $\mathbb{R}$-curve of degree at least 6 has nested ovals—see Corollary 2.7.17. We now give a proof of the Petrovskii inequalities stated in Chapter 2, Theorem 2.7.13.

Theorem 3.3.14 (Petrovskii inequalities) *Let (C, σ) be a non singular projective plane $\mathbb{R}$-curve of even degree $d = 2k$. Let p be the number of positive ovals in $C(\mathbb{R})$ and let n be the number of negative ovals as in Definition 2.7.12. We then have that*

$$p - n \leqslant \frac{3}{2}k(k-1) + 1 \ ; \tag{3.11}$$

$$n - p \leqslant \frac{3}{2}k(k-1) \ . \tag{3.12}$$

We will prove this theorem via a useful refinement of the Smith-Thom inequality, Theorem 3.3.6, based on the Hodge decomposition of cohomology, defined in Section D.3.

Remark 3.3.15 In [Har74] Kharlamov presents the inequalities (3.13) below as a generalisation of the Petrovskii inequalities (3.11) and (3.12) above.

Theorem 3.3.16 (Petrovskii–Oleinik inequalities) *Let (X, σ) be a compact connected Kähler $\mathbb{R}$-variety of even complex dimension $2n$. We then have that*

$$2 - h^{n,n}(X) \leqslant \chi_{top}(X(\mathbb{R})) \leqslant h^{n,n}(X) \ . \tag{3.13}$$

Proof We can deduce this result from the Atiyah-Singer index theorem (Corollary 3.4.15), see [Wil78, Remark 1 after the proof of Proposition 4.2]. This result was first proved (in Russian) in [Har74] using Lefschetz's fixed point theorem (Theorem 3.4.23). We refer to [Wil78, Proposition 4.2] for a proof of this theorem in English. □

In the case of surfaces, the Petrovskii–Oleinik inequalities(3.13) are called the Comessatti inequalities, see [Com28]:

Corollary 3.3.17 (Comessatti inequalities) *Let (X, σ) be a projective non singular $\mathbb{R}$-surface. We then have that*

$$2 - h^{1,1}(X) \leqslant \chi_{top}(X(\mathbb{R})) \leqslant h^{1,1}(X) . \tag{3.14}$$

Proof of Theorem 3.3.14 This proof is based on ideas developed by Arnol'd [Arn71]. Let C be a non singular projective complex plane curve of even degree $d = 2k$. (C is not assumed irreducible, but is assumed to be a reduced effective divisor). We will define a double cover $X \to \mathbb{P}^2$ ramified along C such that if $(C, \sigma_{\mathbb{P}}|_C)$ is an $\mathbb{R}$-curve then X has two real structures lifting $\sigma_{\mathbb{P}}$. We imitate Wilson's construction [Wil78, Section 5, page 64]. Let $P(x_0 : x_1 : x_2) = 0$ be an equation for C in $\mathbb{P}^2(\mathbb{C})$. The homogeneous polynomial P is of even degree $d = 2k$: set

$$Z := \{(x_0 : x_1 : x_2 : x_3) \in \mathbb{P}^3(\mathbb{C}) \mid x_3^{2k} - P(x_0 : x_1 : x_2) = 0\}$$

and let $p_Z : Z \to \mathbb{P}^2(\mathbb{C})$ be the restriction to Z of the projection $(x_0 : x_1 : x_2 : x_3) \mapsto (x_0 : x_1 : x_2)$. The map p_Z is a branched cover of degree d with branching locus C. We define an intermediate covering $Z \to X \to \mathbb{P}^2(\mathbb{C})$ by considering the action on Z of the group μ_d of dth roots of unity given for any $\varepsilon \in \mu_d$ by

$$(x_0 : x_1 : x_2 : x_3) \mapsto (x_0 : x_1 : x_2 : \varepsilon x_3) .$$

The covering $Z \to \mathbb{P}^2(\mathbb{C})$ can then be identified with the quotient map $Z \to Z/\mu_d$ and we set

$$X := Z/\mu_k .$$

The complex surface X thus obtained is a double cover of $\mathbb{P}^2(\mathbb{C})$ branching along C. We denote by $\eta : X \to X$ the covering involution. Assume that the homogeneous polynomial P has real coefficients: the restriction $\sigma_{\mathbb{P}}|_Z$ of the canonical real structure on $\mathbb{P}^3(\mathbb{C})$ is then a real structure on Z which induces a real structure σ_1 on X which in turn lifts the real structure $\sigma_{\mathbb{P}}$ on $\mathbb{P}^2(\mathbb{C})$. The composition $\sigma_2 := \eta \circ \sigma_1 = \sigma_1 \circ \eta$ is another real structure on X which also lifts $\sigma_{\mathbb{P}}$. Note that the passage from σ_1 to σ_2 is equivalent to replacing P by $-P$ in the equation $x_3^{2k} - P(x_0 : x_1 : x_2) = 0$.

Replacing the polynomial P by $-P$ if necessary we can therefore assume that P is negative on the unique non orientable component of the complement of $C(\mathbb{R})$ in $\mathbb{P}^2(\mathbb{R})$. We denote by $F := \{(x_0 : x_1 : x_2) \in \mathbb{P}^2(\mathbb{R}) \mid P(x_0 : x_1 : x_2) \geqslant 0\}$ the surface with boundary consisting of all the points where P is positive or zero. The boundary of F is the zero locus $C(\mathbb{R})$ of P. By construction, the Euler characteristic of F is then $p - n$. The Euler characteristic of the double cover $X(\mathbb{R})$ of F ramified along

$C(\mathbb{R})$ is therefore equal to $2\chi(F) = 2(p-n)$. Moreover, it is possible to express the Hodge numbers of a double cover of the plane as a function of the degree of the branching curve as in Example D.4.5 and in particular we have that

$$h^{1,1}(X) = 3k^2 - 3k + 2\,.$$

We now simply apply the inequalities (3.14). □

3.3.1 Singular Curves

Brusotti's theorem 2.7.10 enables us to give an upper bound for the number of double points which are locally $\mathbb{R}$-analytically isomorphic to $x^2 + y^2 = 0$ that can appear in a planar $\mathbb{R}$-curve of given degree.

Lemma 3.3.18 *For given degree d the number of isolated ordinary double points-by which we mean points that are $\mathbb{R}$-analytically isomorphic to $x^2 + y^2 = 0$—on a real algebraic plane curve of degree d is bounded above by the maximal number of empty ovals that can appear on a non singular real algebraic curve of degree d.*

Remark 3.3.19 See [Cos92] for a generalisation of this lemma to higher dimension.

Corollary 3.3.20 *Let C be a projective plane $\mathbb{R}$-curve of degree d. If $C(\mathbb{R})$ is finite then d is even, $d = 2k$, and*

$$\#(C(\mathbb{R})) \leqslant \frac{3k(k-1)}{2} + 1\,.$$

Proof If the degree is odd then the real locus contains at least one pseudo-line because every line in a pencil of lines centred on a point outside C meets the real locus in at least one point. There is therefore a natural number k such that $d = 2k$. If all the points of $C(\mathbb{R})$ are isolated ordinary double points then the bound given on the number of points in $C(\mathbb{R})$ follows immediately from Lemma 3.3.18 and the inequality of Corollary 2.7.16. The general case requires a little more work: we can use either [Cos92] or [BDIM19]. □

3.4 The Intersection Form on an Even-Dimensional $\mathbb{R}$-Variety

Quadratic forms arise in the theory of non singular complex algebraic varieties because on any compact orientable variety of dimension $4m$ the homology intersection form (or the cohomology intersection form, in which case we should, strictly speaking, call the product a cup-product) in dimension $2m$ is unimodular—see Corollary B.7.7. See Appendix A.6 for the general theory of quadratic forms on free $\mathbb{Z}$-modules. The heart of this section is Rokhlin's famous theorem 3.4.2(3.15) which states that for certain $\mathbb{R}$-varieties there is a congruence relation between the Euler characteristic of the real locus and the index of the intersection form in dimension $2m$.

Let (X, σ) be a non singular projective $\mathbb{R}$-variety of even (complex) dimension $n = 2m$. The underlying differentiable manifold is therefore of real dimension $4m$. Let $Q_{\mathbb{R}}$ be the intersection form on $H^{2m}(X; \mathbb{R})$: it is a unimodular quadratic form by Poincaré duality B.7.7.

Definition 3.4.1 If (a, b) is the *signature* (Definition A.6.10) of the intersection form $Q_{\mathbb{R}}$ then we define the *index* of the form $Q_{\mathbb{R}}$ by

$$\tau := \tau(X) = a - b .$$

Theorem 3.4.2 (Extremal congruences) *Let (X, σ) be a non singular projective $\mathbb{R}$-variety of even dimension $n = 2m$.*

If (X, σ) is an M-variety (Definition 3.3.10) then

$$\chi_{top}(X(\mathbb{R})) \equiv \tau(X) \mod 16 \quad (\textit{Rokhlin}). \tag{3.15}$$

If (X, σ) is an $(M-1)$-variety (Definition 3.3.11) then

$$\chi_{top}(X(\mathbb{R})) \equiv \tau(X) \pm 2 \mod 16 \quad (\textit{Gudkov–Kharlamov–Krakhnov}) \tag{3.16}$$

where $\tau(X)$ is the index of the intersection form on the real vector space $H^{2m}(X; \mathbb{R})$.

Remark 3.4.3 We can go further: see for example [Wel02], unifying and extending several classical congruence results.

We state a series of intermediate results before attacking the proof of this theorem. We start by introducing *Wu classes* which are a key ingredient of the proof. We motivate these objects by studying the real 2-dimensional case. This discussion draws on [MH73, Chapter V].

Lemma 3.4.4 *A compact topological surface without boundary V is orientable if and only if*

$$(x \cdot x) = 0 \quad \textit{for every} \quad x \in H_1(V; \mathbb{Z}_2)$$

where $(x, y) \mapsto (x \cdot y)$ is the intersection form on $H_1(V; \mathbb{Z}_2)$.

Proof We can reduce to the case where V is connected: every homology class $x \in H_1(V; \mathbb{Z}_2)$ can then be represented by a simple closed curve $\gamma \subset V$. Note that the self-intersection number $(x \cdot x)$ is zero if and only if a small neighbourhood $W \subset V$ of γ is orientable. Indeed, if W is orientable then a small homotopy deforms γ to a curve $\gamma' \subset W$ which is disjoint from γ. But if γ does not have an orientable neighbourhood then it must have a neighbourhood W which is a Möbius band. In this case we can deform γ to a curve γ' meeting γ transversally in an odd number of points. The lemma follows. □

The classification of compact connected topological surfaces (Theorem E.1.6) yields the following corollary.

Corollary 3.4.5 *Two compact connected topological surfaces without boundary V and V' are homeomorphic if and only if the quadratic $\mathbb{Z}_2$-modules $H_1(V;\mathbb{Z}_2)$ and $H_1(V';\mathbb{Z}_2)$ are isomorphic.*

Example 3.4.6 If $\mathbb{T}^2$ is a torus, $H_1(\mathbb{T}^2;\mathbb{Z}_2)$ is a two dimensional vector space and the intersection product is given by the matrix $\begin{pmatrix} 0 & 1 \\ 1 & 0 \end{pmatrix}$ with respect to a basis of two classes (e_1, e_2) that represent two non identified edges in a quadrilateral from which the torus is constructed by identifying opposite sides.

Example 3.4.7 For the Klein bottle $\mathbb{K}^2$, $H_1(\mathbb{K}^2;\mathbb{Z}_2)$ is also a two dimensional space generated by classes (e_1, e_2) representing two edges of a quadrilateral. However, this time the intersection matrix is given by $\begin{pmatrix} 0 & 1 \\ 1 & 1 \end{pmatrix}$. Alternatively, consider the basis $(e_1 + e_2, e_2)$: the matrix is then given by $\begin{pmatrix} 1 & 0 \\ 0 & 1 \end{pmatrix}$.

See [MS74, page 131] for the definition and construction of the *Wu classes* $\mathrm{v}_k \in H^k(V;\mathbb{Z}_2)$ of a compact differentiable manifold V. In particular, we have that

Proposition 3.4.8 *Let V be a compact differentiable manifold of even dimension $2n$. There is then a unique element $\mathrm{v}_n \in H^n(V;\mathbb{Z}_2)$ such that for every element $x \in H^n(V;\mathbb{Z}_2)$,*

$$(x \cdot x) = (x \cdot \mathrm{v}_n) \ .$$

The element $\mathrm{v}_n \in H^n(V;\mathbb{Z}_2)$ is the nth Wu class *of V.*

By the cohomological analogue of Lemma 3.4.4 a compact topological surface without boundary V is orientable if and only if the Wu class

$$\mathrm{v}_1(V) \in H^1(X;\mathbb{Z}_2)$$

vanishes. We saw in Proposition 3.1.15 that the orientability of V depended on the vanishing of its first Stiefel–Whitney class

$$\mathrm{w}_1(V) \in H^1(X;\mathbb{Z}_2) \ .$$

The relationship between these two classes is simple: we have that

$$\mathrm{v}_1(V) = \mathrm{w}_1(V) \ .$$

This equality can be deduced from the following proposition, which in its turn follows from Wu's theorem, [MS74, Theorem 11.14] or [Wil78, Lemma 3.8].

Proposition 3.4.9 *Let V be a compact differentiable manifold. For any k the kth Wu class of V is a polynomial in the Stiefel–Whitney classes of V. In particular we have that*

$$\mathrm{v}_1 = \mathrm{w}_1 \quad \textit{and} \quad \mathrm{v}_2 = \mathrm{w}_2 + \mathrm{w}_1 \smile \mathrm{w}_1 \ .$$

Exercise 3.4.10 Recall as in [MS74, pages 90–91] that the *Steenrod squares*

$$\mathrm{Sq}^l \colon H^k(V;\mathbb{Z}_2) \to H^{k+l}(V;\mathbb{Z}_2)$$

are morphisms such that for any $x \in H^k(V;\mathbb{Z}_2)$, $\mathrm{Sq}^0(x) = x$, $\mathrm{Sq}^k(x) = x \smile x$ and $\mathrm{Sq}^l(x) = 0$ for any $l > k$. Using Wu's theorem [MS74, Theorem 11.14]

$$\mathrm{w}_k = \sum_{p+q=k} \mathrm{Sq}^p(\mathrm{v}_q) \ , \tag{3.17}$$

reprove the formulas of Proposition 3.4.9 above. Calculate v_3 as a function of the classes $\mathrm{w}_1, \mathrm{w}_2, \mathrm{w}_3$ and v_4 as a function of the classes $\mathrm{w}_1, \mathrm{w}_2, \mathrm{w}_3, \mathrm{w}_4$ using Wu's formula, [MS74, Problem 8-A] and Cartan's formula [MS74, (4) page 91].

Recall that as in Definition A.6.12 a symmetric bilinear form Q on a free $\mathbb{Z}$-module is *even* (or of *type II*) if and only if for all x, $Q(x,x)$ is even.

Corollary 3.4.11 *Let V be a simply connected oriented differentiable manifold of dimension 4. V then has a spin structure if and only if its degree 2 intersection pairing is even.*

Proof The first Stiefel–Whitney class vanishes because V is orientable so $\mathrm{v}_2(V) = \mathrm{w}_2(V)$ in $H^2(V;\mathbb{Z}_2)$. Moreover, the second Stiefel–Whitney class $\mathrm{w}_2(V)$ vanishes if and only if V has a spin structure and the result follows by Proposition 3.4.8. □

Definition 3.4.12 Let (X,σ) be a projective non singular $\mathbb{R}$-variety of even dimension $n = 2m$ whose intersection form on $H^{2m}(X;\mathbb{R})$ will be denoted $Q_{\mathbb{R}}$. The *index* $\tau(\sigma)$ of the involution σ is the index of the symmetric bilinear form $(x,y) \mapsto Q_{\mathbb{R}}(x,\sigma(y))$.

Remark 3.4.13 The index $\tau(\sigma)$ can equivalently be defined by

$$\tau(\sigma) = \tau_+ - \tau_-$$

where $Q_{\mathbb{R}}$ is the intersection form on $H^{2m}(X;\mathbb{R})$ and we have set $\tau_+ := \tau_+(\sigma) = \tau(Q_{\mathbb{R}}|_{H^{2m}(X;\mathbb{R})^{\sigma}})$ and by $\tau_- := \tau_-(\sigma) = \tau(Q_{\mathbb{R}}|_{H^{2m}(X;\mathbb{R})^{-\sigma}})$.

Theorem 3.4.14 (Atiyah, Singer) *On any oriented compact differentiable manifold of real dimension $4m$ the index of any orientation-preserving involution is equal to the auto-intersection of its fixed locus.*

Proof See [AS68] for the original proof or [JO69] for a more elementary proof. □

Corollary 3.4.15 *Let (X, σ) be a non singular projective $\mathbb{R}$-variety of dimension $n = 2m$. We then have that*

$$\tau(\sigma) = (-1)^m \chi_{top}(X(\mathbb{R})).$$

Proof See Propositions 2.2.27 and 2.2.28 for the differentiable manifold structures on X and $X(\mathbb{R})$. As X has real dimension $4m$, σ^* is orientation preserving by Proposition 2.2.28. The index of the self intersection of $X(\mathbb{R})$ in X is therefore equal by [Hir76, page 132] to the self-intersection of $X(\mathbb{R})$ in its normal bundle $N_{X|X(\mathbb{R})}$ (see also [MS74, page 119]). At any real point multiplication by i in the tangent bundle T_X induces an isomorphism of real vector bundles between $T_{X(\mathbb{R})}$ and the normal bundle $N_{X|X(\mathbb{R})}$. Indeed, consider a point $x \in X(\mathbb{R})$ and let $(u_1, \dots, u_{2m})$ be a basis of the vector space $T_{X(\mathbb{R}),x}$. Since x is a point in the real locus we have $T_{X,x} = T_{X(\mathbb{R}),x} \otimes_{\mathbb{R}} \mathbb{C}$. The $4m$-tuplet $(u_1, iu_1, \dots, u_{2m}, iu_{2m})$ is therefore a basis of $T_{X,x}$ and the $2m$-tuplet $(iu_1, \dots, iu_{2m})$ is a basis for the normal bundle $N_{X|X(\mathbb{R}),x}$. As the natural orientation on the manifold X of real dimension $4m$ is given by $(u_1, iu_1, \dots, u_{2m}, iu_{2m})$, the orientation induced on $N_{X|X(\mathbb{R}),x}$ is obtained via a permutation of sign $(-1)^m$ of the set of vectors $(u_1, iu_1, \dots, u_{2m}, iu_{2m})$. As the Euler characteristic $\chi_{\text{top}}(X(\mathbb{R}))$ is equal to the self-intersection of $X(\mathbb{R})$ in its own tangent bundle $T_{X(\mathbb{R})}$ as in [Hir76, page 13] the result follows. □

Remark 3.4.16 Both the statement and the proof remain valid if we replace "projective" by "compact Kähler" throughout—see Appendix D.

Lemma 3.4.17 *Let (X, σ) be a projective $\mathbb{R}$-variety. If $b_*(X(\mathbb{R}); \mathbb{Z}_2) = b_*(X; \mathbb{Z}_2)$ then σ_* is the identity on $H_*(X; \mathbb{Z}_2)$.*

Proof Immediate by Remark 3.2.4 and Equation (3.4). □

Exercise 3.4.18 Check that the converse of Lemma 3.4.17 is false by showing that the projective plane conic without real points given by the equation $x^2 + y^2 + z^2 = 0$ is a counter-example.

Lemma 3.4.19 *Let (X, σ) be a non singular projective $\mathbb{R}$-variety of even dimension $n = 2m$. The submodules $H^{2m}(X; \mathbb{Z})^G$ and $H^{2m}(X; \mathbb{Z})^{-\sigma}$ are then orthogonal for the intersection form on $H^{2m}(X; \mathbb{Z})$.*

Proof Recall that

$$H^{2m}(X; \mathbb{Z}) = H^{2m}(X; \mathbb{Z})_f \oplus \text{Tor}(H^{2m}(X; \mathbb{Z}))$$

where $H^{2m}(X; \mathbb{Z})_f := H^{2m}(X; \mathbb{Z}) / \text{Tor}(H^{2m}(X; \mathbb{Z}))$ is the free part of $H^{2m}(X; \mathbb{Z})$. By Proposition 3.1.8, the submodules $H^{2m}(X; \mathbb{Z})_f^G$ and $H^{2m}(X; \mathbb{Z})_f^{-\sigma}$ are orthogonal for the form induced by the intersection pairing Q on $H^{2m}(X; \mathbb{Z})$. Moreover, for any $x \in \text{Tor}(H^{2m}(X; \mathbb{Z}))$ and any $y \in H^{2m}(X; \mathbb{Z})$ we have that $Q(x, y) = 0$. Indeed, let

$k \in \mathbb{N}^*$ be such that $kx = 0$ in $H^{2m}(X; \mathbb{Z})$: by linearity in the first variable we then have that $0 = Q(kx, y) = kQ(x, y)$ whence $Q(x, y) = 0$ in the integral ring $\mathbb{Z}$. □

Lemma 3.4.20 *Let (X, σ) be a non singular projective $\mathbb{R}$-variety of even dimension $n = 2m$. We consider the intersection form Q on $H^{2m}(X; \mathbb{Z})$. If m is even then the restriction of Q to $H^{2m}(X; \mathbb{Z})^{-\sigma}$ is even (Definition A.6.12) and if m is odd then the restriction of Q to $H^{2m}(X; \mathbb{Z})^G$ is even.*

Proof By Propositions 3.4.9 and 3.1.16(1) the Wu class $\mathrm{v}_{2m} := \mathrm{v}_{2m}(X) \in H^{2m}(X; \mathbb{Z}_2)$ is a polynomial $P(\mathrm{w})$ in the Stiefel–Whitney classes $\mathrm{w}_{2k}(X)$ which are reduction modulo 2 of the Chern classes $c_k(X)$ by Proposition 3.1.16(2). Since $\sigma^* c_k(X) = (-1)^k c_k(X)$, it follows by Exercise 3.4.21 below that if m is even then v_{2m} is the reduction modulo 2 of an element $y \in H^{2m}(X; \mathbb{Z})^G$. The element y is obtained by considering the same polynomial $P(c)$ in the Chern classes. Consider an element $x \in H^{2m}(X; \mathbb{Z})^{-\sigma}$ and let $x_2 \in H^{2m}(X; \mathbb{Z}_2)$ be its reduction modulo 2. As $H^{2m}(X; \mathbb{Z})^{-\sigma}$ and $H^{2m}(X; \mathbb{Z})^G$ are orthogonal for Q we have that $(x \cdot y) = 0$ so $(x_2 \cdot \mathrm{v}_{2m}) = (x_2 \cdot x_2) = 0$ which implies that $(x \cdot x) = Q(x, x)$ is even. The result for n odd can be proved in a similar way since v_{2m} is the reduction modulo 2 of an element in $H^{2m}(X; \mathbb{Z})^{-\sigma}$.

□

Exercise 3.4.21 Using the fact that every monomial in $P(\mathrm{w})$ belongs to $H^{2m}(X; \mathbb{Z}_2)$, prove that $P(c) \in H^{2m}(X; \mathbb{Z})^G$ if m is even and $P(c) \in H^{2m}(X; \mathbb{Z})^{-\sigma}$ if m is odd.

Proof of Theorem 3.4.2 We will prove Rokhlin's congruence (3.15) and refer to [DK00, 2.7.1] (or [Sil89, II.(2.9)] when $m = 1$) for the proof of the Gudkov–Kharlamov–Krakhnov congruence. We have that $\tau(X) = \tau_+(\sigma) + \tau_-(\sigma)$ so by Corollary 3.4.15:

$$\tau(X) - 2\tau_-(\sigma) = \chi_{top}(X(\mathbb{R})) \quad \text{if } m \text{ is even },$$
$$\tau(X) - 2\tau_+(\sigma) = \chi_{top}(X(\mathbb{R})) \quad \text{if } m \text{ is odd }.$$

If (X, σ) is an M-variety then σ^* is the identity on $H^{2m}(X; \mathbb{Z}_2)$ by Lemma 3.4.17. We now apply Lemma 3.4.20: if m is even then the restriction of Q to $H^{2m}(X; \mathbb{Z})_f^{-\sigma}$ is an even quadratic form and hence $\tau_-(\sigma) \equiv 0 \mod 8$ by Proposition A.6.13. Similarly, if m is odd then the restriction of Q to $H^{2m}(X; \mathbb{Z})_f^{\sigma}$ is an even quadratic form and it follows that $\tau_+(\sigma) \equiv 0 \mod 8$. □

Remark 3.4.22 The presentation above is the same as in Risler, [Ris85, Théorème 2.1]. In [DK00, 2.7.1], Degtyarev and Kharlamov propose a different proof which yields an additional result on $(M - 2)$-varieties. Silhol gives yet another proof for surfaces ($m = 1$) in [Sil89, II.(2.4)].

We conclude this section with a series of useful results and an application to surfaces in $\mathbb{P}^3$.

Lefschetz's formula links the Euler characteristic of the fixed locus of an involution on a triangulable space X to the traces of the endomorphisms that it induces on the homology of X. This formula also appears in the litterature as the "Lefschetz trace formula or Lefschetz index theorem".

Theorem 3.4.23 (Lefschetz formula) *Let (X, σ) be a projective $\mathbb{R}$-variety. We then have that*

$$\chi_{top}(X(\mathbb{R})) = \sum_{k \geqslant 0} (-1)^k \operatorname{tr}\left(\sigma^* \colon H_k(X; \mathbb{Q}) \to H_k(X; \mathbb{Q})\right) . \tag{3.18}$$

Proof Smith's exact sequence (3.6) immediately gives us

$$\chi_{top}(X) = \chi_{top}(X(\mathbb{R})) + 2\chi_{top}(Y, X(\mathbb{R})) .$$

Using the exact sequence of the pair $(Y, X(\mathbb{R}))$ (by Theorem B.3.6 we have that $\chi_{top}(Y, X(\mathbb{R})) - \chi_{top}(Y) + \chi_{top}(X(\mathbb{R})) = 0$) it follows that

$$\chi_{top}(X) = 2\chi_{top}(Y) - \chi_{top}(X(\mathbb{R})) . \tag{3.19}$$

Moreover, taking the coefficient ring to be a field of characteristic different from 2 we have that

$$H_k(X; \mathbb{Q})^\sigma \simeq H_k(Y; \mathbb{Q}) . \tag{3.20}$$

(see [DIK00, Corollary I.1.3.3])

Diagonalising its matrix, we immediately see that the trace of a linear involution τ on a $\mathbb{Q}$-vector space E satisfies

$$\operatorname{tr}(\tau \colon E \to E) = 2 \dim E^\sigma - \dim E .$$

The right hand side of Equation (3.18) becomes

$$2 \sum_{k \geqslant 0} (-1)^k \dim_{\mathbb{Q}} H_k(X; \mathbb{Q})^\sigma - \sum_{k \geqslant 0} (-1)^k \dim_{\mathbb{Q}} H_k(X; \mathbb{Q}) ,$$

or in other words

$$2\chi_{top}(Y) - \chi_{top}(X)$$

by (3.20). The theorem follows by (3.19).

The above proof is taken from [DIK00, I.1.3.5]. We refer to [Mun84, Chapter 2, Section 22] or [Hat02, Theorem 2C.3] for a proof of the Lefschetz–Hopf theorem. □

We now consider a compact Kähler $\mathbb{R}$-variety (X, σ). See Appendix D.3 for the basic properties of the Hodge decomposition on the cohomology of X. Note in particular Corollary D.3.15 which states that

$$\overline{H^{p,q}(X)} = H^{q,p}(X)$$

and Lemma D.3.17 which states that

$$\sigma^* H^{p,q}(X) = H^{q,p}(X) \tag{3.21}$$

where σ^* is the action on $H^*(X;\mathbb{C}) = H^*(X;\mathbb{Q}) \otimes_{\mathbb{Q}} \mathbb{C}$ induced by σ.

Proposition 3.4.24 *Let (X, σ) be a compact Kähler $\mathbb{R}$-variety of dimension n and consider an integer $k \in \{0, \ldots, 2n\}$. We denote by $r_k(X) = \dim H_k(X;\mathbb{Q})^G$ the dimension of the $\mathbb{Q}$-subspace of homology classes that are invariant under $G = \{1, \sigma\}$. The number $r_k(X)$ is then subject to the following constraints depending on the kth Betti number and the Hodge numbers of X:*

$$r_k(X) = \frac{1}{2} b_k(X) \quad \textit{if } k \textit{ is odd} \tag{3.22}$$

and

$$\sum_{\substack{p+q=2n-k \\ p<q}} h^{p,q}(X) \leqslant r_k(X) \leqslant b_k(X) - \sum_{\substack{p+q=2n-k \\ p<q}} h^{p,q}(X) \quad \textit{if } k \textit{ is even.} \tag{3.23}$$

Proof By Corollary 3.1.9, $H_k(X;\mathbb{Z})$ is isomorphic (or anti-isomorphic depending on the parity of n) to $H^{2n-k}(X;\mathbb{Z})$ as an involutive module. We will prove that theorem in the case where n is even: the case where n is odd is similar. In particular, $b_k(X) = \dim_{\mathbb{C}} H^{2n-k}(X;\mathbb{C})$ and $r_k(X) = \dim_{\mathbb{C}} H^{2n-k}(X;\mathbb{C})^G$. Consider the Hodge decomposition $H^{2n-k}(X;\mathbb{C}) = \bigoplus_{p+q=2n-k} H^{p,q}(X)$. By Equation (3.21) quoted above, an element of

$$\bigoplus_{\substack{p+q=2n-k \\ p\neq q}} H^{p,q}(X)$$

is stable under σ^* if and only if it is of the form $\omega + \sigma^*\omega$. In other words, the dimension of the subspace of $\bigoplus_{\{p+q=2n-k,\ p\neq q\}} H^{p,q}(X)$ fixed by σ^* is equal to

$$\sum_{\substack{p+q=2n-k \\ p<q}} h^{p,q}(X) .$$

If k is odd this implies that

$$r_k(X) = \frac{1}{2} b_k(X)$$

and if k is even this implies that

$$r_k(X) \geqslant \sum_{\substack{p+q=2n-k \\ p<q}} h^{p,q}(X) \, .$$

As Lemma D.3.17 also applies to $-\sigma$ a similar argument gives that $\dim_{\mathbb{C}} H^{2n-k}(X; \mathbb{C})^{-\sigma} \geqslant \sum_{\substack{p+q=2n-k \\ p<q}} h^{p,q}(X)$. Moreover, as $H^{2n-k}(X; \mathbb{C})$ is a vector space and σ^* is an involution we have a direct sum decomposition $H^{2n-k}(X; \mathbb{C}) = H^{2n-k}(X; \mathbb{C})^{\sigma} \oplus H^{2n-k}(X; \mathbb{C})^{-\sigma}$ whence it follows that

$$r_k(X) \leqslant b_k(X) - \sum_{\substack{p+q=2n-k \\ p<q}} h^{p,q}(X) \, .$$

□

Proposition 3.4.25 *Let (X, σ) be a non singular projective $\mathbb{R}$-variety of dimension n and consider $G = \mathrm{Gal}(\mathbb{C}|\mathbb{R})$ which acts on X via σ. The topological Euler characteristic of the real locus $X(\mathbb{R})$ then satisfies*

$$\chi_{top}(X(\mathbb{R})) = \sum_{r \ even} (2r_k(X) - b_k(X)) \tag{3.24}$$

where $r_k(X) = \dim H_k(X; \mathbb{Q})^G$.

Proof When (X, σ) is an $\mathbb{R}$-surface, a version of this result was proved by Comessatti [Com28]. We use Silhol's proof [Sil89, Proposition I.(2.1), page 9]. Consider once more the double cover $p \colon X \to Y = X/G$ ramified along $X(\mathbb{R})$ as in Section 3.2. Using a triangulation of X of which $X(\mathbb{R})$ is a sub-triangulation we have that

$$\chi_{top}(X) = 2\chi_{top}(Y) - \chi_{top}(X(\mathbb{R})) \, . \tag{3.25}$$

Moreover, the homology morphism p^* induces an isomorphism

$$H_r(Y; \mathbb{Q}) \simeq H_r(X; \mathbb{Q})^G \tag{3.26}$$

(See [Gro57, page 202] or [Flo60, page 38] in [Bor60]). Note that $\chi_{top}(X) = \sum_{k=0}^{2n} (-1)^k H_k(X; \mathbb{Q})$ and $\chi_{top}(Y) = \sum_{k=0}^{2n} (-1)^k H_k(Y; \mathbb{Q})$. Combining (3.25) and (3.26) above we get that

$$\chi_{top}(X(\mathbb{R})) = \sum_{k=0}^{2n} 2r_k(X) - b_k(X) \, .$$

We now simply apply Proposition 3.4.24 to get (3.24). □

Remark 3.4.26 The identity (3.24) still holds if X is a compact Kähler $\mathbb{R}$-variety.

3.4.1 *Surfaces in* $\mathbb{P}^3$

Applying the various bounds and congruences stated above to degree d hypersurfaces in $\mathbb{P}^3$ we obtain the following bounds.

Theorem 3.4.27 *Let (X, σ) be a non singular $\mathbb{R}$-surface of degree d in $\mathbb{P}^3$. We have the following bounds on the number of connected components in its real locus:*

$$\#\pi_0(X(\mathbb{R})) \leqslant \frac{d(5d^2 - 18d + 25)}{12} - \varepsilon(d)$$

where

$$\begin{cases} \varepsilon(d) = 0 & \text{if } d \equiv 0 \mod 16 \text{ or } d \equiv 1 \mod 4\,; \\ \varepsilon(d) = \frac{1}{2} & \text{if } d \equiv \pm 2 \mod 16\,; \\ \varepsilon(3) = 2\,; & \\ \varepsilon(d) = 1 & \text{if } d \equiv \pm 4 \text{ or } 8 \mod 16 \text{ or } d \neq 3 \text{ and } d \equiv 3 \mod 4\,; \\ \varepsilon(d) = \frac{3}{2} & \text{if } d \equiv \pm 6 \mod 16\,. \end{cases}$$

Proof The proof can be found in [Sil89, Chapter II, Theorem 3.9, Corollary 3.10]. It uses notably the calculation of the Hodge numbers of a non singular hypersurface of degree d in $\mathbb{P}^3$ as in Example D.4.4. □

Remark 3.4.28 As $\varepsilon(2) = \frac{1}{2}$, it follows in particular that a non empty real quadric is connected. As $\varepsilon(3) = 2$, we get that for any non singular cubic $\#\pi_0(X(\mathbb{R})) \leqslant 2$: this upper bound is optimal by Example 3.1.13. As $\varepsilon(4) = 1$ we have that $\#\pi_0(X(\mathbb{R})) \leqslant 10$ for any quartic, which is optimal—see Chapter 4 for more details. The bound given by this theorem for the number of connected components of a quintic is 25: at the time of writing we do not know what the actual maximal number of components of a real quintic is. See [KI96] for the construction of a quintic with 22 components and [Ore01] for the construction of a quintic with 23 components, the best known example at the time of writing.

3.5 Classification of $\mathbb{R}$-Curves and XVIth Hilbert's Problem

In this section we will apply the results obtained above to non singular projective curves. We start by establishing an abstract classification- by which we mean a classification independent of a choice of embedding- and then take a closer look at the classical case of plane curves, ie. curves with the extra information of an embedding into $\mathbb{P}^2$. This classification is more restrictive than the previous one, since not all curves have a non singular embedding into the plane. After characterising non singular projective $\mathbb{R}$-curves which can be embedded into the plane, we will address a more delicate question: what are the possible relative positions of the ovals of a given curve? In other words, for fixed degree, we will try to classify topological pairs $(\mathbb{P}^2(\mathbb{R}), X(\mathbb{R}))$ rather than the pairs $(X, X(\mathbb{R}))$.

3.5.1 Abstract Classification

Let us briefly review our analysis of the situation we met in the second proof of Corollary 3.3.7. Let (X, σ) be a non singular projective $\mathbb{R}$-curve. The involution σ of the topological surface X is orientation reversing so it can be linearised in a neighbourhood of any point in the real locus as a symmetry with respect to an axis. It follows that the quotient $Y = X/\langle\sigma\rangle$ is a compact connected surface whose boundary can be identified with $X(\mathbb{R})$. As the base of the double cover $X \setminus X(\mathbb{R}) \to Y \setminus \partial Y$ is connected, $X \setminus X(\mathbb{R})$ has at most two connected components, and in this case they are exchanged by σ. The topological surfaces X minus all its invariant circles apart from one is still connected. By Riemann's definition of the genus, Definition E.1.2, if $X(\mathbb{R})$ has $g(X) + 1$ circles then $X \setminus X(\mathbb{R})$ is necessarily not connected: the real locus $X(\mathbb{R})$ disconnects or *separates* X. More generally, we propose the following definition.

Definition 3.5.1 (*Klein*) Let (X, σ) be a non singular projective $\mathbb{R}$-curve. The curve (X, σ) is said to be *separating* if $X(\mathbb{R})$ disconnects X and *non separating* otherwise.

We can assign to any non singular projective $\mathbb{R}$-curve (X, σ) a triplet of integers (g, s, a) where $g := g(X)$ is the genus of the compact orientable surface X, $s := s(X) = s(X, \sigma)$ is the number of connected components of $X(\mathbb{R})$ and $a := a(X) = a(X, \sigma)$ is the binary invariant defined by $a := 2 - \#\pi_0(X \setminus X(\mathbb{R}))$. This triplet is a complete homeomorphic equivalence invariant for our surfaces with involutions. We now explain exactly what we mean by this. Following Gabard, [Gab00], we will call a compact connected orientable topological surface without boundary equipped with an orientation-reversing involution a *symmetric surface*. As seen above, a projective $\mathbb{R}$-curve whose algebraic structure has been "forgotten", leaving only the topological structure is naturally a symmetric surface. Klein–Weichold's theorem, see [Gab00, Théorème 2.4], then states that any such symmetric surfaces are homeomorphic via an involution-preserving homeomorphism if and only if they have the same associated triplet. The interested reader will find more information in [Nat99, Section 1] where Natanzon classifies symmetric surfaces and equips them with a Riemann surface structure that is both explicit and simple. (See examples 1.1 and 1.2 in [Nat99].) Since all compact Riemann surfaces are algebraic curves—see Theorem E.2.28—this provides us with a source of real algebraic curves.

As the surface X is orientable $\chi_{top}(X) = 2 - 2g$ as in Proposition E.1.5. As the Euler characteristic of a circle vanishes, the Euler characteristic of the quotient Y is $g - 1$. Moreover, Y is orientable if and only if $a = 0$, ie if and only if the curve is separating. Indeed, $X \setminus X(\mathbb{R}) \to Y \setminus \partial Y$ is the orientation covering of $Y \setminus \partial Y$ which is therefore orientable if and only if $X \setminus X(\mathbb{R})$ is not connected. The boundary of the compact surface Y has s connected components: from the classification of compact surfaces (Theorem E.1.6) we deduce the following constraints:

$$\begin{aligned} &\text{If} \quad a = 0 \quad \text{then} \quad 1 \leqslant s \leqslant g+1 \quad \text{and} \quad g - s \equiv 1 \mod 2\,. \\ &\text{If} \quad a = 1 \quad \text{then} \quad 0 \leqslant s \leqslant g\,. \end{aligned} \tag{3.27}$$

Our classification will be complete once we have proved that conversely for any triplet satisfying these conditions there is a non singular projective $\mathbb{R}$-curve of which it is the invariant.

Theorem 3.5.2 *Let (g, s, a), $a \in \{0, 1\}$ be a triplet of integers. There is a non singular irreducible projective plane $\mathbb{R}$-curve (X, σ) of genus $g(X) = g$ such that $\#\pi_0(X(\mathbb{R})) = s$ and $2 - \#\pi_0(X \setminus X(\mathbb{R})) = a$ if and only if the conditions* (3.27) *above are satisfied.*

Proof See [Gab00, Section 4]. □

We now consider plane curves. If d is the degree of a plane curve then the genus formula $g = \frac{(d-1)(d-2)}{2} = \frac{1}{2}d(d-3)+1$ (Theorem 1.6.17) leads us to consider the triplet (d, s, a) instead of (g, s, a). Non singular projective plane $\mathbb{R}$-curves are subject to two additional constraints beside those coming from the genus formula and Equations (3.27).

Theorem 3.5.3 *Let (d, s, a) be a triplet of integers such that $d > 0$, $s \geqslant 0$, $a \in \{0, 1\}$. There is a non singular irreducible projective plane $\mathbb{R}$-curve (X, σ) of degree $\deg(X) = d$ such that $\#\pi_0(X(\mathbb{R})) = s$ and $2 - \#\pi_0(X \setminus X(\mathbb{R})) = a$ if and only if the following conditions are satisfied*

$$\begin{aligned} &d \equiv 1 \mod 2 \implies s \geqslant 1\,, \\ &a = 0 \implies \left\lfloor \tfrac{d+1}{2} \right\rfloor \leqslant s \leqslant \tfrac{(d-1)(d-2)}{2} + 1 \quad \textit{and} \quad \tfrac{1}{2}d(d-3) - s \equiv 0 \mod 2\,, \\ &a = 1 \implies 0 \leqslant s \leqslant \tfrac{(d-1)(d-2)}{2}\,. \end{aligned}$$

Proof See [Gab00, Théorème 5.2] for a proof that such curves exist. Here, we will restrict ourselves to explaining where the conditions on (d, s, a) come from. The first condition comes from the fact that a plane curve of odd degree always has real points. Indeed, a general real line in the plane meets X is a set of points that is globally stable under σ. As the number of such points is odd, at least one of them is stable under σ. (This is a special case of Proposition 2.6.48). The third condition is simply condition (3.27) where $g = \frac{(d-1)(d-2)}{2}$ by the genus formula (Theorem 1.6.17). The second condition arises on putting together the first condition (3.27) with $g = \frac{1}{2}d(d-3)+1$ and Rokhlin's formula in Theorem 3.5.5 below. See [Mar80, page 59] or [Gab00, Théorème 5.1] for more details. □

3.5.2 First Part of XVIth Hilbert's Problem

En 1891, Hilbert asked the following question: what are the possible relative positions of the ovals of a degree d curve in $\mathbb{P}^2(\mathbb{R})$ up to isotopy for $d = 1, 2, 3, \ldots$. This question about curves (or more generally surfaces in $\mathbb{P}^3(\mathbb{R})$- Hilbert asked this question for surfaces of degree 4 in particular) became in 1900 the first part of what is now known as the XVIth Hilbert's problem. For curves of degree 6 and surfaces of degree 4 the solution was given by Gudkov and Kharlamov in the 70s. One of the most recent versions of this problem is the (asymptotic) solution of Ragsdale's (incorrect) conjecture from 1906 (see Remark 2.7.15), which was found a hundred years later by Brugallé in 2006 [Bru06] using ideas by Arnol'd [Arn71], Viro [Vir80], Itenberg [Ite93] (who gave the first counter-example) and many others.

In this section we present some known constraints on the arrangement of ovals of a curve in the plane and refer to [A'C80], [Gab04, page 50] and [Gab00] for classical curve construction methods, notably due to Hilbert and Harnack. For modern methods, namely Viro's method, "dessins d'enfants" applied to trigonal curves and tropical methods we refer to [Ris93], [Ore03], [BB06] and [IMS09] amongst others. Our first constraint is an immediate consequence of Bézout's theorem 2.7.1.

Theorem 3.5.4 (Hilbert) *Consider a curve of degree $d = 2k$. The number of its ovals contained in a nest (see Definition 2.7.11) or in a disjoint union of two nests is at most k.*

Taken together with Theorem 3.5.3, this constraint yields a complete classification of isotopy types of real plane algebraic curves for $k = 1, 2$. For $k = 1$, the curve is a non singular conic, which is therefore either empty or connected as in Exercise 1.2.68. For $k = 2$, the curve is a quartic so its genus is 3 and the number of ovals is $0, 1, 2, 3$ or 4. By the above theorem there is at most one nest of ovals and in this case the curve has only two nested ovals. Such a curve can be constructed as a small perturbation of the product of equations of two circles with the same centre. (Before perturbation, Bézout's theorem implies that the complex curve has four singular non real points). The sextic $k = 3$ is more interesting and was the object of Hilbert's original question. We get some extra constraints using the following theorem where as in Definition 2.7.12 we denote by p the number of positive ovals (called *even ovals*) and by n the number of negative (or *odd*) ovals of our $\mathbb{R}$-curve. We have the following result.

Theorem 3.5.5 (Gudkov–Rokhlin congruence) *For any M-curve of degree $2k$ we have that*

$$p - n \equiv k^2 \mod 8 . \tag{3.28}$$

This theorem follows from Theorem 3.4.2 applied to the surface obtained as a double cover of the plane branched along our curve as in the proof of Theorem 3.3.14.

Hilbert's question about degree 6 curves was answered by Gudkov in 1969 [Gud69] (see [Gud71] and [Arn71]) using Petrovskii's results [Pet33]. The question on degree 4 surfaces in $\mathbb{P}^3$ (linked to the above by the fact that a degree 4 surface and a double cover of a plane branched along a sextic are both K3 surfaces), was solved by Kharlamov in 1975 [Har76]. The hope of finding a classification for curves of degree $\geqslant 7$ and surfaces of degree $\geqslant 5$ in $\mathbb{P}^3$ subsequently generated intense activity in this area. The most complete result found was Viro's classification of degree 7 curves [Vir80].

3.5.3 Ragsdale's Conjecture

By Corollary 2.7.14, the number p of positive ovals on a non singular real curve of degree $2k$ is bounded above by $7k^2/4 - 9k/4 + 3/2$. A famous incorrect conjecture by Ragsdale (see Remark 2.7.15) states that p must be less than $3k^2/2 - 3k/2 + 1$. We refer the reader to the seminal articles [Ite93, Ite95] and [Bru06] and their bibliographies for the history of the Ragsdale conjecture and the work it has inspired. In [Ite93, Ite95], Itenberg shows that Ragsdale's conjecture is false and in [Bru06], Brugallé proves the existence of a family of non singular real algebraic curves of degree $2k$ such that $p/k^2 \to_{k\to\infty} 7/4$. More recently, Renaudineau gave a constructive proof of the existence of such a family in [Ren17]. Let us also mention the article [Haa95] in which Haas improved the result of [Ite93].

The study of curves in the plane can be generalised to the study of curves on a given surface X. For example, there has been significant progress in the classification of trigonal curves on a Hirzebruch surface X in recent years. We refer the interested reader to [DIK08, Deg12, DIZ14] for more details. We will undertake a systematic study of surfaces in Chapter 4. Meanwhile we present as an example the construction of an M-surface of degree 4 in $\mathbb{P}^3$.

3.5.4 Construction of a Maximal Quartic $\mathbb{R}$-surface[4]

We will illustrate the above theory with the construction of a quartic surface in $\mathbb{P}^3$ which is obtained as the double cover of a quadric surface branched along a curve of bidegree (4, 4). The three-dimensional diagram shows the double cover of a single-sheeted hyperboloid of revolution ramified along the intersection with four hyperplanes in general position as in Figure 3.1. All the figures below are shown in intersection with a Euclidean ball in $\mathbb{R}^3$.

Once we have constructed the singular curve on the hyperboloid, we construct the double cover. We do this by choosing a sign convention, indicated by the choice of colours, and then construct a double cover, which is represented twice, once with the hyperboloid included, and once without the hyperboloid: see Figure 3.2.

The result is a singular surface whose singular points are all ordinary double points.

[4] Illustrations created in collaboration with C. Raffalli in 2001.

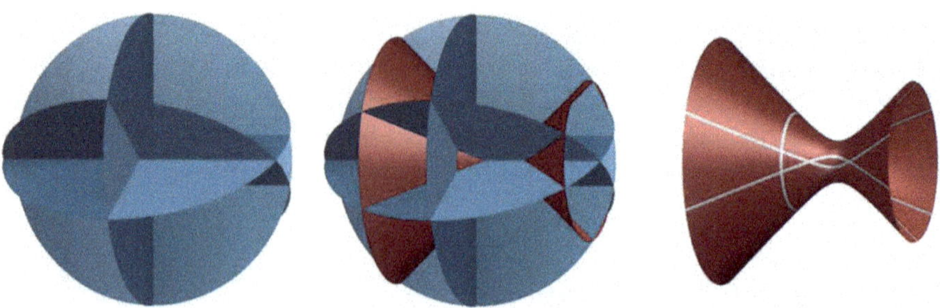

Fig. 3.1 Construction of a maximal $\mathbb{R}$-quartic. From left to right: the four planes; the four planes plus the hyperboloid; a singular curve cut out by the four hyperplanes on the hyperboloid

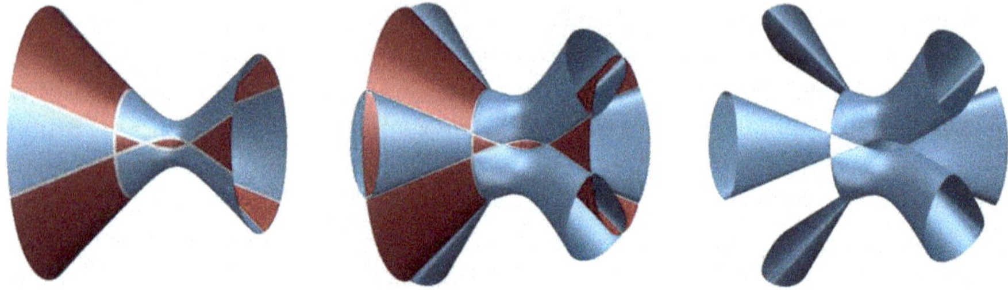

Fig. 3.2 Construction of a maximal $\mathbb{R}$-quartic, bis. From left to right: choice of signs, double cover with hyperboloid, double cover without hyperboloid

In Figure 3.3, we construct a small perturbation of this $(M-1)$-surface. (From the second image onwards, the angle of vision has been changed.) To check that this perturbation is indeed an M-surface, we start by calculating the total Betti number of a quartic in $\mathbb{P}^3$, given that $b^*(X_d) = d(d^2 - 4d + 6)$ for any degree d surface and then calculate the Euler characteristic of the smooth compact connected surface $X(\mathbb{R})$. The Euler characteristic of the double cover is twice the Euler characteristic of the light coloured surfaces cut out on the hyperboloid. Removing the singular points three surfaces remain: each one is isomorphic to a disc of Euler characteristic 1. There are twelve double points so $\chi_{top}(X(\mathbb{R})) = 2(3-12) = -18$: we conclude that the real locus is diffeomorphic to an orientable surface $\mathbb{S}_{10}$ of genus 10. (The real locus is orientable because X is of even degree). Prolonging the deformation a second component which is diffeomorphic to a sphere appears in the centre of the figure. Finally we see that $X(\mathbb{R}) \approx \mathbb{S}^2 \sqcup \mathbb{S}_{10}$ and (X, σ) is therefore an M-surface.

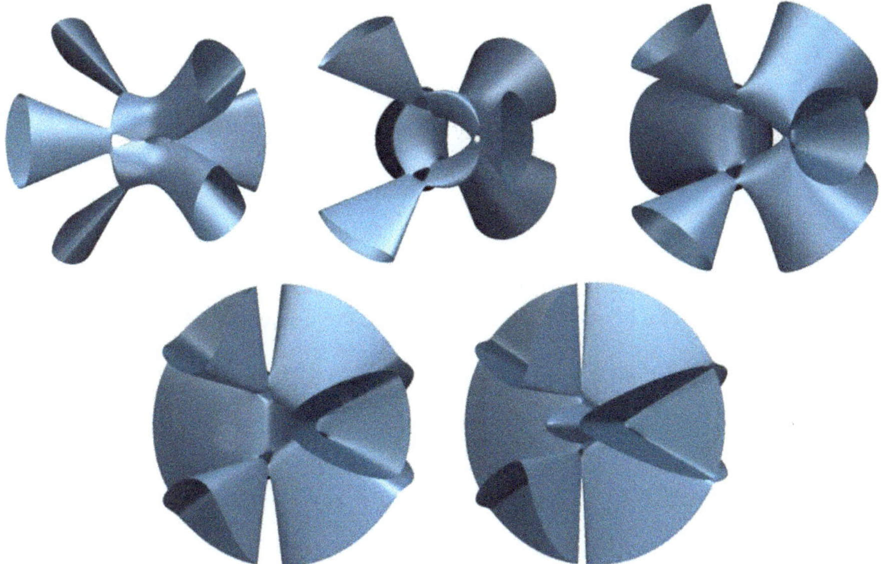

Fig. 3.3 Construction of a maximal $\mathbb{R}$-quartic, conclusion

3.6 Galois-Maximal Varieties

Let (X, σ) be an irreducible projective $\mathbb{R}$-variety, which means that the complex variety X is irreducible, see Definition 2.3.13. Considering the Galois cohomology groups (see Proposition 3.1.4) of the homology of the complex variety X, we obtain refinements of the Smith-Thom inequality (3.8), see (3.30), (3.32) and (3.33).

Lemma 3.6.1 *Let (X, σ) be a projective $\mathbb{R}$-variety. For any $0 \leqslant k \leqslant 2n$, on setting $\lambda_k := \dim_{\mathbb{Z}_2}(1 + \sigma_*)H_k(X; \mathbb{Z}_2)$ we have that*

$$\dim_{\mathbb{Z}_2} H_k(X; \mathbb{Z}_2)^G = \dim_{\mathbb{Z}_2} H_k(X; \mathbb{Z}_2) - \lambda_k ;$$

$$\dim_{\mathbb{Z}_2} H^1(G, H_k(X; \mathbb{Z}_2)) = \dim_{\mathbb{Z}_2} H_k(X; \mathbb{Z}_2) - 2\lambda_k .$$

Proof Simply apply Lemma 3.1.5 to the $\mathbb{Z}_2$-vector space $H_k(X; \mathbb{Z}_2)$ with its involution σ_*. □

Remark 3.6.2 The invariant λ_k defined above on $H_k(X; \mathbb{Z}_2)$ may be different from the invariant defined in Definition 3.1.3 on $H_k(X; \mathbb{Z}) \otimes_{\mathbb{Z}} \mathbb{Z}_2$ if the homology of X contains torsion.

Lemma 3.6.3 *Let (X, σ) be a projective $\mathbb{R}$-variety. For any $0 \leqslant k \leqslant 2n$ we set $a_k := \dim_{\mathbb{Z}_2} \operatorname{Im} \rho_k$. We then have that*

$$\forall k, \quad \lambda_k \leqslant a_k . \tag{3.29}$$

Proof Consider an integer k such that $0 \leqslant k \leqslant 2n$. We have that $(1+\sigma_*)H_k(X;\mathbb{Z}_2) \subset \operatorname{Im}\rho_k$ by definition of ρ so it follows that $\lambda_k \leqslant a_k$. □

Theorem 3.6.4 *Let (X,σ) be a projective irreducible $\mathbb{R}$-variety. We then have that*

$$\sum_{l=1}^{n} b_l(X(\mathbb{R});\mathbb{Z}_2) \leqslant \sum_{k=0}^{2n} \dim H^1\left(G, H_k(X;\mathbb{Z}_2)\right) . \tag{3.30}$$

Definition 3.6.5 A non singular $\mathbb{R}$-variety (X,σ) is said to be *Galois-Maximal* (we will sometimes say that (X,σ) is a *GM-variety*) if the inequality (3.30) is an equality.

Proof (Proof of Theorem 3.6.4) By Lemma 3.6.3 and equality (3.4) we have that

$$b_*(X(\mathbb{R});\mathbb{Z}_2) \leqslant \sum_{k=0}^{2n}\left(b_k(X;\mathbb{Z}_2) - 2\lambda_k\right) . \tag{3.31}$$

and the result follows by Lemma 3.6.1. □

Lemma 3.6.6 *A non singular irreducible projective $\mathbb{R}$-variety (X,σ) is Galois-Maximal if and only if $\forall k$, $\lambda_k = a_k$.*

Proof By Lemma 3.6.1 the variety (X,σ) is Galois-Maximal if and only if

$$\sum_{l=1}^{n} b_l(X(\mathbb{R});\mathbb{Z}_2) = \sum_{k=0}^{2n}\left(b_k(X;\mathbb{Z}_2) - 2\lambda_k\right) .$$

We deduce from Equation (3.4), $\sum_{l=0}^{n} b_l(X(\mathbb{R});\mathbb{Z}_2) = \sum_{k=0}^{2n}\left(b_k(X;\mathbb{Z}_2) - 2a_k\right)$ that (X,σ) is Galois-Maximal if and only if $\sum_{k=0}^{2n}\lambda_k = \sum_{k=0}^{2n} a_k$. We further deduce from Equation (3.29) that this is equivalent to $\forall k$, $\lambda_k = a_k$. □

We have the following upper bounds. (See [Kra83, Theorem 2.3] for a proof using spectral sequences and [Sil89, I.3.13] for a direct proof using the Galois action on the Hodge decomposition.):

$$\sum_{l \text{ even}} \dim_{\mathbb{Z}_2} H_l(X(\mathbb{R});\mathbb{Z}_2) \leqslant \sum_{k=0}^{2n} \dim_{\mathbb{Z}_2} H^2\left(G, H_k(X;\mathbb{Z})\right) ; \tag{3.32}$$

$$\sum_{l \text{ odd}} \dim_{\mathbb{Z}_2} H_l(X(\mathbb{R});\mathbb{Z}_2) \leqslant \sum_{k=0}^{2n} \dim_{\mathbb{Z}_2} H^1\left(G, H_k(X;\mathbb{Z})\right) . \tag{3.33}$$

Note that in inequalities (3.32) and (3.33), the homology groups being considered on X have coefficients in $\mathbb{Z}$ whereas the Galois cohomology groups $H^2(G, H_k(X;\mathbb{Z}))$ and $H^1(G, H_k(X;\mathbb{Z}))$ are $\mathbb{Z}_2$-vector spaces.

Definition 3.6.7 A non singular $\mathbb{R}$-variety (X, σ) is said to be *$\mathbb{Z}$-Galois-Maximal* (or to be a *$\mathbb{Z}GM$-variety*) if the inequalities (3.32) and (3.33) are equalities.

Proposition 3.6.8 *Let (X, σ) be a non singular projective $\mathbb{R}$-variety.*

1. *If (X, σ) is $\mathbb{Z}$-Galois-Maximal then it is Galois-Maximal.*
2. *If the homology of X has no 2-torsion then (X, σ) is Galois-Maximal if and only if it is $\mathbb{Z}$-Galois-Maximal.*

Proof See [Kra83, Proposition 3.6]. □

Example 3.6.9 ($\mathbb{Z}$-Galois-Maximal varieties)

1. Smooth projective curves with non empty real locus are $\mathbb{Z}$-Galois-Maximal (and are therefore Galois-Maximal). See [Sil82] for a proof.
2. Abelian varieties of arbitrary dimension with non empty real locus are $\mathbb{Z}$-Galois-Maximal. See [Kra83] for a proof.
3. All non singular projective surfaces with non empty real locus such that $H_1(X; \mathbb{Z}_2) = 0$ are $\mathbb{Z}$-Galois-Maximal: see Theorem 3.6.11 and Corollary 3.6.12 below.

Example 3.6.10 Let (X, σ) be a projective non singular $\mathbb{R}$-surface such that $H_1(X; \mathbb{Z}) = 0$. (For example, by the Lefschetz hyperplane theorem D.9.2 any surface in $\mathbb{P}^3$ satisfies this condition.) The inequalities (3.32) and (3.33) then give us

$$\#\pi_0(X(\mathbb{R})) \leqslant 1 + \frac{1}{2} \dim_{\mathbb{Z}_2} H^2(G, H_2(X; \mathbb{Z})) \ ;$$

$$b_1(X(\mathbb{R}); \mathbb{Z}_2) \leqslant \dim_{\mathbb{Z}_2} H^1(G, H_2(X; \mathbb{Z})) \ .$$

Indeed, by hypothesis and Poincaré duality, $H_k(X; \mathbb{Z}) = 0$ for $k \in \{1, 3\}$ and as X is globally invariant under σ we have that $H_k(X; \mathbb{Z})^G = H_k(X; \mathbb{Z}) = \mathbb{Z}$ for any $k \in \{0, 4\}$ which implies that $H^2(G, H_k(X; \mathbb{Z})) = \mathbb{Z}_2$ and $H^1(G, H_k(X; \mathbb{Z})) = 0$ for any $k \in \{0, 4\}$

It turns out that in the situation of this example if $X(\mathbb{R}) \neq \varnothing$ then all the above are equalities.

Theorem 3.6.11 (Krasnov) *Let (X, σ) be a non singular projective $\mathbb{R}$-surface such that $X(\mathbb{R}) \neq \varnothing$ and $H_1(X; \mathbb{Z}_2) = 0$. We then have that (X, σ) is Galois-Maximal.*

The original proof of this theorem can be found in [Kra83, page 262]. We give an alternative proof without spectral sequences at the end of this section.

Corollary 3.6.12 *Let (X, σ) be a non singular projective $\mathbb{R}$-surface such that $X(\mathbb{R}) \neq \varnothing$ and $H_1(X; \mathbb{Z}_2) = 0$. We then have that (X, σ) is $\mathbb{Z}$-Galois-Maximal and*

$$\#\pi_0(X(\mathbb{R})) = 1 + \frac{1}{2} \dim_{\mathbb{Z}_2} H^2(G, H_2(X; \mathbb{Z})) \ ;$$

$$b_1(X(\mathbb{R}); \mathbb{Z}_2) = \dim_{\mathbb{Z}_2} H^1(G, H_2(X; \mathbb{Z})) \ .$$

Proof By Theorem 3.6.11 and Proposition 3.6.8 the $\mathbb{R}$-variety (X, σ) is $\mathbb{Z}$-Galois-Maximal and the inequalities of Example 3.6.10 are equalities. □

The following examples show that the hypotheses of Theorem 3.6.11 are necessary.

Example 3.6.13 $(b_1(X) \neq 0)$ Consider an $\mathbb{R}$-curve C of genus $g > 0$ whose real locus has two connected components C_1 and C_2. (The $\mathbb{R}$-curve (C, σ_1) of Example 2.1.29 is an example of this.) Note that $b_1(X) = g(C) \neq 0$. By a theorem of Witt's ([Wit34] or [Kne76a, Kne76b]) there is a rational function $f \in \mathbb{R}(C)^*$ such that $f > 0$ on C_1 and $f < 0$ on C_2 (see also [Sil89, V.(2.3)]). Let $\pi : X \to C$ be the conic bundle given in $\mathbb{A}^2_{x,y} \times C_t$ by the equation

$$x^2 + y^2 = f(t) \ .$$

The space $X(\mathbb{R}) \approx \mathbb{S}^1 \times \mathbb{S}^1$ is then connected: it is a torus lying over $C_1 \approx \mathbb{S}^1$, so $b_*(X(\mathbb{R})) = 4$. Moreover, as $\dim H^1(G, H_4(X; \mathbb{Z}_2)) = \dim H^1(G, H_0(X; \mathbb{Z}_2)) = 1$ and C is a $(M - (g-1))$-curve, by [Sil89, V.4, page 108] we have that $\dim H^1 (G, H_1(X; \mathbb{Z}_2)) = \dim H^1(G, H_1(C; \mathbb{Z}_2)) = 2g - 2(g-1)$ from which it follows that

$$\dim H^1(G, H_1(X; \mathbb{Z}_2)) = \dim H^1(G, H_3(X; \mathbb{Z}_2)) = 2 \ .$$

We then have that $\sum_{k=0}^4 \dim H^1(G, H_k(X; \mathbb{Z}_2)) \geqslant 6$ and (X, σ) is not Galois-Maximal. We can generalise this example as in [Sil89, V.4] and [vH00]: let (X, σ) be an $\mathbb{R}$-surface with a conic bundle structure $\pi : X \to C$ over a projective $\mathbb{R}$-curve (C, σ_C). We suppose that the real structures on X and C are compatible with π (i.e. π is an $\mathbb{R}$-morphism: $\sigma_C \circ \pi = \pi \circ \sigma$). We can then prove that (X, σ) is $\mathbb{Z}$-Galois-Maximal if and only if the number of connected components of the real locus $X(\mathbb{R})$ is equal to the number of connected components of the real locus $C(\mathbb{R})$ of the base curve.

Example 3.6.14 $(b_1(X) = 0)$ See Section 4.5 for the definition and main properties of real Enriques surfaces. In particular, the fundamental group of a complex Enriques surface X is equal to $\mathbb{Z}_2$ so $b_1(X) = 0$ and $b_1(X; \mathbb{Z}_2) = 1$. We will see in Section 4.5 that there are real Enriques surfaces with non empty real locus which are not Galois-maximal (see Theorem 4.5.20), such as those whose real locus is connected and orientable (by Theorem 4.5.16, $X(\mathbb{R})$ must then be diffeomorphic to $\mathbb{S}^2$ or $\mathbb{T}^2$). We can do better than this: it turns out that all possible cases of behaviour with respect to Galois maximality can be attained by real Enriques surfaces (Theorem 4.5.20). In other words, there are real Enriques surfaces which are $\mathbb{Z}$-Galois-maximal, others that are Galois-maximal but not $\mathbb{Z}$-Galois-maximal and yet others that are not Galois-maximal.

We end this subsection with a result confirming a conjecture due to R. Silhol in 1989. See [Sil89, Remark I.4.5] for more details. We saw in Chapter 2 Theorem 2.6.32 which gives a sufficient condition for an invariant linear class to be represented by an invariant divisor. We will now give a similar result for another class group of divisors, the Néron-Severi group $\mathrm{NS}(X)$ (see Definition 2.6.34).

Theorem 3.6.15 (van Hamel 1998) *Let (X, σ) be a non singular irreducible projective algebraic $\mathbb{R}$-variety. If X is $\mathbb{Z}$-Galois-maximal then for any divisor D algebraically equivalent to $\sigma(D)$ there is a divisor D' algebraically equivalent to D such that $D' = \sigma(D')$. In other words*[5]*:*

$$\mathrm{Div}(X)^G / \mathrm{Div}^0(X)^G = \mathrm{NS}(X)^G \ .$$

Proof The proof is tricky and uses equivariant cohomology. See [vH00, Cor. IV.5.2] for more details. □

Returning to Example 3.6.13 we now shows that the $\mathbb{Z}$-Galois maximal hypothesis cannot be weakened.

Example 3.6.16 ($\mathrm{Div}(X)^G / \mathrm{Div}^0(X)^G \neq \mathrm{NS}(X)^G$) Let $X \to C$ be a conic bundle over a curve with non empty real part $X(\mathbb{R})$ which nevertheless has fewer connected components that the real part $C(\mathbb{R})$ of the base curve. The variety X is then a non singular irreducible projective algebraic $\mathbb{R}$-variety which is not $\mathbb{Z}$-Galois-Maximal and satisfies:

$$\mathrm{Div}(X)^G / \mathcal{P}(X)^G = \mathrm{Pic}(X)^G,$$

but

$$\mathrm{Div}(X)^G / \mathrm{Div}^0(X)^G \neq \mathrm{NS}(X)^G \ .$$

See [vH00, Example III.9.5] for more details.

3.6.1 σ-Representable Classes and Proof of Theorem 3.6.11

We saw in Chapter 2 that if the real locus of an $\mathbb{R}$-variety (X, σ) is non empty then any linear class of divisors on X invariant under σ is *representable* by a divisor on X which is itself invariant under σ (Theorem 2.6.32). By analogy we introduce the notion of an invariant topological class which is *σ-representable*, which will help us characterise Galois-maximal varieties (Proposition 3.6.19). Moreover, using these techniques we will give a proof of Krasnov's theorem 3.6.11 without spectral sequences.

Throughout this paragraph (X, σ) will be a projective algebraic $\mathbb{R}$-variety of dimension n. We use the conventions of Section 3.2: $\widetilde{X}$ will be a finite simplicial complex underlying the compact topological space X such that if the real structure σ fixes a simplex s of $\widetilde{X}$ globally then it fixes each of the vertices of s. We denote the

[5] Scheme theoretically: if X is a scheme over $\mathbb{R}$ satisfying the hypotheses of the theorem then $\mathrm{NS}(X) = \mathrm{NS}(X_{\mathbb{C}})^G$.

subcomplex fixed by σ by $\widetilde{X}^G$ and $C(\widetilde{X}; \mathbb{Z}_2)$ and $C(\widetilde{X}^G; \mathbb{Z}_2)$ will be the chain groups with coefficients in $\mathbb{Z}_2$. We will freely identify the homology groups $H_k(C(\widetilde{X}; \mathbb{Z}_2))$ with the groups $H_k(X; \mathbb{Z}_2)$ and $H_k(C(\widetilde{X}^G; \mathbb{Z}_2))$ with $H_k(X(\mathbb{R}); \mathbb{Z}_2))$.

Definition 3.6.17 Let (X, σ) be an $\mathbb{R}$-variety. An invariant class $\alpha \in H_k(X; \mathbb{Z}_2)^G$ is said to be *σ-representable* if there is an invariant k-cycle $c \in C(\widetilde{X}; \mathbb{Z}_2)^G$ representing α.

Lemma 3.6.18 *We use the same notation as in the spectral sequences* (3.1) *and* (3.2) *with $L = \varnothing$. Consider $\alpha \in H_k(X; \mathbb{Z}_2)$: the following are equivalent :*

- $\rho_k(\alpha) = 0$ *in* $H_k(\rho C(\widetilde{X}; \mathbb{Z}_2))$;
- *α is an invariant σ-representable class.*

Proof If α is an invariant σ-representable class and c is an invariant representative it is clear that,

$$\rho(c) = (1 + \sigma)(c) \equiv 0 \mod 2 \quad \text{whence} \quad \rho_k(\alpha) = 0 \, .$$

Conversely, let $\alpha \in H_k(X; \mathbb{Z}_2)$ be such that $\rho_k(\alpha) = 0$ in $H_k(\rho C(\widetilde{X}; \mathbb{Z}_2))$ and let $c \in C_k(\widetilde{X}; \mathbb{Z}_2)$ be a k-cycle representing α. There is then a $(k+1)$-chain $b \in \rho C_{k+1}(\widetilde{X}; \mathbb{Z}_2)$ such that

$$\rho(c) = \partial b \, .$$

There is therefore a chain $\tilde{b} \in C_{k+1}(\widetilde{X}; \mathbb{Z}_2)$ such that

$$\rho(\tilde{b}) = b \quad \text{and} \quad \rho(c) = \partial b = \rho(\partial \tilde{b}) \, .$$

The k-cycle $c' = \partial \tilde{b} - c = \partial \tilde{b} + c$ also represents α and satisfies $\rho(c') = 0$. By the exact sequence (3.1) we have that $c' \in \rho C(\widetilde{X}; \mathbb{Z}_2) \oplus C(\widetilde{X}^G; \mathbb{Z}_2)$ and the invariance of c' follows. □

Proposition 3.6.19 *Let (X, σ) be a $\mathbb{R}$-variety. The following properties are equivalent.*

- *(X, σ) is Galois-Maximal ;*
- *For any $0 \leqslant k \leqslant 2n$, any homology class $\alpha \in H_k(X; \mathbb{Z}_2)$ invariant under σ can be represented by a cycle invariant under σ.*

Remark 3.6.20 It is interesting to compare the above result with Theorem 3.6.15 in the complex codimension 1 case. This theorem characterises the invariant *algebraic* classes which are representable by invariant divisors.

Proof Consider an integer k such that $0 \leqslant k \leqslant 2n$. By Lemma 3.6.18 we have that $\ker \rho_k \subset H_k(X; \mathbb{Z}_2)^G$. Recall that on setting $a_k = \dim_{\mathbb{Z}_2} \operatorname{Im} \rho_k$ and $\lambda_k = \dim_{\mathbb{Z}_2}(1 + \sigma_*)H_k(X; \mathbb{Z}_2)$ Lemma 3.6.3 implies that $\lambda_k \leqslant a_k$ and Lemma 3.6.6 implies that $\lambda_k = a_k$ for all k if and only if (X, σ) is Galois-Maximal. We start by supposing that every invariant class is σ-representable: we then have that

$$\ker \rho_k = H_k(X; \mathbb{Z}_2)^G$$

by Lemma 3.6.18. It follows that $a_k = \dim \operatorname{Im} \rho_k = \lambda_k$ by Lemma 3.6.1 and hence the $\mathbb{R}$-variety (X, σ) is Galois-Maximal.

Conversely, assume that (X, σ) is Galois-Maximal. For any $k \in \{0, \dots, 2n\}$ we then have that $\dim \operatorname{Im} \rho_k = \lambda_k$ from which since $\ker \rho_k \subset H_k(X; \mathbb{Z}_2)^G$ we can deduce that $\dim \ker \rho_k = \dim H_k(X; \mathbb{Z}_2)^G$. Using Lemma 3.6.18 once more, we see that every invariant class is σ-representable. □

Proposition 3.6.19 is the key element of the elementary proof of Theorem 3.6.11. The original proof ([Kra83, page 262]) is based on the degeneration of the Grothendieck spectral sequence. An English version of this proof can be found in [Sil89, A1.7]. We start with a lemma which is essentially due to Hirzebruch and which will be used several times in the rest of this book.

Lemma 3.6.21 *Let (X, σ) be a non singular projective algebraic $\mathbb{R}$-surface. We set $G = \{1, \sigma\} \simeq \mathbb{Z}_2$ and we let $Y = X/G$ be the topological quotient of X by G. The fundamental class of $X(\mathbb{R})$ in $H_2(Y; \mathbb{Z}_2)$ then vanishes.*

Proof The complex surface X with its Euclidean topology is a compact oriented differentiable manifold of dimension 4 on which G acts by orientation preserving diffeomorphism (see Proposition 2.2.28). The subset of fixed points $X(\mathbb{R})$ is a compact differentiable submanifold all of whose connected components are of codimension 2 in X; the projection $p\colon X \to Y$ is then a branched double cover of Y whose branching locus is $p(X(\mathbb{R}))$ [Hir69, Section 1]. If $X(\mathbb{R})$ is orientable then the normal bundle $\mathcal{N}$ of $p(X(\mathbb{R}))$ in Y is the tensor square of the normal bundle $N_{X|X(\mathbb{R})}$ of $X(\mathbb{R})$ in X: all these objets are differentiable bundles of complex lines (see Hirzebruch [Hir69, pages 259–260]). The first Chern class of the bundle $\mathcal{N}$ is therefore divisible by 2 in $H^2(Y; \mathbb{Z})$. The fundamental class of $X(\mathbb{R})$ in $H_2(Y; \mathbb{Z})$- ie. the class of $p(X(\mathbb{R}))$- is the Poincaré dual of $c_1(\mathcal{N})$. This class is therefore also divisible by 2. On the other hand, if $X(\mathbb{R})$ is not orientable then a second application of [Hir69, pages 259–260] shows that its fundamental class vanishes in Y modulo 2. Reducing modulo 2 it follows that in every case the fundamental class of $X(\mathbb{R})$ in $H_2(Y; \mathbb{Z}_2)$ vanishes. □

Lemma 3.6.22 *Using the same notation as in the Smith exact sequence* (3.2)*, let (X, σ) be a non singular projective $\mathbb{R}$-surface such that $X(\mathbb{R}) \neq \varnothing$ and $H_1(X; \mathbb{Z}_2) = 0$. We then have that*

$$H_3(Y, X(\mathbb{R}); \mathbb{Z}_2) \simeq \mathbb{Z}_2 \ .$$

Moreover if β_0 is the unique non zero class in $H_3(\rho C(\widetilde{X}; \mathbb{Z}_2))$ then the second component of $\gamma_3(\beta_0)$ in $H_2(\rho C(\widetilde{X}; \mathbb{Z}_2)) \oplus H_2(X(\mathbb{R}); \mathbb{Z}_2)$ is non trivial. (See the exact sequence (3.2)*).*

Proof The real locus $X(\mathbb{R})$ is a differentiable submanifold of X of dimension 2 and in particular $b_k(X(\mathbb{R}); \mathbb{Z}_2) = 0$ for any $k > 2$. The hypothesis $H_1(X; \mathbb{Z}_2) = 0$ implies that $H^3(X;\mathbb{Z}_2)=0$ by Poincaré duality and hence $H_3(X;\mathbb{Z}_2)\simeq H^3(X;\mathbb{Z}_2)=0$. The exact long sequence (3.2) can be decomposed as follows.

$$0\to H_4(\rho C(\widetilde{X}))\to H_4(X;\mathbb{Z}_2)\to H_4(\rho C(\widetilde{X}))\to H_3(\rho C(\widetilde{X}))\to 0 \tag{3.34}$$

$$\begin{aligned}0\to H_3(\rho C(\widetilde{X}))\xrightarrow{\gamma_3} H_2(\rho C(\widetilde{X}))\oplus H_2(X(\mathbb{R});\mathbb{Z}_2)\xrightarrow{i_2} H_2(X;\mathbb{Z}_2)\xrightarrow{\rho_2}\\ H_2(\rho C(\widetilde{X}))\to H_1(\rho C(\widetilde{X}))\oplus H_1(X(\mathbb{R});\mathbb{Z}_2)\to 0\end{aligned} \tag{3.35}$$

By convention the complex surface X is connected so by (3.34) we have that

$$\dim H_4(\rho C(\widetilde{X};\mathbb{Z}_2))=\dim H_4(X;\mathbb{Z}_2)=\dim H_3(\rho C(\widetilde{X};\mathbb{Z}_2))=1\ .$$

The group $H_r(\rho C(\widetilde{X};\mathbb{Z}_2))$ is isomorphic to $H_r(Y,X(\mathbb{R});\mathbb{Z}_2)$ (Proposition 3.2.5) and as this isomorphism is natural it is possible to identify the maps γ_r and Δ_r in the exact sequences (3.2) and (3.6). As above, we denote by β_0 the unique non zero class in $H_3(Y,X(\mathbb{R});\mathbb{Z}_2)$. We deduce from the exact sequence (3.7)

$$H_3(Y,X(\mathbb{R});\mathbb{Z}_2)\xrightarrow{\delta_3} H_2(X(\mathbb{R});\mathbb{Z}_2)\to H_2(Y;\mathbb{Z}_2)$$

and Lemma 3.6.21 that the fundamental class of $X(\mathbb{R})$ in $H_in2(X(\mathbb{R});\mathbb{Z}_2)$ is the image of β_0 under δ_3. As $X(\mathbb{R})$ is non empty its fundamental class in $H_2(X(\mathbb{R});\mathbb{Z}_2)$ is non zero, moreover δ_3 is the second component of $\Delta_3\colon H_3(Y,X(\mathbb{R}))\to H_2(Y,X(\mathbb{R}))\oplus H_2(X(\mathbb{R}))$, which is the same thing as the second component of $\gamma_3\colon H_3(\rho C(\widetilde{X}))\to H_2(\rho C(\widetilde{X}))\oplus H_2(X(\mathbb{R}))$, and as $H_3(Y,X(\mathbb{R});\mathbb{Z}_2)\simeq H_3(\rho C(\widetilde{X};\mathbb{Z}_2))$ the lemma follows. □

Proof of Theorem 3.6.11 For any $k\in\{0,1,3,4\}$ we have that $H_k(X;\mathbb{Z}_2)^G=H_k(X;\mathbb{Z}_2)$ and it follows that $\lambda_k=a_k$ since by hypothesis we have that $b_1(X;\mathbb{Z}_2)=b_3(X;\mathbb{Z}_2)=0$. The failure of X to be Galois-Maximal is therefore due to the behaviour of the group $H_2(X;\mathbb{Z}_2)$. Suppose that (X,σ) is not Galois-Maximal. It then follows from Proposition 3.6.19 that there is an invariant class $\alpha\in H_2(X;\mathbb{Z}_2)^G$ which is not σ-representable.

We now use the fact that in the exact sequence (3.35) the target space of ρ_2 is also the target space of the first component of γ_3. Lemma 3.6.18 implies that $\rho_2(\alpha)\neq 0$ in $H_2(\rho C(\widetilde{X};\mathbb{Z}_2))$. We denote by $\alpha'=\rho_2(\alpha)$ this non zero class in $H_2(\rho C(\widetilde{X};\mathbb{Z}_2))$. By definition of ρ if c is a representative of α which is not invariant then by hypothesis α' is the class of $c+\sigma(c)$ in $H_2(\rho C(\widetilde{X};\mathbb{Z}_2))$. As α is invariant we have that $i_2(\alpha'\oplus 0)=\alpha+\sigma(\alpha)=2\alpha=0$ in $H_2(X;\mathbb{Z}_2)$. As the class β_0 is the only non zero class in $H_3(\rho C(\widetilde{X};\mathbb{Z}_2))$, it follows from the exactness of the sequence (3.35) at $H_2(\rho C(\widetilde{X};\mathbb{Z}_2))$ that α' is the image under β_0 of γ_3. This contradicts Lemma 3.6.22 which implies that the image of γ_3 is not contained in $H_2(\rho C(\widetilde{X};\mathbb{Z}_2))$. □

3.7 Algebraic Cycles

3.7.1 Fundamental Class

A connected compact manifold without boundary V of dimension n has a fundamental $\mathbb{Z}_2$ homology class $[V] \in H_n(V; \mathbb{Z}_2)$. If the variety V is also *oriented* (see Definition B.5.3) then it has an oriented fundamental $\mathbb{Z}$-class $[V] \in H_n(V; \mathbb{Z})$, see Remark B.5.8.

A (non singular) compact complex analytic variety V of complex dimension n therefore has a fundamental homology class $[V] \in H_{2n}(V; \mathbb{Z})$. Similarly, a (non singular) compact real analytic variety L of dimension n has a fundamental homology class $[L] \in H_n(L; \mathbb{Z}_2)$.

In [BH61], Borel and Haefliger describe how to define a fundamental class $[V] \in H_{2n}(V; \mathbb{Z})$ in the general case where V is a complex analytic space with singularities and a fundamental class $[L] \in H_n(L; \mathbb{Z}_2)$ when L is a locally real analytic space. For more details of the construction of this fundamental class using a semi-algebraic triangulation (see Appendix B.2) we refer to [BCR98, Theorem 11.1.1 and Proposition 11.3.1].

Definition 3.7.1 Let X be a complex analytic space of (complex) dimension n and let Y be a compact analytic subspace of (complex) dimension k. The *homology class represented by* Y (or *class of* Y), denoted by $[Y] \in H_{2k}(X; \mathbb{Z})$, is the image of the fundamental homology class $[Y] \in H_{2k}(Y; \mathbb{Z})$ under the morphism $i_*: H_{2k}(Y; \mathbb{Z}) \to H_{2k}(X; \mathbb{Z})$ induced by the inclusion map $i: Y \hookrightarrow X$.

If X is non singular the *cohomology class represented by* Y, denoted η_Y, will be the image in $H_c^{2n-2k}(X; \mathbb{Z})$ of the class $[Y] \in H_{2k}(X; \mathbb{Z})$ under the Poincaré duality morphism (see Corollary 3.1.9 for the case where X is compact and Theorem B.7.1 for the general case)

$$D_X^{-1}: H_{2k}(X; \mathbb{Z}_2) \to H_c^{2n-2k}(X; \mathbb{Z}_2) .$$

Similarly, let $L \subset V$ be a locally real analytic subspace of dimension k in a locally real analytic space of dimension n. If L is compact the *the homology class represented by* L, denoted $[L] \in H_k(V; \mathbb{Z}_2)$, will be the image of the fundamental homology class $[L] \in H_k(L; \mathbb{Z}_2)$ under the morphism induced by inclusion $L \hookrightarrow V$.

If V is non singular the *cohomology class represented by* L, denoted η_V, will be the image in $H_c^{n-k}(V; \mathbb{Z}_2)$ of $[L] \in H_k(V; \mathbb{Z}_2)$ under the Poincaré duality morphism

$$D_V^{-1}: H_k(V; \mathbb{Z}_2) \to H_c^{n-k}(V; \mathbb{Z}_2) .$$

3.7.2 Algebraic Cycles

Let (X, σ) be a projective $\mathbb{R}$-surface. Both the complex variety X and its real locus $X(\mathbb{R})$ have a Euclidean topology. We consider the subgroup

$$H_1^{\text{alg}}(X(\mathbb{R}); \mathbb{Z}_2) \subset H_1(X(\mathbb{R}); \mathbb{Z}_2)$$

of homology classes which can be represented by the real locus of an algebraic $\mathbb{R}$-curve. One of the main questions in this area is to determine this group and apply this knowledge to classification problems and topology.

More generally, we consider a projective $\mathbb{R}$-variety of dimension n, (X, σ), and an irreducible subvariety $Y \subset X$ which is stable under σ (so that $(Y, \sigma|_Y)$ is an $\mathbb{R}$-subvariety) of codimension k in X. If $Y(\mathbb{R})$ is of codimension k in $X(\mathbb{R})$ then we have a homology class $[Y(\mathbb{R})] \in H_{n-k}(X(\mathbb{R}); \mathbb{Z}_2)$ where n is the dimension of X. Let $H_{n-k}^{\text{alg}}(X(\mathbb{R}); \mathbb{Z}_2)$ be the subgroup in $H_{n-k}(X(\mathbb{R}); \mathbb{Z}_2)$ generated by these classes. If X is non singular we denote by $H^k_{\text{alg}}(X(\mathbb{R}); \mathbb{Z}_2)$ the subgroup generated by the Poincaré duals of these classes.

Definition 3.7.2 The group $H_{n-k}^{\text{alg}}(X(\mathbb{R}); \mathbb{Z}_2)$ is called the group of $(n-k)$*-algebraic cycles on* $X(\mathbb{R})$ or the group of *algebraic cycles of codimension* k *in* $X(\mathbb{R})$. (Note that for non singular X we generally call the elements of the group $H^k_{\text{alg}}(X(\mathbb{R}); \mathbb{Z}_2)$ "algebraic cycles" rather than "algebraic cocycles").

Remark 3.7.3 This is a slight abuse of notation, as an element of $H_{n-k}^{\text{alg}}(X(\mathbb{R}); \mathbb{Z}_2)$ is actually a homology class rather than a cycle.

Let (X, σ) be a projective $\mathbb{R}$-variety. We set

$$H_*^{\text{alg}}(X(\mathbb{R}); \mathbb{Z}_2) = \bigoplus_{k \geqslant 0} H_k^{\text{alg}}(X(\mathbb{R}); \mathbb{Z}_2) \ ;$$

and if X is non singular we set

$$H^*_{\text{alg}}(X(\mathbb{R}); \mathbb{Z}_2) = \bigoplus_{k \geqslant 0} H^k_{\text{alg}}(X(\mathbb{R}); \mathbb{Z}_2) \ .$$

Theorem 3.7.4 *Let* (X, σ) *and* (Y, τ) *be projective* $\mathbb{R}$*-varieties and let* $f : X \to Y$ *be a morphism of* $\mathbb{R}$*-varieties. We then have that*

$$f_*(H_*^{\text{alg}}(X(\mathbb{R}); \mathbb{Z}_2)) \subset H_*^{\text{alg}}(Y(\mathbb{R}); \mathbb{Z}_2) \ ;$$

and if X *and* Y *are non singular,*

$$f^*(H^*_{\text{alg}}(Y(\mathbb{R}); \mathbb{Z}_2)) \subset H^*_{\text{alg}}(X(\mathbb{R}); \mathbb{Z}_2) \ .$$

Proof This follows immediately from results found in [BH61, Section 5]. See also [BCR98, Theorem 11.3.4]. □

Theorem 3.7.5 *Let (X, σ) be a non singular projective $\mathbb{R}$-variety. The direct sum*

$$H^*_{\mathrm{alg}}(X(\mathbb{R}); \mathbb{Z}_2) = \bigoplus_{k \geqslant 0} H^k_{\mathrm{alg}}(X(\mathbb{R}); \mathbb{Z}_2)$$

is a graded ring with respect to cup-product.

Proof This follows immediately from [BH61, Section 5]. See also [BCR98, Theorem 11.3.5]. □

Definition 3.7.6 (Totally algebraic variety) A real algebraic variety V such that

$$H_*^{\mathrm{alg}}(V; \mathbb{Z}_2) = H_*(V; \mathbb{Z}_2)$$

is said to be *totally algebraic*. Similarly, an $\mathbb{R}$-variety (X, σ) such that

$$H_*^{\mathrm{alg}}(X(\mathbb{R}); \mathbb{Z}_2) = H_*(X(\mathbb{R}); \mathbb{Z}_2)$$

is said to be *totally algebraic*.

Example 3.7.7 Projective spaces $\mathbb{P}^n(\mathbb{R})$ and Grassmannians (see below) $\mathbb{G}_{n,k}(\mathbb{R})$ are totally algebraic

$$H_*^{\mathrm{alg}}(\mathbb{P}^n(\mathbb{R}); \mathbb{Z}_2) = H_*(\mathbb{P}^n(\mathbb{R}); \mathbb{Z}_2)\ ;$$
$$H_*^{\mathrm{alg}}(\mathbb{G}_{n,k}(\mathbb{R}); \mathbb{Z}_2) = H_*(\mathbb{G}_{n,k}(\mathbb{R}); \mathbb{Z}_2)\ .$$

See [BCR98, Proposition 11.3.3].

Definition 3.7.8 (*Grassmannian*) Let K be a field and let $n \geqslant k$ be natural numbers. The *Grassmannian* (or *Grassmann variety*) $\mathbb{G}_{n,k}(K)$ is the set of k-dimensional subspaces of K^n.

Remark 3.7.9 The Grassmannian is a generalisation of projective space. For any natural number n, $\mathbb{P}^n(K) = \mathbb{G}_{n+1,1}(K)$.

Proposition 3.7.10 *Let n and k be natural numbers such that $n \geqslant k$. The Grassmanian $\mathbb{G}_{n,k}(\mathbb{R})$ is a complete non singular real affine algebraic variety.*

Proof See [BCR98, Proposition 3.4.5]. □

Proposition 3.7.11 *Let n and k be natural numbers such that $n \geqslant k$. The real algebraic varieties $\mathbb{G}_{n,k}(\mathbb{R})$ and $\mathbb{G}_{n,n-k}(\mathbb{R})$ are isomorphic.*

Proof See [BCR98, Propositions 3.4.3, 3.4.4 et 3.4.11]. □

Beyond the fact that characterising the homology classes coming from algebraic varieties is interesting in itself, the group of algebraic cycles intervenes in the following question: how well can a $\mathcal{C}^\infty$ hypersurfaces be approximated by algebraic hypersurfaces? (See [BCR98, Section 12.4] for more details). A closed $\mathcal{C}^\infty$ manifold M in a real algebraic variety V *has an algebraic approximation in* V if for every open neighbourhood Ω of the inclusion $M \hookrightarrow V$ in $\mathcal{C}^\infty(M, V)$ (with the $\mathcal{C}^\infty$ topology, see Remark 5.2.2), there is an $h \in \Omega$ such that $h(M)$ is a non singular algebraic subset of V. We have the following important result.

Theorem 3.7.12 *Let (X, σ) be a non singular quasi-projective $\mathbb{R}$-variety of dimension n such that $X(\mathbb{R})$ is compact and non empty (i. e. $X(\mathbb{R})$ is complete, see Definition 1.4.11) and let $M \subset X(\mathbb{R})$ be a compact $\mathcal{C}^\infty$ hypersurface. The following conditions are then equivalent.*

- *The fundamental class $[M]$ of M belongs to $H_{n-1}^{\mathrm{alg}}(X(\mathbb{R}); \mathbb{Z}_2)$.*
- *M has an algebraic approximation in $X(\mathbb{R})$.*
- *There is a $\mathcal{C}^\infty$ diffeotopy of $X(\mathbb{R})$ arbitrarily close to the identity sending M to a non singular algebraic subset of codimension 1 in $X(\mathbb{R})$.*

Proof See [BCR98, Theorem 12.4.11]. □

3.7.3 Cycle Map

Let (X, σ) be an $\mathbb{R}$-variety such that $X(\mathbb{R}) \neq \varnothing$. We denote by $H_{2k}^{\mathbb{R}-\mathrm{alg}}(X; \mathbb{Z})$ the subgroup of $H_{2k}(X; \mathbb{Z})$ generated by classes that can be represented by complex algebraic subvarieties of dimension k fixed by σ. We can define a *cycle map*

$$\psi_X \colon H_{2k}^{\mathbb{R}-\mathrm{alg}}(X; \mathbb{Z}) \to H_k^{\mathrm{alg}}(X(\mathbb{R}); \mathbb{Z}_2) \tag{3.36}$$

as follows.

Consider an element $\alpha \in H_{2k}^{\mathbb{R}-\mathrm{alg}}(X; \mathbb{Z})$. By hypothesis there exist irreducible complex algebraic subvarieties D_j, $j = 1, \ldots, s$ of dimension k such that $\sigma D_j = D_j$ and integers $n_j \in \mathbb{Z}$ such that $\alpha = \sum_{j=1}^{s} n_j [D_j]$. If $\dim_{\mathbb{R}} D_j(\mathbb{R}) = k$, then the image of D_j under ψ_X is the class $[D_j(\mathbb{R})]$ in $H_k(X(\mathbb{R}); \mathbb{Z}_2)$ and otherwise $\psi_X(D_j) = 0$. The image of α under ψ_X is then the linear combination $\sum_{j=1}^{s} n_j [D_j(\mathbb{R})]$ where the sum is only taken over the indices j such that $\dim_{\mathbb{R}} D_j(\mathbb{R}) = k$.

For a special class of algebraic varieties we have the following result.

Lemma 3.7.13 ([BH61, Section 5.15, Proposition, page 496]) *Let (X, σ) be a non singular projective $\mathbb{R}$-variety. If every homology class modulo 2 on X (resp. on $X(\mathbb{R})$) (with not necessarily compact support) can be represented by a σ-invariant algebraic cycle on X (resp. a real algebraic cycle on $X(\mathbb{R})$) then the inverse of the cycle map*

$$H_*(X(\mathbb{R}); \mathbb{Z}_2) \to H_*(X; \mathbb{Z}_2)$$

induces a ring morphism that doubles dimensions.

Example 3.7.14 A basic example of this phenomenon is projective space. In this case, the cycle map

$$H_{2k}(\mathbb{P}^n(\mathbb{C}); \mathbb{Z}) \otimes_{\mathbb{Z}} \mathbb{Z}_2 \simeq H_{2k}(\mathbb{P}^n(\mathbb{C}); \mathbb{Z}_2) \longrightarrow H_k(\mathbb{P}^n(\mathbb{R}); \mathbb{Z}_2)$$

is an isomorphism of $\mathbb{Z}_2$-modules.

3.7.4 Applications to $\mathbb{R}$-Surfaces

We end this section by showing that any rational $\mathbb{R}$-surface is totally algebraic. We will continue the study of totally algebraic $\mathbb{R}$-surfaces in Chapter 4.

When (X, σ) is a non singular projective $\mathbb{R}$-surface the $\mathbb{Z}_2$-vector spaces $H^1(X(\mathbb{R}); \mathbb{Z}_2)$ et $H_1(X(\mathbb{R}); \mathbb{Z}_2)$ are isomorphic by Poincaré duality. We denote their dimension by $b^1(X(\mathbb{R}))$. Similarly, we write

$$b^1_{\text{alg}}(X(\mathbb{R})) = \dim_{\mathbb{Z}_2} H^1_{\text{alg}}(X(\mathbb{R}); \mathbb{Z}_2).$$

Any algebraic $\mathbb{R}$-surface X is therefore totally algebraic (Definition 3.7.6) if and only if

$$b^1_{\text{alg}}(X(\mathbb{R})) = b^1(X(\mathbb{R})) .$$

There are natural bounds on b^1_{alg} coming from complex geometry. For example, when X is simply connected, $b^1_{\text{alg}}(X(\mathbb{R}))$ is bounded above by the Hodge number $h^{1,1}(X) = \dim_{\mathbb{C}} H^1(X, \Omega^1_X)$. There are other upper bounds coming from the real structure σ, such as the obvious upper bound $b^1_{\text{alg}}(X(\mathbb{R})) \leqslant b^1(X(\mathbb{R}))$.

We note for future reference that the decomposition

$$H^1(X(\mathbb{R}); \mathbb{Z}_2) = \bigoplus_{V \subset X(\mathbb{R})} H^1(V; \mathbb{Z}_2)$$

where V runs over the set of connected components of $X(\mathbb{R})$ is orthogonal for the intersection form: see Section B.7 for more details, particularly Corollary B.7.7.

Surfaces have a cycle map that is more sophisticated that the one defined above.

Proposition 3.7.15 *Let (X, σ) be a non singular projective $\mathbb{R}$-surface. There is then a surjective map*

$$\varphi_X\colon \operatorname{Pic}(X)^G \to H^1_{\text{alg}}(X(\mathbb{R}); \mathbb{Z}_2) \tag{3.37}$$

defined essentially by mapping a real algebraic curve to the fundamental class of its real part.

Proof We refer to [Sil89, Chapter III]. If (X, σ) has a non empty real locus we know that every linear class of divisors that is invariant under σ can be represented by an invariant divisor. The canonical map $\mathrm{Div}(X)^G \to \mathrm{Pic}(X)^G$ is therefore surjective. As in (3.36) we associate to any irreducible divisor the fundamental class of its real locus if this real locus has codimension 1 and zero otherwise. This yields a morphism $\mathrm{Div}(X)^G \to H_1(X(\mathbb{R}); \mathbb{Z}_2)$. We can prove that the image under this map of a principal divisor to be zero. We then compose with the Poincaré duality map $D_{X(\mathbb{R})}^{-1}\colon H_1(X(\mathbb{R}); \mathbb{Z}_2) \to H^1(X(\mathbb{R}); \mathbb{Z}_2)$, which yields a well-defined morphism $\varphi_X\colon \mathrm{Pic}(X)^G \to H^1(X(\mathbb{R}); \mathbb{Z}_2)$. Conversely, the surjectivity of the map φ_X on $H^1_{\mathrm{alg}}(X(\mathbb{R}); \mathbb{Z}_2)$ can be proved by complexification of algebraic cycles on $X(\mathbb{R})$. (When the real locus is empty φ_X is zero by both convention and necessity). □

We recall (see Definition 3.1.14) that $\mathrm{w}_1(V)$ denotes the first Stiefel–Whitney class of the tangent bundle of the compact differentiable variety V. Let K_X be a canonical divisor on X. We will use the following properties of φ_X:

Proposition 3.7.16 *Let (X, σ) be a non singular projective $\mathbb{R}$-surface. The morphism φ_X defined above has the following properties*

$$\forall D \in \mathrm{Div}(X)^\sigma,\ \forall D' \in \mathrm{Div}(X)^\sigma,\quad \varphi_X(D)\cdot\varphi_X(D') \equiv (D\cdot D') \mod 2 \tag{3.38a}$$

$$w_1(X(\mathbb{R})) - \varphi_X(K_X) \in \varphi_X(\mathrm{Pic}^0(X)^G)\ . \tag{3.38b}$$

Proof In (3.38a) the left hand side uses the intersection form on the vector space $H^1(X(\mathbb{R}); \mathbb{Z}_2)$ and the right hand side uses the intersection form on $\mathrm{Pic}(X)$. The result follows on noting that this intersection form is invariant—see Proposition 4.1.16.

To prove (3.38b) we start by assuming that $H^1(X; \mathbb{Q}) = 0$. The first Chern class map $c_1\colon \mathrm{Pic}(X) \to H^2(X; \mathbb{Z})$ is then injective. If $X(\mathbb{R})$ is non empty $\mathrm{Div}(X)^G \to \mathrm{Pic}(X)^G$ is surjective and $\mathrm{Pic}(X)^G$ can be included in $H_2(X; \mathbb{Z})$ on composition with the Poincaré duality map $D_X\colon H^2(X; \mathbb{Z}) \to H_2(X; \mathbb{Z})$. The morphism φ_X can then be identified with ψ_X:

$$\psi_X \circ D_X \circ c_1 = D_{X(\mathbb{R})} \circ \varphi_X\ .$$

We now interpret $\mathrm{Pic}(X)$ as a group of isomorphism classes of line bundles. Since $c_1(T_X) = -c_1(\mathcal{K}_X)$, note that the restriction to the real locus of the tangent bundle of X is isomorphic to the complexification of the tangent bundle of $X(\mathbb{R})$:

$$T_X|_{X(\mathbb{R})} \simeq T_{X(\mathbb{R})} \otimes \mathbb{C}\ .$$

By definition of ψ_X it follows that

$$\mathrm{w}_1(X(\mathbb{R})) = D_{X(\mathbb{R})}^{-1}(\psi_X(D_X(c_1(\mathcal{K}_X)))) = \varphi_X(K_X)\ .$$

When $H^1(X; \mathbb{Q}) \neq 0$ the image of $D_X(c_1(\mathcal{K}_X))$ under ψ_X is only defined up to addition of an element in $\varphi_X(\mathrm{Pic}^0(X)^G)$. □

Proposition 3.7.17 *Let (X,σ) be a non singular projective $\mathbb{R}$-surface. If the real locus $X(\mathbb{R})$ has a non orientable connected component then*

$$b^1_{\mathrm{alg}}(X(\mathbb{R})) \geqslant 1 \;.$$

Proof This result, first proved in [BKS82], is a combination of Proposition 3.1.15 and the fact that $\mathrm{w}_1(X(\mathbb{R})) \in H^1_{\mathrm{alg}}(X(\mathbb{R});\mathbb{Z}_2)$ because of Property (3.38b). □

Theorem 3.7.18 *Let X be a non singular projective $\mathbb{R}$-surface whose geometric genus $p_g(X)$ is zero (Definition 4.1.1) such that $H_1(X;\mathbb{Z}_2)=0$. We then have that*

$$H^1_{\mathrm{alg}}(X(\mathbb{R});\mathbb{Z}_2) = H^1(X(\mathbb{R});\mathbb{Z}_2)\;.$$

We apply this theorem to rational surfaces below. We refer the interested reader to the survey article [BCP11] for the case of surfaces of general type and geometric genus 0.

Remark 3.7.19 The assumption that the homology of the complex surface X does not contain 2-torsion is necessary. There are non singular projective $\mathbb{R}$-surfaces such that $p_g(X)=q(X)=0$ but $H^1_{\mathrm{alg}}(X(\mathbb{R});\mathbb{Z}_2) \subsetneq H^1(X(\mathbb{R});\mathbb{Z}_2)$. See Theorem 4.5.17 for more details.

Proof By the final hypothesis $H_1(X;\mathbb{Z})_f=0$ from which it follows that $H^1(X;\mathbb{Q})=0$. (See the universal coefficients theorem, B.4.3). We therefore have that $p_g(X)=q(X)=0$ and by the Lefschetz theorem D.9.3 it follows that $\mathrm{Pic}(X)\simeq \mathrm{NS}(X)\simeq H^{1,1}(X)\cap H^2(X;\mathbb{Z}) = H^2(X;\mathbb{Z})$. Moreover we have that

$$\dim_{\mathbb{Z}_2} H^2(G,\mathrm{Pic}(X)) = \dim_{\mathbb{Z}_2} H^1\,(G,H_2(X;\mathbb{Z}))$$

by Proposition D.6.5. The homology of X has no 2-torsion by the universal coefficients theorem B.4.3 and by Corollary 3.6.12 we have that $\dim_{\mathbb{Z}_2} H_1(X(\mathbb{R});\mathbb{Z}_2) = \dim_{\mathbb{Z}_2} H^1\,(G,H_2(X;\mathbb{Z}))$. It remains to show that

$$\dim_{\mathbb{Z}_2} H^1_{\mathrm{alg}}(X(\mathbb{R});\mathbb{Z}_2) = \dim_{\mathbb{Z}_2} H^2(G,\mathrm{Pic}(X))\;.$$

We recall that $H^2(G,\mathrm{Pic}(X)) = \mathrm{Pic}(X)^G/(1+\sigma^*)\,\mathrm{Pic}(X)$ by Proposition 3.1.4. Set $r := \mathrm{rk}\,\mathrm{Pic}(X)^G$ and $\lambda := \dim_{\mathbb{Z}_2}(1+\sigma^*)\,\mathrm{Pic}(X)\otimes_{\mathbb{Z}}\mathbb{Z}_2$. By Lemma 3.1.1 we can find a basis $(d_1,\ldots,d_{\rho(X)})$ of the free $\mathbb{Z}$-module $\mathrm{Pic}(X)$ whose first r elements form a basis of $\mathrm{Pic}(X)^G$ such that

$$\begin{aligned} d_j &\in (1+\sigma^*)\,\mathrm{Pic}(X) \quad \text{for } j=1,\ldots,\lambda\;;\\ d_j &\notin (1+\sigma^*)\,\mathrm{Pic}(X) \quad \text{for } j=\lambda+1,\ldots,r\;. \end{aligned} \tag{3.39}$$

For any a and b in $\mathrm{Pic}(X)$ we have that

$$(a\cdot b) = (\sigma^* a\cdot\sigma^* b)$$

by Proposition 4.1.16. It follows that for any $j = 1, \dots, \lambda$ and for any $d \in \mathrm{Pic}(X)^G$ we have that

$$(d_j \cdot d) = (d' \cdot d) + (\sigma^* d' \cdot d) = 2(d' \cdot d) \equiv 0 \mod 2\,. \tag{3.40}$$

We write the matrix of the intersection form restricted to $\mathrm{Pic}(X)^G$ in the basis $(d_1, \dots, d_r)$ mentioned above: calculating the determinant and using Equation (3.40) we see that for all $j = \lambda + 1, \dots, r$, there is a k such that

$$(d_j \cdot d_k) \equiv 1 \mod 2$$

and it follows that

$$(\varphi(d_j) \cdot \varphi(d_k)) = 1$$

where $\varphi\colon \mathrm{Pic}(X)^G \to H^1_{\mathrm{alg}}(X(\mathbb{R}); \mathbb{Z}_2)$ is the morphism defined in (3.37). This implies that for all j, $\lambda < j \leqslant r$, $\varphi(d_j) \neq 0$. If we replace d_{j_1} by $d_{j_1} + \cdots + d_{j_s}$ (for distinct j_i such that $\lambda < j_i \leqslant r$, we obtain another basis of $\mathrm{Pic}(X)^G$ satisfying the hypotheses (3.39). For the same reasons as above, $\varphi(d_{j_1} + \cdots + d_{j_s}) \neq 0$ or in other words, the $\varphi(d_j)$, $j = \lambda + 1, \dots, r$ are linearly independent in $H^1_{\mathrm{alg}}(X(\mathbb{R}); \mathbb{Z}_2)$. We conclude by noting that since $\dim_{\mathbb{Z}_2} H^1_{\mathrm{alg}}(X(\mathbb{R}); \mathbb{Z}_2) \leqslant \dim_{\mathbb{Z}_2} H^2(G, \mathrm{Pic}(X)) = r - \lambda$, we get that $\dim_{\mathbb{Z}_2} H^1_{\mathrm{alg}}(X(\mathbb{R}); \mathbb{Z}_2) = \dim_{\mathbb{Z}_2} H^2(G, \mathrm{Pic}(X))$. □

Corollary 3.7.20 (Silhol) *Any geometrically rational $\mathbb{R}$-surface is totally algebraic.*

Remark 3.7.21 The proof of Theorem 3.7.18 proposed here is a generalisation of [Sil89, III.(3.4)]. Theorem 3.7.18 can also be found, in a similar generality but with a different proof, in [vH00, Chapter IV, Corollary 4.4 and Chapter III, Lemma 8.9].

3.8 Solutions to exercises of Chapter 3

3.3.4

1. The fundamental group of $\mathbb{RP}^2$ is $\pi_1(\mathbb{RP}^2) \simeq \mathbb{Z}_2$ (see Proposition 2.7.6). By Hurewicz's theorem B.3.9, $H_1(\mathbb{RP}^2; \mathbb{Z}) \simeq \mathbb{Z}_2$ since it is the abelianisation of $\pi_1(\mathbb{RP}^2)$. As the degree zero homology is torsion free, the universal coefficients theorem B.4.3 implies that $H_1(\mathbb{RP}^2; \mathbb{Z}_2) \simeq H_1(\mathbb{RP}^2; \mathbb{Z}) \otimes \mathbb{Z}_2 \simeq \mathbb{Z}_2$.
2. This follows from a similar calculation applied to the fundamental group $\pi_1(\mathbb{S}^1 \times \mathbb{S}^1) \simeq \mathbb{Z} \oplus \mathbb{Z}$ or by direct calculation using the Künneth formula. See Exercise B.6.5 for more details.
3. Use the fact that $\pi_1(\mathbb{K}^2) \simeq \mathbb{Z}_2 \oplus \mathbb{Z}$.

3.3.12 We calculate the total Betti numbers $b^*(C; \mathbb{Z}_2) = 2 + 2g$ and $b^*(C(\mathbb{R}); \mathbb{Z}_2) = 2s$ and it follows that $b^*(C; \mathbb{Z}_2) - b^*(C(\mathbb{R}); \mathbb{Z}_2) = 2(g + 1 - s)$.

3.4.10 Apply Equation (3.17) for $k = 1$

$$\mathrm{w}_1 = \mathrm{Sq}^0(\mathrm{v}_1) + \mathrm{Sq}^1(\mathrm{v}_0) = \mathrm{Sq}^0(\mathrm{v}_1) = \mathrm{v}_1 \ .$$

As $\mathrm{Sq}^p(\mathrm{v}_q) = 0$ whenever $p > q$, Formula (3.17) is simply a sum

$$\mathrm{w}_k = \sum_{p=0,\dots,\lfloor\frac{k}{2}\rfloor} \mathrm{Sq}^p(\mathrm{v}_{k-p}) \ .$$

For $k = 2$ we get that $\mathrm{w}_2 = \mathrm{Sq}^0(\mathrm{v}_2) + \mathrm{Sq}^1(\mathrm{v}_1) = \mathrm{v}_2 + \mathrm{v}_1 \smile \mathrm{v}_1$, and since $\mathrm{v}_1 = \mathrm{w}_1$ on setting $\mathrm{w}_k\,\mathrm{w}_l := \mathrm{w}_k \smile \mathrm{w}_l$ it follows that

$$\mathrm{v}_2 = \mathrm{w}_2 + \mathrm{w}_1^2 \ .$$

We calculate v_3 using Wu's formula [MS74, Problem 8-A]

$$\mathrm{Sq}^p(\mathrm{w}_k) = \sum_{l=0}^{p} \binom{k-p+l-1}{l} \mathrm{w}_{p-l} \smile \mathrm{w}_{k+l}$$

which gives us

$$\mathrm{Sq}^1(\mathrm{w}_2) = \mathrm{w}_1 \smile \mathrm{w}_2 + \mathrm{w}_3 \tag{3.41}$$

and Cartan's formula [MS74, (4) page 91] which gives us $\mathrm{Sq}^1(\mathrm{w}_1 \smile \mathrm{w}_1) = \mathrm{Sq}^0(\mathrm{w}_1) \smile \mathrm{Sq}^1(\mathrm{w}_1) + \mathrm{Sq}^1(\mathrm{w}_1) \smile \mathrm{Sq}^0(\mathrm{w}_1) = 2\,\mathrm{w}_1 \smile \mathrm{w}_1 \smile \mathrm{w}_1 = 0$. It follows that $\mathrm{w}_3 = \mathrm{Sq}^0(\mathrm{v}_3) + \mathrm{Sq}^1(\mathrm{v}_2) = \mathrm{v}_3 + \mathrm{Sq}^1(\mathrm{w}_2 + \mathrm{w}_1 \smile \mathrm{w}_1) = \mathrm{v}_3 + \mathrm{Sq}^1(\mathrm{w}_2) + \mathrm{Sq}^1(\mathrm{w}_1 \smile \mathrm{w}_1)$ and hence

$$\mathrm{v}_3 = \mathrm{w}_1\,\mathrm{w}_2 \ .$$

For v_4 we use $\mathrm{v}_2 \smile \mathrm{v}_2 = (\mathrm{w}_2 + \mathrm{w}_1 \smile \mathrm{w}_1) \smile (\mathrm{w}_2 + \mathrm{w}_1 \smile \mathrm{w}_1) = \mathrm{w}_2 \smile \mathrm{w}_2 + \mathrm{w}_1 \smile \mathrm{w}_1 \smile \mathrm{w}_1 \smile \mathrm{w}_1$, Cartan's formula and equation (3.41): $\mathrm{Sq}^1(\mathrm{w}_1 \smile \mathrm{w}_2) = \mathrm{Sq}^0(\mathrm{w}_1) \smile \mathrm{Sq}^1(\mathrm{w}_2) + \mathrm{Sq}^1(\mathrm{w}_1) \smile \mathrm{Sq}^0(\mathrm{w}_2) = \mathrm{w}_1 \smile (\mathrm{w}_1 \smile \mathrm{w}_2 + \mathrm{w}_3) + \mathrm{w}_1 \smile \mathrm{w}_1 \smile \mathrm{w}_2 = \mathrm{w}_1 \smile \mathrm{w}_3$. We then have that $\mathrm{w}_4 = \mathrm{Sq}^0(\mathrm{v}_4) + \mathrm{Sq}^1(\mathrm{v}_3) + \mathrm{Sq}^2(\mathrm{v}_2) = \mathrm{v}_4 + \mathrm{Sq}^1(\mathrm{w}_1 \smile \mathrm{w}_2) + \mathrm{v}_2 \smile \mathrm{v}_2$ and finally we get

$$\mathrm{v}_4 = \mathrm{w}_4 + \mathrm{w}_1\,\mathrm{w}_3 + \mathrm{w}_2^2 + \mathrm{w}_1^4 \ .$$

3.4.18 Let $X \subset \mathbb{P}^2(\mathbb{C})_{x:y:z}$ be the $\mathbb{R}$-curve of equation $x^2 + y^2 + z^2 = 0$. Its real locus $X(\mathbb{R}) = X \cap \mathbb{P}^2(\mathbb{R})$ is empty. The topological surface underlying X is connected and of genus zero and hence $H_0(X;\mathbb{Z}_2) = H_1(X;\mathbb{Z}_2) = H_2(X;\mathbb{Z}_2) = \mathbb{Z}_2$. In particular, the $\mathbb{Z}_2$-linear map $\sigma_i : H_i(X;\mathbb{Z}_2) \to H_i(X;\mathbb{Z}_2), i = 0, 1, 2$ is the identity, but $b_*(X;\mathbb{Z}_2) = 3$ whereas $b_*(X(\mathbb{R});\mathbb{Z}_2) = 0$.

3.4.21 By hypothesis, the $2m$th Wu class is a polynomial in m variables $\mathrm{w}_{2k} := \mathrm{w}_{2k}(X), k = 1, \ldots, m$, and each monomial in these variables belongs to $H^{2m}(X; \mathbb{Z}_2)$. We therefore have that

$$P(\mathrm{w}) = \sum_{i=(i_1,\ldots,i_m)|\sum 2i_l\alpha_{i_l}=2m} a_i \prod_{l=1}^{m} \mathrm{w}_{2i_l}^{\alpha_{i_l}}$$

from which it follows on setting $c_k := c_k(X), k = 1, \ldots, m$ that

$$P(c) = \sum_{i=(i_1,\ldots,i_m)|\sum 2i_l\alpha_{i_l}=2m} a_i \prod_{l=1}^{m} c_{i_l}^{\alpha_{i_l}} .$$

Since $\sigma^* c_k = (-1)^k c_k$ we get that

$$\sigma^* P(c) = \sum_{i=(i_1,\ldots,i_m)|\sum 2i_l\alpha_{i_l}=2m} a_i \prod_{l=1}^{m} (-1)^{i_l\alpha_{i_l}} c_{i_l}^{\alpha_{i_l}}$$

from which the result follows because for every $i = (i_1, \ldots, i_m)$, $\sum i_l\alpha_{i_l} = m$.

Chapter 4
Surfaces

This chapter contains a partial classification of real algebraic surfaces. Some of the results presented here are classical, others are more recent: we have tried to provide a panorama without attempting to be exhaustive, the selection criteria being the author's personal preferences. This chapter provides a review of the geometry of real and complex surfaces: our leitmotif is an attempt to describe as far as possible the topological types and deformation classes of real algebraic surfaces, and whether each family of surfaces thus described contains any totally algebraic elements.

In an ideal world we would find, as for algebraically closed base fields, a discrete invariant (i.e. a multi-integer) classifying the possible topological types of real varieties, plus, for each value attained by this discrete invariant, a continuous subinvariant, called a *moduli space*.[1] A perfect classification would establish a bijection between irreducible moduli spaces and possible values of the multi-integral invariant.

The first natural integral invariant is the dimension. In dimension 1 the topological classification, established by Klein, was described in Section 3.5: we now review it as motivation for higher dimensional theory. Any non singular projective $\mathbb{R}$-curve (X, σ) is associated to a triplet of integers (g, s, a) where $g := g(X)$ is the genus of the orientable compact surface X, $s := s(X, \sigma)$ is the number of connected components of $X(\mathbb{R})$ and $a := a(X, \sigma)$ is the binary invariant determined by $a := 2 - \#\pi_0(X \setminus X(\mathbb{R}))$. These invariants must satisfy the following conditions.

1. If $a = 0$—i.e., if the curve is *separating* (Definition 3.5.1) then $1 \leqslant s \leqslant g + 1$ and $g - s \equiv 1 \mod 2$.
2. If $a = 1$ then $0 \leqslant s \leqslant g$.

We can give a discrete topological classification of $\mathbb{R}$-curves as follows: for any triplet of integers[2] (g, s, a), $a \in \{0, 1\}$, satisfying the above conditions, there is a

[1] From a set theoretic point of view a "moduli space" is simply a set parameterising possible structures up to isomorphism: in this book we will settle for this point of view. It is rarely simple to equip such a set with an appropriate "space" structure.

[2] Which are necessarily positive or zero.

F. Mangolte, *Real Algebraic Varieties*, Springer Monographs in Mathematics,
https://doi.org/10.1007/978-3-030-43104-4_4

non singular projective $\mathbb{R}$-curve (X, σ) realising it, or in other words such that $g(X) = g$, $s(X, \sigma) = s$ and $a(X, \sigma) = a$. A refined topological classification of $\mathbb{R}$-curves follows because two $\mathbb{R}$-curves (X, σ) and (Y, τ) are deformation equivalent if and only if they have the same triplets: $(g(X), s(X), a(X)) = (g(Y), s(Y), a(Y))$. See [Gab00] for more details.

From dimension 2 onwards it frequently becomes difficult to give such a precise classification, even for special classes of surfaces. We generally start our investigations by classifying real loci up to homeomorphism, as in Section 4.2. Even when we manage to identify a suitable multi-integer for the classification of a particular type of surface and endow the corresponding moduli space with a natural structure, it is not usually clear whether the number of irreducible components of this moduli space is finite. When it is finite, it is often difficult to calculate the number of its irreducible or connected components.

In this chapter we will list

1. All known classifications of real loci of $\mathbb{R}$-surfaces.
2. All known classifications of $\mathbb{R}$-surfaces up to isomorphism.
3. The cases in which the "quasi-simplicity" problem- a real version of the Def=Diff problem, see Question 4.3.29—is solved.

Example 4.0.1 In this chapter we will study various "classes" of $\mathbb{R}$-surfaces: the word "class" is deliberately vague. For example, we will classify topological types of

- Geometrically rational $\mathbb{R}$-surfaces (Definition 4.4.1): the "class" is then a $\mathbb{C}$-birational equivalence class;
- Rational $\mathbb{R}$-surfaces (Definition 4.4.1): the "class" is then an $\mathbb{R}$-birational equivalence class;
- Real Enriques surfaces, (Definition 4.5.13) resp. real K3 surfaces (Definition 4.5.3): in this case we consider that $\mathbb{R}$-surfaces (X, σ) and (Y, τ) belong to the same class if and only if the complex surfaces X and Y belong to the unique irreducible family of complex deformations (Definition 4.3.25) of Enriques surfaces, resp. K3 surfaces;
- Real elliptic surfaces (Definition 4.6.1): the $\mathbb{R}$-surfaces (X, σ) and (Y, τ) belong to the same "class" if the complex surfaces X and Y belong to one of the irreducible families (of which there are an infinite number) of complex deformations of elliptic surfaces.
- Real Jacobian elliptic surfaces of irregularity zero and fixed holomorphic Euler characteristic: in this case the "class" is once again a unique irreducible family of complex Jacobian elliptic surfaces.

4.1 Curves and Divisors on Complex Surfaces

Section 2.6 of Chapter 2 deals with divisors on varieties of arbitrary dimension. Recall in particular that on a non singular irreducible complex variety X there is a one-to-one correspondance between Cartier divisors and Weil divisors and the linear equivalence groups $\mathrm{Cl}(X)$ and $\mathrm{CaCl}(X)$ are isomorphic. To any divisor D on X represented by $(U_i, f_i)_i$ we associate a line bundle $\mathcal{O}_X(D)$ defined by $\mathcal{O}_X(D)|_{U_i} = f_i^{-1}\mathcal{O}_X|_{U_i}$ as in Definition 2.6.11. If the variety X is quasi-projective and non singular then the map $D \mapsto \mathcal{O}_X(D)$ induces an isomorphism

$$\mathrm{Cl}(X) \simeq \mathrm{Pic}(X) ,$$

as in Corollary 2.6.17.

On a surface, prime divisors are just irreducible curves and divisors are linear combinations of irreducible curves with integral coefficients. When this linear combination has positive coefficients the divisor is said to be *effective*. Many authors consider that a *curve on a surface* is simply an effective divisor on this surface: this recalls our plane curves of Section 1.6 which were allowed to be reducible or non-reduced. Recall that as in Definition 2.6.26 the *canonical divisor* K_X of a complex surface X is a[3] divisor associated to the *canonical bundle* $\mathcal{K}_X = \det \Omega_X$. In particular, we have that $\mathcal{O}_X(K_X) = \bigwedge^2 \Omega_X = \Omega_X^2$. On an $\mathbb{R}$-surface (X, σ) the canonical bundle is an $\mathbb{R}$-bundle, ${}^\sigma\mathcal{K}_X = \mathcal{K}_X$. Recall that by the Cartan–Serre theorem D.1.3 the $\mathbb{C}$-vector spaces $H^i(X, \mathcal{F})$ of the cohomology of a coherent sheaf $\mathcal{F}$ are finite dimensional.

Definition 4.1.1 Let X be a non singular complex projective surface, or more generally a compact Kähler surface.

1. The *geometric genus* of X is defined to be $p_g(X) := \dim H^2(X, \mathcal{O}_X)$.
2. The *irregularity* of X is defined to be $q(X) := \dim H^1(X, \mathcal{O}_X)$.
3. The *holomorphic Euler characteristic* of X is defined to be

$$\chi(\mathcal{O}_X) = 1 - q(X) + p_g(X) .$$

4. The *Hodge numbers* of X are defined by $h^{a,b}(X) := \dim H^b(X, \Omega_X^a)$.

Proposition 4.1.2 *Let X be a compact Kähler surface. We have the following identities.*

1. $p_g(X) = \dim H^0(X, \mathcal{K}_X) = h^{2,0}(X) = h^{0,2}(X)$;
2. $q(X) = h^{1,0}(X) = h^{0,1}(X) = h^{3,0}(X) = h^{0,3}(X)$.

Proof We refer to Appendix D for the proofs. By Hodge symmetry we have that $h^{p,q} = h^{q,p}$. As $H^0(X, \mathcal{K}_X) = H^0(X, \Omega_X^2)$ we have that $h^{2,0}(X) = \dim H^0(X, \mathcal{K}_X)$.

[3]We remind the reader that it is customary to call this object "the" canonical divisor despite the fact that it is only defined up to linear equivalence.

We could also have used Serre duality, which gives us $H^2(X, \mathcal{O}_X) = H^0(X, \mathcal{K}_X)$. We complete the proof using Poincaré duality with complex coefficients which gives us $b_k = b_{4-k}$. As this duality is compatible with the Hodge decomposition we get that $h^{p,q} = h^{2-p,2-q}$. □

Definition 4.1.3 Let X be a non singular irreducible complex projective surface. For any $m \geqslant 1$ the number $P_m(X) := \dim H^0(X, \mathcal{K}_X^{\otimes m})$ is called the *mth plurigenus* of X: in particular, $P_1(X) = p_g(X)$. The *canonical dimension* $\kappa(X)$, also called the *Kodaira dimension*, is defined to be the Iitaka dimension of the canonical divisor

$$\kappa(X) := \begin{cases} -\infty & \text{if } P_m(X) = 0 \text{ for any } m \geqslant 1\ ; \\ k \geqslant 0 & \text{the smallest integer such that the sequence } \left\{\frac{P_m(X)}{m^k}\right\}_m \text{ is bounded.} \end{cases}$$

If φ_{mK} denotes the rational map from X to a projective space associated to the linear system $|mK|$ then $\kappa(X)$ is the maximal dimension of the images $\varphi_{mK}(X)$ for $m \geqslant 1$.

It turns out that the Kodaira dimension can be defined for any complex compact analytic variety. For a surface X, $\kappa(X)$ can be $-\infty$, 0, 1 or 2—see Definition D.4.8, Proposition D.4.9 and Remark D.4.10. In what follows, we will consider each possible Kodaira dimensions in turn.

Recall that a projective variety is said to be of *general type* if and only if its canonical bundle is big, or equivalently if $\kappa(X) = \dim X$. See Definitions 2.6.22 and 2.6.29 for more details.

Definition 4.1.4 A complex projective surface X (resp. a projective $\mathbb{R}$-surface (X, σ)) is said to be *of general type* if $\kappa(X) = 2$ and *of special type* if $\kappa(X) < 2$.

Remark 4.1.5 The Kodaira dimension of a scheme is invariant under base change so for any projective $\mathbb{R}$-scheme X we have that $\kappa(X) = \kappa(X \times_{\operatorname{Spec}\mathbb{R}} \operatorname{Spec}\mathbb{C})$.

4.1.1 Intersection Form

The free $\mathbb{Z}$-module $\operatorname{Div}(X)$ generated by curves on a non singular projective surface X has a symmetric bilinear form endowing $\operatorname{Cl}(X)$ with a quadratic module structure.

We start by generalising Definition 1.6.11 of the intersection multiplicity of two plane curves in $\mathbb{P}^2(\mathbb{C})$ to curves in an arbitrary non singular surface.

Definition 4.1.6 Let X be a non singular complex quasi-projective variety, let C_1 and C_2 be two distinct irreducible curves in X and let P be a point in X. If $P \in C_1 \cap C_2$ and f_i is an equation for C_i $(i = 1, 2)$ in the local ring $\mathcal{O}_{X,P}$ of X at P then we set

$$(C_1 \cdot C_2)_P := \dim_{\mathbb{C}} \mathcal{O}_{X,P}/(f_1, f_2)\ .$$

If $P \notin C_1 \cap C_2$ then we set $(C_1 \cdot C_2)_P := 0$ The number thus defined is called the *intersection multiplicity* of the curves C_1 and C_2 at the point P.

If $(C_1 \cdot C_2)_P = 1$ then we say that the curves C_1 and C_2 are *transverse* (or *meet transversely*) at P.

Exercise 4.1.7 1. Prove that if $P \in C_1 \cap C_2$ then the ring $\mathcal{O}_{X,P}/(f_1, f_2)$ is a finite-dimensional complex vector space. (Use the Nullstellensatz.)
2. Prove that $(C_1 \cdot C_2)_P = 1$ if and only if f_1 and f_2 generate the maximal ideal $\mathfrak{m}_P$ (i.e. if and only if f_1 and f_2 form a local system of parameters of X in a neighbourhood of P—see Definition 1.5.47).

Definition 4.1.8 Let X be a non singular complex projective surface and let C_1, C_2 be distinct irreducible curves on X. We set

$$(C_1 \cdot C_2) := \sum_{P \in X} (C_1 \cdot C_2)_P = \sum_{P \in C_1 \cap C_2} (C_1 \cdot C_2)_P \ .$$

This is called the *intersection number* of the curves C_1 and C_2.

Theorem 4.1.9 *Let X be a non singular complex projective surface. There is a unique symmetric bilinear form*

$$\mathrm{Div}(X) \times \mathrm{Div}(X) \longrightarrow \mathbb{Z}, \quad (A, B) \longmapsto (A \cdot B)$$

with the following properties:

- *If A and B are non singular curves who meet transversely then $(A \cdot B) = \#(A \cap B)$;*
- *if A and A' are linearly equivalent then $(A \cdot B) = (A' \cdot B)$ for any divisor B on X.*

Proof See [Bea78, I.4]. □

Definition 4.1.10 Let X be a non singular complex projective surface. It follows from Theorem 4.1.9 that there is a symmetric bilinear form on the $\mathbb{Z}$-module $\mathrm{Cl}(X)$, the *intersection form*

$$\mathrm{Cl}(X) \times \mathrm{Cl}(X) \longrightarrow \mathbb{Z}, \quad (A, B) \longmapsto (A \cdot B) \ .$$

If A and B are divisors on X then we call $(A \cdot B)$ the *intersection number* of A and B and we denote by $(A^2) = (A \cdot A)$ the *self-intersection number* of A.

When there is no risk of confusion we will sometimes abusively denote the intersection number $(A \cdot B)$ by $A \cdot B$.

Proposition 4.1.11 *Let X be a non singular complex projective surface and let $\mathcal{L}_1$ and $\mathcal{L}_2$ be line bundles on X. We set*

$$(\mathcal{L}_1 \cdot \mathcal{L}_2) = \chi(\mathcal{O}_X) - \chi(\mathcal{L}_1^{-1}) - \chi(\mathcal{L}_2^{-1}) + \chi(\mathcal{L}_1^{-1} \otimes \mathcal{L}_2^{-1}) \ .$$

The map

$$\mathrm{Pic}(X) \times \mathrm{Pic}(X) \longrightarrow \mathbb{Z}, \quad (\mathcal{L}_1, \mathcal{L}_2) \longmapsto (\mathcal{L}_1 \cdot \mathcal{L}_2)$$

is then a symmetric bilinear form on the $\mathbb{Z}$*-module* $\mathrm{Pic}(X)$ *and the isomorphism of* $\mathbb{Z}$*-modules* $\mathrm{Cl}(X) \simeq \mathrm{Pic}(X)$ *induced by* $D \mapsto \mathcal{O}_X(D)$ *is an isometry for the symmetric bilinear forms on* $\mathrm{Cl}(X)$ *and* $\mathrm{Pic}(X)$*. In other words, if A and B are two divisors on X then*

$$(\mathcal{O}_X(A) \cdot \mathcal{O}_X(B)) = (A \cdot B) .$$

Proof See [Bea78, Théorème I.4]. □

The restriction of a line bundle to a projective curve has a well-defined degree.

Proposition 4.1.12 *Let C be a non singular irreducible projective curve on X and let $\mathcal{L}$ be a line bundle on X. We then have that*

$$(\mathcal{O}_X(C) \cdot \mathcal{L}) = \deg(\mathcal{L}|_C) .$$

Proof See [Bea78, Lemme I.6]. □

Example 4.1.13 1. If $X = \mathbb{P}^2(\mathbb{C})$ then $\mathrm{Pic}(X) = \mathbb{Z}$ is generated by the class of a line (see Exercise 2.6.5). Any curve of degree d on X is linearly equivalent to a divisor dH where H is a line. Let C and C' be two curves of respective degrees d and d' and let L, L' be two distinct lines. Since $C \sim dL$ and $C' \sim d'L'$ we recover Bézout's theorem

$$(C \cdot C') = (dL \cdot d'L') = dd'(L \cdot L') = dd' .$$

2. If $X = \mathbb{P}^1(\mathbb{C}) \times \mathbb{P}^1(\mathbb{C})$ then $\mathrm{Pic}(X) = \mathbb{Z} \times \mathbb{Z}$ is generated by the classes $F_1 = \{0\} \times \mathbb{P}^1(\mathbb{C})$ and $F_2 = \mathbb{P}^1(\mathbb{C}) \times \{0\}$. The multiplication table is given by $(F_1^2) = (F_2^2) = 0$ and $(F_1 \cdot F_2) = 1$. A curve on X is determined by a bihomogeneous polynomial in four variables. Let C, C' be two curves of bidegrees (d_1, d_2) and (d_1', d_2'): we then have that

$$(C \cdot C') = (d_1 F_1 + d_2 F_2) \cdot (d_1' F_1 + d_2' F_2) = d_1 d_2' + d_1' d_2 .$$

Using the first Chern class map $c_1 \colon \mathrm{Pic}(X) \to H^2(X; \mathbb{Z})$ (see Appendix D for more details) we can link the intersection form to the cup-product (see Section B.7):

Proposition 4.1.14 *Let X be a non singular complex variety of dimension n and let Y be a non singular* compact *complex subvariety of codimension* 1*. The fundamental class of Y in $H_{2n-2}(X; \mathbb{Z})$ is then the Poincaré dual of $c_1(\mathcal{O}_X(Y)) \in H_c^2(X; \mathbb{Z})$.*

Proof See [Hir66, Theorem 4.9.1]. □

Proposition 4.1.15 *Let X be a non singular complex projective surface and let D and D' be divisors on X. We then have that $c_1(\mathcal{O}_X(D)) \in H^2(X; \mathbb{Z})$, $c_1(\mathcal{O}_X(D')) \in H^2(X; \mathbb{Z})$ and*

$$(D \cdot D') = c_1(\mathcal{O}_X(D)) \smile c_1(\mathcal{O}_X(D')) .$$

Proof See [*Ibid.*]. □

Proposition 4.1.16 *The intersection form on a non singular projective $\mathbb{R}$-surface (X, σ) is compatible with the real structure. In other words*

$$\forall \mathcal{L}, \mathcal{L}' \in \mathrm{Pic}(X), (\mathcal{L} \cdot \mathcal{L}') = ({}^{\sigma}\mathcal{L} \cdot {}^{\sigma}\mathcal{L}')$$

and

$$\forall A, B \in \mathrm{Cl}(X), (A \cdot B) = (\sigma A \cdot \sigma B) .$$

Proof In order to apply Proposition 4.1.11 recall that

$$\chi(\mathcal{L}) = \sum (-1)^k \dim_{\mathbb{C}} H^k(X, \mathcal{L}) .$$

The first equation now follows from a simple application of Proposition 2.2.2. We then use Proposition 2.6.30 to obtain the second equation on linear divisor classes.

We note that by Section 3.7, this result also follows from Corollary 3.1.9 in singular cohomology. See [Sil89, II.1] for more details if necessary. □

Throughout the rest of this chapter we will freely identify $\mathrm{Pic}(X)$ and $\mathrm{Cl}(X)$ whenever X is a non singular projective surface.

Theorem 4.1.17 (Serre duality) *Let X be a non singular projective surface and let $\mathcal{L}$ be a line bundle on X. We then have that*

$$H^k(X, \mathcal{L}) \simeq H^{2-k}(X, \mathcal{K}_X \otimes \mathcal{L}^{-1})$$

and in particular

$$\chi(\mathcal{L}) = \chi(\mathcal{K}_X \otimes \mathcal{L}^{-1}) .$$

Proof See [Bea78, Théorème I.11]. □

Theorem 4.1.18 (Riemann–Roch formula for surfaces) *Let X be a non singular projective surface and let D be a divisor on X. We then have that*

$$\chi(\mathcal{O}_X(D)) = \frac{1}{2} D \cdot (D - K_X) + \chi(\mathcal{O}_X) .$$

Proof For any divisor A on X we have that $(\mathcal{O}_X(-A))^{-1} = \mathcal{O}_X(A)$. By Proposition 4.1.11 we therefore have that

$$(-D)\cdot(D-K_X) = \\ \chi(\mathcal{O}_X)-\chi(\mathcal{O}_X(D))-\chi(\mathcal{O}_X(K_X-D))+\chi(\mathcal{O}_X(D)\otimes\mathcal{O}_X(K_X-D))\,.$$

Using Serre duality applied to $\mathcal{O}_X(K_X-D)$ we get that

$$(-D)\cdot(D-K_X)=\chi(\mathcal{O}_X)-\chi(\mathcal{O}_X(D))-\chi(\mathcal{O}_X(D))+\chi(\mathcal{O}_X(K_X))$$

and the required formula follows on applying Serre duality to the canonical bundle $\chi(\mathcal{O}_X(K_X))=\chi(\mathcal{O}_X)$. □

The holomorphic Euler characteristic of a complex surface X is linked to its topological Euler characteristic $\chi_{\text{top}}(X)=\sum_{k=0}^{4}(-1)^k\dim_{\mathbb{Q}} H_k(X;\mathbb{Q})$ by the following formula.

Theorem 4.1.19 (Noether's formula) *Let X be a non singular complex projective surface. We then have that*

$$\chi(\mathcal{O}_X)=\frac{1}{12}(K_X^2+\chi_{top}(X))\,.$$

This formula is often written in terms of Chern numbers $c_1^2(X)=(K_X^2)$ and $c_2(X)=\chi_{\text{top}}(X)$, which yields

$$\chi(\mathcal{O}_X)=\frac{1}{12}(c_1^2(X)+c_2(X))\,.$$

Proof See [GH78, III.5]. □

Definition 4.1.20 Let (X,σ) be a non singular projective $\mathbb{R}$-surface. We recall (Definition 2.6.34, Theorem 2.6.35 and Definition 2.6.36) that $\text{NS}(X)=\text{Pic}(X)/\text{Pic}^0(X)$ is the *Néron–Severi group* of the complex surface X, that $\rho(X)=\text{rk}(\text{Pic}(X)/\text{Pic}^0(X))$ is the *Picard number* of X and if $X(\mathbb{R})$ is non empty then $\rho_{\mathbb{R}}(X)=\text{rk}(\text{Pic}(X)^G/\text{Pic}^0(X)^G)$ is the *real Picard number* of the $\mathbb{R}$-surface (X,σ).

Remark 4.1.21 By definition we have that $\rho_{\mathbb{R}}(X)\leqslant\rho(X)$. By Proposition 2.6.37, if $q(X)=0$ then $\rho(X)=\text{rk}\,\text{Pic}(X)$ and moreover if $X(\mathbb{R})$ is non empty then $\rho_{\mathbb{R}}(X)=\text{rk}\,\text{Pic}(X)^G$ by Theorem 2.6.32.

Definition 4.1.22 Let X be a non singular complex projective surface and let A and B be divisors on X. We denote by $A\equiv B$ the *numerical equivalence* relation: $A\equiv B$ if and only if $(A\cdot C)=(B\cdot C)$ for any effective divisor C on X. We denote by

$$\text{Num}(X):=\text{Div}(X)/\equiv$$

the quotient group.

Proposition 4.1.23 *For any non singular complex projective surface X we have that*

$$\mathrm{Num}(X) \simeq \mathrm{NS}(X)/\operatorname{Tor}(\mathrm{NS}(X)) .$$

Proof See [GH78, Chapter V]. □

Theorem 4.1.24 (Hodge index) *Let X be a non singular projective surface and let H be an ample divisor on X. If D is a divisor on X such that $D \cdot H = 0$ then $(D^2) \leqslant 0$ with equality if and only if $D \equiv 0$.*

Proof We recall that for any divisor D on X we denote by $h^k(D)$ the dimension of the space $H^k(X, \mathcal{O}_X(D))$.

We claim that if D is a divisor on X such that $(D^2) > 0$ then either $h^0(mD) \neq 0$ or $h^0(-mD) \neq 0$ for large enough m.

It follows that either mD or $-mD$ is equivalent to a non zero effective divisor for large enough m which implies that $H \cdot D > 0$ or $H \cdot D < 0$ and the first part of the theorem follows.

We now prove the claim. Let D be a divisor on X such that $(D^2) > 0$. By Riemann–Roch $\chi(\mathcal{O}_X(mD))$ is then equivalent to $\frac{m^2}{2}(D^2)$ as m tends to infinity. As $h^0(mD) + h^2(mD) \geqslant \chi(\mathcal{O}_X(mD))$ either $h^0(mD)$ or $h^2(mD)$ tends to infinity as m tends to infinity. By Serre duality $h^2(mD) = h^0(K_X - mD)$. Using the same argument replacing D by $-D$, we conclude that either $h^0(-mD)$ or $h^0(K_X + mD)$ tends to infinity as m tends to infinity, but $h^0(K_X - mD)$ and $h^0(K_X + mD)$ cannot both tend to infinity. Indeed, if $s \in H^0(X, \mathcal{O}_X(K_X - mD))$ then multiplication by s defines an inclusion $H^0(X, \mathcal{O}_X(K_X + mD)) \hookrightarrow H^0(X, \mathcal{O}_X(2K_X))$. It follows that either $h^0(mD) \neq 0$ or $h^0(-mD) \neq 0$ for large enough m.

To prove the second claim assume that $D \cdot H = 0$ and $(D^2) = 0$. Assume by contradiction that $D \cdot C > 0$ for some effective divisor C. Let $\lambda = \frac{p}{q} \in \mathbb{Q}$ be defined by $(C - \lambda H) \cdot H = 0$. The divisor $mD + q(C - \lambda H)$ then does not satisfy $(mD + q(C - \lambda H))^2 \leqslant 0$ for large enough m despite the fact that $(mD + q(C - \lambda H)) \cdot H = 0$. □

Corollary 4.1.25 *Let X be a non singular complex projective curve. The index of the intersection form (Definition 3.4.1) is then given by*

$$\tau(X) = 2 + 2h^{0,2}(X) - h^{1,1}(X) = 2 + 4p_g(X) - b_2(X) .$$

Proof By the Hodge index theorem the restriction of the intersection form Q to $H^{1,1}(X)$ has signature $(1, h^{1,1}(X) - 1)$. As the restriction of Q to $H^{2,0}(X) \oplus H^{0,2}(X)$ is definite positive the signature of Q is equal to $(h^{2,0}(X) + h^{0,2}(X) + 1, h^{1,1}(X) - 1)$. The result now follows from the identities $h^{2,0}(X) = h^{0,2}(X) = p_g(X)$ and $b_2(X) = h^{2,0}(X) + h^{1,1}(X) + h^{0,2}(X)$. □

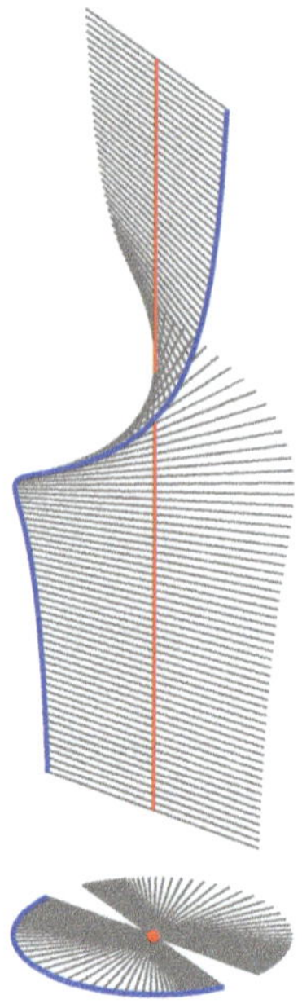

Fig. 4.1 Blow up: the exceptional curve is represented by the vertical line

4.1.2 Blow-Up

See Appendix F or [Bea78, II.1] for more details.

Let X be a complex projective surface and let $P \in X$ be a non singular point. There is then a surface $\widetilde{X}$ and a birational morphism $\pi \colon \widetilde{X} \to X$, unique up to isomorphism, such that

1. The restriction of π to $\widetilde{X} \setminus \pi^{-1}(P) \to X \setminus \{P\}$ is an isomorphism.
2. $E_P := \pi^{-1}(P)$ is isomorphic to $\mathbb{P}^1(\mathbb{C})$,
3. The variety $\widetilde{X}$ is non singular along the divisor E_P.

Definition 4.1.26 The morphism π is called the *blow up of* X *at* P (or *centred at* P, the surface $B_P X := \widetilde{X}$ is called the *blow up* of X at P and the curve $E_P := \pi^{-1}(P)$ is called the *exceptional curve* of the blow up.

Example 4.1.27 (*Blow up of a point in the affine plane*) Applying Definition F.2.1 and restricting ourselves to an affine neighbourhood of $(0, 0)$, we see that the blow up $B_{(0,0)}\mathbb{A}^2$ of $\mathbb{A}^2$ at the point $(0, 0)$ is the quadric hypersurface defined in $\mathbb{A}^2 \times \mathbb{P}^1$ by

$$B_{(0,0)}\mathbb{A}^2 = \{((x, y), [u : v]) \in \mathbb{A}^2_{x,y} \times \mathbb{P}^1_{u:v} \mid uy = vx\} .$$

See Figure 4.1.[4]

[4]Figure created by Daniel Naie.

Example 4.1.28 (*Blow up of a point in the projective plane*) The blow up $B_{(0:0:1)}\mathbb{P}^2$ of $\mathbb{P}^2$ at $P = (0:0:1)$ is the algebraic surface $\widetilde{\mathbb{P}^2}$ defined locally over a neighbourhood $U = (z \neq 0)$ of P by

$$B_P U := \{((x, y), [u:v]) \in U_{x,y} \times \mathbb{P}^1_{u:v} \mid uy = vx\}.$$

More generally, the blow up of the projective plane $\mathbb{P}^2_{x:y:z}$ at a point $P = (a:b:1)$ in the open affine set $(z \neq 0)$ is given by

$$B_{(a:b:1)}\mathbb{P}^2 := \{([x:y:z], [u:v]) \in \mathbb{P}^2_{x:y:z} \times \mathbb{P}^1_{u:v} \mid u(y - bz) - v(x - az) = 0\},$$

and in particular

$$B_{(0:0:1)}\mathbb{P}^2 := \{([x:y:z], [u:v]) \in \mathbb{P}^2_{x:y:z} \times \mathbb{P}^1_{u:v} \mid uy - vx = 0\}.$$

Remark 4.1.29 If X is a complex analytic space we can deduce a description of the blow up of X in a non singular point from the examples above. We simply carry out the blow up in a chart sending an open neighbourhood of P to an open set in $\mathbb{C}^2$. Note that when working with the Zariski topology we cannot generally use this "local" description of blow ups, since a surface containing a dense open subset isomorphic to a non empty Zariski open subset of $\mathbb{A}^2$ must be rational (Definition 4.4.1).

Proposition 4.1.30 *Let X be a non singular complex projective surface, let $\pi\colon \widetilde{X} \to X$ be the blow up of X at a point P and let E be the exceptional curve of π.*

1. *The map* $\operatorname{Pic}(X) \oplus \mathbb{Z} \to \operatorname{Pic}(\widetilde{X})$ *defined by* $(A, n) \mapsto \pi^*A + nE$ *is an isomorphism.*
2. *Let A and B be divisors on X. We then have that*

$$(\pi^*A \cdot \pi^*B) = (A \cdot B), \qquad (E \cdot \pi^*A) = 0, \qquad (E)^2 = -1\,.$$

3. *We have that* $\operatorname{NS}(\widetilde{X}) \simeq \operatorname{NS}(X) \oplus \mathbb{Z}[E]$.
4. *We have that* $K_{\widetilde{X}} = \pi^*K_X + E$.

Proof See [Bea78, II.3]. □

Remark 4.1.31 We can also blow up singular points. In Example 4.7.6 we calculate the blow up of a surface at an ordinary double point.

4.1.3 Adjunction Formula

Let $f\colon C \to X$ be the embedding of an effective divisor in a surface X and let D be a divisor on X. We set

$$\mathcal{O}_C(D) := f^*(\mathcal{O}_X(D))$$

and in particular $\mathcal{O}_C(C) = \mathcal{O}_X(C)|_C$. If C is non singular then $\mathcal{O}_C(C)$ is the normal bundle of C in X.

Theorem 4.1.32 (Non singular variety) *Let X be a non singular complex variety and let Y be a non singular complex subvariety of codimension* 1. *We then have that*

$$\mathcal{K}_Y = \mathcal{K}_X \otimes \mathcal{O}_X(Y)|_Y .$$

Proof See [BHPVdV04, Theorem I.6.3]. □

Remark 4.1.33 The canonical bundles $\mathcal{K}_X$ and $\mathcal{K}_Y$ (see Appendix D) are defined for any non singular varieties and subvarieties.

Corollary 4.1.34 (Non singular curves on a surface) *Let X be a complex surface (which is assumed non singular but not necessarily connected or compact) and let C be a non singular curve on X. The canonical sheaf on C is then given by*

$$\mathcal{K}_C = \mathcal{K}_X \otimes \mathcal{O}_C(C) .$$

Remark 4.1.35 When C is singular but X is non singular the right hand side of the previous formula is well defined and gives rise to a sheaf on C

$$\omega_C := \mathcal{K}_X \otimes \mathcal{O}_C(C) .$$

This definition appears to depend on the embedding of C in X but this turns out not in fact to be the case. The sheaf ω_C is known as the *dualising sheaf* of C. See [BHPVdV04, II.1] and [Har77, III.7] for more details.

4.1.4 Genus of an Embedded Curve

This subsection draws on [BHPVdV04, II.11]. Let C be a non singular connected complex curve: the *geometric genus* of C is the genus of the underlying topological surface (see Definition E.1.2). More generally, if C is a reduced and irreducible complex curve and $\nu\colon \widetilde{C} \to C$ is its normalisation (see Example 1.5.38) then the normalisation $\widetilde{C}$ is a connected non singular curve.

Definition 4.1.36 The *geometric genus* $g(C)$ of a reduced and irreducible complex algebraic curve C is defined to be the topological genus of its normalisation.

$$g(C) := g(\widetilde{C}) .$$

It is a birational invariant.

Definition 4.1.37 The *arithmetic genus* $p_a(C)$ of a complex algebraic curve C (which is assumed neither reduced nor irreducible) is defined by:

$$p_a(C) := 1 - \chi(\mathcal{O}_C) .$$

Remark 4.1.38 If C is non singular, irreducible and reduced then $p_a(C) = g(C)$.

Remark 4.1.39 Assume that C is irreducible and reduced and is embedded in a non singular surface X. The following then hold.

1. The arithmetic genus of C is equal to the geometric genus $g(C')$ of a non singular curve C' obtained by perturbing C in the surface X whenever such a perturbation is possible. (The curve $C' \subset X$ is then linearly equivalent to C.)
2. By Remark 4.1.35 we have that

$$p_a(C) = 1 - \chi(\mathcal{O}_C) = 1 + \chi(\omega_C) .$$

Definition 4.1.40 Let C be a curve on a non singular complex surface and let P be a point of multiplicity r_P on C. The point P is said to be an *ordinary multiple point* of C if and only if it is locally analytically isomorphic to a singularity of the form $\prod_{k=1,\dots,r_P}(x - \varepsilon^k y) = 0$ where ε is a primitive r_Pth root of unity.

Lemma 4.1.41 *Let P be an ordinary multiple point of multiplicity r_P of an irreducible curve C and let $\widetilde{C}$ be the strict transform of C on the blow up of X centred at P. We then have that*

$$\widetilde{C}^2 = C^2 - r_P^2 .$$

Proof By [Bea78, Lemme II.2] we have that $\pi^* C = \widetilde{C} + r E_P$, so the formula follows from Proposition 4.1.30. □

Definition 4.1.42 Let X be a surface, let $P \in X$ be a non singular point and let $X' \to X$ be the blow up of X centred at P with exceptional curve $E_P \subset X'$. Any point $Q \in E_P$ is said to be an *infinitely close* point of P. More generally, if $\pi\colon X'' \to X$ is a sequence of blow ups then any point $Q \in X''$ such that $\pi(Q) = P$ is said to be an *infinitely close point* of P.

Let $C \subset X$ be a reduced curve on a non singular surface. We set $\delta_P(C) = \sum \frac{1}{2} r_Q(r_Q - 1)$ where the sum is taken over all infinitely close points Q of P including P itself. In particular, if P is an ordinary multiple point of multiplicity r on C then $\delta_P(C) = \frac{1}{2} r_P(r_P - 1)$: an ordinary double point counts for 1, a triple point counts for 3 and a quadruple point counts for 6 . See [Har77, Chapitre V, exercice 3.7] for more details.

Proposition 4.1.43 *Let $C \subset X$ be an irreducible reduced curve on a non singular surface and let $\nu\colon \widetilde{C} \to C$ be the normalisation of C. We then have that*

$$p_a(C) = g(\widetilde{C}) + \delta(C)$$

where $\delta(C) = \sum_{P \in C} \delta_P(C)$. *If* C *is non singular then* $\delta(C) = 0$.

Proof See [Har77, Chapitre V, exemple 3.9.2]. □

Theorem 4.1.44 (Adjunction for singular curves) *Let* X *be a complex analytic surface which is assumed to be non singular but not necessarily connected or compact and let* C *be a compact curve on* X *which is assumed neither reduced nor irreducible. We then have that*

$$2p_a(C) - 2 = \deg(\mathcal{K}_X \otimes \mathcal{O}_C(C)) .$$

If moreover X *is compact then the intersection form is well defined and the above equality can be written as*

$$2p_a(C) - 2 = C \cdot (K_X + C) . \tag{4.1}$$

Exercise 4.1.45 Let $(C, \sigma_{\mathbb{P}}|_C)$ be a projective plane reduced and irreducible $\mathbb{R}$-curve of degree 4. Prove that if C is rational—i.e. $g(C) = 0$—then at least one of its singular points is real, i.e. $\operatorname{Sing}(C) \cap C(\mathbb{R}) \neq \varnothing$.

Here is another application of adjunction to $\mathbb{R}$-curves, taken from [KM16, Proposition 23].

Proposition 4.1.46 *Let* $C \subset \mathbb{P}^1 \times \mathbb{P}^1$ *be a rational* $\mathbb{R}$*-curve whose real locus* $C(\mathbb{R})$ *is non singular,* $C(\mathbb{R}) \cap \operatorname{Sing}(C) = \varnothing$. *The fundamental class* $[C(\mathbb{R})] \in H_1(\mathbb{T}^2, \mathbb{Z}_2)$ *is then non vanishing.*

Proof Let $\{E_1, E_2\}$ be a basis of $H_2(\mathbb{P}^1(\mathbb{C}) \times \mathbb{P}^1(\mathbb{C}); \mathbb{Z})$ such that $(E_k)^2 = 0$ and $E_1 \cdot E_2 = 1$. The fundamental class of the complex curve C is therefore equal to $a_1 E_1 + a_2 E_2$ where a_1 and a_2 are natural numbers. The fundamental class of the canonical divisor is given by $K_X = -2E_1 - 2E_2$. The adjunction formula then gives us

$$2p_a(C) - 2 = \big(a_1 E_1 + a_2 E_2\big) \cdot \big((a_1 - 2)E_1 + (a_2 - 2)E_2\big) = a_1(a_2 - 2) + a_2(a_1 - 2),$$

so that

$$p_a(C) = (a_1 - 1)(a_2 - 1) . \tag{4.2}$$

As C is stable under σ we have moreover that

$$a_k = (C \cdot E_{3-k}) \equiv \big(C(\mathbb{R}) \cdot E_{3-k}(\mathbb{R})\big) \mod 2$$

for any $k \in \{1, 2\}$. If the class $[C(\mathbb{R})] \in H_1(\mathbb{T}^2, \mathbb{Z}_2)$ were zero then a_1, a_2 would both be even and $p_a(C)$ would be odd. Since C is rational $g(\widetilde{C}) = 0$ so C would then have an odd number of singular points one of which would be real. □

Using the adjunction formula (4.1) we define the *virtual genus* of a divisor D on a compact surface by

$$p_v(D) := \frac{1}{2}\big(D \cdot (K_X + D)\big) + 1 . \tag{4.3}$$

If $D = A + B$ where A and B are effective divisors then we have that

$$p_v(D) := p_a(A) + p_a(B) + A \cdot B - 1 \tag{4.4}$$

and

$$p_v(-D) = D^2 - p_a(D) + 2 . \tag{4.5}$$

See [Har77, Chapitre V, exercice 1.3] for more details.

Proposition 4.1.47 ([BHPVdV04, II.11.c]) *Let C be a reduced connected curve on a surface X. We then have that $p_v(C) = p_a(C) \geqslant 0$.*

Exercise 4.1.48 Let $C = C_1 + C_2$ be the union of two non singular disjoint rational curves. We then have that $\chi_{top}(C) = 4$ and $p_v(C) = -1$.

4.2 Examples of $\mathbb{R}$-Surfaces

We start by recalling the definitions of some special types of surfaces of negative Kodaira dimension. $\kappa = -\infty$.

Definition 4.2.1 (*Hirzebruch surfaces* [Hir51]) A complex surface X is a *Hirzebruch surface of index n*, denoted $\mathbb{F}_n$, if it is the total space of a locally trivial $\mathbb{P}^1(\mathbb{C})$ bundle over $\mathbb{P}^1(\mathbb{C})$ and $\mathbb{F}_n = \mathbb{P}_{\mathbb{P}^1}(\mathcal{O}_{\mathbb{P}^1} \oplus \mathcal{O}_{\mathbb{P}^1}(n))$, by which we mean that $\mathbb{F}_n$ is the projectivisation of the 2-dimensional vector bundle $\mathcal{O}_{\mathbb{P}^1} \oplus \mathcal{O}_{\mathbb{P}^1}(n)$ over $\mathbb{P}^1$.

By convention, the *real Hirzebruch surface of index n* is obtained by equipping $\mathbb{F}_n$ with the canonical real structure induced by $\sigma_{\mathbb{P}}$.

Proposition 4.2.2 *If $n > 0$ then the curve $E_\infty = \mathbb{P}_{\mathbb{P}^1}(\mathcal{O}_{\mathbb{P}^1}(n))$ is an exceptional section of the line bundle $\mathbb{F}_n \to \mathbb{P}^1$ whose self-intersection is $(E_\infty^2) = -n$.*

Proof See [BHPVdV04, Propositions 4.1 et 4.2, p. 141] and [Bea78, Chapitre III]. □

Remark 4.2.3 The surface $\mathbb{F}_n$ is obtained by gluing the local charts $\mathbb{P}^1_{u:v} \times \mathbb{A}^1_t$ and $\mathbb{P}^1_{u_1:v_1} \times \mathbb{A}^1_{t_1}$ over the open sets $\{t \neq 0\}$ and $\{t_1 \neq 0\}$ via the map

$$((u : v), t) \mapsto ((u_1 : v_1), t_1)$$

where $t_1 = \frac{1}{t}$ and $uv_1 = t^n u_1 v$.

Remark 4.2.4 Hirzebruch surfaces have negative κ dimension because the general fibre of $\mathbb{F}_n \to \mathbb{P}^1$ has negative κ dimension.

Exercise 4.2.5 1. Prove that the Hirzebruch surface $\mathbb{F}_1$ is isomorphic to the blow up of $\mathbb{P}^2(\mathbb{C})$ in a point.
2. Prove that if n is odd then $\mathbb{F}_n$ has only one equivalence class of real structures.
3. Prove that if $n = 2k$ then $\mathbb{F}_{2k}$ has a second class of real structures whose real locus is empty.

Example 4.2.6 (*Conic bundles over* $\mathbb{P}^1$) Historically a *conic bundle* over a field K was a surface given by an equation of the form

$$x^2 + axy + by^2 = f(t)$$

with $a, b \in K$ and $f \in K[t]$. When $K = \mathbb{R}$, on reducing the quadratic form on the left, we can always reduce to the case of an equation of the form

$$x^2 - ay^2 = f(t)$$

where $a = -1, 0, 1$. Completing the affine surface defined above gives us a variety X_0 whose equation is

$$x^2 - ay^2 - f(t)z^2 = 0$$

in $\mathbb{P}^2_{x:y:z} \times \mathbb{A}^1_t$.

Recall that we can define the round up $\lceil x \rceil$ of a real number x using the round down: $\lceil x \rceil = -\lfloor -x \rfloor$. We denote by $m := \left\lceil \frac{\deg f}{2} \right\rceil$ so the degree of f is equal to $2m$ or $2m - 1$. Set $f_1 := t^{2m} f(\frac{1}{t})$ (classically f_1 is called the *reciprocal polynomial* of f) and glue the surface X_1 of equation $x_1^2 - ay_1^2 - f_1(t_1)z_1^2 = 0$ in $\mathbb{P}^2_{x_1:y_1:z_1} \times \mathbb{A}^1_{t_1}$ to the surface X_0 along the open sets $\{t \neq 0\}$ and $\{t_1 \neq 0\}$ using the isomorphism

$$((x : y : z), t) \longmapsto ((x_1 : y_1 : z_1), t_1) = ((x : y : zt^m), \frac{1}{t}) .$$

If $a \neq 0$ and f has simple roots then X is a non singular projective surface and the map $\pi \colon X \to \mathbb{P}^1$ defined by $\pi \colon ((x : y : z), t) \mapsto t$ on X_0 and $\pi \colon ((x_1 : y_1 : z_1), t_1) \mapsto t_1$ on X_1 turns X into a conic bundle over $\mathbb{P}^1$.

If moreover the degree of f is even we can avoid having a root at infinity and we can choose the sign of the dominant coefficient in such a way that the real locus of the completed surface is diffeomorphic to that of the initial surface X_0.

Remark 4.2.7 If the degree of f is odd there is at least one fibre of the form $x^2 - y^2 = 0$ consisting of two real lines meeting in a point.

Example 4.2.8 (*Topology of conic bundles*) Let X be the projective completion of the conic bundle of the equation

$$x^2 + y^2 = f(t)$$

where $f \in \mathbb{R}[t]$ is a polynomial of even degree $2s$ which is negative at infinity and has exactly $2s$ distinct real simple zeros, for example $f(t) = -\prod_{i=1,\dots,2s}(t-i)$ for $s \geqslant 2$. It follows immediately that $X(\mathbb{R})$ is compact and has s connected components. The variety $X(\mathbb{R})$ is a disjoint union of s spheres.

The examples above can be generalised to conic bundles over a curve of arbitrary genus.

Definition 4.2.9 (*Conic bundles*) A *conic bundle* is a pair (X, π) where X is a complex surface and $\pi\colon X \to B$ is a morphism to a non singular complex curve such that every fibre is isomorphic to a possibly singular or non reduced plane conic (see Exercise 1.2.68). A *real conic bundle* is a pair $((X, \sigma), \pi)$ where (X, σ) is an $\mathbb{R}$-surface and $\pi\colon X \to B$ is a morphism of $\mathbb{R}$-varieties to an $\mathbb{R}$-curve (B, σ_B) such that every fibre is isomorphic as a complex curve to a plane conic.

Remark 4.2.10 A Hirzebruch surface is a conic bundle whose fibres are all non singular.

Exercise 4.2.11 (*Conic bundle*)

1. Prove that any surface with a conic bundle structure has negative κ dimension.
2. Prove that the total space X of a conic bundle $\pi\colon X \to B$ with at least one irreducible fibre is non singular if and only if all the fibres of π are reduced.
3. Prove that any conic bundle over $\mathbb{P}^1$ with reduced complex fibres is the blow up of a Hirzebruch surface in a finite number of points.
4. Give an example of a real conic bundle whose real locus is not connected in the Euclidean topology.

Definition 4.2.12 (*Del Pezzo surfaces*) A complex surface X is said to be a *del Pezzo surface* if and only if its anti-canonical bundle $-K_X$ is ample. The *degree* of the del Pezzo surface X is then defined to be the integer (K_X^2). A *real del Pezzo surface* is an $\mathbb{R}$-surface (X, σ) such that X is a del Pezzo surface.

We refer the interested reader to Demazure's survey [DPT80, pp. 21–69] for a study of the multicanonical morphisms of del Pezzo surfaces and their generalisations, the *weak del Pezzo surfaces*, whose anti-canonical divisor $-K_X$ is only assumed *nef* and *big*.

Exercise 4.2.13 1. Prove that del Pezzo surfaces have negative κ dimension.
2. Give an example of a del Pezzo surface with a conic bundle structure. (See [BM11] for a characterisation of such surfaces.)
3. Prove that $\mathbb{P}^1 \times \mathbb{P}^1$ is a degree 8 del Pezzo surface.
4. Prove that a double cover of the projective plane branched along a non singular quartic curve is a del Pezzo surface of degree 2.

4.2.1 Topological Surfaces: Conventions and Notations

A topological surface is a topological manifold of dimension 2. Recall that any topological manifold of dimension 2 has a unique $\mathcal{C}^\infty$ differentiable manifold structure (see [Hir76, Chapter 9] for more details) and any homeomorphism between topological manifolds can be approximated by $\mathcal{C}^\infty$ diffeomorphisms. We will therefore always assume that any topological surface comes equipped with this differentiable structure and our *topological surfaces* will be differentiable manifolds of real dimension 2. It will therefore make sense to talk about *diffeomorphisms* between topological surfaces, for example.

Throughout this section our topological surfaces will be assumed *compact*:

Convention 4.2.14 *A* topological surface *is a compact topological manifold without boundary of dimension* 2.

Notation 4.2.15 *If A and B are topological surfaces, we will write $A \approx B$ if A and B are $\mathcal{C}^\infty$-diffeomorphic. $A \sqcup B$ will be the disjoint union of A and B and $A\#B$ will be their connected sum as in Definition B.5.12 and Remark B.5.13. $\sqcup^s A$ will denote the disjoint union of s copies of A and $\#^k A$ will denote the connected sum of k copies of A. By convention, $\sqcup^0 A = \varnothing$ and $\#^0 A = \mathbb{S}^2$. We denote by*

1. $\mathbb{S}^2$ *the sphere of dimension* 2;
2. $\mathbb{T}^2 \approx \mathbb{S}^1 \times \mathbb{S}^1$ *the torus of dimension* 2;
3. $\mathbb{S}_g = \#^g\mathbb{T}^2$ *the orientable topological surface of genus* $g \geqslant 0$. *In particular,* $\mathbb{S}_0 = \mathbb{S}^2$ *and* $\mathbb{S}_1 = \mathbb{T}^2$;
4. $\mathbb{RP}^2 \approx \mathbb{S}^2/\mathbb{Z}_2$ *the real projective plane;*
5. $\mathbb{K}^2$ *the Klein bottle;*
6. $\mathbb{V}_g = \#^g\mathbb{RP}^2$ *the non orientable surface of genus*[5] $g > 0$. *In particular* $\mathbb{V}_1 = \mathbb{RP}^2$ *and* $\mathbb{V}_2 = \mathbb{K}^2$.

Remark 4.2.16 For $g > 0$, $\mathbb{V}_g$ is the non orientable surface of topological Euler characteristic $2 - g$ since $\mathbb{V}_g$ is the connected sum of g copies of $\mathbb{RP}^2 = \mathbb{V}_1$. For convenience we extend the notation $\mathbb{V}_g$ to the case $g = 0$: $\mathbb{V}_0 = \mathbb{S}_0 = \mathbb{S}^2$.

Exercise 4.2.17 Prove that the real locus of a Hirzebruch surface $\mathbb{F}_n$ equipped with its canonical real structure is diffeomorphic to the torus $\mathbb{T}^2$ if n is even and to the Klein bottle $\mathbb{K}^2$ if n is odd.

Example 4.2.18 See Appendix F for more details. If (X, σ) is an $\mathbb{R}$-surface and $P \in X(\mathbb{R})$ is real then by the universal property of blow ups (Corollary F.2.6) σ lifts to a real structure on the blow up $B_P X$ and E_P is an $\mathbb{R}$-curve for this structure. The real locus therefore satisfies $(B_P X)(\mathbb{R}) = B_P(X(\mathbb{R}))$ and we can denote it by $B_P X(\mathbb{R})$ without risking confusion. Topologically, blow up corresponds to the

[5] As there are at least two incompatible definitions of the genus of a non orientable surface in the literature, let us specify that we will use Riemann's original definition (see Definition E.1.2): the genus $g := g(S)$ of a compact surface S of Euler characteristic $e := e(S)$ is given by $g := 2 - e$ if S is non orientable and by $g = \frac{2-e}{2}$ if S is orientable.

following surgery of the real locus: we remove from $X(\mathbb{R})$ a disc centred at P (whose boundary is a circle) and we glue along this circle a Möbius band (whose boundary is also a circle) to get $B_P X(\mathbb{R})$. In other words:

$$B_P X(\mathbb{R}) \approx X(\mathbb{R}) \# \mathbb{RP}^2 \ .$$

Example 4.2.19 (*Real algebraic models of compact surfaces*) We now present real algebraic models (see Introduction) of all compact topological surfaces. In other words, for any finite family of integers $g_i \geqslant 0$ and $g'_j \geqslant 0$ we give an example of an $\mathbb{R}$-surface whose real locus is diffeomorphic to a disjoint union of the $\mathbb{S}_{g_i}$s and $\mathbb{V}_{g'_i}$s. We will explain the rationale behind the choice of these particular algebraic models in subsequent sections. Note that these algebraic models are not all of negative Kodaira dimension because Theorem 4.4.14 gives constraints on the topology of the real locus of such a variety. On the other hand, for each of these real algebraic models (X, σ) the complex surface X is simply connected in the Euclidean topology.

1. The real locus of (X, σ) is connected and non empty.
   (a) The real projective plane $\mathbb{RP}^2 \approx \mathbb{P}^2(\mathbb{R})$, $X = \mathbb{P}^2(\mathbb{C})$.
   (b) The quadric sphere in $\mathbb{R}^3_{x,y,z}$

$$\mathbb{S}^2 \approx \mathcal{Z}(x^2 + y^2 + z^2 - 1)$$

   whose projective completion is the quadric sphere in $\mathbb{P}^3(\mathbb{R})$

$$\mathbb{S}^2 \approx Q_{3,1}(\mathbb{R})$$

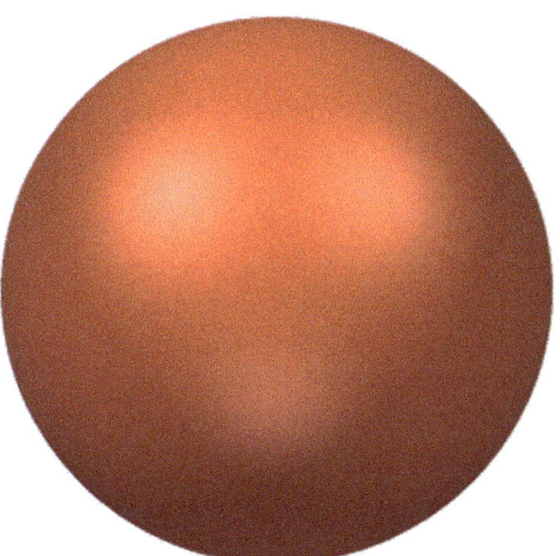

Fig. 4.2 Quadric sphere $\mathbb{S}^2$

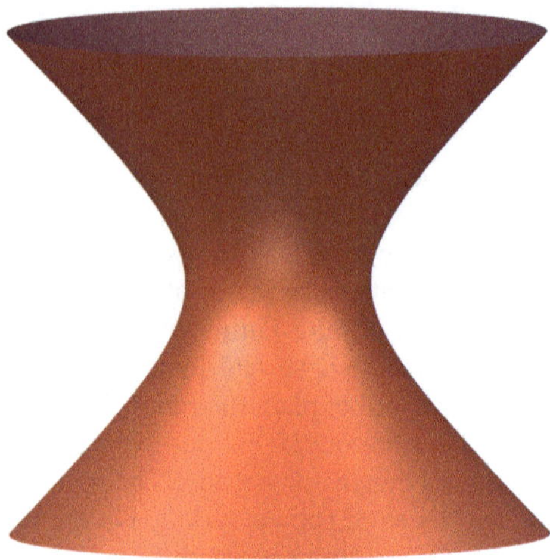

Fig. 4.3 Hyperboloid in $\mathbb{R}^3$ whose projective completion is the quadric torus $Q_{2,2}(\mathbb{R}) \subset \mathbb{P}^3(\mathbb{R})$

where

$$X = Q_{3,1} := \mathcal{Z}(x^2 + y^2 + z^2 - w^2) \subset \mathbb{P}^3_{w:x:y:z}(\mathbb{C}) .$$

(c) The quadric torus

$$\mathbb{T}^2 \approx Q_{2,2}(\mathbb{R}) \subset \mathbb{P}^3(\mathbb{R}) ,$$

where $X = Q_{2,2} := \mathcal{Z}(x^2 + y^2 - z^2 - w^2) \subset \mathbb{P}^3(\mathbb{C})$. This is the projective completion of the hyperboloid of revolution $\mathcal{Z}(x^2 + y^2 - z^2 - 1) \subset \mathbb{R}^3$.

(d) The Klein bottle is a blow up of the projective plane $\mathbb{K}^2 \approx B_P\mathbb{P}^2(\mathbb{R})$ at a point $P \in \mathbb{P}^2(\mathbb{R})$—see Example 4.2.18. The blow-up of the projective plane at a point is also a Hirzebruch surface of index 1 and $\mathbb{K}^2 \approx \mathbb{F}_1(\mathbb{R})$. See Exercise 4.2.11.

(e) The non orientable surface of genus g can be obtained from the blow up of the projective plane in $g - 1$ points

$$\mathbb{V}_g \approx B_{P_1,\ldots,P_{g-1}}\mathbb{P}^2(\mathbb{R})$$

where $P_1, \ldots, P_{g-1} \in \mathbb{P}^2(\mathbb{R})$ as in Example 4.2.18.

(f) The orientable surface $\mathbb{S}_g$ of genus $g \leqslant 10$ can be obtained as the real locus of a K3 surface; see Section 4.5 for more details.

(g) The orientable surface $\mathbb{S}_g$ of arbitrary genus g can be obtained as the real locus of a proper elliptic surface over $\mathbb{P}^1$: see Section 4.6 for more details.

2. The real locus of (X, σ) is empty or not connected.

(a) The empty set is the real locus of the quadric

$$\varnothing = Q_{4,0}(\mathbb{R}) \subset \mathbb{P}^3(\mathbb{R})$$

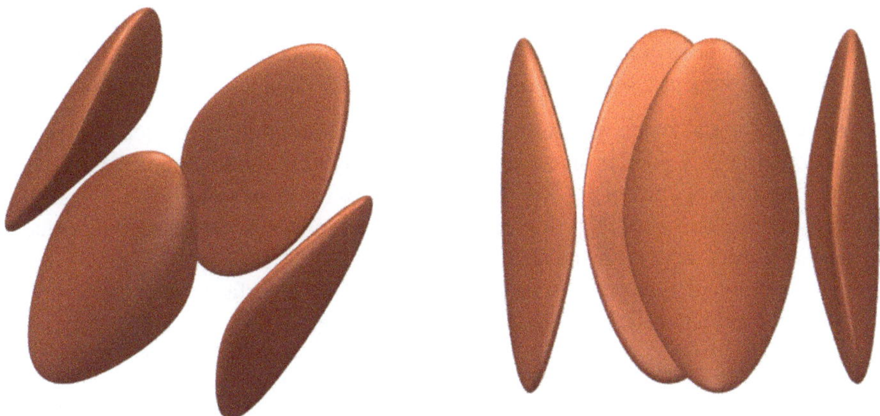

Fig. 4.4 A del Pezzo surface of degree 2 with four connected components

where

$$X = Q_{4,0} := \mathcal{Z}(x^2 + y^2 + z^2 + w^2) \subset \mathbb{P}^3(\mathbb{C}) .$$

(b) For any $s \geqslant 2$ the disjoint union of s spheres can be obtained as the real locus of the projective completion of the conic bundle

$$\mathcal{Z}(x^2 + y^2 - f(t)) \subset \mathbb{R}^3$$

where $f(t) = -\prod_{i=1,\ldots,2s}(t - i)$. See Example 4.2.8 for more details.

(c) The disjoint union of a finite number $s \geqslant 2$ of spheres and non orientable surfaces can be obtained as the real locus of the surface described in Example (2b) by blowing up real points.

(d) The disjoint union of four spheres is the real locus of a certain del Pezzo surface $(X, \sigma_{\mathbb{P}}|_X)$ of degree 2 (the exact values of the coefficients correspond to the diagram in Figure 4.4). Specifically, X is a projective completion in a weighted projective space $\mathbb{P}(1, 1, 1, 2)$ (i.e. the quotient of $\mathbb{C}^4_{z_0,\ldots,z_3}$ by the $\mathbb{C}^*$ action given by $(z_0, \ldots, z_3) \mapsto (\lambda z_0, \lambda z_1, \lambda z_2, \lambda^2 z_3)$) of the affine surface of equation

$$z^2 + 8x^4 + 20x^2y^2 - 24x^2 + 8y^4 - 24y^2 + 16,25 = 0 . \tag{4.6}$$

Note that by Proposition 2.3.22 this surface is a geometrically rational non rational $\mathbb{R}$-surface as in Definition 2.3.18. It is also a minimal $\mathbb{R}$-surface, see Definition 4.3.10.

(e) The disjoint union of a finite number of orientable and non orientable surfaces can be obtained as the real locus of a surface obtained from an elliptic surface fibered over $\mathbb{P}^1$ as in Section 4.6 by blowing up real points.

Remark 4.2.20 All compact topological surfaces therefore have a real algebraic model (X, σ) whose complex surface X is simply connected in the Euclidean topology. We now present a selection of models for which X is not simply connected.

1. All topological types which do not contain an orientable connected component of genus strictly greater than 1 can be realised by a real conic bundle over a curve B of non zero genus $g(B)$. See Theorem 4.4.14 for more details.
2. Only a finite number of topological types can be realised by real Enriques surfaces. See Theorem 4.5.16 for more details.
3. All topological types can be realised by real elliptic surfaces over a curve B of non zero genus $g(B)$: see Section 4.6 for more details.

4.3 $\mathbb{R}$-Minimal Surfaces

We refer the interested reader to [Kol01a, Section 2]—containing most of the preprint [Kol97]—for a presentation of minimal surfaces based on Mori theory.

Definition 4.3.1 The inverse operation of a blow-up is called a *contraction*: see Appendix F for more details. A contraction $\pi: X \to Y$ is an $\mathbb{R}$-*contraction* if and only if the birational morphism π is an $\mathbb{R}$-morphism.

Of course, not every curve can be contracted to a non singular point, despite the fact that it is possible to blow up any non singular point.

Definition 4.3.2 (*$(-n)$-curves*)

1. A (-1)-*curve* L on a non singular complex projective surface X is a curve isomorphic to $\mathbb{P}^1(\mathbb{C})$ whose self intersection $L \cdot L$ is -1. In particular, L is rational, irreducible and non singular.
2. A (-1)-*real curve* L on a non singular projective $\mathbb{R}$-surface (X, σ) is a complex (-1)-curve which is stable under σ. We equip any such curve with the restriction of σ.
3. More generally, for any natural number n, a $(-n)$-*curve* L on X is a curve isomorphic to $\mathbb{P}^1$ such that $L \cdot L = -n$.

Consider a point $P \in X$ and let $\pi_P: B_P X \to X$ be the blow up of X centred at P. It follows from Proposition 4.1.30 that the exceptional line $E_P := \pi^{-1}(P)$ is a (-1)-curve.

Exercise 4.3.3 Prove that any (-1)-curve C on a surface X satisfies $(K_X \cdot C) = -1$.

We have the following criterion for curve contractions on a complex surface:

Theorem 4.3.4 (Castelnuovo's criterion) *Suppose that Y is a non singular complex projective surface and that $E \subset Y$ is a (-1)-curve. There is then a projective surface X and a morphism $\pi: Y \to X$ such that $P = \pi(E)$ is a non singular point of X and π is the blow up of X centred on P.*

We refer to [Bea78, II.17] for a proof. More generally, Grauert's theorem enables us to contract curves to not necessarily non singular points.

Theorem 4.3.5 (Grauert) *Let $E \subset Y$ be a connected reduced projective curve on a non singular complex projective surface Y and let $E = \sqcup E_i$ be its decomposition into irreducible components.*

There is then a (not necessarily projective) normal algebraic surface X and a birational map $\pi\colon Y \to X$ such that $P = \pi(E)$ is a point of X and the restriction of π to $Y \setminus E \to X \setminus P$ is an isomorphism if and only if the matrix $(E_i \cdot E_j)_{i,j}$ is negative definite.

See [BHPVdV04, Theorem III.2.1] for more details.

Corollary 4.3.6 *For any non singular $\mathbb{R}$-surface (Y, τ) it follows from Castelnuovo's criterion that any (-1)-real curve or any pair of disjoint conjugate (-1)-curves can be contracted to a non singular $\mathbb{R}$-surface (X, σ).*

Proof See [Sil89, II.6.2]. □

Example 4.3.7 (*See Example* 4.2.18) In the first case of Corollary 4.3.6 we have that $Y(\mathbb{R}) \approx X(\mathbb{R})\#\mathbb{RP}^2$ and in the second case $Y(\mathbb{R}) \approx X(\mathbb{R})$.

Proposition 4.3.8 (Strong factorisation) *Any birational map between non singular complex projective algebraic surfaces factorises as a sequence of blow ups and contractions of (-1)-curves. More precisely, if $f\colon X \dashrightarrow Y$ is a birational map then there is a non singular complex algebraic surface Z and birational morphisms $\pi_1\colon Z \to X$ and $\pi_2\colon Z \to Y$ such that the diagram below is commutative.*

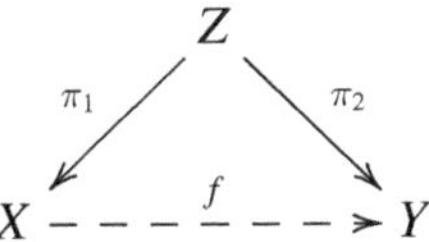

Proof See [Bea78, II.12]. □

Corollary 4.3.9 (Strong factorisation for ℝ-surfaces) *Any birational real map between non singular projective $\mathbb{R}$-surfaces factorises as a sequence of blow ups of real points, blow ups of pairs of conjugate points, contractions of real (-1)-curves and contractions of disjoint conjugate pairs of (-1)-curves. More precisely, if $f\colon (X, \sigma) \dashrightarrow (Y, \tau)$ is a real birational map then there is a non singular projective algebraic $\mathbb{R}$-surface (Z, σ_Z) and $\mathbb{R}$-birational morphisms $\pi_1\colon Z \to X$ and $\pi_2\colon Z \to Y$ such that the diagram below commutes*

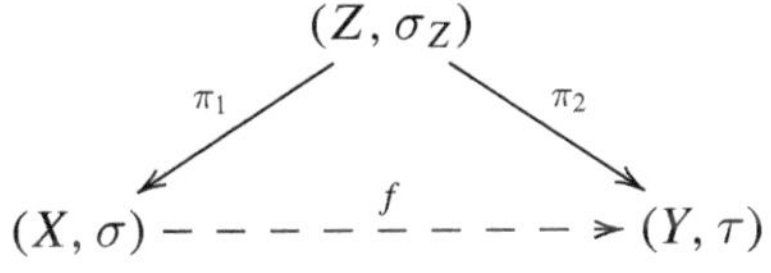

Proof See [Sil89, II.6.4]. □

Definition 4.3.10 (*Minimal surfaces*)

1. A non singular complex surface X is said to be *minimal* if and only if it has no contraction to a non singular surface.
2. A non singular $\mathbb{R}$-surface (X, σ) is said to be *minimal* if and only if it has no $\mathbb{R}$-contraction to a non singular $\mathbb{R}$-surface.

Remark 4.3.11 Riemannian geometers also study *minimal surfaces*, by which they mean (compact) surfaces with a certain boundary whose area is minimal amongst (compact) surfaces with the same boundary in a given Riemannian manifold: the best known example of this phenomenon is soap bubbles. Obviously, there is no link between these two types of minimal surface.

Remark 4.3.12 By Theorem 4.3.4, a non singular complex surface X is minimal if and only if it contains no (-1)-curves. By Corollary 4.3.6, a non singular $\mathbb{R}$-surface is minimal if and only if it contains neither a real (-1)-curve nor a pair of disjoint conjugate (-1)-curves.

Remark 4.3.13 If the complex surface X is minimal then (X, σ) is $\mathbb{R}$-minimal, but the converse is false: this can be seen by considering an $\mathbb{R}$-surface with two conjugate (-1)-curves which meet in a real point. For example, the irreducible components of a singular fibre with equation of the form $x^2 + y^2 = 0$ in a conic bundle are non real conjugate (-1)-curves meeting in a real point $(0, 0)$. The complex surface is not minimal because we can contract one of the (-1)-curves, but there is no contraction to a non singular surface which respects the real structure.

Exercise 4.3.14 We use the same notations as in Example 4.2.6. If the polynomial f is of odd degree then Remark 4.2.7 implies that there is a fibre containing two real (-1)-curves. In particular, the $\mathbb{R}$-surface is not minimal.

Exercise 4.3.15 (*Continuation of Exercise* 4.2.11) Prove that if the total space X of a conic bundle over $\mathbb{P}^1$ is a non singular projective surface then it has $8 - K_X^2$ singular fibres.

Definition 4.3.16 Let (X, σ) be a non singular projective $\mathbb{R}$-surface, let (B, σ_B) be a non singular $\mathbb{R}$-curve and let g be a natural number.

1. We say that a morphism of complex varieties $\pi\colon X \to B$ is a *genus g bundle* if its general fibre is a non singular projective curve of genus g or in other words if there is a non empty Zariski open set U in B such that $\forall x \in U$, $\pi^{-1}(x)$ is isomorphic to a projective non singular curve of genus g.
2. An $\mathbb{R}$-morphism $\pi\colon (X, \sigma) \to (B, \sigma_B)$ is said to be a *real genus g bundle* if the map of complex varieties $\pi\colon X \to B$ is a genus g bundle.
3. We say that the fibered complex surface (X, π) (resp. the fibration π) is *minimal* if no fibre of π contains a (-1)-curve. We sometimes say that (X, π) is *relatively minimal* to underline the fact that this minimality is relative to the morphism.

4. We say that the fibered ℝ-surface $((X, \sigma), \pi)$ is *minimal* (or *relatively minimal*) if no fibre of π contains either a real (-1)-curve or a pair of disjoint conjugate (-1)-curves.

Exercise 4.3.17 Let $\pi\colon X \to B$ be a complex surface which is a genus g bundle. Suppose that (X, π) is relatively minimal and the surface X is not minimal. As no fibre of π contains a (-1)-curve there is a *horizontal* (-1)-curve E, by which we mean that the image $\pi(E)$ is not a point. In this case B is a rational curve and $g = 0$. In particular, the complex surface X is rational.

Exercise 4.3.18 Let $\pi\colon X \to B$ be a complex surface with a genus g bundle structure. Suppose that some fibre of π contains a pair of non disjoint (-1)-curves. Prove that we then have that $g = 0$.

Exercise 4.3.19 1. Prove that any conic bundle is a surface with a genus 0 bundle structure.
2. Prove that any projective non singular ℝ-surface with a genus 0 bundle structure is birationally equivalent to a real conic bundle. (See [Sil89, Corollary V.2.7].)

Proposition 4.3.20 *Let (X, σ) be a non singular projective ℝ-surface with a real genus g bundle structure $\pi\colon X \to B$. We assume moreover that all the fibres of π are geometrically connected.*

1. *If $g \geqslant 1$ then the ℝ-surface (X, σ) is relatively minimal if and only if the complex surface X is relatively minimal.*
2. *If $g = 0$ and (X, σ) is relatively minimal then any fibre F of π containing a (-1)-curve E is necessarily of the form $F = E + \sigma E$ with $E \cdot \sigma E = 1$.*

Proof See [Man67, Man86] or [Sil89, V.1.6]. □

Corollary 4.3.21 *1. A complex conic bundle (X, π) is* minimal *if and only if π has no singular fibres.*
2. An real conic bundle $((X, \sigma), \pi)$ is minimal *if and only if all its singular fibres are real (i.e. lie over $B(\mathbb{R})$) and no irreducible component of a singular fibre of π is a real (-1)-curve.*

Proof All fibres are geometrically connected and a conic has at most two irreducible components. □

Recall that we denote by G the Galois group $\mathrm{Gal}(\mathbb{C}|\mathbb{R})$ acting non trivially on X via the real structure. If X has an ℝ-bundle structure $\pi\colon X \to B$ over a projective ℝ-curve, then we denote by $\mathrm{Pic}(X/B)$ or $\mathrm{Pic}(X/\pi)$, the *relative Picard group*[6]

$$\mathrm{Pic}(X/B) = \mathrm{Pic}(X)/\pi^*(\mathrm{Pic}(B)) \ .$$

[6]The corresponding scheme theoretic object is $\mathrm{Pic}_{X/B}(B)$.

Proposition 4.3.22 *A real del Pezzo surface* (X, σ) *is minimal if and only if* $\mathrm{Pic}(X)^G = \mathbb{Z}$. *A real conic bundle* $(X, \sigma) \to (B, \sigma_B)$ *is minimal if and only if* $\mathrm{Pic}(X/B)^G = \mathbb{Z}$.

Proof Exercise. □

Theorem 4.3.23 (Minimal $\mathbb{R}$-surfaces) *Let* (X, σ) *be a non singular minimal projective* $\mathbb{R}$*-surface. The variety* (X, σ) *is then isomorphic to exactly one* $\mathbb{R}$*-surface from the following list.*

• $\kappa(X) = -\infty$ *(section 4.4)*

1. $(\mathbb{P}^2(\mathbb{C}), \sigma_{\mathbb{P}})$;
2. $(Q_{3,1}, \sigma_{\mathbb{P}}|_{Q_{3,1}})$;
3. $(\mathbb{F}_n, \sigma_{\mathbb{F}_n})$ *where* $\mathbb{F}_n$ *is a real Hirzebruch surface such that* $n \neq 1$; *(in cases* 1, 2 *and* 3, (X, σ) *is rational. See Section 4.4 for more details)*
4. (X, σ) *such that* X *is a complex rational surface and* $X(\mathbb{R}) = \emptyset$, *namely* $Q_{4,0}$, $Q_{3,0} \times \mathbb{P}^1$, *or a surface which is a real conic bundle over a conic with empty real locus* $\pi\colon X \to (\mathbb{P}^1(\mathbb{C}), \sigma_{\mathbb{P}}{}')$ *where* $\sigma_{\mathbb{P}}{}'$ *is the involution of* $\mathbb{P}^1(\mathbb{C})$ *defined by* $(x_0 : x_1) \mapsto (-\overline{x_1} : \overline{x_0})$ *as in Remark 2.1.41;*
5. *a del Pezzo surface of degree* 1 *or* 2 *such that* $\rho_{\mathbb{R}}(X) = 1$;
6. *an real conic bundle* $\pi\colon X \to (\mathbb{P}^1, \sigma_{\mathbb{P}})$ *with an even number of singular fibres* $2r \geqslant 4$ *such that* $\rho_{\mathbb{R}}(X/\pi) = 1$ *or in other words* $\rho_{\mathbb{R}}(X) = 2$; *(in cases* 4, 5 *and* 6, (X, σ) *is geometrically rational but not rational. See Section 4.4 for more details)*
7. *An real conic bundle* $\pi\colon X \to B$ *such that* $g(B) > 0$ *and such that* $\rho_{\mathbb{R}}(X/\pi) = 1$ *or in other words* $\rho_{\mathbb{R}}(X) = 2$; *(in case* 7, (X, σ) *is a uniruled surface which is not geometrically rational. See Section 4.4 for more details)*

• $\kappa(X) = 0$ *(See Section 4.5)*

8. *X is a K3 surface, see Definition 4.5.3;*
9. *X is an Enriques surface see Definition 4.5.13;*
10. *X is an abelian surface, see Definition 4.5.22;*
11. *X is a bi-elliptic surface, see Definition 4.5.28.*

• $\kappa(X) = 1$ *(See Section 4.6)*

12. *X is a properly elliptic surface, see Definition 4.6.10.*

• $\kappa(X) = 2$ *(section 4.7)*

13. *X is a surface of general type, see Definition 4.1.4.*

Proof See [Kol01a, Theorem 30]. □

The above theorem is the basis for the classification of real and complex projective algebraic surfaces. See [BHPVdV04, Chapter VI] for the classification of compact complex analytic surfaces. The classification of projective surfaces in positive characteristic was carried out in a series of articles [Mum69, BM77, BM76, BH75]—see [Bǎd01] for a summary.

Exercise 4.3.24 (*ℝ-elementary transformations*) Let (X, σ) be a Hirzebruch surface of index n and let $P \in X$ be a real point. The blow up of X centred at P transforms the fibre through P into a real (-1)-curve which can then be contracted to a new non singular ℝ-surface X'. Prove that if $n > 0$ then X' is a Hirzebruch surface of index $n+1$ (resp. $n-1$) if P lies on the exceptional section E_∞ (resp. if P does not lie on this section). If $n = 0$ there is no exceptional section and the surface obtained from this transformation is $\mathbb{F}_1$ for any P. The ℝ-surfaces X and X' are birationally equivalent and minimal.

Similarly, let P and $\sigma(P)$ be two conjugate non real points on X and let X'' be the surface obtained by the elementary ℝ-transformation consisting of blowing up the two points and then contracting the conjugate non real (-1)-curves thus obtained. Calculate the index of the Hirzebruch surface obtained, distinguishing the cases where the two points do or do not lie on the exceptional section.

4.3.1 Deformation Families

In the following sections we study the topology of real algebraic surfaces and their deformation families. We also state some theorems on the group of algebraic cycles H^1_{alg}, defined in Section 3.7, which enable us to compare the behaviour of various different families of surfaces. See [Man97, MvH98, Man00, Man03] for more details.

Definition 4.3.25 (*Deformations*)

1. • A complex analytic variety Y is a *deformation* of a complex variety X if and only if there is a complex analytic variety $\mathcal{M}$, a proper holomorphic submersion
$$\pi \colon \mathcal{M} \to D = \{z \in \mathbb{C} \mid |z| < 1\}$$
and a point $z_0 \in D$ such that $X = \pi^{-1}(0)$ and $Y = \pi^{-1}(z_0)$.
   • An analytic ℝ-variety (Y, τ) is a *deformation* of an analytic ℝ-variety (X, σ) if and only if there is an analytic ℝ-variety $(\mathcal{M}, \sigma_\mathcal{M})$, a proper holomorphic submersion $\pi \colon \mathcal{M} \to D$ and a point $z_0 \in [-1, 1] = D(\mathbb{R})$ such that $\sigma_D \circ \varphi = \varphi \circ \sigma_\mathcal{M}$, $X = \pi^{-1}(0)$ and $Y = \pi^{-1}(z_0)$. In particular we have that $\sigma = \sigma_\mathcal{M}|_X$ and $\tau = \sigma_\mathcal{M}|_Y$.
2. Two varieties X and Y are said to be *deformation equivalent* if and only if there is a finite family of varieties Z_i $i = 1 \ldots l$ such that $Z_1 = X$, $Z_l = Y$ and for every i, Z_{i+1} is either a deformation of Z_i or isomorphic to Z_i.

Remark 4.3.26 The definition of deformation equivalence given above is justified by the fact that two varieties that are deformation equivalent do not necessarily belong to the the same non singular deformation family. For example, Horikawa proved in [Hor75] that the space of numerical quintics has two irreducible components of dimension 40 which meet along a subspace of dimension 39.

Definition 4.3.27 Two complex surfaces X and Y are said to belong to the same *complex family* if and only if X and Y are fibres of a proper holomorphic submersion (sometimes called a *large deformation*)

$$\pi: \mathcal{M} \to B$$

over an *irreducible* complex analytic variety B.

Two $\mathbb{R}$-surfaces (X, σ) and (Y, τ) belong to the same *real family* if and only if they are fibres of some equivariant large deformation whose base has *connected* real locus.

Theorem 4.3.28 (Ehresmann's fibration theorem) *Let $f: \mathcal{M} \to B$ be a differentiable map between manifolds: f, $\mathcal{M}$ and B are supposed at least C^2 and at most C^∞. If f is a surjective proper submersion then f is a locally trivial fibration (Definition C.3.5).*

Proof See [Ehr51, Ehr95]. □

The Galois group $G = \mathrm{Gal}(\mathbb{C}|\mathbb{R})$ acts on X (resp. Y, resp. $\mathcal{M}$) by involution σ (resp. τ, resp. $\sigma_{\mathcal{M}}$). Adapting the proof of Ehresmann's fibration theorem, we can prove that if two $\mathbb{R}$-varieties (X, σ) and (Y, τ) are deformation equivalent then X is diffeomorphic to Y via a G-equivariant diffeomorphism [Dim85, Lemma 4] and in particular $X(\mathbb{R})$ is diffeomorphic to $Y(\mathbb{R})$: the converse is false and there are many known examples where it fails.

In general, every complex deformation family corresponds to many real families. For example, there is a unique complex family of Enriques surfaces, but over 200 real families of real Enriques surfaces. See [DIK00] for more details.

Question 4.3.29 *(Def=Diff) The Def=Diff problem is the following: if two complex surfaces X and Y are diffeomorphic, are they necessarily deformation equivalent?*

See [Man01] for the proof that this is not always the case. See [KK02], [Cat03], [Cat08] for other examples. The precise real version of this question was given by Kharlamov.

Definition 4.3.30 (*Quasi-simplicity of $\mathbb{R}$-surfaces*) An $\mathbb{R}$-surface (X, σ) is said to be *quasi-simple* if and only if any $\mathbb{R}$-surface (Y, τ) such that there is a G-equivariant diffeomorphism $(X, \sigma) \to (Y, \tau)$ is deformation equivalent to (X, σ) whenever the complex surface Y is deformation equivalent to X.

The following definition is useful for expressing classifications of topological types of real surfaces appearing in a given class of complex surfaces.

Definition 4.3.31 (*Morse simplification*) Given a compact topological surface S without boundary, (which is neither assumed connected nor orientable) a *topological Morse simplification* of S is a Morse transformation that decreases the total Betti number by two. There are two types of Morse simplifications:

- removing a spherical component $\mathbb{S}^2 \to \varnothing$,
- contracting a handle $\mathbb{S}_{g+1} \to \mathbb{S}_g$ or $\mathbb{V}_{q+2} \to \mathbb{V}_q$.

Definition 4.3.32 (*Topological type, extremal topological type*) A *topological type* is a class of $\mathbb{R}$-surfaces with diffeomorphic real loci. Given a class of complex surfaces (see Example 4.0.1) a topological type is said to be *extremal* if it cannot be obtained by topological Morse simplification from a topological type belonging to the same class of complex surfaces.

Example 4.3.33 In the diagrams shown in Figures 4.11 and 4.12, the extremal topological types are those corresponding to points with no ascending adjacent edge.

Remark 4.3.34 There is a stronger version of topological types in the literature which states that two $\mathbb{R}$-surfaces (X, σ) and (Y, τ) are of the same topological type if and only if there is an equivariant diffeomorphism $(X, \sigma) \to (Y, \tau)$. It is then immediate that $X(\mathbb{R})$ is diffeomorphic to $Y(\mathbb{R})$, but the converse is false. See [DIK00] for more details.

Remark 4.3.35 When the surfaces being considered belong to the same complex deformation class the reader should be aware that Morse simplification is abstract in the sense that the existence of a continuous deformation realising the topological transformation is not guaranteed. It is simply a practical definition helping us list topological types. In certain special cases it is however possible to realise Morse transformations by explicit deformation: see Theorem 4.6.13 for more details.

4.4 Uniruled and Rational Surfaces ($\kappa = -\infty$)

The aim of this section is to classify topological types of rational $\mathbb{R}$-surfaces and more generally classify $\mathbb{R}$-surfaces of negative Kodaira dimension using the topological type of their real locus. The complete classification of non singular projective surfaces of negative Kodaira dimension is Theorem 4.4.14. The main intermediate result is Theorem 4.4.15, sometimes described as a generalisation of Comessatti's theorem 4.4.16, which bounds the genus of an orientable surface contained in the real locus of a rational surface. There are basically two different approaches to the proof of this classification. One is based on reduction to minimal surfaces followed by case by case analysis as in Theorem 4.3.23. The other is based on the action of the Galois group on the cohomology ring of X, and it is this second proof that will be presented in this section. Both methods have their advantages.

4.4.1 Rational $\mathbb{R}$-Surfaces

Let us mention two survey articles on rational $\mathbb{R}$-surfaces[7]: [Man17a] (continuing [Hui11]) dealing with topological classification and [BM14] dealing with birational geometry. We now specialise Definition 2.3.18 to surfaces.

Definition 4.4.1 (*Rational and uniruled $\mathbb{R}$-surfaces*) Let (X, σ) be an algebraic $\mathbb{R}$-surface.

1. The $\mathbb{R}$-surface (X, σ) is said to be *rational* or *$\mathbb{R}$-rational* if and only if it is birationally equivalent to the $\mathbb{R}$-projective plane $(\mathbb{P}^2(\mathbb{C}), \sigma_{\mathbb{P}})$, by which we mean there is a birational map of $\mathbb{R}$-surfaces

$$(X, \sigma) \dashrightarrow (\mathbb{P}^2(\mathbb{C}), \sigma_{\mathbb{P}}) .$$

The real algebraic surface $X(\mathbb{R})$ is then rational (Definition 1.3.37).
2. The $\mathbb{R}$-surface (X, σ) is said to be *geometrically rational* or *$\mathbb{C}$-rational* if and only if it is $\mathbb{C}$-birationally equivalent to the projective plane $\mathbb{P}^2(\mathbb{C})$, or in other words if and only if there is a birational map of complex surfaces

$$X \dashrightarrow \mathbb{P}^2(\mathbb{C}) .$$

The real algebraic surface $X(\mathbb{R})$ is then geometrically rational if it is Zariski dense in X, since the complex algebraic surface X, which is a complexification of $X(\mathbb{R})$, is then rational (Definition 1.3.37).
3. The $\mathbb{R}$-surface (X, σ) is *uniruled* if and only if it is dominated by a cylinder of dimension 2, or in other words if and only if there exists an $\mathbb{R}$-curve (Y, τ) and a rational map of $\mathbb{R}$-varieties

$$(Y \times \mathbb{P}^1, \tau \times \sigma_{\mathbb{P}}) \dashrightarrow (X, \sigma)$$

whose image is dense in the Zariski topology.

Remark 4.4.2 Unlike rationality, uniruledness is invariant under change of base field: an $\mathbb{R}$-surface (X, σ) is uniruled if and only if the complex surface X is uniruled. See [Deb01, §4.1, Remark 4.2(5)] for more details.

Remark 4.4.3 If X is uniruled then $\kappa(X) = -\infty$ because X is dominated by a ruled variety. In dimension 2 the converse holds and in fact a stronger result turns out to be true: any surface X such that $\kappa(X) = -\infty$ is birationally ruled, i.e. birationally equivalent to a cylinder $Y \times \mathbb{P}^1$ of dimension 2. See [Bea78, Exemple VII.3 and Chapitre III] for more details. A complex surface is therefore uniruled if and only if

[7] Several authors, such as [Sil89] or [DK02], consider that (X, σ) is a rational surface whenever the complex surface X is rational: this can be confusing, and in this case we will say that (X, σ) is geometrically rational or $\mathbb{C}$-rational. See Definition 4.4.1 for more details.

it is birationally ruled and such surfaces are often said to be "ruled" in the literature. The notion of "ruled surface" becomes difficult to handle over the real numbers: a conic bundle can by $\mathbb{C}$-birationally equivalent to a ruled surface without being $\mathbb{R}$-birationally equivalent to it—we prove in Proposition 4.4.10 that this is the case in Example 4.2.8 whenever $s \geqslant 2$. This example is however uniruled over both $\mathbb{R}$ and $\mathbb{C}$.

Remark 4.4.4 It follows from Definition 1.3.37 and Proposition 4.3.8 that a complex surface is rational if and only if it is obtained by applying a sequence of blow ups and contractions of (-1)-curves to the complex projective plane. An $\mathbb{R}$-surface (X, σ) is therefore geometrically rational if and only if it can be obtained from the projective plane by a sequence of not necessarily real blow ups and contractions. The function field $K(X)$ of the complex surface X is then isomorphic to the field of rational functions $\mathbb{C}(X_1, X_2)$. If we also require that these blow ups and contractions should be real[8] then the $\mathbb{R}$-surface (X, σ) is rational. In this case the $\mathbb{R}$-algebra of restrictions to $X(\mathbb{R})$ of elements of $K(X)$ is isomorphic as an $\mathbb{R}$-algebra to the field of rational fractions $\mathbb{R}(X_1, X_2)$.

Remark 4.4.5 In Definition 4.4.1 the complex surface X is not assumed to be complete, projective or non singular. We state a classification theorem for non singular projective X (Theorem 4.4.14) below. We will return to singular varieties at the end of this section and we will deal with affine varieties in Chapter 5 (Definition 5.5.2).

Remark 4.4.6 Let F be a real algebraic surface and let (X, σ) be a complexification of F. The surface F is then rational if and only if (X, σ) is $\mathbb{R}$-rational.

Remark 4.4.7 By definition of $\mathbb{R}$-morphisms, any rational $\mathbb{R}$-surface is geometrically rational. Similarly, as the product $\mathbb{R}$-surface $\mathbb{P}^1 \times \mathbb{P}^1$ is birationally equivalent to the $\mathbb{R}$-surface $\mathbb{P}^2$, it is immediate that any geometrically rational $\mathbb{R}$-surface is uniruled.

Theorem 4.4.8 *Let (X, σ) be a non singular projective $\mathbb{R}$-surface such that $\kappa(X) = -\infty$. The following then hold.*

1. *X is uniruled and (X, σ) is birationally equivalent to a real conic bundle $\pi\colon X \to B$ with $g(B) = q(X)$;*
2. *(X, σ) is geometrically rational if and only if $q(X) = 0$;*
3. *(X, σ) is rational if and only if $q(X) = 0$ and $X(\mathbb{R})$ is connected and non empty.*

Proof 1. By Remark 4.4.3, any complex surface such that $\kappa(X) = -\infty$ is uniruled (and indeed birationally ruled). Comessatti proved in [Com12] that any $\mathbb{R}$-surface (X, σ) such that X is birationally ruled is $\mathbb{R}$-birationally equivalent to an real conic bundle $\pi\colon X \to B$ such that $g(B) = q(X)$. See [Sil89, Chapter V] for more details.

---

[8] By "real" we mean "globally real": in other words, if P is the centre of a blow up then so is $\overline{P}$ and if E is contracted then so is $\overline{E}$.

2. This follows from the classification of complex surfaces. See [Bea78, Chapitre IV] for more details.
3. It remains to prove that if the complex surface X is rational then (X, σ) is rational if and only if $X(\mathbb{R})$ is connected and non empty. As the number of connected components of the real locus is invariant under birational maps defined over $\mathbb{R}$ this condition is necessary. It is sufficient by [Sil89, Corollary IV.6.5].

□

Corollary 4.4.9 *Let (X, σ) be a uniruled non geometrically rational $\mathbb{R}$-surface. The surface (X, σ) is then birationally equivalent to a conic bundle $\pi\colon X \to B$ such that $g(B) = q(X) > 0$.*

We refer the interested reader to [Com12, Com14, Isk65, Isk67, Man67, Man86, Sil89, Kol97, Kol01a, DK02, Wel03, BM11] for classical and recent results on the classification of conic bundles.

Recall that $\mathbb{V}_g = \#^g\mathbb{RP}^2$ denotes the non orientable surface of genus g whose Euler characteristic is $2 - g$. For example, $g(\mathbb{RP}^2) = 1$ and $g(\mathbb{K}^2) = 2$. We now list the real algebraic models of Example 4.2.19 which are uniruled.

Proposition 4.4.10 (Examples of uniruled surfaces) *We will now classify the real algebraic models (X, σ) described in Example 4.2.19 for each topological type of the real locus.*

1. *(X, σ) is an $\mathbb{R}$-rational surface.*
   (a) *The real projective plane $X = \mathbb{P}^2(\mathbb{C})$, $\sigma = \sigma_{\mathbb{P}}$, $X(\mathbb{R}) = \mathbb{P}^2(\mathbb{R})$.*
   (b) *The quadric sphere $X = Q_{3,1} \subset \mathbb{P}^3(\mathbb{C})$, $\sigma = \sigma_{\mathbb{P}}|_{Q_{3,1}}$, $X(\mathbb{R}) = \mathbb{S}^2$.*
   (c) *The quadric torus $X = Q_{2,2}$, $\sigma = \sigma_{\mathbb{P}}|_{Q_{2,2}}$, ($(X, \sigma)$ is isomorphic to the $\mathbb{R}$-surface $(\mathbb{P}^1(\mathbb{C}) \times \mathbb{P}^1(\mathbb{C}), \sigma_{\mathbb{P}} \times \sigma_{\mathbb{P}})$ and $Q_{2,2}(\mathbb{R}) = \mathbb{T}^2$.*
   (d) *The torus considered as the real locus of even index Hirzebruch surfaces with their canonical real structure. $\mathbb{F}_{2k}(\mathbb{R}) \approx \mathbb{T}^2$.*
   (e) *The Klein bottle considered as the real locus of the blow up of $\mathbb{P}^2(\mathbb{R})$ at a point $B_P\mathbb{P}^2(\mathbb{R}) = \mathbb{K}^2$.*
   (f) *The Klein bottle considered as the real locus of the Hirzebruch surfaces of odd index $\mathbb{F}_{2k+1}(\mathbb{R}) = \mathbb{K}^2$.*
   (g) *The non orientable surface of genus $g > 0$ obtained by blowing up $\mathbb{P}^2(\mathbb{R})$: $B_{P_1,\ldots,P_{g-1}}\mathbb{P}^2(\mathbb{R}) = \mathbb{V}_g$ for any $P_1, \ldots, P_{g-1} \in \mathbb{P}^2(\mathbb{R})$.*
   (h) *The non orientable surface $g > 0$ obtained by blowing up the quadric sphere: $B_{P_1,\ldots,P_g}Q_{3,1}(\mathbb{R}) = \mathbb{V}_g$ for any $P_1, \ldots, P_g \in Q_{3,1}(\mathbb{R})$.*
2. *(X, σ) is geometrically rational but not rational.*
   (a) *The empty set considered as the real locus of the quadric $X = Q_{4,0}$, $\sigma = \sigma_{\mathbb{P}}|_{Q_{4,0}}$, $X(\mathbb{R}) = \varnothing$.*
   (b) *The disjoint union of $s > 1$ spheres considered as the real locus of the projective completion of the conic bundle*

$$\mathcal{Z}\left(x^2 + y^2 + \prod_{i=1,\dots,2s} (t - i)\right) \subset \mathbb{R}^3 .$$

(c) *The disjoint union of a finite union of $s > 1$ spheres and non orientable surfaces considered as the real locus of the surface obtained by blowing up real points in the above example.*

(d) *The disjoint union of four spheres considered as the real locus of the degree 2 del Pezzo surface of equation* (4.6), *Sect. 4.2.1.*

3. *(X, σ) is a uniruled non geometrically rational variety.*

(a) *The disjoint union of a finite number $s \geqslant 0$ of spheres, toruses and Klein bottles considered as the real locus of conic bundle over a curve of non zero geometric genus.*

(b) *The disjoint union of a finite number $s \geqslant 0$ of spheres, toruses and non orientable surface of arbitrary genus considered as the real locus of a surface produced by blowing up real points in the previous example.*

Proof of Proposition 4.4.10 1. By Corollary 4.3.9 an $\mathbb{R}$-surface (X, σ) is $\mathbb{R}$-rational if and only if there is a sequence of blow ups of real points or conjugate pairs of points and contractions of real (-1)-curves or pairs of disjoint conjugate (-1)-curves producing this surface from the real projective plane.

(a) The $\mathbb{R}$-surface $(\mathbb{P}^2(\mathbb{C}), \sigma_{\mathbb{P}})$ is rational by definition.

(b) The quadric surface $Q_{3,1}$ in projective space $\mathbb{P}^3$ is rational. For any real point P in $Q_{3,1}$ let $T_P Q_{3,1} \subset \mathbb{P}^3(\mathbb{R})$ be the real projectivisation of the tangent plane to $Q_{3,1}$ at P. The stereographic projection $Q_{3,1} \setminus T_P Q_{3,1} \to \mathbb{A}^2$ is then an isomorphism of $\mathbb{R}$-surfaces. For example if P is the north pole $N = [1 : 0 : 0 : 1]$ let $\pi_N \colon Q_{3,1} \to \mathbb{P}^2_{U:V:W}$ be the rational map given by

$$\pi_N \colon [w : x : y : z] \dashrightarrow [x : y : w - z] .$$

The restriction of π_N is then the stereographic projection of $Q_{3,1} \setminus T_N Q_{3,1}$ onto its image $\pi_N(Q_{3,1} \setminus T_N Q_{3,1}) = \{w \neq 0\} \simeq \mathbb{A}^2$.
(The inverse rational map $\pi_N^{-1} \colon \mathbb{P}^2 \dashrightarrow Q_{3,1}$ is given by $[U : V : W] \dashrightarrow [U^2 + V^2 + W^2 : 2UW : 2VW : U^2 + V^2 - W^2]$.).
The rational map π_N can be decomposed as the blow up of $Q_{3,1}$ in N, followed by the contraction of the birational transform of the curve $z = w$ (or in other words the intersection of $Q_{3,1}$ with the tangent plane $T_N Q_{3,1}$), which is the union of two non real conjugate lines. The rational map π_N^{-1} can be decomposed as the blow up of two non real conjugate points $[1 : \pm i : 0]$ followed by the contraction of the birational transform of the line $z = 0$.
The surface $Q_{3,1}$ is therefore birational to the surface Y obtained as follows. Let $P, \overline{P}$ be a pair of non real conjugate points in $\mathbb{P}^2(\mathbb{C})$ and let $L := L_{P,\overline{P}}$ be the line passing through these two points. Note that $(L, \sigma_{\mathbb{P}}|_L)$ is an $\mathbb{R}$-line. The self-intersection number of the strict transform $\widetilde{L}$ of L in the blown up

surface $\widetilde{X} = B_{P,\overline{P}}\mathbb{P}^2(\mathbb{C})$ is

$$(\widetilde{L}^2) = (L^2) - 2 = -1$$

and it follows that there is a contraction $c\colon \widetilde{X} \to Y$ to a non singular $\mathbb{R}$-surface whose real locus is a sphere. Indeed, by construction $\widetilde{X}(\mathbb{R}) \approx X(\mathbb{R}) = \mathbb{RP}^2$ and the contraction c remplaces a Moebius band by a disc so $Y(\mathbb{R}) \approx \mathbb{S}^2$.

(c) We carry out the same construction using two distinct real points P, Q in $\mathbb{P}^2(\mathbb{C})$. We obtain $B_{P,Q}\mathbb{P}^2(\mathbb{R}) \approx \mathbb{V}_2$ and on contracting the $\mathbb{R}$-line $L_{P,Q}$ which is a real (-1)-curve we get a real locus $Y(\mathbb{R})$ diffeomorphic to a torus $\mathbb{T}^2$. Indeed, by construction $Y(\mathbb{R})$ is diffeomorphic to $\mathbb{T}^2$ or $\mathbb{K}^2$. Moreover, the complex surface thus obtained is isomorphic to $\mathbb{P}^1(\mathbb{C}) \times \mathbb{P}^1(\mathbb{C})$ and we know (see Exercise 2.1.42) that the real locus of any real structure on $\mathbb{P}^1(\mathbb{C}) \times \mathbb{P}^1(\mathbb{C})$ is $\varnothing$, $\mathbb{S}^2$ or $\mathbb{T}^2$.

(d) Simply note that the complex surface $\mathbb{F}_{2k}$ is obtained from $\mathbb{F}_0 = \mathbb{P}^1(\mathbb{C}) \times \mathbb{P}^1(\mathbb{C})$ by a succession of $2k$ elementary transformations. Choosing k elementary $\mathbb{R}$-transformations based at k pairs of non real conjugate points as in Exercise 4.3.24 we get that $\mathbb{F}_{2k}(\mathbb{R}) \approx \mathbb{F}_0(\mathbb{R}) \approx \mathbb{T}^2$.

(e) Let P be a real point in $\mathbb{P}^2(\mathbb{C})$. By Example 4.2.18 we know that $B_P\mathbb{P}^2(\mathbb{R}) \approx \mathbb{F}_1(\mathbb{R}) \approx \mathbb{K}^2$.

(f) We use the same construction as in (1d) starting with $\mathbb{F}_{2k+1}(\mathbb{R}) \approx \mathbb{F}_1(\mathbb{R}) \approx \mathbb{K}^2$.

(g) Consider points $P_1, \ldots, P_{g-1} \in \mathbb{P}^2(\mathbb{R})$. By Example 4.2.18 we have that

$$B_{P_1,\ldots,P_{g-1}}\mathbb{P}^2(\mathbb{R}) \approx \mathbb{V}_g\ .$$

(h) Consider points $P_1, \ldots, P_g \in Q_{3,1}(\mathbb{R})$. By Example 4.2.18 we have that

$$B_{P_1,\ldots,P_g}Q_{3,1}(\mathbb{R}) = B_{P_1,\ldots,P_g}\mathbb{S}^2 \approx \mathbb{V}_g\ .$$

2. Let (X, σ) be a geometrically rational $\mathbb{R}$-surface. By Proposition 2.3.22, if the $\mathbb{R}$-surface (X, σ) is $\mathbb{R}$-rational then $X(\mathbb{R})$ is connected and non empty.

(a) Any complex quadric surface is birational to $\mathbb{P}^2(\mathbb{C})$ so the $\mathbb{R}$-surface $(Q_{4,0}, \sigma_{\mathbb{P}}|_{Q_{4,0}})$ is $\mathbb{C}$-rational but as its real locus is empty it is not rational.

(b) As X is a conic bundle over $\mathbb{P}^1$ it is a complex rational surface and (X, σ) is therefore geometrically rational. The number of connected components is at least half the number of simple singular fibres of the conic bundle, i.e. half the number of simple roots of f. By hypothesis this gives us $\#\pi_0(X(\mathbb{R})) > 1$ so by Proposition 2.3.22 the $\mathbb{R}$-surface (X, σ) is not rational.

(c) Blowing up a real point does not change the number of connected components.

(d) As any complex del Pezzo surface is rational (X, σ) is geometrically rational. By construction $\#\pi_0(X(\mathbb{R})) > 1$ so the $\mathbb{R}$-surface (X, σ) is not rational.

3. (a) Let $X \to B$ be such a fibration. By [Deb01, Remarks 4.2(5), p. 87] the complex surface X is uniruled but not rational because $q(X) = g(B) > 0$.
   (b) Idem.

□

Theorem 4.4.11 (Real locus of a $\mathbb{C}$-rational surface) *Let (X, σ) be a non singular projective geometrically rational $\mathbb{R}$-surface.*

1. *The $\mathbb{R}$-surface (X, σ) is rational if and only if $X(\mathbb{R})$ is connected and non empty. When moreover (X, σ) is also minimal $X(\mathbb{R})$ is diffeomorphic to one of the following surfaces: the real projective plane $\mathbb{RP}^2$, a sphere $\mathbb{S}^2$, a torus $\mathbb{T}^2$ or a Klein bottle $\mathbb{K}^2$. In this last case, X is a Hirzebruch surface $\mathbb{F}_n$ of odd index $n = 2k + 1 > 1$.*
2. *When (X, σ) is a minimal real del Pezzo surface of degree 1, $X(\mathbb{R})$ is diffeomorphic to the disjoint union of a real projective planes and 4 spheres. If (X, σ) is a minimal real del Pezzo surface of degree 2, then $X(\mathbb{R})$ is diffeomorphic to the disjoint union of 4 spheres.*
3. *If (X, σ) is non rational and has a minimal real conic bundle structure with $2s$ singular fibres then $X(\mathbb{R})$ is diffeomorphic to a disjoint union of s spheres, $s \geqslant 2$.*

Proof As most of these statements have been proved earlier, we refer the reader to [Man14] or [Man17a] for the missing pieces. See also [Rus02] which includes a complete classification of minimal del Pezzo surfaces of degree 1 and 2 based on Silhol's construction in [Sil89, § VI.4]. □

Exercise 4.4.12 If (X, σ) has a minimal real conic bundle structure prove that $\#\pi_0(X(\mathbb{R})) = 4 - \frac{1}{2}K_X^2$ (see Exercise 4.3.15).

Remark 4.4.13 If (X, σ) is a geometrically rational minimal $\mathbb{R}$-surface such that $X(\mathbb{R}) = \varnothing$ then X is a Hirzebruch surface of even index.

The main result of this section is the following theorem which summarises and completes previous results.

Theorem 4.4.14 (Topology of the real locus when $\kappa(X) = -\infty$) *Let (X, σ) be a non singular projective $\mathbb{R}$-surface of negative Kodaira dimension. We equip the real locus $X(\mathbb{R})$ with its Euclidean topology and consider topological surfaces up to homeomorphism. We denote by $s := \#\pi_0(X(\mathbb{R}))$ the number of connected components of the real locus.*

1. *If (X, σ) is rational then $s = 1$ and $X(\mathbb{R})$ is homeomorphic to one of the following compact connected surfaces:*
   (a) *The torus $\mathbb{T}^2$;*
   (b) *The sphere $\mathbb{S}^2$;*
   (c) *A non orientable surface $\mathbb{V}_g$ for some $g \in \mathbb{N}$.*
2. *If (X, σ) is geometrically rational, or in other words if the complex surface X is rational, then $s \in \mathbb{N}$ can be arbitrary and $X(\mathbb{R})$ is homeomorphic to one of the following compact topological spaces (which are all surfaces except for $\varnothing$):*

(a) *The empty set* $\varnothing$;
(b) *A torus* $\mathbb{T}^2$;
(c) *A disjoint union of spheres and non orientable surfaces*

$$\sqcup^l \mathbb{S}^2 \sqcup \mathbb{V}_{g_1} \sqcup \cdots \sqcup \mathbb{V}_{g_{s-l}}$$

where $l, g_1, \ldots, g_{s-l} \in \mathbb{N}^*$.

3. *If* (X, σ) *is uniruled (or in other words if the complex surface* X *is geometrically ruled) of irregularity* $q := q(X)$ *then* $s \in \mathbb{N}$ *is arbitrary and* $X(\mathbb{R})$ *is homeomorphic to one of the following compact topological spaces (which are all surfaces apart from* $\varnothing$*):*

(a) *The empty set* $\varnothing$;
(b) *A disjoint union of* $q + 1$ *toruses*

$$\sqcup^{q+1} \mathbb{T}^2 \ ;$$

(c) *A disjoint union of toruses, spheres and non orientable surfaces*

$$\sqcup^t \mathbb{T}^2 \sqcup^l \mathbb{S}^2 \sqcup \mathbb{V}_{g_1} \sqcup \cdots \sqcup \mathbb{V}_{g_{s-t-l}}$$

where $t < q + 1$ *and* $l, g_1, \ldots, g_{s-t-l} \in \mathbb{N}$.

4. *Conversely, any topological surface in the first list has a rational algebraic model, any topological surface in the second list has a geometrically rational algebraic model and any topological surface in the third list has a uniruled algebraic model.*

Before attacking the proof of this theorem we give a series of intermediate results.

Some authors call the orientable topological surface of genus g a *surface with* g *holes.* The main result in the classification of uniruled $\mathbb{R}$-surfaces is that orientable components of such surfaces have at most one hole. The theorem below is actually more general because there are surfaces of general type whose geometric genus is zero. We refer the interested reader to the previously cited review article [BCP11] which studies surfaces of general type with geometric genus equal to 0.

Theorem 4.4.15 *Let* (X, σ) *be a non singular projective* $\mathbb{R}$*-surface such that* $p_g(X) = 0$. *Any* orientable *component of the real locus* $X(\mathbb{R})$ *is then diffeomorphic to the sphere* $\mathbb{S}^2$ *or the torus* $\mathbb{T}^2$.

The following result is due to Comessatti [Com14].

Corollary 4.4.16 (Comessatti's theorem) *Let* (X, σ) *be an non singular projective* $\mathbb{R}$*-surface which is rational over the real numbers. Its real locus is then non empty and connected and if* $X(\mathbb{R})$ *is orientable then it is diffeomorphic to a sphere* $\mathbb{S}^2$ *or a torus* $\mathbb{T}^2$.

Proof By Theorem 1.5.55 the number of connected components of the real locus is invariant under birational maps of $\mathbb{R}$-surfaces. By hypothesis the $\mathbb{R}$-surface (X, σ) is birational to the $\mathbb{R}$-surface $(\mathbb{P}^2(\mathbb{C}), \sigma_{\mathbb{P}})$ and it follows that

$$\#\pi_0(X(\mathbb{R})) = \#\pi_0(\mathbb{P}^2(\mathbb{R})) = 1 .$$

As the geometric genus is a birational invariant and $p_g(\mathbb{P}^2(\mathbb{C})) = 0$, Theorem 4.4.15 completes the proof. □

The proof of Theorem 4.4.15 given here is a "modern" proof. Comessatti's original proof [Com14] of Corollary 4.4.16 starts by reducing to the case of minimal surfaces and then enumerating the possible topological types of real loci of minimal geometrically rational surfaces.

Lemma 4.4.17 *Let (X, σ) be a non singular projective $\mathbb{R}$-surface and let $V \subset X(\mathbb{R})$ be an* orientable *connected component of its real locus. The fundamental class (Definition 3.7.1) $\alpha \in H^2(X; \mathbb{Z})$ of V is then σ^*-invariant and its square $(\alpha \cdot \alpha)$ is minus the topological Euler characteristic of V:*

$$(\alpha \cdot \alpha) = -\chi_{top}(V) . \tag{4.7}$$

Proof The fact that α is σ^*-invariant is immediate. We prove (4.7) as in [Sil89, p. 71].

See Propositions 2.2.27 and 2.2.28 for the differentiable manifold structures on X and $X(\mathbb{R})$. Since V is orientable the product $(\alpha \cdot \alpha)$ is the self intersection of the variety V in X which by [Hir76, p. 132] is the same thing as the self intersection of V in its normal bundle $N_{X|V}$. (See also [MS74, p. 119].) At any real point multiplication by i in the tangent space T_X yields an orientation reversing isomorphism between $T_{X(\mathbb{R})}$ and the normal bundle $N_{X|X(\mathbb{R})}$. Indeed, consider a point $x \in X(\mathbb{R})$ and let (u_1, u_2) be a basis for the vector space $T_{X(\mathbb{R}),x}$. Since x is a point in the real locus we have that $T_{X,x} = T_{X(\mathbb{R}),x} \otimes_{\mathbb{R}} \mathbb{C}$. The quadruplet (u_1, iu_1, u_2, iu_2) is therefore a basis for the vector space $T_{X,x}$ and the pair (iu_1, iu_2) is a basis for the normal vector space $N_{X|X(\mathbb{R}),x}$. Since the natural orientation of the differentiable manifold X of real dimension 4 is given by (u_1, iu_1, u_2, iu_2) the induced orientation on $N_{X|X(\mathbb{R}),x}$ is given by (iu_2, iu_1). As the Euler characteristic $\chi_{\text{top}}(V)$ is equal to the self intersection of V in its tangent bundle T_V (see [Hir76, p. 13]) the result follows. □

Remark 4.4.18 Both the statement and the proof remain valid if we replace "projective" by "compact Kähler". See Appendix D for more details.

Lemma 4.4.19 *Let (X, σ) be a non singular projective $\mathbb{R}$-surface. The intersection form is then negative definite on the σ^*-invariant part of the real vector space $H^{1,1}(X) \cap H^2(X; \mathbb{R})$.*

Proof The Hodge index theorem 4.1.24 implies that the intersection form restricted to the subspace $H^{1,1}(X) \cap H^2(X; \mathbb{R})$ is Lorentzian, by which we mean that it has signature $(1, h^{1,1}(X) - 1)$. As the surface is projective, Proposition 2.6.43 implies that it has a real embedding in a projective space $\varphi\colon X \hookrightarrow \mathbb{P}^N(\mathbb{C})$. As the surface is non singular, Bertini's theorem D.9.1 implies is has a non singular hyperplane section H': we denote by h the fundamental class of the $\mathbb{R}$-curve $(H, \sigma|_H)$ where $H = \varphi^*(H')$. The class h is then σ^*-anti-invariant. Indeed, by Proposition 2.2.28, the anti-holomorphic involution σ is orientation preserving on the differentiable manifold X of real dimension 4, but orientation reversing on the submanifold H of real dimension 2. The eigenspaces of the involution σ^* are orthogonal and the eigenspace of eigenvalue 1 is therefore orthogonal to the line generated by the class of H. □

Corollary 4.4.20 *Let (X, σ) be a non singular projective $\mathbb{R}$-surface: if the geometric genus $p_g(X)$ vanishes then the intersection form restricted to the σ^*-invariant part of the real vector space $H^2(X; \mathbb{R})$ is negative definite.*

Proof Simply recall the Hodge decomposition (see Appendix D) $H^2(X; \mathbb{C}) = H^{2,0}(X) \oplus H^{1,1}(X) \oplus H^{0,2}(X)$ and $p_g(X) = h^{0,2}(X) = h^{2,0}(X)$ which gives us $H^{1,1}(X) = H^2(X; \mathbb{C})$. □

Remark 4.4.21 Once again, the same statement with almost the same proof (replacing "hyperplane section" by "Kähler class") remains valid if we replace "projective" by "compact Kähler". This is not however an actual generalisation because any compact Kähler manifold with $p_g = 0$ is projective. See [BHPVdV04, Chapter VI] for more details.

Proof of Theorem 4.4.15 Let $V \subset X(\mathbb{R})$ be an orientable connected component of the real locus. Lemma 4.4.17 and Corollary 4.4.20 imply that $\chi_{\text{top}}(V) \geqslant 0$. It follows that V is diffeomorphic to the sphere or a torus. □

We will prove Theorem 4.4.14 using the following refinement of Theorem 4.4.15.

Proposition 4.4.22 *Let (X, σ) be a non singular projective $\mathbb{R}$-surface such that $p_g(X) = 0$. Let $q := q(X)$ be its irregularity, let $s = \#\pi_0(X(\mathbb{R}))$ be the number of connected components of its real locus and let $t \leqslant s$ be the number of connected components of $X(\mathbb{R})$ which are diffeomorphic to a torus $\mathbb{T}^2$.*

1. *If $q = 0$ and the homology of X has no 2-torsion then*
$$t \leqslant 1$$
and if $t = 1$, then $s = t$ and
$$X(\mathbb{R}) \approx \mathbb{T}^2 .$$
2. *If $\kappa(X) = -\infty$ then*
$$t \leqslant q + 1$$
and if $t = q + 1$ then $s = t$ and
$$X(\mathbb{R}) \approx \sqcup^{q+1} \mathbb{T}^2 .$$

Remark 4.4.23 Note that the hypotheses of Proposition 4.4.22(1) cannot be weakened.

1. If X is a K3 surface (see Section 4.5) then $q(X) = \pi_1(X) = 0$ and $p_g(X) = 1$. Looking at Figure 4.11 of this section we see that if $g \leqslant 10$ then there is a real K3 surface whose real locus is diffeomorphic to a orientable surface of degree g. Similarly, there is a real K3 surface whose real locus is diffeomorphic to the disjoint union of a torus and several spheres and there is a real K3 surface whose real locus is diffeomorphic to the disjoint union of two toruses.
2. If X is an Enriques surface (see Section 4.5) then $p_g(X) = q(X) = 0$ and $\pi_1(X) = \mathbb{Z}_2$. Checking the list provided in Theorem 4.5.16 we see that there is a real Enriques surface whose real locus is diffeomorphic to the disjoint union of a torus and two Klein bottles and there is a real Enriques surface whose real locus is diffeomorphic to the disjoint union of two toruses.

On the other hand, we can weaken the hypothesis "$\kappa(X) = -\infty$" in Proposition 4.4.22(2) by replacing it by "$\kappa(X) \neq 1$ and the homology of X has no 2-torsion" as in Complement 4.4.24.

Proof of Proposition 4.4.22 This proof is based on an argument by Risler [Ris85, p. 161] quoted by Silhol [Sil89, p. 72]. As in Section 3.2, let $Y = X/G$ be the topological quotient of X by the involution and let $p \colon X \to Y$ be the canonical surjection. Note that the spaces Y and $Y \setminus X(\mathbb{R})$ are topological manifolds which have a $\mathcal{C}^\infty$ structure. It follows from Remark 1.5.28 that Y is a differentiable manifold of dimension 4 and from Proposition 2.2.27 that $X(\mathbb{R})$ is a differentiable manifold of real dimension 2. The subvariety $X(\mathbb{R})$ in Y is therefore of real codimension 2 in Y.

Recall that in the exact sequence (3.6) of Theorem 3.2.6,

$$\begin{aligned} \cdots \to H_r(Y, X(\mathbb{R}); \mathbb{Z}_2) \oplus H_r(X(\mathbb{R}); \mathbb{Z}_2) \to H_r(X; \mathbb{Z}_2) \to \\ H_r(Y, X(\mathbb{R}); \mathbb{Z}_2) \xrightarrow{\Delta_r} H_{r-1}(Y, X(\mathbb{R}); \mathbb{Z}_2) \oplus H_{r-1}(X(\mathbb{R}); \mathbb{Z}_2) \to \cdots \end{aligned} \tag{4.8}$$

the second component of Δ_r is the boundary map δ_r of the homology sequence associated to the pair $(Y, X(\mathbb{R}))$:

$$H_r(Y, X(\mathbb{R}); \mathbb{Z}_2) \xrightarrow{\delta_r} H_{r-1}(X(\mathbb{R}); \mathbb{Z}_2) \to H_{r-1}(Y; \mathbb{Z}_2)\,. \tag{4.9}$$

Since the homology has no 2-torsion we have that $b_1(X; \mathbb{Z}_2) = b_3(X; \mathbb{Z}_2) = b^1(X) = 2q(X)$. Since $H_4(X(\mathbb{R}); \mathbb{Z}_2) = H_3(X(\mathbb{R}); \mathbb{Z}_2) = \{0\}$ the exact sequence (4.8) yields

$$\begin{aligned} 0 \to H_4(Y, X(\mathbb{R}); \mathbb{Z}_2) \to H_4(X; \mathbb{Z}_2) \to H_4(Y, X(\mathbb{R}); \mathbb{Z}_2) \to \\ \to H_3(Y, X(\mathbb{R}); \mathbb{Z}_2) \to H_3(X; \mathbb{Z}_2)\,. \end{aligned} \tag{4.10}$$

Since $H_4(X;\mathbb{Z}_2) \simeq \mathbb{Z}_2$, we can deduce from the first line that $H_4(Y, X(\mathbb{R});\mathbb{Z}_2) \simeq \mathbb{Z}_2$ and that $1 \leqslant \dim_{\mathbb{Z}_2} H_3(Y, X(\mathbb{R});\mathbb{Z}_2) \leqslant 2q+1$. The part of the exact sequence (4.9) that we need is:

$$H_3(Y, X(\mathbb{R});\mathbb{Z}_2) \xrightarrow{\delta_2} H_2(X(\mathbb{R});\mathbb{Z}_2) \xrightarrow{i_2} H_2(Y;\mathbb{Z}_2)\ .$$

Using the above calculation it follows that

$$\dim_{\mathbb{Z}_2} \ker i_2 \leqslant \dim_{\mathbb{Z}_2} H_3(Y, X(\mathbb{R});\mathbb{Z}_2) \leqslant 2q+1\ . \tag{4.11}$$

Let $\{V_r\}_{r=1,\dots,s}$ be the connected components of $X(\mathbb{R})$. The group morphism $H_2(X(\mathbb{R});\mathbb{Z}_2) \xrightarrow{i_2} H_2(Y;\mathbb{Z}_2)$ sends the fundamental homology class of V_r in $H_2(X(\mathbb{R});\mathbb{Z}_2)$ to its fundamental homology class in $H_2(Y;\mathbb{Z}_2)$.

We know by Lemma 4.4.17 that the fundamental homology class in $H_2(X;\mathbb{Z})$ of a connected component diffeomorphic to $\mathbb{T}^2$ is isotropic for the intersection form and it follows from Corollary 4.4.20 that it is zero. It follows that its fundamental class in $H_2(Y;\mathbb{Z})$ and hence in $H_2(Y;\mathbb{Z}_2)$ vanishes.

The fundamental classes in $H_2(X(\mathbb{R});\mathbb{Z}_2)$ of the connected components of $X(\mathbb{R})$ diffeomorphic to $\mathbb{T}^2$ are linearly independent and their images under i_2 all vanish. It follows from (4.11) that $t \leqslant 2q+1$.

We conclude by noting that as $p\colon X \to Y$ is a double covered ramified along $X(\mathbb{R})$ the fundamental class of $X(\mathbb{R})$ in $H_2(Y;\mathbb{Z}_2)$ vanishes. For example, if $X(\mathbb{R})$ is orientable then its fundamental class in $H_2(Y;\mathbb{Z})$ is 2-divisible (see Lemma 3.6.21 which applies because X is non singular).

Permuting terms if necessary we can assume that $V_1, \dots, V_t$ are the connected components of $X(\mathbb{R})$ diffeomorphic to $\mathbb{T}^2$. If $V_1 \sqcup \dots \sqcup V_t \subsetneq X(\mathbb{R})$ then the fundamental classes of $V_1, \dots, V_t$ and $X(\mathbb{R})$ in $H_2(X(\mathbb{R});\mathbb{Z}_2)$ are linearly independent and by (4.11) we get that $t < 2q+1$ because the fundamental class of $X(\mathbb{R})$ belongs to $\ker i_2$. This result applied to $q=0$, proves the first part of the proposition.

To prove the second part of the proposition we recall that a non singular complex projective surface such that $\kappa(X) = -\infty$ is uniruled and its homology has no 2-torsion. Indeed, any such surface is birationally equivalent to a genus 0 fibration with non singular fibres which has a section. (See [Bea78, Exemple VII.3 and Chapitre III] for more details.) The existence of such a section implies the homology is torsion free and this property is invariant under birational maps. We can therefore apply the first part of the proposition to prove the result when $q=0$.

By Theorem 4.4.8 if X is uniruled and $q>0$ then X is a conic bundle $\pi\colon X \to C$ defined over $\mathbb{R}$ over a curve C of genus $g(C) = q(X) \geqslant 1$. Analysing the singular fibres of π shows that they cannot meet a torus and we then apply Harnack's inequality (3.3.7): $t \leqslant \#\pi_0(C(\mathbb{R})) \leqslant g(C)+1 = q+1$. □

Proof of Theorem 4.4.14 Any surface X of negative Kodaira dimension has geometric genus $p_g = 0$. Indeed, $\kappa(X) = -\infty$ means that all positive multiples of the canonical bundle have no global sections and in particular $p_g(X) = \dim H^2(X, \mathcal{O}_X)$

$= \dim H^0(X, \Omega_X) = 0$. Theorem 4.4.15 therefore applies and implies that the only orientable surfaces that can appear are $\mathbb{S}^2$ and $\mathbb{T}^2$. Point (1) follows from Comessatti's Theorem 4.4.16. The upper bound on the number of toruses in (2) and (3) follows from Proposition 4.4.22 since p_g and q are birational invariants of complex surfaces. If (X, σ) is geometrically rational then $p_g(X) = p_g(\mathbb{P}^2(\mathbb{C})) = 0$ and $q(X) = q(\mathbb{P}^2(\mathbb{C})) = 0$ (in general, uniruled surfaces have zero geometric genus but non zero irregularity). Finally (4) follows from Proposition 4.4.10. □

Complement 4.4.24 *Let (X, σ) be a non singular projective $\mathbb{R}$-minimal $\mathbb{R}$-surface.*

1. *If $p_g(X) = 0$, $q(X) > 0$ and X is not uniruled then $q(X) = 1$, $K_X^2 = 0$, $b_2(X) = 2$ and $\kappa(X) \in \{0, 1\}$.*
2. *If X is bi-elliptic and the homology of X has no 2-torsion then $t < 3$, and if $t = 2$ then $t = s$, see [CF03, Remark 7.3], [Suw69].*
3. *If the homology of X has no 2-torsion and $t \geqslant 3$ then $\kappa(X) = 1$ (i.e. X is a properly elliptic surface).*

Proof 1. Note that under these hypotheses, X must be of special type. Indeed if X is minimal and of general type $c_1^2(X) > 0$ and it follows that $\chi(\mathcal{O}_X) \geqslant 1$ by the Noether formula 4.1.19 and hence $q(X) = 0$. If X is not uniruled then $\chi_{\text{top}}(X) = c_2(X) \geqslant 0$ [Bea78, Theorem X.4] and then $\chi(\mathcal{O}_X) \geqslant 0$ by Noether: it follows that if $p_g(X) = 0$ then $q(X) \leqslant 1$. By minimality $c_1^2(X) = 0$ and by Noether's formula $c_2(X) = 0$. As $b_1(X) = b_3(X) = 2q(X)$ it follows that $b_2(X) = 2$. (See [Bea78, VI.1 et VI.2] for any alternative proof.)

2. There are bi-elliptic surfaces whose real locus is made up of three or four toruses—see Theorem 4.5.30—but in this case the homology of X contains 2-torsion [CF03, Remark 7.3], [Suw69].

3. The classification of complex compact surfaces [BHPVdV04, Chapter VI, Table 10] tells us that only bi-elliptic surfaces can satisfy $\kappa(X) = 0$, $p_g(X) = 0$ and $q(X) = 1$. As such surfaces were dealt with in the previous question we have that $\kappa(X) = 1$ or in other words X is a properly elliptic surface.

We have not had the time to construct explicit examples contradicting the conclusion of Proposition 4.4.22 when the hypotheses are weakened but we propose two sketch constructions.

1. By [BHPVdV04, Théorème III.18.2], if a fibration $\pi: X \to C$ has singular fibres they are all of the form mE where E is a non singular elliptic curve. If $t > \#\pi_0(C(\mathbb{R}))$ it follows from Silhol's classification of singular real fibres [Sil84], [Sil89, Chapitre VII] that π has an even number of fibres with even multiplicity m.
2. We could also use [Bea78, Théorème VI.13, case II non bi-elliptic]: under our hypotheses, X is necessarily a quotient of the form $(B \times F)/H$ where F is a non singular elliptic curve, B is a non singular curve of genus at least 2, H is a finite group acting faithfully on both B and F, B/H is elliptic, F/H is rational and H acts freely on $B \times F$.

□

4.4.2 *Singular Surfaces and Parabolas*

In this subsection based on [Kol99b, CM08, CM09] we give a classification of possible topological types of singular geometrically rational Du Val $\mathbb{R}$-surfaces 4.4.30 (see Definition 1.3.37 for the definition of geometrically rational). We will do this using the orbifold structure with conic points (4.4.31) on the connected components of the topological normalisation (4.4.35). This will give us a generalisation of Comessatti's theorem (4.4.36). Another consequence, which was initially our main motivation in this section, is the proof of three conjectures of Kollár's on rationally connected varieties. See Theorem 6.2.11 in Chapter 6.

Du Val Surfaces

We start by recalling a definition of Artin's, [Art66] (see also [Har77, p. 250]).

Definition 4.4.25 Let X be a normal complex surface defined over $\mathbb{C}$ and let P be a singular point of X. We say that P is a *rational singularity* if and only if there is a resolution $\pi\colon \widetilde{X} \to X$ of P such that $R^q\pi_*(\mathcal{O}_{\widetilde{X}}) = 0$ for all $q \geqslant 0$ where $(R^q\pi_*(\mathcal{O}_{\widetilde{X}}))$ denotes the q-th direct image of the sheaf $\mathcal{O}_{\widetilde{X}}$.

Theorem 4.4.26 *Let X be a normal complex surface and let P be a singular point of X. The following properties are equivalent.*

1. *P is rational of embedding dimension* 3.
2. *P is rational of multiplicity* 2*: we say it is a* rational double point.
3. *P is of multiplicity* 2 *and can be resolved by a sequence of blow ups of points.*
4. *The minimal resolution of P has a configuration of exceptional curves of type A_n, D_n, E_6, E_7 or E_8. (See below for more details.)*

Proof See [Slo80, p. 71]. □

Remark 4.4.27 There are double points on surfaces which are not rational—for example, $z^6 + y^2 + x^3 = 0$ is a elliptic double point—but any double point on a surface has embedding dimension 3, see [Lau71, p. 7].

Definition 4.4.28 If one of these four equivalent properties is satisfied then P is said to be a *rational double point of type A_n, $n \geqslant 1$, D_n, $n \geqslant 4$ or E_n, $n =$* 6, 7, 8. Over $\mathbb{C}$ we have the following characteristic equations [BHPVdV04, p. 87]:

$$\begin{aligned}
A_n (n \geqslant 1): &\quad z^2 + x^2 + y^{n+1} = 0\\
D_n (n \geqslant 4): &\quad z^2 + y(x^2 + y^{n-2}) = 0\\
E_6: &\quad z^2 + x^3 + y^4 = 0\\
E_7: &\quad z^2 + x(x^2 + y^3) = 0\\
E_8: &\quad z^2 + x^3 + y^5 = 0\,.
\end{aligned}$$

Remark 4.4.29 Rational double points on surfaces are the same thing as canonical singularities. These singularities are quotients of $\mathbb{C}^2$ by finite subgroups of $\mathbf{SL}_2(\mathbb{C})$. We also call them Du Val singularities.

Definition 4.4.30 A projective surface is said to be *Du Val* if and only if its only singularities are rational double points.

Over $\mathbb{C}$, Du Val singularities are classified in Definition 4.4.28 below: there are the *cyclic* singularities A_n, $n \geqslant 1$, the *dihedral* singularities D_n, $n \geqslant 4$, the *tetrahedral* singularity E_6, the *octahedral* singularity E_7 and the *icosahedral* singularity E_8. There are many other types of singularities over $\mathbb{R}$ and in this section we will only present two series of cyclic singularities. We refer to [CM08, section 1 and example 1.3] for more details.

A real surface X is said to have a singularity of type $A_n^{\pm}$ at a point $P \in X(\mathbb{R})$ if in some neighbourhoood of P X is $\mathbb{R}$-analytically isomorphic to

$$x^2 \pm y^2 - z^{n+1} = 0, \ n \geqslant 1 \ .$$

The grey part of Figure 4.5 represents the zone in $\mathbb{R}^2_{z,x}$ where $z^{n+1} - x^2$ is positive. The surface X which is locally a double cover of the plane branched over the curve $z^{n+1} - x^2 = 0$ only has real points over this zone.

Note that all these singularities are non isomorphic except for A_1^+ and A_1^-.

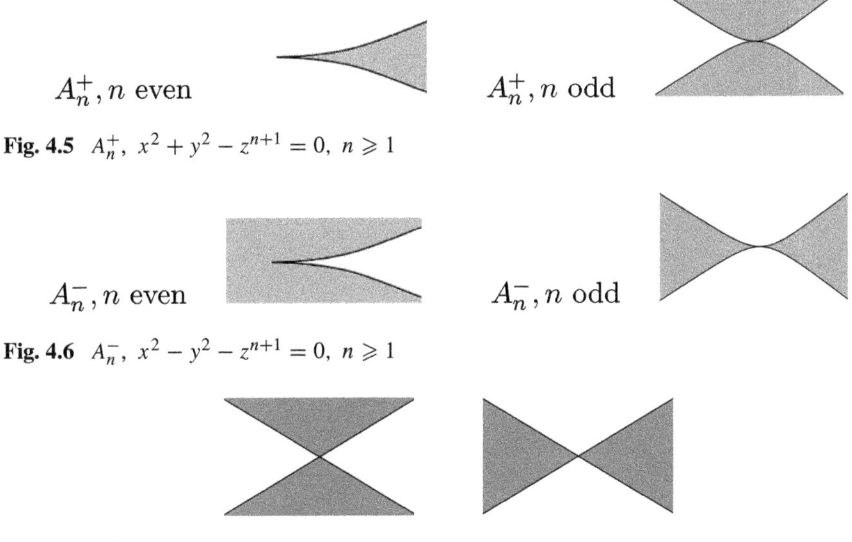

Fig. 4.5 A_n^+, $x^2 + y^2 - z^{n+1} = 0$, $n \geqslant 1$

Fig. 4.6 A_n^-, $x^2 - y^2 - z^{n+1} = 0$, $n \geqslant 1$

Fig. 4.7 $A_1^+ \simeq A_1^-$

4.4.3 Generalisation of Comessatti's Theorem

Orbifolds of Dimension 2

The term *orbifold* derives from the expression *n-manifold* denoting a topological space M equipped with a family of charts $(\widetilde{U}, \phi)$ where $\widetilde{U}$ is an open set and ϕ is a homeomorphism onto an open set $U \subset \mathbb{R}^n$.

An n-orbifold is a space equipped with an atlas whose charts $\phi\colon \widetilde{U} \to U \subset \mathbb{R}^n$ are finite branched covers: in the case where all the maps ϕ are of degree 1 the orbifold is simply a manifold. More precisely, every open chart of an orbifold is equipped with a G-action for some finite group G and ϕ factorises through a homeomorphism $G\backslash\widetilde{U} \to U$. See [BMP03, Chapter 2] for more details.

Definition 4.4.31 If G is cyclic and acts by rotation of angle $2\pi/m$ its unique fixed point is said to be a *conic point* of index m.

Orbifolds are not always homeomorphic to a manifolds, except in dimension 2 where any orbifold M is homeomorphic to a topological manifold denoted $|M|$. See [Sco83, §2], for example, for more details.

Definition 4.4.32 Let p and q be coprime integers, $(p, q) = 1$. We denote by $\mathbb{S}(p, q)$ the orbifold whose underlying smooth surface is $|\mathbb{S}(p, q)| = \mathbb{S}^2$ with two conic points whose indices are respectively p and q.

Let M be a compact 2 dimensional orbifold with a global finite covering map of degree d from a smooth surface $\widetilde{M} \to M$. The orbifold Euler characteristic is then defined by

$$\chi(M) := \frac{1}{d}\chi(\widetilde{M}) \in \mathbb{Q} .$$

Let M be a 2-orbifold with k conic points of angles $2\pi/m_j$, $j = 1, \ldots, k$ and let $|M|$ be the smooth surface underlying M. We then have that

$$\chi(M) = \chi(|M|) - \sum_{j=1}^{k}(1 - \frac{1}{m_j}) .$$

Definition 4.4.33 The orbifold M is said to be *spherical* (resp. *Euclidean*) if and only if $\chi(M) > 0$ (resp. $\chi(M) = 0$).

Proposition 4.4.34 *The orbifold M is spherical or Euclidean if and only if $|M|$ is spherical and $\sum_{j=1}^{k}(1 - \frac{1}{m_j}) \leqslant 2$ or $|M|$ is Euclidean and $k = 0$.*

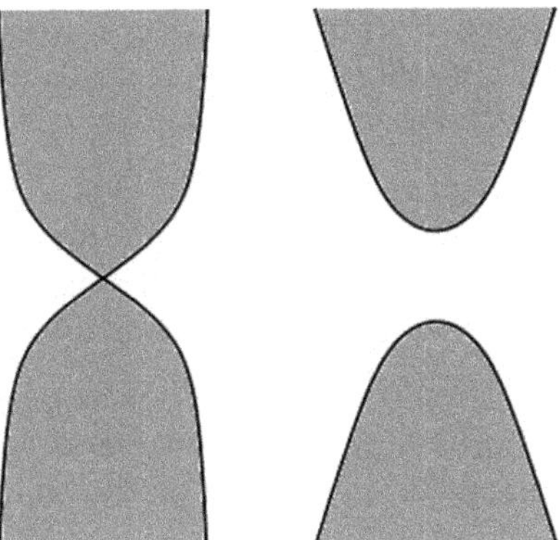

Fig. 4.8 M and $\overline{M}$ in a neighbourhood of a singular point of type $A_n^{\pm}$, n odd

Topological Normalisation

Kollár [Kol99a] introduced an operation imitating "branch separation" in algebraic geometry (Definition 1.5.37) in order to handle the situation where the real locus of an $\mathbb{R}$-variety is singular.

Definition 4.4.35 Let V be a simplicial complex whose singular locus $\mathrm{Sing}(V)$ is finite, where here $\mathrm{Sing}(V)$ is defined as being the set of points $x \in V$ whose star (the union of all the simplexes in V having x as a vertex) is not homeomorphic to a disc. The topological normalisation $\overline{\nu}\colon \overline{V} \to V$ is the unique continuous proper map such that

1. $\overline{\nu}$ is a homeomorphism over $V \setminus \mathrm{Sing}(V)$,
2. If $P \in \mathrm{Sing}(V)$ then the fibre $\overline{\nu}^{-1}(P)$ is in bijection with the set of local connected components of V in a neighbourhood of P.

Let X be a geometrically rational algebraic $\mathbb{R}$-surface and let $M \subset \overline{X(\mathbb{R})}$ be a connected component of the topological normalisation of its real locus. If X is non singular and the smooth surface M is orientable then Comessatti's theorem 4.4.16 implies that M is a sphere or a torus. The generalisation below was proved in [CM09].

If X is Du Val then we equip M with an orbifold structure with conic points (Definition 4.4.38).

Theorem 4.4.36 *Let X be a geometrically rational algebraic $\mathbb{R}$-surface and let $M \subset \overline{X(\mathbb{R})}$ be a connected component of the topological normalisation of its real locus. If X is Du Val and the orbifold M is orientable then M is spherical or Euclidean.*

This result is a corollary of Theorem 4.4.39 below. Before stating this theorem we need one more technical definition When $X(\mathbb{R})$ is two dimensional the normalisation $\overline{X(\mathbb{R})}$ is a topological manifold and if $P \in X(\mathbb{R})$ is a singular point of type $A_n^{\pm}$ with n odd then $\overline{X(\mathbb{R})}$ has two locally connected components in a neighbourhood of P.

Definition 4.4.37 The point P is *globally non separating* if the two locally connected components in a neighbourhood of P are in the same connected component of $\overline{X(\mathbb{R})}$ and *globally separating* otherwise.

Let X be a Du Val $\mathbb{R}$-surface and let $\overline{\nu}\colon \overline{X(\mathbb{R})} \to X(\mathbb{R})$ be the topological normalisation of the real locus. We denote by Σ_X the set of real singular points which are either of type A_n^- with n even or globally separating and of type A_n^- with n odd. We denote by $\mathcal{P}_X := \operatorname{Sing}(X) \setminus \Sigma_X$ the set of all other singular points.

Definition 4.4.38 Let $M \subset \overline{X(\mathbb{R})}$ be a connected component of the topological normalisation of the real locus of a Du Val $\mathbb{R}$-surface. We equip M with an orbifold structure whose conic points of index m correspond to the singular points of type $A_m^{\pm}$ contained in $\mathcal{P}_X \cap \overline{\nu}(M)$.

We denote by $k(M)$ the cardinality $\#\{\overline{\nu}^{-1}(\mathcal{P}_X) \cap M\}$ and for $i = 1 \dots k(M)$ we let $m_i(M)$ be the index of a point in $\mathcal{P}_X \cap \overline{\nu}(M)$.

Theorem 4.4.39 *Let X be a Du Val $\mathbb{R}$-surface and let $M \subset \overline{X(\mathbb{R})}$ be a possibly non orientable connected component of the topological normalisation of its real locus. If X is geometrically rational then*

- $k(M) \leqslant 4$,
- $\sum_{i=1}^{k}(1 - \frac{1}{m_i+1}) \leqslant 2$,
- $|M| = \mathbb{S}^1 \times \mathbb{S}^1 \implies k(M) = 0$.

Proof See [CM08, Corollary 0.2, Theorem 0.3] and [CM09, Theorem 0.2] for a full proof: here we only discuss the inequality $k(M) \leqslant 4$. The heart of the proof is a reduction to the case of certain double covers of the quadratic cone branched along singular curves of degree 6. A clever counting argument then enables us to complete the proof.

The minimal model programme enables us to reduce to the case where X is a del Pezzo (Definition 4.2.12) Du Val surface of degree 1. See Lemma [CM08, Lemma 1.8] for more details.

The anticanonical model of X is a branched double cover $q : X \to Q$ of a quadric cone $Q \subset \mathbb{P}^3(\mathbb{C})$ whose branching locus is the union of the summit of the cone and a cubic section B not passing through the summit. See [DPT80, Exposé V] for more details. Note that the pull back under q of the summit of the cone is a non singular point. Let X' be the singular elliptic surface obtained from X by blowing up this non singular point.[9]

We recall that $\overline{\nu}\colon \overline{X'(\mathbb{R})} \to X'(\mathbb{R})$ is the topological normalisation of the real locus. We therefore want to prove that for any connected component $M \subset \overline{X'(\mathbb{R})}$ we have that

$$\#(\overline{\nu}^{-1}(\mathcal{P}_{X'}) \cap M) \leqslant 4 .$$

The surface X' is a ramified double cover of a Hirzebruch surface $\mathbb{F}_2$ whose branching curve is the union of the unique section of negative self-intersection Σ_∞

[9] Exercise: prove that X' is an elliptic surface.

of the fibration $\mathbb{F}_2 \to \mathbb{P}^1(\mathbb{C})$ and a trisection B which does not meet Σ_∞. The cone Q is isomorphic to the weighted projective space $\mathbb{P}(1, 1, 2)$ which we equip with coordinates (x_0, x_1, y_2) and X is therefore the hypersurface in $\mathbb{P}(1, 1, 2, 3)$ with coordinates (x_0, x_1, y_2, z) defined by the equation

$$z^2 = y_2^3 + p_4(x_0, x_1)y_2 + q_6(x_0, x_1) .$$

We now describe a plane model for Q in which the hyperplane sections of Q embedded in $\mathbb{P}^3$ via $H^0(Q, \mathcal{O}_Q(2))$ correspond to the parabolas tangent to the line at infinity $L_\infty = \{w = 0\}$ at the point $O := \{w = x = 0\}$ in the projective plane equipped with coordinates (x, y, w). In other words, we blow up at O and then at the infinitely close point O' in O corresponding to tangency to the line at infinity L_∞, and denote by $\widetilde{Q}$ the surface thus obtained. Let E and E' be the total transformations of O and O' and note that $E = E' + E''$ where E'' is a (-2)-curve. The linear system $H^0(\widetilde{Q}, \mathcal{O}_{\widetilde{Q}}(2H - E - E'))$ sends $\widetilde{Q}$ birationally to the quadric cone $Q \subset \mathbb{P}^3(\mathbb{C})$ contracting both the strict transform $\widetilde{L_\infty}$ of the line L_∞ and the curve E'' to points. Since $\widetilde{L_\infty}$ and E'' are disjoint the contraction of $\widetilde{L_\infty}$ gives a Hirzebruch surface $\mathbb{F}_2$ whose (-2)-section Σ_∞ is the image of E''. We write this using coordinates (x, y, w) on $\mathbb{P}^2$: $H^0(Q, \mathcal{O}_Q(1))$ then corresponds to $H^0(\widetilde{Q}, \mathcal{O}_{\widetilde{Q}}(H - E))$ which is generated by w, x and $y_2 := yw$ extends w^2, wx, x^2 to a basis of $H^0(\widetilde{Q}, \mathcal{O}_{\widetilde{Q}}(2H - E - E')) \simeq H^0(Q, \mathcal{O}_Q(2))$. The morphism $\widetilde{Q} \to \mathbb{P}(1, 1, 2)$ is therefore given by $x_0 := w$, $x_1 := x$, $y_2 := yw$.

The elliptic surface X' is the double cover of $\mathbb{F}_2$ branched over Σ_∞ and over the curve B corresponding to the curve in Q of equation $y^3 + p_4(x_0, x_1)y + q_6(x_0, x_1) = 0$. The curve B therefore corresponds to the plane curve of equation $w^3y^3 + p_4(w, x)yw + q_6(w, x) = 0$ whose affine part has equation

$$y^3 + p_4(1, x)y + q_6(1, x) = 0 . \tag{4.12}$$

Note that a parabola of this form, by which we mean a curve of the form $C \in (2H - E - E')$ is disjoint from E'' (which contracts onto the summit of the cone) unless it degenerates as two lines passing through the point O. In particular, we can always modify the coordinates in the affine plane so that C is sent to the line $y = 0$.

In order to describe the geometry at infinity of parabolas of this form, recall that the surface $\mathbb{F}_2$ is covered by two open sets isomorphic to $\mathbb{C} \times \mathbb{P}^1(\mathbb{C})$. On one of these open charts we have affine coordinates $\frac{x}{w} \in \mathbb{C}$ and homogeneous coordinates $(w : y) \in \mathbb{P}^1(\mathbb{C})$, whereas on the other chart the coordinates are $\frac{w}{x} \in \mathbb{C}$ and $(\frac{x^2}{w} : y) \in \mathbb{P}^1(\mathbb{C})$ (or alternatively $\frac{x^2}{w}/w = (\frac{x}{w})^2$). The section at infinity Σ_∞ corresponds to the curve $E'' \subset \widetilde{X}$ and is defined by $w = 0$ or $\frac{x^2}{w} = 0$ depending on the chart. A parabola $yw = a_0w^2 + a_1xw + a_2x^2$ is therefore given by an equation

$$\frac{1}{\eta} = a_0 + a_1\frac{x}{w} + a_2(\frac{x}{w})^2$$

in the affine chart of coordinates $(\frac{x}{w} : \eta := \frac{w}{y})$. Using these coordinates at infinity it becomes easy to see when a domain "meets" $\mathbb{F}_2$ at infinity.

We now seek the normal form of equation (4.12). The singular points of $X'(\mathbb{R})$ are in one to one correspondence with singular points of $B(\mathbb{R})$. There are different cases to consider corresponding to different numbers of connected components of the trisection B.

Here we will restrict ourselves to the case where the trisection has three irreducible components and we refer to [CM08] for the other cases. We aim to prove that every connected component of the topological normalisation of each of the two double covers branched along B has at most 4 singular points. We start by noting that as B is real at least one of the irreducible components is real. Equation (4.12) becomes

$$(y - \alpha(x))(y - \beta(x))(y - \gamma(x)) = 0$$

and changing the real coordinates on $Q = \mathbb{P}(1, 1, 2)$ if necessary we can assume that $\gamma = 0$. The case $\beta = \overline{\alpha}$ where there are two irreducible complex conjugate components only gives us 2 singular points: $\operatorname{Re}\alpha(x) = 0$, $y = \operatorname{Im}\alpha(x)$. We can therefore assume that all three irreducible components are real. Equation (4.12) then becomes $(y - \alpha)(y - \beta)y = 0$ where $\alpha(x) = \alpha_0 + \alpha_1 x + \alpha_2 x^2$ and $\beta(x) = \beta_0 + \beta_1 x + \beta_2 x^2$ are polynomials of degree 2.

Case Without Tangency

Suppose that none of the parabolas are tangent to each other. Since we can permute the three curves we can assume the smallest one is at infinity. (The smallest curve is the one that has the smallest value of a_2 when we write their equations in the form $y = a_0 + a_1 x + a_2 x^2$.) Changing coordinates we get a curve given by the equation $y = 0$ and two convex parabolas, i.e. parabolas for which $\alpha_2 > 0$ and $\beta_2 > 0$.

The 6 intersection points are distinct and are given by

$$y = \alpha(x)\beta(x) = 0\ , \qquad \alpha(x) = \beta(x) = y\ .$$

The curve B is real so if one of these singular points is not real then the number of real singular points is bounded by 4. We suppose therefore that the 6 singular points are real and set

$$\begin{cases} \alpha(x) = \alpha_2(x - a_1)(x - a_2),\ a_1 < a_2\ ; \\ \beta(x) = \beta_2(x - b_1)(x - b_2)\ . \end{cases} \tag{4.13}$$

Multiplying y by β_2 if necessary we can assume that $\beta_2 = 1$. Furthermore we can reduce to the case $0 < \alpha_2 < 1$ by exchanging the roles of α and β if necessary. Using a translation along the x axis we can assume that $b_1 = -b_2$. Equation (4.13) then becomes

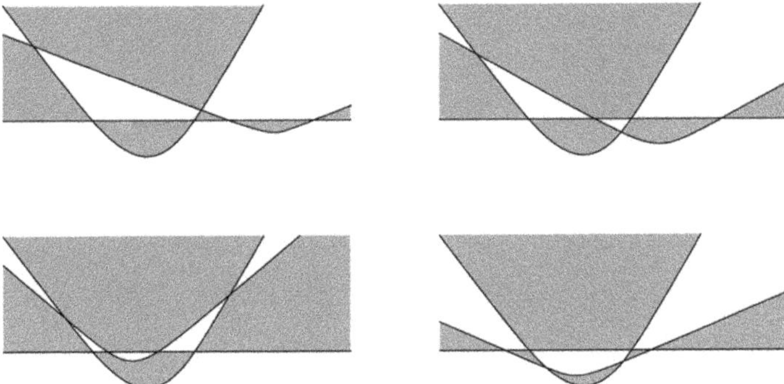

Fig. 4.9 6 A_1 points

$$\begin{cases} \alpha(x) = \alpha_2(x - a_1)(x - a_2), \ a_1 < a_2, \ 0 < \alpha_2 < 1 \ ; \\ \beta(x) = (x^2 - b^2), \ 0 < b \ . \end{cases}$$

Up to symmetry $x \leftrightarrow -x$ this leaves us with 4 possibilities.

$$b < a_1 \ , \ -b < a_1 < b < a_2 \ , \ a_1 < -b < b < a_2 \ , \ -b < a_1 < a_2 < b \ .$$

This configuration is shown in Figure 4.9. In order to help the reader visualise the situation we invite them to count the double points in each connected component of the complement of B. To do this, note that two connected components are connected at infinity if and only their boundary contains two non bounded curves belonging to the same pair of parabolas.

Case with Tangency

A detailed study similar to the one carried out above enables us to reduce to the five cases shown in Figure 4.10. □

Non Orientable Components

All non orientable surfaces can be realised as components of real non singular rational surfaces by blowing up real points in the real projective plane. Similarly, it is easy to construct hyperbolic non orientable orbifolds.

When X is geometrically rational minimal and non singular Comessatti's theorem implies that M is spherical or euclidean—in fact minimality implies that M is diffeomorphic to $\mathbb{S}^2$ or $\mathbb{RP}^2$ (which are both spherical), $\mathbb{S}^1 \times \mathbb{S}^1$ or $\mathbb{K}^2$ (which are both Euclidean). The singular case is rather different.

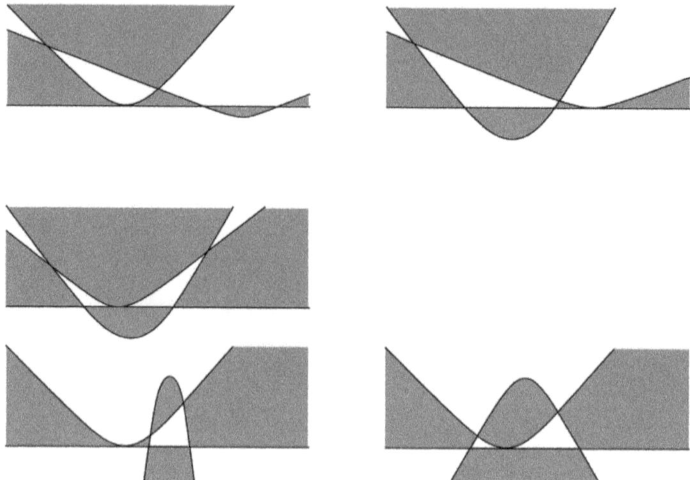

Fig. 4.10 4 A_1 points, 1 A_2 point

Theorem 4.4.40 *There is a geometrically rational minimal Du Val* $\mathbb{R}$*-surface* X *which has a component* $M \subset \overline{X(\mathbb{R})}$ *which is a hyperbolic orbifold.*

Proof See [CM09, Theorem 0.4]. □

4.5 K3, Enriques, Abelian and Bi-elliptic Surfaces ($\kappa = 0$)

The four classes of non singular minimal complex projective surfaces with Kodaira dimension $\kappa = 0$ can be distinguished by their geometric genus p_g and their irregularity q. We can show that if the Kodaira dimension of a surface X vanishes then there is a strictly positive integer m such that $mK_X \sim 0$.

Let m_0 be the smallest strictly positive integer such that $m_0 K_X \sim 0$.

Theorem 4.5.1 *Let* X *be a non singular minimal complex projective surface with* $\kappa(X) = 0$. *There are then four possibilities for the pair* $(p_g(X), q(X))$*:*

1. $p_g(X) = 1, q(X) = 0$*: we then have that* $m_0 = 1$ *and* X *is a* projective K3 surface *(Definition 4.5.3).*
2. $p_g(X) = 0$, $q(X) = 0$*: we then have that* $m_0 = 2$ *and* X *is an* Enriques surface *(Definition 4.5.13).*
3. $p_g(X) = 1$, $q(X) = 2$*: we then have that* $m_0 = 1$ *and* X *is an* abelian surface *(Definition 4.5.22).*
4. $p_g(X) = 0, q(X) = 1$*: we then have that* $m_0 \in \{2, 3, 4, 6\}$ *and* X *is a* bi-elliptic surface *(Definition 4.5.28).*

Proof See [Bea78, Liste VI.20 and Théorème VIII.2]. □

Corollary 4.5.2 *Any non singular minimal complex projective surface X with $\kappa(X) = 0$ satisfies $4K_X \sim 0$ or $6K_X \sim 0$.*

4.5.1 K3 Surfaces

The book [X85] is a standard reference for complex K3 surfaces: we refer to [Sil89, Chapter VIII] for real K3 surfaces. The K3 surfaces in Theorem 4.5.1 are assumed to be *projective*. More generally, a K3 surface is a compact non singular complex analytic surface with trivial canonical divisor and vanishing first Betti number.

Definition 4.5.3 Let X be a non singular compact complex analytic surface. X is said to be a *K3 surface* if and only if $K_X \sim 0$ and $b_1(X) = 0$. A *real K3 surface* is an $\mathbb{R}$-surface (X, σ) such that X is a K3 surface.

Proposition 4.5.4 *Let X be a K3 surface. X is then minimal: moreover $\kappa(X) = 0$, $p_g(X) = 1$, $q(X) = 0$ and X is simply connected.*

Proof See [X85], [BHPVdV04, Chapitre VIII]. □

K3 surfaces are not all projective but by a fundamental theorem due to Siu they are all Kähler. See [X85, Exposé XII] for more details.

Theorem 4.5.5 (Kharlamov 1975) *There are 66 topological types of real K3 surfaces. Each of them can be obtained by topological Morse simplification (Definition 4.3.31) from one of the 6 extremal types listed below. Conversely, any type obtained in this way can be realised as the real locus of a real K3 surface.*

The 6 extremal types are:

1. *M-surfaces, $b^*(X(\mathbb{R}); \mathbb{Z}_2) = 24$, $\chi(X(\mathbb{R})) = -16, 0, 16$,*

$$\mathbb{S}_{10} \sqcup \mathbb{S}^2, \quad \mathbb{S}_6 \sqcup 5\mathbb{S}^2, \quad \mathbb{S}_2 \sqcup 9\mathbb{S}^2 \ ;$$

2. *$(M-2)$-surfaces, $b^*(X(\mathbb{R}); \mathbb{Z}_2) = 20$, $\chi(X(\mathbb{R})) = \pm 8$,*

$$\mathbb{S}_7 \sqcup 2\mathbb{S}^2, \quad \mathbb{S}_3 \sqcup 6\mathbb{S}^2 \ ;$$

3. *Pair of toruses,*

$$\mathbb{T}^2 \sqcup \mathbb{T}^2 \ .$$

Proof See [Har76], [Sil89, Chapter VIII]. □

Figure 4.11 shows all pairs $(\chi(X(\mathbb{R})), b^*(X(\mathbb{R})); \mathbb{Z}_2)$ which can be realised by real K3 surfaces. We deduce from this figure the possible topological types of real K3 surfaces using the following proposition.

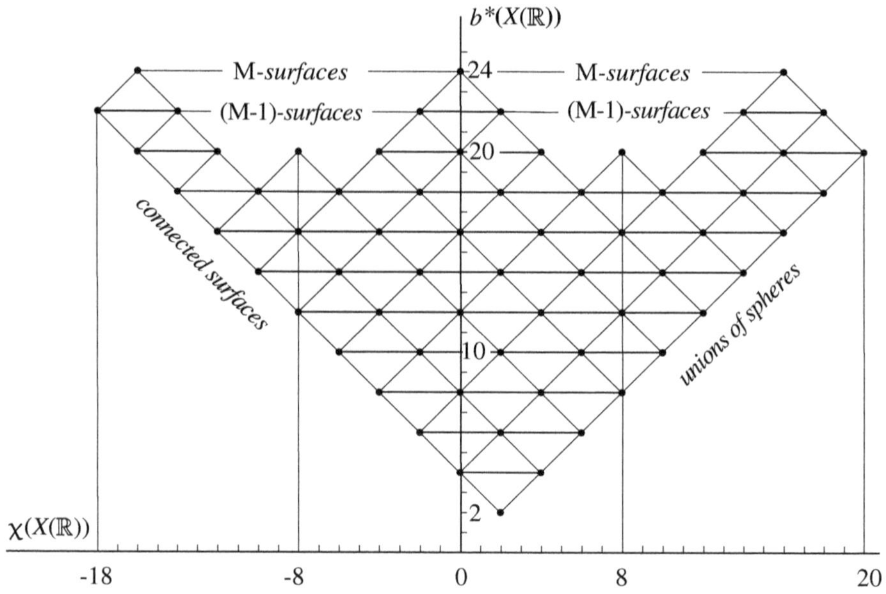

Fig. 4.11 Topological types of real K3 surfaces with non empty real locus

Proposition 4.5.6 *Let (X, σ) be a real K3 surface. If $X(\mathbb{R})$ is non empty then it is an orientable topological surface. Moreover $X(\mathbb{R})$ has at most one connected component whose Euler characteristic is $\leqslant 0$ (the other components are therefore all diffeomorphic to the sphere) unless $X(\mathbb{R})$ is the disjoint union of two toruses.*

Proof Let (X, σ) be a real K3 surface whose real locus is non empty. Note first that $X(\mathbb{R})$ is orientable. Indeed, by definition $c_1(X) = b_1(X) = 0$ so by Proposition 3.1.16 $w_2(X) = 0$. Moreover we know by Proposition 4.5.4 that $\pi_1(X) = 0$ which implies in particular that $b_1(X; \mathbb{Z}_2) = 0$. Theorem 3.1.18 therefore applies.

As in the proof of Theorem 3.2.6 we set $Y = X/G$ and let $p: X \to Y$ be the associated branched double cover. Since the homology of X has no 2-torsion and $q(X) = 0$ we have that $H_3(Y, X(\mathbb{R}); \mathbb{Z}_2) \simeq \mathbb{Z}_2$ by Lemma 3.6.22 and as in the first part of the proof of Proposition 4.4.22 this yields an exact sequence arising from (4.9)

$$0 \to \mathbb{Z}_2 \stackrel{\delta_2}{\to} H_2(X(\mathbb{R}); \mathbb{Z}_2) \stackrel{i_2}{\to} H_2(Y; \mathbb{Z}_2) . \tag{4.14}$$

Let $\{V_r\}_{r=1,\dots,s}$ be the connected components of $X(\mathbb{R})$ which are therefore compact orientable topological surfaces by previous results. For any $r = 1, \dots, s$, we denote by α_r the fundamental class of V_r in $H_2(X; \mathbb{Z})$ and by β_r the image of α_r in $H_2(Y; \mathbb{Z}_2)$ under the map p_* induced by p. Obviously we have that $\alpha_r \cdot \alpha_{r'} = 0$ if $r \neq r'$ and α_r is in the invariant part of $H_2(X; \mathbb{Z})$.

By construction the group morphism $H_2(X(\mathbb{R}); \mathbb{Z}_2) \xrightarrow{i_2} H_2(Y; \mathbb{Z}_2)$ sends the fundamental homology class of V_r in $H_2(X(\mathbb{R}); \mathbb{Z}_2)$ to $\beta_r \in H_2(Y; \mathbb{Z}_2)$. Recall that as $p\colon X \to Y$ is a double cover ramified along $X(\mathbb{R})$ the fundamental class of $X(\mathbb{R})$ in $H_2(Y; \mathbb{Z}_2)$ vanishes by Lemma 3.6.21. This class is equal to the sum $\beta_1 + \beta_2 + \cdots + \beta_s$ which therefore vanishes. Moreover, the relationship $\beta_1 + \beta_2 + \cdots + \beta_s = 0$ is the only relationship between the classes β_r because of the exact sequence (4.14). There are therefore two possibilities for the classes α_r:

1. the classes α_r are linearly independent;
2. there is exactly one relationship between the classes α_r which is of the form $\sum_{r=1}^{s} d_r\alpha_r$ with $\prod_{r=1}^{s} d_r \neq 0$.

Using Lemma 4.4.19 and the equalities $h^{2,0} = 1$ and $\sigma^* H^{2,0} = H^{0,2}$, we can prove that the positive index of the intersection form restricted to the invariant part of $H_2(X; \mathbb{Z})$ is equal to 1. Moreover, $\alpha_r \cdot \alpha_r = -\chi_{\mathrm{top}}(V_r)$ by Lemma 4.4.17.

In the first case it follows that there is at most one component which has negative or zero Euler characteristic.

In the second case it follows from the fact that $\sum_{r=1}^{s} d_r\alpha_r = 0$ that $\alpha_r \cdot \alpha_r = 0$ for all r because $\alpha_r \cdot \alpha_{r'} = 0$ whenever $r \neq r'$. The fact that the positive index of the intersection form restricted to $H_2(X; \mathbb{Z})^G$ is 1 implies that $X(\mathbb{R})$ contains at most two connected components diffeomorphic to $\mathbb{T}^2$ and that if α_1 and α_2 are two such components then they are linearly dependent. □

Remark 4.5.7 There is a more sophisticated proof of the orientability of the real locus of a real K3 surface which proceeds as follows: we start by noting that by Proposition 3.1.10 the real locus of a real K3 surface of degree 4 in $\mathbb{P}^3(\mathbb{C})$ is orientable and then we apply the following theorem [Har76], [Sil89, Chapter VIII]:

Theorem 4.5.8 *Any real K3 surface is a deformation (Definition 4.3.25) of a quartic* $\mathbb{R}$*-surface in* $\mathbb{P}^3$.

See [Sil89, Corollary VIII.4.2] for more details.

Let (X, σ) be a real K3 surface whose real locus is non empty. Let s be the number of connected components of $X(\mathbb{R})$ and let g be the sum of genuses of connected components of $X(\mathbb{R})$. If $g > 0$ and $X(\mathbb{R})$ is not a union of two toruses then g is the genus of the unique component which is not diffeomorphic to a sphere. We then have that

$$s = \frac{b^*(X(\mathbb{R}); \mathbb{Z}_2) + \chi(X(\mathbb{R}))}{4}, \qquad g = \frac{b^*(X(\mathbb{R}); \mathbb{Z}_2) - \chi(X(\mathbb{R}))}{4}.$$

The topological type of $X(\mathbb{R}) \neq \varnothing$ is therefore determined by the value of the pair $(\chi(X(\mathbb{R})), b^*(X(\mathbb{R})); \mathbb{Z}_2)$ except for the pair $(0, 8)$ which corresponds to two topological types realised by real K3 surfaces:

$$\mathbb{T}^2 \sqcup \mathbb{T}^2 \quad \text{and} \quad \mathbb{S}_2 \sqcup \mathbb{S}^2 .$$

4.5.2 Algebraic Cycles on K3 Surfaces

In the moduli space of complex K3 surfaces, the isomorphism classes of complex surfaces of given Picard number ρ form a countable union of subspaces of dimension $20 - \rho$. See [GH78, p. 594] for more details. We now prove a similar result for real K3 surfaces: it turns out that in the moduli space of real K3 surfaces satisfying certain conditions explained below, the isomorphism classes of real K3 surfaces such that $b^1_{\text{alg}} \geqslant k$ form a countable union of subspaces of dimension $20 - k$.

Example 4.5.9 (*Quartic surface in $\mathbb{P}^3$ such that $b^1_{\text{alg}} = 0$*) Consider the surface $\mathbb{S}^1 \times \mathbb{S}^1$ realised as a quartic in $\mathbb{P}^3(\mathbb{R})$, for example as the real locus of the surface $X \subset \mathbb{P}^3(\mathbb{C})$ whose equation with real coefficients is $16(x_1^2 + x_2^2) - (x_1^2 + x_2^2 + x_3^2 + 3x_0^2)^2 = 0$ as in Example 2.6.38. The surface in question has non real singularities: perturbing the equation slightly so that the real locus is still a torus we get a general non singular $\mathbb{R}$-surface X'. We then have that $\rho(X) = 1$ by Noether's theorem (see [Del73, 1.2.1]) and as X' is a non singular quartic it is a K3 surface. The hyperplane section generates a non trivial algebraic cycle in $H^2(G, \text{Pic}(X))$, but does not generate a real algebraic cycle—if the intersection of this plane with the torus is not empty then it contains two homologous circles or a circle homologous to 0. See [BKS82] for more details.

There is a unique complex family of complex K3 surfaces [X85], but there are 75 real families of real K3-surfaces [DIK00].

Let (X, σ) be an $\mathbb{R}$-surface. The number b^1_{alg} is not generally invariant under real deformation. If X is a K3 surface then

$$b^1_{\text{alg}}(X(\mathbb{R}); \mathbb{Z}_2) \leqslant b^1(X(\mathbb{R}); \mathbb{Z}_2) \leqslant h^{1,1}(X) = 20 .$$

Theorem 4.5.10 *Let (X, σ) be a real K3 surface which is not an M-surface. For any subgroup K in $H^1(X(\mathbb{R}); \mathbb{Z}_2)$ there is a real deformation Y of X and an isometry*

$$u \colon H^1(X(\mathbb{R}); \mathbb{Z}_2) \to H^1(Y(\mathbb{R}); \mathbb{Z}_2)$$

such that

$$u(K) = H^1_{\text{alg}}(Y(\mathbb{R}); \mathbb{Z}_2) .$$

When X is a *general* K3 surface we have that

$$b^1_{\text{alg}}(X(\mathbb{R}); \mathbb{Z}_2) \leqslant 1 ,$$

but Theorem 4.5.10 enables us to realise any value of b^1_{alg} authorised by topological constraints by specialisation.

Corollary 4.5.11 *Let (X, σ) be a real K3 surface. The following then hold.*

1. *For any integer* $1 \leqslant k < b^1(X(\mathbb{R}); \mathbb{Z}_2)$ *there is a real deformation* Y *of* X *such that*
$$b^1_{\mathrm{alg}}(Y(\mathbb{R}); \mathbb{Z}_2) = k \ .$$
2. *If* X *is not an* M*-surface there is also a real deformation* Y *of* X *such that*
$$b^1_{\mathrm{alg}}(Y(\mathbb{R}); \mathbb{Z}_2) = b^1(X(\mathbb{R}); \mathbb{Z}_2) \ .$$

For completion's sake, we note that for any maximal real K3 surface there cannot be a real deformation to a totally real algebraic surface because for any such surface the inequality $b^1_{\mathrm{alg}}(X(\mathbb{R}); \mathbb{Z}_2) < b^1(X(\mathbb{R}); \mathbb{Z}_2)$ holds by the following proposition.

Proposition 4.5.12 *Let* (X, σ) *be a compact Kähler* $\mathbb{R}$*-surface such that* $H_1(X; \mathbb{Z}_2) = 0$. *We have that*

$$b^1_{\mathrm{alg}}(X(\mathbb{R}); \mathbb{Z}_2) \leqslant b^1(X(\mathbb{R}); \mathbb{Z}_2) - (p_g(X) - a)$$

where $a = \frac{1}{2}(b^*(X; \mathbb{Z}_2) - b^*(X(\mathbb{R}); \mathbb{Z}_2))$, *so that* (X, σ) *is an* $(M - a)$*-surface (Definition 3.3.11).*

Proof See [Man97, Proposition 3.2]. □

Proof of Theorem 4.5.10 We sketch the proof of Theorem 4.5.10: see [Man97] for the complete proof. Let X be a K3 surface. We know by [X85, Exposé IV] that $H^2(X; \mathbb{Z})$ equipped with the cup product is isomorphic to a certain free $\mathbb{Z}$-module L with an integral even non degenerate symmetric bilinear form Q of signature $(3, 19)$. Let $f\colon H^2(X; \mathbb{Z}) \to L$ be an isometry. We say that the pair (X, f) is a *marked K3 surface*. Consider the Hodge decomposition (see Appendix D)

$$H^2(X, \mathbb{C}) = H^{2,0}(X) \oplus H^{1,1}(X) \oplus H^{0,2}(X) \ .$$

Here we have that $h^{2,0} = \dim H^{2,0}(X) = 1$ and $h^{1,1} = \dim H^{1,1}(X) = 20$. Let (X, f) be a marked K3 surface. We consider $P \subset L_{\mathbb{R}} = L \otimes \mathbb{R}$, the image under $f_{\mathbb{R}}$ of the subspace $H^2(X; \mathbb{R}) \cap (H^{2,0}(X) \oplus H^{0,2}(X))$ in $H^2(X; \mathbb{R})$. Since $H^{2,0}(X)$ is of complex dimension 1, P is of real dimension 2 in $L_{\mathbb{R}}$. We choose an orientation of P such that for any holomorphic 2-form $\omega \in H^{2,0}(X)$ the basis $(\Re(\omega), \Im(\omega))$ is direct. This oriented plane P is called the *period* of the marked K3 surface (X, f).

Let (X, σ) be a real K3 surface. The real structure induces an involution σ of (L, Q). For simplicity's sake we consider the case where X is not an M-surface and set $K = H^1(X(\mathbb{R}); \mathbb{Z})$. Following ideas due to Nikulin [Nik83] we show that in this case there is a primitive submodule $M \subset L^{-\sigma}$ whose quotient $M/((1 - \sigma)L \cap M)$ has rank $b^1(X(\mathbb{R}); \mathbb{Z}_2)$ and whose orthogonal $M^{\perp}_{\mathbb{R}}$ meets the cone of periods of real K3 surfaces deformation equivalent to X. For any real K3 surface (Y, g) whose period is orthogonal to M we have that $g^{-1}(M) \subset H^2_{\mathrm{alg}}(Y; \mathbb{Z})$ by the Lefschetz theorem on $(1, 1)$-cycles (Theorem D.9.3).

Consider an element $\alpha \in g^{-1}(M)$. By construction, α is anti-invariant for the real structure on Y and as Y is simply connected it is representable by a real divisor. The first Chern class induces an isomorphism $\mathrm{Pic}(X)^\sigma \longrightarrow H^2_{\mathrm{alg}}(X;\mathbb{Z})^{-\sigma}$. We complete the proof by establishing that $H^1(Y(\mathbb{R});\mathbb{Z}_2)$ is equal to $\varphi \circ c_1^{-1} \circ g^{-1}(M)$. □

4.5.3 Enriques Surfaces

Definition 4.5.13 Let X be a non singular compact complex analytic surface. X is said to be an *Enriques surface* if and only if $q(X) = 0$, $K_X \not\sim 0$ and $2K_X \sim 0$. A *real Enriques surface* is an $\mathbb{R}$-surface (X, σ) such that X is an Enriques surface.

Remark 4.5.14 Unlike K3 surfaces, all Enriques surfaces are projective.

Proposition 4.5.15 *Let X be an Enriques surface. X is then a minimal projective surface such that $\kappa(X) = 0$ and $p_g(X) = 0$.*

Proof See [BHPVdV04, Chapitre VIII]. □

There are 87 topological types of real Enriques surfaces [DK96b]. The theorem below completes the classification started by Nikulin in [Nik96].

In the list below $X(\mathbb{R})$ is the real locus of a real Enriques surface realising a given topological type. Recall that by definition of an $(M-a)$-surface, the modulo 2 Betti numbers satisfy the relationship $2a = \sum_{i=0}^4 b_i(X;\mathbb{Z}_2) - \sum_{i=0}^2 b_i(X(\mathbb{R});\mathbb{Z}_2)$.

Theorem 4.5.16 (Degtyarev, Kharlamov 1996) *There are 87 topological types of real Enriques surfaces. Each of them can be obtained by topological Morse simplification (Definition 4.3.31) from one of the 22 extremal types listed below. Conversely, except for $6\mathbb{S}^2$ and $\mathbb{T}^2 \sqcup 5\mathbb{S}^2$ any type obtained in this way can be realised as a real Enriques surface.*

The 22 extremal types are the following.

1. *M-surfaces,*
   (a) $\chi(X(\mathbb{R})) = 8$,
   $4\mathbb{RP}^2 \sqcup 2\mathbb{S}^2$, $\mathbb{V}_3 \sqcup \mathbb{RP}^2 \sqcup 4\mathbb{S}^2$, $\mathbb{V}_4 \sqcup 5\mathbb{S}^2$,
   $\mathbb{K}^2 \sqcup 2\mathbb{RP}^2 \sqcup 3\mathbb{S}^2$, $2\mathbb{K}^2 \sqcup 4\mathbb{S}^2$, $\mathbb{K}^2 \sqcup \mathbb{T}^2 \sqcup 4\mathbb{S}^2$,
   (b) $\chi(X(\mathbb{R})) = -8$,
   $\{\mathbb{V}_l \sqcup \mathbb{V}_{12-l}\}_{l=1\dots6}$, $\mathbb{V}_{10} \sqcup \mathbb{T}^2$;
2. *$(M-2)$-surfaces with $\chi(X(\mathbb{R})) = 0$,*
   $\mathbb{V}_4 \sqcup 2\mathbb{RP}^2$, $\mathbb{V}_6 \sqcup 2\mathbb{S}^2$, $\mathbb{V}_3 \sqcup \mathbb{K}^2 \sqcup \mathbb{RP}^2$, $\mathbb{V}_4 \sqcup \mathbb{T}^2 \sqcup \mathbb{S}^2$,
   $\mathbb{V}_5 \sqcup \mathbb{RP}^2 \sqcup \mathbb{S}^2$, $2\mathbb{V}_3 \sqcup \mathbb{S}^2$, $\mathbb{V}_4 \sqcup \mathbb{K}^2 \sqcup \mathbb{S}^2$, $2\mathbb{K}^2 \sqcup \mathbb{T}^2$;
3. *Pair of toruses $\mathbb{T}^2 \sqcup \mathbb{T}^2$.*

Proof See [Nik96, DK96b, DK96a]. □

4.5.4 Algebraic Cycles on Enriques Surfaces

There is a unique complex family of Enriques surfaces, but several hundred real families. See [DIK00] for more details. Let (X, σ) be a real Enriques surface. As $b^2(X) = 10$ and $b^1(X; \mathbb{Z}_2) = 1$ we have that $b^1_{\text{alg}}(X(\mathbb{R}); \mathbb{Z}_2) \leqslant b^1(X(\mathbb{R}); \mathbb{Z}_2) \leqslant 12$ by Inequality (3.8) of Theorem 3.3.6.

Unlike K3 surfaces, the number b^1_{alg} is invariant under real deformation of Enriques surfaces. The theorem below characterises the group of algebraic cycles of a real Enriques surface topologically.

Theorem 4.5.17 *Let (X, σ) be a real Enriques surface with $X(\mathbb{R}) \neq \varnothing$. If all the connected components of the real part $X(\mathbb{R})$ are orientable we have that*

$$H^1_{\text{alg}}(X(\mathbb{R}); \mathbb{Z}_2) = H^1(X(\mathbb{R}); \mathbb{Z}_2) .$$

Otherwise

$$\dim H^1_{\text{alg}}(X(\mathbb{R}); \mathbb{Z}_2) = \dim H^1(X(\mathbb{R}); \mathbb{Z}_2) - 1 .$$

Corollary 4.5.18 *A real Enriques surface (X, σ) is totally algebraic if and only if $X(\mathbb{R})$ is empty or orientable.*

Corollary 4.5.19 *There are real families of Enriques surfaces such that $b^1_{\text{alg}}(X(\mathbb{R}); \mathbb{Z}_2) < b^1(X(\mathbb{R}); \mathbb{Z}_2)$ for all members of the family. In particular, if X is a real Enriques surface with maximal first Betti number, $b^1(X(\mathbb{R}); \mathbb{Z}_2) = 12$, there is no real deformation of X which is totally algebraic.*

Proof We will give a proof of the first part of Theorem 4.5.17: we refer to [MvH98] for a complete proof. For technical reasons we prove this theorem in homology. In other words, we will establish the following equivalence.

$$H_1^{\text{alg}}(X(\mathbb{R}); \mathbb{Z}_2) = H_1(X(\mathbb{R}); \mathbb{Z}_2) \iff X(\mathbb{R}) \text{ is orientable.}$$

As the canonical divisor of an Enriques theorem is 2-torsion, the condition is necessary by Theorem 4.5.21 below. The converse is tricky. If Y is a K3 surface then Y is simply connected and there is a surjective morphism

$$H_2(Y; \mathbb{Z})^{-\sigma} \longrightarrow H_1(Y(\mathbb{R}); \mathbb{Z}_2) .$$

This morphism is not well defined for an Enriques surface X because its fundamental group is $\pi_1(X) = \mathbb{Z}_2$. On the other hand, when X is an Enriques surface we can always define a morphism of equivariant cohomology [MvH98, Sec. 4]

$$\alpha^X : H_2(X; G, \mathbb{Z}(1)) \longrightarrow H_1(X(\mathbb{R}); \mathbb{Z}_2)$$

whose image is precisely the group $H_1^{\text{alg}}(X(\mathbb{R}); \mathbb{Z}_2)$. Any Enriques surface is a quotient of a K3 surface by a holomorphic involution without fixed points [Bea78,

Proposition VIII.17]. Let Y be a complex K3 surface such that X is the quotient of Y by a holomorphic involution η without fixed points. The real structure on X naturally lifts to two real structures σ_1 and $\sigma_2 = \eta \circ \sigma_1$ on Y which commute with each other [Sil89, Theorem A_8.6]. The real part $X(\mathbb{R})$ is covered by the union of the real parts $Y_1(\mathbb{R}) = Y^{\sigma_1}$ and $Y_2(\mathbb{R}) = Y^{\sigma_2}$. For any $j \in \{1, 2\}$ let X_j be the disjoint union of components of $X(\mathbb{R})$ covered by $Y_j(\mathbb{R})$. This gives us a natural decomposition of the real part of an Enriques surface in "halves"

$$X(\mathbb{R}) = X_1 \sqcup X_2 \ .$$

Recall that all the connected components of the real locus of a K3 surface are orientable. Let M be a connected component of one of the halves X_j. If M is orientable then it is covered by two components of $Y_j(\mathbb{R})$ which are exchanged by η. If M is non orientable then M is covered by a unique component of $Y_j(\mathbb{R})$ which is the orientation covering of M. This gives us a morphism

$$H_1(Y_1(\mathbb{R}); \mathbb{Z}_2) \oplus H_1(Y_2(\mathbb{R}); \mathbb{Z}_2) \longrightarrow H_1(X(\mathbb{R}); \mathbb{Z}_2)$$

which is surjective whenever $X(\mathbb{R})$ is orientable. This morphism gives us a commutative diagram.

$$\begin{array}{ccc} H_2(Y_1; G, \mathbb{Z}(1)) \oplus H_2(Y_2; G, \mathbb{Z}(1)) & \longrightarrow & H_2(X; G, \mathbb{Z}(1)) \\ \Big\downarrow {\scriptstyle \alpha^{Y_1} \oplus \alpha^{Y_2}} & & \Big\downarrow {\scriptstyle \alpha^X} \\ H_1(Y_1(\mathbb{R}); \mathbb{Z}_2) \oplus H_1(Y_2(\mathbb{R}); \mathbb{Z}_2) & \longrightarrow & H_1(X(\mathbb{R}); \mathbb{Z}_2) \end{array} \tag{4.15}$$

The morphisms α^{Y_1} and α^{Y_2} are surjective because Y is simply connected and hence α^X is surjective whenever $X(\mathbb{R})$ is orientable. □

The decomposition of the real locus in halves can also be used to characterise Galois-Maximality (Definition 3.6.5) of real Enriques surfaces

Theorem 4.5.20 *Let (X, σ) be a real Enriques surface of non empty real locus $X(\mathbb{R}) = X_1 \sqcup X_2$.*

1. *Suppose that both the halves X_1 and X_2 are non empty. The surface X is then Galois-Maximal. Moreover, X is $\mathbb{Z}$-Galois-Maximal if and only if $X(\mathbb{R})$ is non orientable*
2. *Suppose one of the halves X_1 or X_2 is empty. The surface X is then Galois-Maximal if and only if $X(\mathbb{R})$ is non orientable. Moreover, X is $\mathbb{Z}$-Galois-Maximal if and only if $X(\mathbb{R})$ has at least one component which is of odd Euler characteristic.*

Proof See [MvH98]. □

All cases of Galois-Maximality (see Example 3.6.14) are realised by Enriques surfaces: on inspecting the proof of Theorem 4.5.16 (see [DK96b, §5] or [DK96a]) we see that there exist examples of Enriques surfaces for each of the cases listed in the previous theorem.

Theorem 4.5.21 *Let $d \geqslant 2$ be an integer. An algebraic $\mathbb{R}$-surface X whose canonical bundle K_X is d-torsion can only be totally algebraic if its real part $X(\mathbb{R})$ is empty or orientable.*

Proof Let X be a totally algebraic $\mathbb{R}$-surface, by which we mean that

$$H^1_{\text{alg}}(X(\mathbb{R}); \mathbb{Z}_2) = H^1(X(\mathbb{R}); \mathbb{Z}_2).$$

We use the properties (3.38) of $\varphi_X \colon \operatorname{Pic}(X)^\sigma \to H^1_{\text{alg}}(X(\mathbb{R}); \mathbb{Z}_2)$ discussed in Proposition 3.7.16. Let D be a divisor whose class in the Néron–Severi group $\operatorname{NS}(X)$ has a trivial multiple: we then have that $(D \cdot D') = 0$ for any divisor D'. When D is real we have that $\varphi_X(D) = 0$ in $H^1(X(\mathbb{R}); \mathbb{Z}_2)$ since by hypothesis any cohomology class $u \in H^1(X(\mathbb{R}); \mathbb{Z}_2)$ is the image under φ_X of a real divisor D' so $\varphi_X(D) \cdot u = 0$. As the intersection form is non degenerate on $H^1(X(\mathbb{R}); \mathbb{Z}_2)$ it follows that $\varphi_X(D) = 0$. Now if $X(\mathbb{R}) \neq \varnothing$ we can assume that K_X is real by Theorem 2.6.32. As K_X is torsion in $\operatorname{NS}(X)$ we have that $\varphi_X(K_X) = 0$ whence $\mathrm{w}_1(X(\mathbb{R})) = 0$ and therefore $X(\mathbb{R})$ is orientable. □

4.5.5 Abelian Surfaces

A detailed study of real abelian varieties is available in Comessatti's articles [Com25, Com26]. Their moduli spaces and their compactifications are described in [Sil89, Chap IV] and [Sil92].

Definition 4.5.22 A *complex torus* of dimension g is a quotient of $\mathbb{C}^g$ by a sub-$\mathbb{Z}$-module $\Lambda \subset \mathbb{C}^g$ of maximal rank $2g$ (also called a *lattice*). An *abelian variety* is a projective complex torus, or in other words a complex torus with an ample divisor. A *real abelian surface* is a complex torus of dimension 2 equipped with a real structure and an embedding into projective space. The embedding can be assumed equivariant by Theorem 2.6.44.

Remark 4.5.23 Complex toruses are Kähler because they inherit a Kähler metric from $\mathbb{C}^g$. On the other hand, like K3 surfaces, complex toruses of dimension 2 (or more) are not always projective.

Example 4.5.24 (*Complex toruses associated to a variety*) The Picard variety $\operatorname{Pic}^0(X)$ of a compact Kähler variety X of irregularity $q > 0$ is a complex torus of dimension q. (See Definition D.6.6 for the definition of the Picard variety.) If X is projective it is a projective variety by Proposition D.6.7. If X is a complex torus then $\operatorname{Pic}^0(X)$ is isomorphic to X.

The Albanese variety $\operatorname{Alb}(X)$ of a compact Kähler variety X of irregularity $q > 0$ is a complex torus of dimension q. (See Definition D.6.10 for the definition of the

Albanese variety.) It is an abelian variety if X is projective [Voi02, Corollaire 12.12]. If X is a complex torus, $\mathrm{Alb}(X)$ is isomorphic to X.

The Jacobian $\mathrm{Jac}(C)$ of a compact complex curve C of genus g is an abelian variety of dimension g. (See Definition E.4.1 for the definition of the Jacobian.) If C is a curve of genus 1 then $\mathrm{Jac}(C)$ is a curve isomorphic to C.

Proposition 4.5.25 *Let X be an abelian surface. X is then a minimal projective surface such that $\kappa(X) = 0$, $p_g(X) = 1$, $q(X) = 2$ and $K_X \sim 0$.*

Proof See Theorem 4.5.1 and [Bea78, Liste VI.20 and Théorème VIII.2]. □

Theorem 4.5.26 *There are 4 topological types of real abelian surfaces.*

$$\varnothing, \quad \mathbb{T}^2, \quad 2\mathbb{T}^2, \quad 4\mathbb{T}^2 .$$

Proof See [Sil89, Chapter IV]. □

4.5.6 Algebraic Cycles on Abelian Surfaces

Like K3 surfaces, there is a unique complex family of complex abelian surfaces whereas real abelian surfaces are divided into several real families and b^1_{alg} is not invariant under real deformation.

Let (X, σ) be a real abelian surface. By Theorem 4.5.26 we have that $b^1(X(\mathbb{R})) \le 8$. Moreover, using [Kuc96, Theorem 2.1] we also have that $b^1_{\mathrm{alg}}(X(\mathbb{R})) \le 5$.

As for K3 surfaces we can identify certain topological constraints and prove that once these constraints are satisfied we can always deform an abelian surfaces so as to obtain a given $b^1_{\mathrm{alg}}(X(\mathbb{R}))$.

In particular, we can show that the real part of a totally algebraic real abelian surface is either connected or empty—see [Hui94]—and that a real abelian surface with connected real locus can always be deformed to a totally algebraic real abelian surface.

Proposition 4.5.27 *Let X be a real abelian surface with a real point. We then have that*

$$H^1_{\mathrm{alg}}(X(\mathbb{R}); \mathbb{Z}_2) = H^1(X(\mathbb{R}); \mathbb{Z}_2) \implies X(\mathbb{R}) \approx \mathbb{T}^2 .$$

4.5.7 Bi-elliptic Surfaces

Definition 4.5.28 Let X be a non singular compact complex analytic surface. We say that X is a *bi-elliptic surface*[10] if there are two elliptic curves E and F and

[10]Classically, bi-elliptic surfaces were called *hyperelliptic: we refer to [Bea78, VI.19] for an explanation of the terminology used here.*

a finite group H acting by translation on F and by automorphism on E such that $E/H = \mathbb{P}^1$ and X is the quotient of the product $E \times F$ by the product action of H. A *real bi-elliptic surface* is an $\mathbb{R}$-surface (X, σ) such that X is a bi-elliptic surface.

Proposition 4.5.29 *Let X be a bi-elliptic surface. X is then a projective minimal surface such that $\kappa(X) = 0$, $p_g(X) = 0$, $q(X) = 1$ and $m_0 K_X \sim 0$ for some $m_0 \in \{2, 3, 4, 6\}$.*

Proof See Theorem 4.5.1 and [Bea78, Liste VI.20 and Théorème VIII.2]. □

For any bi-elliptic surface X, the *Albanese map* (Definition D.6.13)

$$\alpha\colon X = (E \times F)/H \to \mathrm{Alb}(X) = F/H$$

is an elliptic fibration that is locally but not globally trivial. The fibres of α are all isomorphic to E over $\mathbb{C}$. When X is an $\mathbb{R}$-surface the Albanese fibration α is an $\mathbb{R}$-fibration and the curves F/H and E are real elliptic curves. The real locus of a non singular real elliptic curve is either empty or consists of one or two ovals, It is immediate that the number of connected components of the real locus of X satisfies $0 \leqslant \#\pi_0 X(\mathbb{R}) \leqslant 4$ and every connected component is homeomorphic to a torus $\mathbb{T}^2$ or a Klein bottle $\mathbb{K}^2$. This gives us a list of 15 potential topological types of $X(\mathbb{R})$. Catanese and Frediani [CF03] determined the eleven topological types that are actually possible as a corollary of their description of the moduli space of real bi-elliptic surfaces. If α is further assumed to have a real section then only seven topological types can be realised.

Theorem 4.5.30 (Catanese, Frediani 2003) *There are 11 topological types of real bi-elliptic surfaces.*

1. $\varnothing$, $\mathbb{T}^2$, $2\mathbb{T}^2$, $3\mathbb{T}^2$, $4\mathbb{T}^2$,
2. $\mathbb{K}^2$, $2\mathbb{K}^2$, $3\mathbb{K}^2$, $4\mathbb{K}^2$,
3. $\mathbb{T}^2 \sqcup \mathbb{K}^2$, $\mathbb{T}^2 \sqcup 2\mathbb{K}^2$.

Proof See [CF03]. □

Theorem 4.5.31 *There are 7 topological types of real bi-elliptic surfaces whose Albanese fibration has a real section.*

1. $\mathbb{T}^2$, $2\mathbb{T}^2$, $3\mathbb{T}^2$, $4\mathbb{T}^2$,
2. $2\mathbb{K}^2$, $3\mathbb{K}^2$, $4\mathbb{K}^2$.

Proof See [Man03, Théorème 2.3]. □

4.5.8 Algebraic Cycles on Bi-elliptic Surfaces

The canonical divisor K_X of a bi-elliptic surface is torsion: we denote its order by $d_X \in \{2, 3, 4, 6\}$.

There are exactly seven complex families of bi-elliptic surfaces and each of them corresponds to several real families. We refer to [CF03] for more details. Once again, the number b^1_{alg} is invariant under deformation and the article [Man03] contains a topological characterisation of totally algebraic real bi-elliptic surfaces.

Theorem 4.5.32 *Let X be a real bi-elliptic surfaces with a real point.*

1. *If $H^1_{\mathrm{alg}}(X(\mathbb{R});\mathbb{Z}_2) = H^1(X(\mathbb{R});\mathbb{Z}_2)$ then $X(\mathbb{R})$ is homeomorphic to a torus. If moreover d_X is even then α has a real section.*
2. *Suppose that $X(\mathbb{R})$ is homeomorphic to a torus. If d_X is odd or α has a real section then*

$$H^1_{\mathrm{alg}}(X(\mathbb{R});\mathbb{Z}_2) = H^1(X(\mathbb{R});\mathbb{Z}_2)\ .$$

Proof We give a partial proof of this result and refer to [Man03] for the complete proof. Let (X,σ) be a real bi-elliptic surface. We denote by

$$\pi\colon X \to E/H \simeq \mathbb{P}^1$$

the second elliptic fibration whose only singular fibres are multiple fibres $m_t L_t$ where L_t is a non singular elliptic curve. The Néron–Severi group $\mathrm{NS}(X)$ is generated by a fibre X_x of α and by the reductions L_t of the multiple fibres of π. Let $m_t L_t$ and $m_{t'} L_{t'}$ be two multiple real fibres of π and denote by d the gcd of m_t and $m_{t'}$. Assume that $d \geqslant 2$: the divisor $D = (m_t/d)L_t - (m_{t'}/d)L_{t'}$ is then d-torsion in $\mathrm{NS}(X)$. By the proof of Theorem 4.5.21 we then have that $\varphi_X(D) = 0$. Permuting t and t' if necessary we can assume that m_t/d is odd and in this case

$$\varphi_X((m_t/d)L_t) = \varphi_X(L_t)\ .$$

There are now two possibilities: either $\varphi_X(L_t) = \varphi_X(L_{t'})$ or $\varphi_X(L_t) = 0$. Studying the seven possible configurations of multiples fibres, we deduce that the image under φ_X of the subgroup of $\mathrm{NS}(X)$ generated by the real curves L_t is of dimension $\leqslant 1$.

Suppose now that (X,σ) is totally algebraic. In this case we have that $\varphi_X(\mathrm{Pic}^0(X)^\sigma) = \{0\}$ in $H^1_{\mathrm{alg}}(X(\mathbb{R});\mathbb{Z}_2)$ by [Kuc96, Th. 2.1]. It follows that there is a well defined morphism

$$\mathrm{NS}(X)^\sigma \to H^1_{\mathrm{alg}}(X(\mathbb{R});\mathbb{Z}_2)$$

on $\mathrm{NS}(X)^\sigma$ which is surjective onto $H^1_{\mathrm{alg}}(X(\mathbb{R});\mathbb{Z}_2)$. This gives an upper bound $\dim H^1_{\mathrm{alg}}(X(\mathbb{R});\mathbb{Z}_2) \leqslant 2$. By hypothesis, the dimension of the space $H^1(X(\mathbb{R});\mathbb{Z}_2)$ satisfies the same inequality so $X(\mathbb{R})$ is connected. Moreover, the canonical divisor K_X of a bi-elliptic surface is d_X-torsion for some $d_X \in \{2,3,4,6\}$. By Theorem 4.5.21, if the real part of $X(\mathbb{R})$ is non empty then it is orientable and thus homeomorphic to a torus. We refer to the original article [Man03] for a proof of the converse. □

4.5.9 *Summary: Algebraic Cycles on Surfaces with $\kappa \leqslant 0$*

Gathering the results in this section on surfaces of Kodaira dimension $\kappa(X) = 0$ and checking them directly (exercise for the reader) for surfaces of Kodaira dimension $\kappa(X) = -\infty$ (by [Kuc96, Theorem 2.1] we then have to choose an involution such that $\varphi_X(\mathrm{Pic}^0(X)^\sigma) = \{0\}$ in $H^1_{\mathrm{alg}}(X(\mathbb{R}); \mathbb{Z}_2)$) we note that when X is an algebraic surface of one of the following types: rational, uniruled, abelian, K3 or Enriques, we can always find an algebraic surface Y in the same complex family as X and a non empty real structure on Y which is totally algebraic. (This turns out to also hold for regular elliptic surfaces: see Theorem 4.6.16 for more details.) On the other hand, there are two complex families of bi-elliptic surfaces containing $\mathbb{R}$-surfaces whose real part is diffeomorphic to a torus and which are never totally algebraic:

Theorem 4.5.33 *Let E, F be elliptic curves and let H be the group $\mathbb{Z}_2 \oplus \mathbb{Z}_2$ or $\mathbb{Z}_4 \oplus \mathbb{Z}_2$. For any complex algebraic surface Y which is deformation equivalent to the bi-elliptic surface $X = (E \times F)/H$ and for any real structure on Y with real points we have that*

$$H^1_{\mathrm{alg}}(Y(\mathbb{R}); \mathbb{Z}_2) \neq H^1(Y(\mathbb{R}); \mathbb{Z}_2) .$$

Proof See [Man03, Corollaire 3.3]. □

In each of the five other complex families of bi-elliptic surfaces there is an $\mathbb{R}$-surface X such that $X(\mathbb{R})$ is homeomorphic to a torus and $H^1_{\mathrm{alg}}(Y(\mathbb{R}); \mathbb{Z}_2) = H^1(Y(\mathbb{R}); \mathbb{Z}_2)$.

Corollary 4.5.34 *Except for the surfaces of Theorem 4.5.33, every complex family of surfaces with non positive Kodaira dimension contains an $\mathbb{R}$-surface Y with a real point such that $H^1_{\mathrm{alg}}(Y(\mathbb{R}); \mathbb{Z}_2) = H^1(Y(\mathbb{R}); \mathbb{Z}_2)$.*

4.6 Elliptic Surfaces ($\kappa \leqslant 1$)

Definition 4.6.1 A non singular complex analytic surface X is said to be *elliptic* if and only if there is a non singular complex curve Δ and a proper surjective holomorphic map $\pi\colon X \longrightarrow \Delta$ such that the fibre $X_u = \pi^{-1}(u)$ is a non singular curve of genus 1 for almost all points $u \in \Delta$.

Remark 4.6.2 As the complex variety X is non singular and of dimension 2 all the fibres of π are of dimension 1. (This equidimensionality of fibres no longer holds in higher dimension: see [Uen73] for more details.) Moreover, since π is proper its fibres are compact and X is compact if and only if Δ is compact.

Some elliptic surfaces are projective and hence algebraic by Chow's theorem D.5.1. The algebraic definition is as follows.

Definition 4.6.3 A non singular complex projective algebraic surface X is said to be *elliptic* if there is a non singular complex projective algebraic curve Δ and a regular surjective map $\pi\colon X \longrightarrow \Delta$ whose general fibre is a non singular complex projective algebraic curve of genus 1.

Remark 4.6.4 The image under π of the set of its singular fibres is Zariski closed in Δ so the number of singular fibres of π is finite.

Definition 4.6.5 An $\mathbb{R}$-surface (X, σ) is said to be *real elliptic* if there is a real elliptic fibration $\pi\colon X \longrightarrow \Delta$, i.e. the curve Δ has a real structure σ_Δ and $\pi \circ \sigma = \sigma_\Delta \circ \pi$.

Remark 4.6.6 (*Scheme theoretic definition*) We give the corresponding scheme theoretic definition for the sake of completeness. An *elliptic fibration* of a geometrically integral non singular surface X is a faithfully flat morphism $\pi\colon X \longrightarrow C$ to a non singular curve whose generic fibre $X_{K(C)}$ is isomorphic to a non singular curve of genus 1 over the function field $K(C)$ of rational functions on C. A closed general fibre $X_{\kappa(c)} = \pi^{-1}(c)$ of π is isomorphic to a non singular curve of genus 1 over the residue field $\kappa(c)$ of the point $c \in C$. A closed schematic fibre $\pi^{-1}(c)$ which is not isomorphic to a non singular curve of genus 1 over $\kappa(c)$ is said to be *degenerate*.

Proposition 4.6.7 *If X is an elliptic surface then $\kappa(X) \leqslant 1$.*

Proof See [BHPVdV04, Theorem V.12.5]. □

Proposition 4.6.8 *Let X be a non singular compact complex analytic surface such that $\kappa(X) = 1$. The surface X is then canonically equipped with an elliptic fibration which is the only elliptic fibration on X.*

Proof See [BHPVdV04, § VI.3, case $a(X) = 2$, $\kappa(X) = 1$]. □

Remark 4.6.9 If $\kappa(X) = 1$ the variety X has an elliptic fibration given by the morphism φ_{mK_X} associated to a multiple of the canonical divisor. In particular, if (X, σ) is an $\mathbb{R}$-surface then this fibration is a real elliptic fibration by Proposition 2.6.31.

Definition 4.6.10 A non singular compact complex analytic surface X (resp. $\mathbb{R}$-surface (X, σ)) is a *properly elliptic surface* (resp. a *properly real elliptic surface*) if $\kappa(X) = 1$.

An algebraic surface is said to be *regular* or *of zero irregularity* if $H^1(X, \mathcal{O}_X) = \{0\}$. When X is an elliptic surface this implies that the base curve has genus 0 and the fibration has at least one singular fibre. Conversely, let X be an elliptic fibration and suppose that $\pi\colon X \longrightarrow \mathbb{P}^1$ has at least one singular fibre. The surface X is then of zero irregularity. Recall (Definition 4.3.16) that the fibration π is said to be *minimal* if and only if non of its fibres contains a (-1)-curve (by which we mean, generalising Definition 4.3.2 to non singular analytic surfaces, an irreducible non singular rational curve of self intersection -1). When a relatively minimal elliptic fibration $\pi\colon X \to \mathbb{P}^1$ has a section $s\colon \mathbb{P}^1 \to X$ we say that X is a

Jacobian elliptic surface.[11] All of these definitions make sense over $\mathbb{R}$: a real elliptic surface is an elliptic surface whose fibration morphism commutes with the real structures on X and $\mathbb{P}^1$ and is said to be a real Jacobian elliptic surface if π has a real section.

We recall two results on complex elliptic surfaces which will be useful in the rest of this section.

Lemma 4.6.11 *Two relatively minimal complex elliptic surfaces of zero irregularity without multiple fibres are equivalent by deformation if and only if their holomorphic Euler characteristic are equal.*

Proof See [Kas77]. □

Theorem 4.6.12 *Two regular elliptic surfaces without multiple fibres X and Y are deformation equivalent if and only if*

1. *the minimal models X' of X and Y' of Y are deformation equivalent;*
2. *$\eta_X = \eta_Y$, where η_Z, $Z = X, Y$ is the minimal number of blow ups required to produce Z from Z'.*

Proof See [Kod64]. □

The possible singular fibres of a real elliptic surface were classified by Silhol—see [Sil84] and [Sil89, Chapitre VII] for more details. The following theorem lists all possible topological types of real Jacobian elliptic surfaces in each complex family of complex Jacobian elliptic families.

Theorem 4.6.13 *Let $k \geqslant 1$ be an integer. The possible extremal topological types of real Jacobian elliptic surfaces of zero irregularity and holomorphic Euler characteristic $\chi(\mathcal{O}_X) = k$ are:*

1. *M-surfaces, $a = k + 4\lambda - 1$, $l = 5k - 4\lambda$, $\lambda = 0, 1, \ldots, k$,*
   - *$\mathbb{S}_l \sqcup a\,\mathbb{S}^2$, k even*
   - *$\mathbb{V}_{2l} \sqcup a\mathbb{S}^2$, k odd.*
2. *$(M-2)$-surfaces, $a = k + 4\lambda$, $l = 5k - 4\lambda - 3$, $\lambda = 0, 1, \ldots, k-1$,*
   - *$\mathbb{S}_l \sqcup a\,\mathbb{S}^2$, k even or*
   - *$\mathbb{V}_{2l} \sqcup a\mathbb{S}^2$, k odd.*
3. *$\chi(X(\mathbb{R})) = 0$,*
   - *pair of toruses $\mathbb{K}^2 \sqcup \mathbb{K}^2$, k even or*
   - *pair of Klein bottles $\mathbb{T}^2 \sqcup \mathbb{T}^2$, k odd.*

---

[11] This terminology comes from the fact that in this case the fibrations $X \to \mathbb{P}^1$ and $\mathrm{Jac}(X) \to \mathbb{P}^1$ are isomorphic. See [BHPVdV04, V.9] for more details.

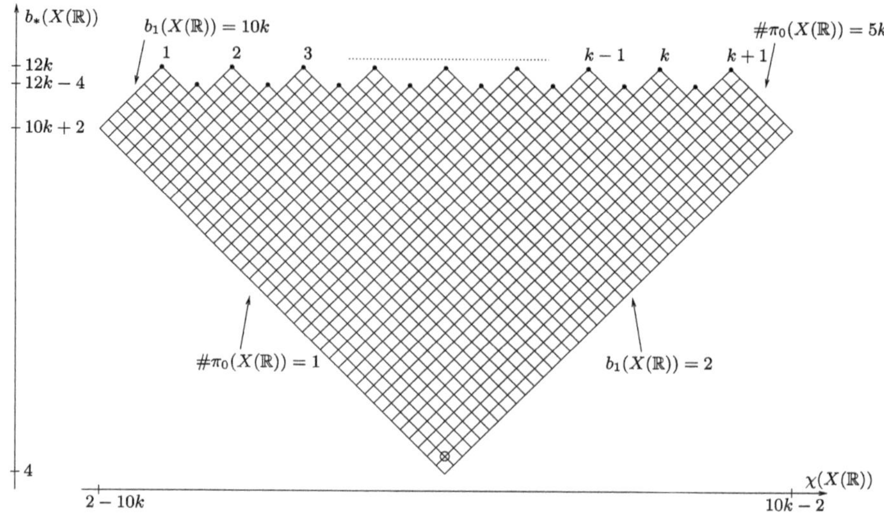

Fig. 4.12 Topological types of real Jacobian regular elliptic surfaces of holomorphic Euler characteristic k

Let X be a real Jacobian elliptic surface of Euler characteristic $\chi(\mathcal{O}_X) = k$. The topological type of $X(\mathbb{R})$ is then obtained by applying Morse simplification to one of the types listed above.

Conversely, any topological type obtained by Morse simplification applied to one of the above types which has total Betti number at least 2 can be realised as the real part of a real Jacobian elliptic surface X with Euler characteristic $\chi(\mathcal{O}_X) = k$.

Proof See [BM07]. □

Note that in the definition of a Jacobian surface we have assumed the elliptic fibration is relatively minimal. The analogue of Theorem 4.6.13 without this hypothesis, i.e. the classification of topological types of real elliptic surfaces of zero irregularity with at least one real section contained in a given family of complex deformations, follows directly from Theorem 4.6.13. Indeed, by Theorem 4.6.12, two complex elliptic surfaces of zero irregularity without multiple fibres are deformation equivalent if and only if their holomorphic Euler characteristic are equal *and* their canonical fibres have the same degree.

To realise a given topological type in a certain complex family—with $k = \chi(\mathcal{O}_X)$ and $K_X^2 = -m < 0$, for example—consider a Jacobian elliptic surface Y of holomorphic Euler characteristic k. By definition we then have that $K_Y^2 = 0$. Let X be the surface obtained by blowing up a set of m points globally fixed by the real structure. We then have that $K_X^2 = -m$. Each blow up at a real point produces a connected sum with an $\mathbb{RP}^2$, as in Example 4.2.18. Conversely, the topological type of any real elliptic surface of zero irregularity with a real section can be obtained in this way.

4.6.1 Algebraic Cycles on Elliptic Surfaces such that $q = 0$

Unlike surfaces of zero Kodaira dimension (K3, abelian, Enriques and bi-elliptic surfaces) there is an infinite number of complex families of regular elliptic surfaces and for each complex family there are several real families.

In general it is fairly difficult to find in a given family of complex surfaces real algebraic surfaces with "large" first Betti number $b^1(X(\mathbb{R}); \mathbb{Z}_2)$. For example, we do not yet know whether there is a surface of degree 5 in $\mathbb{P}^3(\mathbb{R})$ with first Betti number equal to 47 (which is a known upper bound on the Betti number of such surfaces. See Section 4.7 for more details).

Proposition 4.6.14 (Kharlamov) *Any regular real elliptic surface (X, σ) without multiple fibres satisfies the Ragsdale–Viro inequality*

$$b^1(X(\mathbb{R}); \mathbb{Z}_2) \leqslant h^{1,1}(X) . \tag{4.16}$$

This result and an idea of its proof were communicated to us by V. Kharlamov in 1997. We do not know of any published proof other than [AM08], which we reproduce below.

Proof of Proposition 4.6.14 Set $b_*(X(\mathbb{R}); \mathbb{Z}_2) = \sum_{k=0}^{k} b_k(X(\mathbb{R}); \mathbb{Z}_2)$ and $b_*(X; \mathbb{Z}_2) = \sum_{k=0}^{2k} b_k(X; \mathbb{Z}_2)$. When $\pi\colon X \to \mathbb{P}^1$ has no multiple fibres it is easy to check (using the classification of possible singular fibres given in [Sil89, Chapitre VII]) that

$$b_1(X(\mathbb{R}); \mathbb{Z}_2) \leqslant b_1(\mathrm{Jac}(X)(\mathbb{R}); \mathbb{Z}_2)$$

where $\mathrm{Jac}(X) \to \mathbb{P}^1$ is the Jacobian bundle associated to $X \to \mathbb{P}^1$ as in [BHPVdV04, V.9]. By construction this fibration is a real elliptic surface with a real section such that

$$h^{1,1}(\mathrm{Jac}(X)) = h^{1,1}(X) .$$

We may therefore assume without loss of generality that π has a real section. The real structure σ induces an involution, also denoted by σ, on $H_2(X, \mathbb{Z})$. Consider the following homological invariants:

The rank of the submodule invariant under σ,

$$r_2 = \mathrm{rk}\, H_2(X, \mathbb{Z})^\sigma = \mathrm{rk}\ker(1 - \sigma)$$

the Comessatti characteristic

$$\lambda = \mathrm{rk}\,((1 + \sigma)H_2(X, \mathbb{Z})) = \mathrm{rk}\,\mathrm{Im}(1 + \sigma) .$$

As the fibration $\pi\colon X \to \mathbb{P}^1$ has a section it does not have multiple fibres and the Betti numbers $b_1(X)$ and $b_3(X)$ vanish. By Theorem 3.6.11 the surface (X, σ) is therefore *Galois-Maximal* and for this reason (see also Corollary 3.6.12) the Comessatti characteristic corresponds to

$$2\lambda = b_*(X; \mathbb{Z}_2) - b_*(X(\mathbb{R}); \mathbb{Z}_2)$$

and the first Betti number of $X(\mathbb{R})$ corresponds to

$$b_1(X(\mathbb{R}); \mathbb{Z}_2) = b_2(X) - r_2 - \lambda \,. \tag{4.17}$$

If there is no real non singular fibre of π then $X(\mathbb{R})$ is the union of two toruses or two Klein bottles because π has a real section. In this case inequality (4.16) holds. If π has at least one singular real fibre then $X(\mathbb{R})$ has exactly one connected component which is not simply connected and a finite number of other components which are all homeomorphic to spheres. Let s be the number of spherical components: the sum of the Betti numbers of $X(\mathbb{R})$ is then $b_*(X(\mathbb{R}); \mathbb{Z}_2) = 2 + 2s + b_1(X(\mathbb{R}); \mathbb{Z}_2)$ and the Comessatti characteristic is given by

$$\lambda = r_2 - 2s \,. \tag{4.18}$$

From Lemmas 4.4.17 and 4.4.19 it follows that s is a lower bound for the dimension of the invariant part of $H^{1,1}(X)$. As moreover $\sigma(H^{2,0}(X)) = H^{0,2}(X)$ by Lemma D.3.17 and $h^{2,0}(X) = h^{0,2}(X)$ we deduce the following lower bound for r_2:

$$h^{2,0}(X) + s \leqslant r_2 \,.$$

It follows from Equation (4.18) that $h^{2,0}(X) - s \leqslant \lambda$ and equality (4.17) implies that

$$b_1(X(\mathbb{R}); \mathbb{Z}_2) \leqslant b_2(X) - 2h^{2,0}(X) \,.$$

□

For every complex family of regular elliptic surfaces without multiple fibres every $\mathbb{R}$-surface (X, σ) satisfies $b^1(X(\mathbb{R}); \mathbb{Z}_2) \leqslant h^{1,1}(X)$ (Proposition 4.6.14). We prove below that in every complex family of regular elliptic surfaces without multiple fibres there is at least one subfamily of $\mathbb{R}$-surfaces such that $b^1 = h^{1,1}$. Moreover, in each of these real families there is at least one $\mathbb{R}$-surface such that $b^1_{\text{alg}}(X(\mathbb{R}); \mathbb{Z}_2) = b^1(X(\mathbb{R}); \mathbb{Z}_2)$. These two results are proved using the same construction. This establishes that the Ragsdale–Viro inequality $b^1(X(\mathbb{R}); \mathbb{Z}_2) \leqslant h^{1,1}(X)$ is optimal for regular real elliptic surfaces without multiple fibres.

By the Hodge decomposition theorem, for any regular relatively minimal elliptic surface X we have that $h^{1,1}(X) = 10\chi(\mathcal{O}_X)$ so when X is also without multiple fibres

$$b^1(X(\mathbb{R}); \mathbb{Z}_2) \leqslant 10\chi(\mathcal{O}_X) \,. \tag{4.19}$$

Theorem 4.6.15 *For any $k > 0$ there is a regular relatively minimal real elliptic surface (X, σ) such that:*

$$\chi(\mathcal{O}_X) = k, \quad b^1(X(\mathbb{R}); \mathbb{Z}_2) = 10k \,.$$

Proof We give a sketch of the proof and refer to [Man00] for the full proof. The surfaces appearing in the above statement are said to be modular. A modular surface is constructed from a finite index subgroup Γ of the modular group $\mathbf{PSL}_2(\mathbb{Z}) = \mathbf{SL}_2(\mathbb{Z})/\{\pm 1\}$. We adapt the classical construction [Shi71, Shi72b] to the real case and then use the real classification of possible singular fibres of an elliptic fibration given by Silhol [Sil84] to obtain the conditions that must hold on the group Γ for the real surface to have maximal homology in rank 1.

Consider a finite index subgroup $\Gamma \subset \mathbf{PSL}_2(\mathbb{Z})$. As $\mathbf{PSL}_2(\mathbb{Z})$ is a subgroup of $\mathbf{PSL}_2(\mathbb{R})$, the group Γ is a discrete subgroup of the group of isometries of the hyperbolic plane $\mathbb{H}$. It is therefore a Fuchsian group and the quotient $\Delta'_\Gamma = \mathbb{H}/\Gamma$ is a complex curve whose non compacity arises from parabolic classes or cusps. A natural compactification of this space can be obtained on noting that Γ acts on $\mathbb{P}^1(\mathbb{Q})$ considered as a subspace of the boundary of $\mathbb{H} = \{z \in \mathbb{C}/\ \Im(z) > 0\}$: we then consider the compact complex curve

$$\Delta_\Gamma = (\mathbb{H} \cup \mathbb{P}^1(\mathbb{Q}))/\Gamma\ .$$

We then use the fact that Γ is not only a group of isometries of $\mathbb{H}$ but also a group of automorphisms of elliptic curves, which allows us to construct a natural fibration in genus one curves over the open set Δ'_Γ in Δ_Γ. There are then several different ways of extending this fibration over a cusp point P. To determine the complex type of the singular fibre over P it is enough to identify an element of the stabiliser of $P \in \Delta_\Gamma$. This gives us a monodromy representation

$$\rho\colon \pi_1(\Delta'_\Gamma) \to \mathbf{PSL}_2(\mathbb{Z})\ .$$

For every lifting $\rho'\colon \pi_1(\Delta'_\Gamma) \to \mathbf{SL}_2(\mathbb{Z})$ of ρ we obtain an elliptic surface with singular fibres whose complex types are in a prescribed list.

The action of the group $\mathbf{PSL}_2(\mathbb{R})$ on $\mathbb{H}$ is denoted $z \mapsto A.z$ where $A.z = \frac{az+b}{cz+d}$ if A is represented by $\begin{pmatrix} a & b \\ c & d \end{pmatrix}$, $ad - bc = 1$. The involution $\sigma_\mathbb{H}\colon z \mapsto -\bar{z}$ on $\mathbb{H}$ is anti-holomorphic. We set $S\colon \mathbf{SL}_2(\mathbb{R}) \to \mathbf{SL}_2(\mathbb{R})$, $\begin{pmatrix} a & b \\ c & d \end{pmatrix} \mapsto \begin{pmatrix} a & -b \\ -c & d \end{pmatrix}$. The map S then induces an involution on $\mathbf{PSL}_2(\mathbb{R})$ which is also denoted by S. For any $z \in \mathbb{H}$ and $A \in \mathbf{PSL}_2(\mathbb{R})$ we have that $\sigma_\mathbb{H}(A.\sigma_\mathbb{H}(z)) = S(A).z$.

Let Γ be a Fuchsian group (i.e. a discrete subgroup of $\mathbf{PSL}_2(\mathbb{R})$). The involution $\sigma_\mathbb{H}$ induces a real structure on the quotient $\mathbb{H}/\Gamma$ if and only if $\sigma_\mathbb{H}\Gamma = \Gamma\sigma_\mathbb{H}$, i.e. if and only if Γ is stable under S.

In general we cannot entirely control the real types of the singular fibres arising in this construction. When this is possible, we obtain some necessary conditions on Γ and then exhibit a sequence of groups which satisfy these necessary conditions.

For any $k \in \mathbb{N}^*$, let Γ_k be the arithmetic group whose fundamental domain is shown in Figure 4.13. For every group Γ_k there is a real modular elliptic surface $X_k \longrightarrow \Delta_{\Gamma_k} \simeq \mathbb{P}^1$ satisfying the conditions of Theorem 4.6.15. For every k, the real

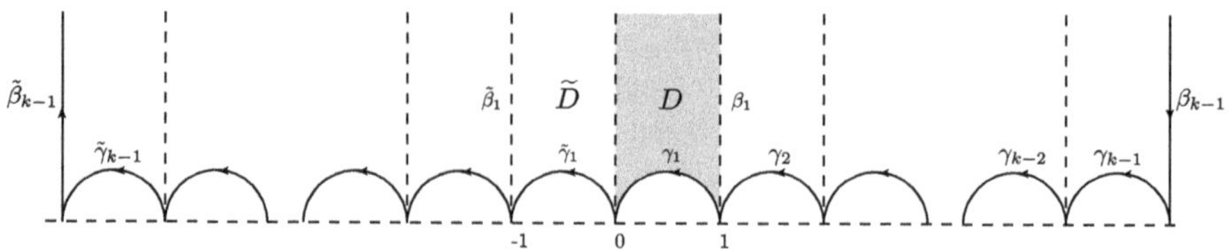

Fig. 4.13 Fundamental domain of the group Γ_k

part $X_k(\mathbb{R})$ is connected. When k is even, $X_k(\mathbb{R})$ is diffeomorphic to the orientable real surface of genus $\frac{5k}{2}$: when k is odd, $X_k(\mathbb{R})$ is diffeomorphic to the non orientable surface of Euler characteristic $2 - 10k$. □

Theorem 4.6.16 *Every regular complex elliptic surface $Y \to \mathbb{P}^1$ without multiple fibres can be deformed over $\mathbb{C}$ to an elliptic surface X with a real structure such that*

$$b^1_{\mathrm{alg}}(X(\mathbb{R}); \mathbb{Z}_2) = b^1(X(\mathbb{R}); \mathbb{Z}_2) = h^{1,1}(X) .$$

Proof Any regular elliptic surface with no multiple fibres $Y \longrightarrow \mathbb{P}^1$ has a relatively minimal model $Y' \longrightarrow \mathbb{P}^1$ which is an elliptic surface. By Lemma 4.6.11, relatively minimal regular elliptic surfaces without multiple fibres are classified by their holomorphic Euler characteristic. By Theorem 4.6.15, there is a relatively minimal regular elliptic surface X' without multiple fibres such that $\chi(\mathcal{O}_{X'}) = \chi(\mathcal{O}_{Y'})$ and $b^1_{\mathrm{alg}}(X'(\mathbb{R}); \mathbb{Z}_2) = b^1(X'(\mathbb{R}); \mathbb{Z}_2) = h^{1,1}(X')$. Now, if V is a real surface and $W \to V$ is a blow up at a point of $V(\mathbb{R})$, then W is a real surface and

$$h^{1,1}(W) = h^{1,1}(V) + 1, \quad b^1_{\mathrm{alg}}(W(\mathbb{R}); \mathbb{Z}_2) = b^1_{\mathrm{alg}}(V(\mathbb{R}); \mathbb{Z}_2) + 1 .$$

The surface Y is obtained from Y' by a finite number of blow ups of points. The theorem follows by Theorem 4.6.12. □

Remark 4.6.17 Unlike elliptic surfaces, a surface fibred in genus 2 curves can be of general type. The real theory of such surfaces is much less well developed than the real genus 1 theory but an initial step towards their classification has nevertheless been made, namely a classification of the possible singular fibres of a pencil of genus 2 curves established in [AM15].

4.7 Surfaces of General Type ($\kappa = 2$)

In this section we will construct some interesting examples of $\mathbb{R}$-surfaces of general type. In particular, we will study their algebraic cycles.

For any non singular complex projective algebraic surface X the image of the map $\mathrm{Pic}(X) \to H^2(X; \mathbb{C}) = H^{2,0}(X) \oplus H^{1,1}(X) \oplus H^{0,2}(X)$ is contained in $H^{1,1}(X)$ (see

Appendix D) so the Picard number $\rho(X)$ is bounded above by the Hodge number $h^{1,1}(X)$. Let X_d be a non singular surface in $\mathbb{P}^3(\mathbb{C})$ of degree $d := \deg(X_d)$. We then have that $\rho(X_d) \leqslant h^{1,1}(X_d) = \frac{d}{3}(2d^2 - 6d + 7)$ (see Example D.4.4).

If $d \leqslant 3$, X_d is a rational surface and $\rho(X_d) = h^{1,1}(X_d)$. If $d > 3$ and X_d is very general amongst degree d surfaces then its Picard number satisfies $\rho(X_d) = 1$ by Noether's theorem [Del73, 1.2.1] and it is fairly difficult to construct surfaces with Picard number close to $h^{1,1}(X_d)$.

If $d = 4$, X_d is a K3 surface, and we have a classification of such surfaces, see Theorem 4.5.10. If $d \geqslant 5$, X is a surface of general type (see [Bom73]) and only a few sporadic examples are understood.

Surfaces X which have "large" Picard number, or in other words for which ρ is close to $h^{1,1}$, are exceptional. (See [Man94], [KI96], [Bih01a] for a study of surfaces in $\mathbb{P}^3$ of degree 5 and their deformations; see [Bih01b] and [Ren15] for more information on surfaces of degree 6.) On the other hand, $\mathbb{R}$-surfaces which have "lots" of algebraic cycles are even rarer. As we can see, it is extremely interesting to construct surfaces with such properties.

Algebraic cycles can be constructed by blowing up singular points on surfaces (namely the irreducible components of the exceptional divisor). Persson uses this method to construct examples of surfaces with maximal Picard number. We refer to [Per82] for more details.

In this section we show that in certain cases we can prove useful results in real algebraic geometry using this method, essentially thanks to Lemma 4.7.7 and an appropriate language for blow-ups, see Appendix F.

Definition 4.7.1 A compact complex analytic surface X which has the same numerical invariants as a degree 5 surface in $\mathbb{P}^3$ is called a *numerical quintic*.

Remark 4.7.2 By [Hor75], for example, any surface X such that

$$(c_1^2(X), h^{0,2}(X)) = (5, 4)$$

is a numerical quintic.

After studying real resolutions of singular points on a real surface we present two Examples 4.7.13 and 4.7.14, of numerical quintics. If X is a surface of degree 5 in $\mathbb{P}^3(\mathbb{C})$ then $h^{1,1}(X) = 45$ as in Example D.4.4, and we have an upper bound

$$\operatorname{rk} \operatorname{Pic}(X) \leqslant 45 \ .$$

In [Per82], Persson gives an example of a non singular numerical quintic for which $\rho = 43$ but this example is relatively uninteresting over the real numbers as we will see in Example 4.7.14. Specifically, the homology of its real locus contains only 31 classes generated by algebraic cycles. In Example 4.7.13, we revisit an example of Hirzebruch's of a quintic with a non singular model with $\rho = 41$. We will see that the real locus of this model has homology of dimension 41 entirely generated by algebraic cycles.

Remark 4.7.3 In 2011, Mathias Schütt [Sch11] constructed a quintic such that $\rho = 45$ which is a quotient of a Fermat surface. (We refer the interested reader to [Shi81] or more generally [Bea14] for more information on Fermat surfaces.) Schütt's surface has equation

$$yzw^3 + xyz^3 + wxy^3 + zwx^3 = 0$$

in $\mathbb{P}^3$ and is clearly defined over $\mathbb{R}$. It would be interesting to calculate b^1 and b^1_{alg} of the real locus of this surface. (See [Sch15] for the construction of complex quintic surfaces with Picard number between 1 and 45.)

4.7.1 Resolution of Singular Points and Double Covers

In this subsection we consider real resolutions of rational double points (Definition 4.4.28) and we prove Lemma 4.7.7 on which we will rely in the rest of the section.

The two examples of surfaces of general type studied in the last part of this chapter are obtained by resolution of singularities of special surfaces. Each exceptional curve obtained by blowing up a point generates a complex algebraic cycle whose real locus is not always easy to understand.

Example 4.7.4 A common phenomenon is illustrated by the surface $X_1 \subset \mathbb{R}^3_{x,y,z}$ of equation $z^2 = x^4 - x^2 - y^4 - y^2$ constituted of two spheres meeting in an ordinary double point. (There are two other singular points in the complex locus which will be unimportant for our purposes.) Blowing up this point $\widetilde{X_1} \to X_1$ we get a smooth sphere $\widetilde{X_1}(\mathbb{R})$. There is therefore a complex algebraic cycle invariant under the real structure which does not give rise to any non trivial class in $H_1(\widetilde{X_1}(\mathbb{R}); \mathbb{Z}_2)$.

Example 4.7.5 Our second example illustrates a slightly different phenomenon. The surface $X_2 \subset \mathbb{R}^3_{x,y,z}$ of equation $z^2 = (x^2 + (y-1)^2 - 4)(x^2 + (y+1)^2 - 4)$ contains two spheres meeting in two ordinary double points. Blowing up these points $\widetilde{X_2} \to X_2$ we get a smooth torus $\widetilde{X_2}(\mathbb{R})$ and the two exceptional curves (which are -2-curves, see Definition 4.3.2) yield the same class in $H_1(\widetilde{X_2}(\mathbb{R}); \mathbb{Z}_2)$ despite the fact that they generate distinct classes in the homology of the complex variety.

It is this second phenomenon that arises in Example 4.7.14. On the other hand, in Example 4.7.13, all the invariant algebraic cycles of the complex variety generate non trivial classes in $H_1(\widetilde{S}(\mathbb{R}); \mathbb{Z}_2)$ because the initial singular surface is homeomorphic to $\mathbb{P}^2(\mathbb{R})$.

Note that, unlike the complex case, a hyperplane section is not always homologically non trivial in the real locus.

Example 4.7.6 (*Resolution of a double point on a surface*) Consider a surface X and a double point P on this surface. As P is a double point its minimal embedding dimension is 3, see Remark 4.4.27. In other words, locally analytically we can assume that there is an open set V in K^3, $K = \mathbb{R}$ or $\mathbb{C}$, centred in $(0, 0, 0)$ in which the equation of X is of the form $z^2 = f(x, y)$ and $P = (0, 0, 0)$. The blow up $\widetilde{V}$ of V in 0 is the set of pairs (a, ξ) in $V \times \mathbb{P}^2(K)$ satisfying the equations

$$x\xi_2 = y\xi_1 \quad \text{and} \quad x\xi_3 = z\xi_1$$

where (x, y, z) are coordinates on V and $[\xi_1, \xi_2, \xi_3]$ are the homogeneous coordinates on $\mathbb{P}^2(K)$.

We can cover the blow up $\widetilde{V}$ of V by charts,

$$U_i = \{(a, \xi) \in \widetilde{V} : \xi_i \neq 0\}, \qquad i = 1, 2, 3\ ;$$

with coordinates (u_i, v_i, w_i) defined by:

$$\begin{aligned} u_1 &= x, & v_1 &= \frac{\xi_2}{\xi_1}, & w_1 &= \frac{\xi_3}{\xi_1} & &\text{on } U_1\ , \\ u_2 &= \frac{\xi_1}{\xi_2}, & v_2 &= y, & w_2 &= \frac{\xi_3}{\xi_2} & &\text{on } U_2\ , \\ u_3 &= \frac{\xi_1}{\xi_3}, & v_3 &= \frac{\xi_2}{\xi_3}, & w_3 &= z & &\text{on } U_3\ . \end{aligned}$$

We can lift X via the blow up map $\pi \colon \widetilde{X} \to X$ to the charts U_i defined above. In U_1 the equation of $\pi^*(X)$ is of the form

$$u_1^2 w_1^2 = u_1^m f_1(u_1, v_1)$$

where m is the multiplicity of f in 0. The strict transform $\widetilde{X}$ of X therefore has equation $w_1^2 = u_1^{m-2} f_1(u_1, v_1)$ in this chart In U_2 the equation is $w_2^2 = v_2^{m-2} f_2(u_2, v_2)$ and in U_3 it is $1 = w_3^{m-2} f_3(u_3, v_3, w_3)$.

In what follows we will be particularly interested in rational double points (Definition 4.4.28) since their resolution will generate algebraic cycles in the complex variety without changing the numerical invariants of the surface. (See [Art62] or [Slo80, p. 70] for more details.) Recall that singular surfaces all of whose singularities are rational double points are called Du Val surfaces (see Definition 4.4.30).

We want to determine the topology of the real locus of the non singular surface $\widetilde{X}$ obtained by blow up and calculate the rank of $H_1^{\mathrm{alg}}(\widetilde{X}(\mathbb{R}); \mathbb{Z}_2)$.

Lemma 4.7.7 *Let X be an $\mathbb{R}$-surface with non empty real locus on which we consider a singular point P belonging to $X(\mathbb{R})$. Let $\pi_P \colon \widetilde{X} \to X$ be an $\mathbb{R}$-resolution (by which we mean that π_P commutes with the real structures on X and $\widetilde{X}$). If the real part $L(\mathbb{R})$ of the exceptional divisor is non empty then we have that*

$$\chi_{top}(\widetilde{X}(\mathbb{R})) = \chi_{top}(X(\mathbb{R})) + \chi_{top}(L(\mathbb{R})) - 1$$

where χ_{top} is the topological Euler characteristic.

Proof To simplify notations we set $V := X(\mathbb{R})$ and $W := L(\mathbb{R})$. We then have that $\widetilde{V} = \widetilde{X}(\mathbb{R})$.

We consider the exact sequence of cohomology with compact support (B.5) from Proposition B.6.8 applied to the compact pair $(\widetilde{V}, W)$:

$$\cdots \to H_c^k(\widetilde{V} \setminus W; \mathbb{Z}_2) \to H^k(\widetilde{V}; \mathbb{Z}_2) \to H^k(W; \mathbb{Z}_2) \to H_c^{k+1}(\widetilde{V} \setminus W; \mathbb{Z}_2) \to \cdots \tag{4.20}$$

We have that

1. $\forall k \geqslant 0, \quad H_c^k(\widetilde{V} \setminus W; \mathbb{Z}_2) = H_c^k(V \setminus \{P\}; \mathbb{Z}_2)$;
2. $\dim H_c^0(V \setminus \{P\}; \mathbb{Z}_2) = \dim H^0(V; \mathbb{Z}_2) - 1$;
3. $\forall k > 0, \quad H_c^k(V \setminus \{P\}; \mathbb{Z}_2) = H^k(V; \mathbb{Z}_2)$;
4. $H^2(W; \mathbb{Z}_2) = 0$.

By definition of π there is a neighbourhood U of P in V and a neighbourhood $\widetilde{U}$ of W in $\widetilde{X}$ such that π_P is biholomorphic to $\widetilde{U} \setminus W$ over $U \setminus \{P\}$. As $\widetilde{V} \setminus W$ and $V \setminus \{P\}$ are homeomorphic (1) follows. Statement (2) follows from the definition of cohomology with compact support. Indeed, if we denote by V_1 the connected component of V containing P we have that $H_c^0(V_1 \setminus \{P\}; \mathbb{Z}_2) = 0$. Statement (3) then follows from the exact sequence (4.20) applied to the pair $(V, \{P\})$, and finally (4) is simply a restatement of the fact that W has dimension 1.

We then simply write that the alternating sum of dimensions of $\mathbb{Z}_2$-vector spaces in the exact sequence is zero to obtain

$$\chi_{top}(\widetilde{V}) = \chi_{top}(V) + \chi_{top}(W) - 1 .$$

□

Corollary 4.7.8 *If $L(\mathbb{R})$ is connected and the hypotheses of the previous lemma hold then we have that*

$$\chi_{top}(\widetilde{X}(\mathbb{R})) = \chi_{top}(X(\mathbb{R})) - \dim H^1(L(\mathbb{R}); \mathbb{Z}_2) .$$

4.7.2 Resolutions of Real Double Covers

Let W be a non singular $\mathbb{R}$-surface whose real locus is connected and non empty and let C be an $\mathbb{R}$-curve on W without multiple components. We assume there is a divisor B on W such that $C \in |2B|$: we will then say that C is an *even curve*. Let X be the double cover of W branched along C and let $\widetilde{X}$ be the canonical resolution of X (see below). We have the choice between two real structures on the surfaces X and $\widetilde{X}$. If W is a rational surface and locally $P(x, y)$ is a polynomial defining C the choice of real structure corresponds to a choice of sign: $z^2 = \pm P(x, y)$. Having made this choice, we study the real locus $\widetilde{X}(\mathbb{R})$ of $\widetilde{X}$.

Remark 4.7.9 We check that

1. The surface X is singular if and only if the curve C is singular.
2. The surface X is projective if and only if W is projective, see [BHPVdV04, p. 182].
3. If the complex surface W is simply connected then the complex surface X is simply connected if and only if the complex curve C is connected.

There is a special method for resolving the singularities of a double cover, namely *canonical resolution*, which is sometimes more efficient than direct resolution by blow up. It has however the disadvantage of not always giving a minimal resolution, as we will see with the line passing through the quadruple points in Example 4.7.14. However, we will prove below that this method always yields a minimal resolution for rational double points.

Definition 4.7.10 Let W be a non singular complex projective algebraic surface and let $C \subset W$ be an even curve without multiple components. The canonical resolution of (W, C) is defined to be the pair $(\widetilde{W}, \widetilde{C})$ defined recursively as follows:

- $(W_0, C_0) = (W, C)$;
- At step (W_k, C_k):

If C_k is non singular, we set $(\widetilde{W}, \widetilde{C}) = (W_k, C_k)$. If C_k is singular we choose a singular point P on C_k. We denote by $\pi\colon W_{k+1} \to W_k$ the blow up of W centred at P and by L the corresponding exceptional divisor and we set $C_{k+1} = \pi^*(C_k) - 2\lfloor m/2 \rfloor L$ where m is the multiplicity of C_k at P. (We note that L is a component of C_{k+1} if and only if m is odd.)

We can show that this definition makes sense, or in other words that the processus eventually stops and the order of the blow-ups does not affect the result. See [Per81, p. 10] for more details.

Consider the canonical resolution of a double cover X defined by a pair (W, C). The equation of C in an open affine subset of K^2 is $f(x, y) = 0$: after blowing up $(0, 0)$ in K^2 we obtain (with some obvious modifications of the notations introduced in Example 4.7.6) $u_1^m f_1(u_1, v_1) = 0$ and $u_2^m f_2(u_2, v_2) = 0$. We then take our new branching locus to be $C_1 = 2\lfloor m/2 \rfloor L$ and consider the double cover. The equations of this double cover are:

$$z^2 = u_i f_i(u_i, v_i) \text{ if } m \text{ is odd},$$
$$z^2 = f_i(u_i, v_i) \text{ if } m \text{ is even}.$$

Comparing with the calculations of Example 4.7.6 we see that this method is equivalent to resolving a singularity by blow up and resolution of the branching curve if $m = 2, 3$.

Lemma 4.7.11 *If (X, σ) is a singular $\mathbb{R}$-surface the resolution $\widetilde{X} \to X$ of a rational double point P belonging to $X(\mathbb{R})$ does not change the number of connected* *components of $X(\mathbb{R})$. Moreover, if P is a point of type A_n, n odd or D_n, n even then we have that*

$$\chi_{top}(\widetilde{X}(\mathbb{R})) = \chi_{top}(X(\mathbb{R})) - n \ .$$

Proof 1. Let L be the exceptional divisor generated by resolution of P. We need to determine what happens to the real locus of L, or in other words what happens to P. Connectedness of $\widetilde{X}(\mathbb{R})$ depends on connectedness of $L(\mathbb{R})$.

2. When we blow up a singularity of type D_n, E_6, E_7 or E_8 the branching locus remains connected since the corresponding singularity of the branching locus is then triple. Moreover, when the singularity is of type D_n with even n the real locus has exactly n double points of type A_1.
3. A singularity of type $A_n, n \geqslant 2$ becomes a singularity of type A_{n-2} after blow up; the branching locus remains singular and connected until we reach A_1 and A_2 whose respective equations (over $\mathbb{C}$) are $z^2 = x^2 - y^2$ and $z^2 = y^2 - x^3$.
4. The blow up of A_1 gives us $w_1^2 = 1 - v_1^2$ (in U_1 for example) and $L(\mathbb{R})$ is then the conic of equation $u_1 = 0$, $w_1^2 + v_1^2 = 1$. This resolution therefore turns a point into a connected curve and the strict transform $\widetilde{X}(\mathbb{R})$ therefore has the same number of connected components as $X(\mathbb{R})$. Moreover, in this case $\chi_{\text{top}}(L(\mathbb{R})) = 0$.
5. A point of type A_2 has two possible equations over $\mathbb{R}$: $z^2 = y^2 - x^3$ and $z^2 = x^3 - y^2$ which give rise after blow up to equations $w_1^2 = v_1^2 - u_1$ and $w_1^2 = u_1 - v_1^2$ respectively. The curve $L(\mathbb{R})$ (corresponding to $u_1 = 0$) consists of two lines, $w_1 = v_1$ and $w_1 = -v_1$, in the first case and the isolated point $w_1 = v_1 = 0$ in the second case.

For A_n with n odd the locus $L(\mathbb{R})$ is connected and contains n double points of type A_1. We complete the proof by applying Lemma 4.7.7 to each blow up of a point of typeA_1. □

Proposition 4.7.12 *Let (W, C) be a pair where W is a non singular compact $\mathbb{R}$-surface and $C \subset W$ is a possibly reducible $\mathbb{R}$-curve. If the real part of C is connected and all its singularities are of type A_n, D_n, E_6, E_7 or E_8, then the double cover $\widetilde{X}(\mathbb{R})$ of the canonical resolution of (W, C) is connected for one of the two real structures lifting the real structure on W.*

Proof 1. If $W(\mathbb{R})$ and $C(\mathbb{R})$ are connected it is clear that the double cover $X(\mathbb{R})$ defined by (W, C) is connected independently of the choice of real structure.

2. The resolution of a singularity that is not in $C(\mathbb{R})$ does not alter the connectedness of $W(\mathbb{R})$ or $C(\mathbb{R})$.
3. Since the singularities of the real part of C are of type A_n, D_n, E_6, E_7 or E_8 the corresponding singularities of X are rational double points of type A_n, D_n, E_6, E_7 or E_8.

By Theorem 4.4.26, we know that the canonical resolution of this type of singularity is equivalent to a sequence of blow ups of points on X, and moreover all the intermediate singularities are also rational double points. The proposition follows from Lemma 4.7.11. □

The first example we present is a non singular model of a certain degree 5 surface in $\mathbb{P}^3$.

Example 4.7.13 (*A real quintic such that* $b_1^{\mathrm{alg}} = b_1$) This example is based on a construction that Persson attributes to Hirzebruch in [Per82, Introduction]. A non singular complex surface X in $\mathbb{P}^3(\mathbb{C})$ is connected and simply connected and by the Lefschetz hyperplane theorem D.9.2:

$$\pi_0(X) \simeq \pi_0(\mathbb{P}^3(\mathbb{C})) ,$$
$$\pi_1(\mathbb{P}^3(\mathbb{C})) \to \pi_1(X) \to 0 .$$

It follows that the Picard group $\mathrm{Pic}(X)$ is a free, finitely generated $\mathbb{Z}$-module of rank ρ

Consider a curve C formed of five lines in general position in $\mathbb{P}^2(\mathbb{C})$. Let $f: \mathcal{S} \to \mathbb{P}^2(\mathbb{C})$ be the cyclic covering of order 5 of $\mathbb{P}^2(\mathbb{C})$ branched along C. The surface $\mathcal{S}$ is then a quintic in $\mathbb{P}^3(\mathbb{C})$ which has ten singularities, corresponding to the intersections of the lines. Let P be one of these singular points. In an affine neighbourhood of P we can write the equation of $\mathcal{S}$ in the form

$$x^5 = z^2 - y^2 ,$$

which implies that P is a rational double point of type A_4 (see Definition 4.4.28). We know that each rational double point can be resolved by successive blow ups and the non singular model thus obtained has the same numerical invariants as a quintic. In particular, the bound on the Picard number remains valid.

On the other hand, the resolution of a singular point of type A_n increases the number of algebraic cycles by exactly n (i.e. the number of irreducible components of the exceptional divisor of the resolution). This yields a non singular surface $\widetilde{\mathcal{S}}$ such that $\rho(\widetilde{\mathcal{S}}) = 41$. The Picard group of $\widetilde{\mathcal{S}}$ is generated by the hyperplane section and 40 cycles arising from resolutions of singularities.

Choosing the lines in C to be real lines in general position in $\mathbb{P}^2(\mathbb{R})$ we obtain a real surface. We will use the topology of $\mathcal{S}(\mathbb{R})$ to calculate the topology of $\widetilde{\mathcal{S}}(\mathbb{R})$: in other words, we will study the behaviour of the real locus under resolution of singularities.

Restricting the covering map f to $\mathcal{S}(\mathbb{R})$ we get a homeomorphism from $\mathcal{S}(\mathbb{R})$ to $\mathbb{P}^2(\mathbb{R})$. As the equation of $\mathcal{S}$ is of the form $t^5 = Q(x, y, z)$ and as f is given by $f(x, y, z, t) = (x, y, z)$, it is easy to see that there is only one real point in each fibre of f. The surface $\mathcal{S}(\mathbb{R})$ is therefore connected and $\dim H_1(\mathcal{S}(\mathbb{R}); \mathbb{Z}_2) = 1$. Moreover, its only homology class is algebraic, as it is simply the pull back via the map

$$f^*: H_1(\mathbb{P}^2(\mathbb{R}); \mathbb{Z}_2) \to H_1(\mathcal{S}(\mathbb{R}); \mathbb{Z}_2)$$

of the hyperplane section of $\mathbb{P}^2(\mathbb{R})$.

We now consider the resolution of a point P of type A_4 on a surface X whose local equation is $z^2 = x^5 + y^2$. We use the notations of Example 4.7.6. Let $P \in V \subset X$ be a neighbourhood of P. After a first blow up $\widetilde{V} \to V$ we get equations

$$w_1^2 = u_1^3 + v_1^2 \quad \text{on } U_1 \quad \text{and} \quad w_2^2 = u_2^5 v_2^3 + 1 \quad \text{on } U_2 \; .$$

The real part of the exceptional curve is connected and consists of two lines (whose equations are $w_1 = v_1$ and $w_1 = -v_1$ in U_1 respectively) meeting in a singular point Q of $B_P X \supset \widetilde{V} = B_P V$.

We now blow up Q and we denote by $\widetilde{X} := B_Q(B_P(X))$ the surface thus obtained. Changing notation slightly for this second blow up, the equation of $\widetilde{X}$ in $\widetilde{U}_1$ is of the form

$$\widetilde{w_1}^2 = \widetilde{u_1} + \widetilde{v_1}^2 \; .$$

The real locus of the exceptional divisor L is again connected with two irreducible components. Moreover, the intersection point $\widetilde{Q}$ is not a singular point of the surface. To summarise,

- two blow ups are needed to resolve P;
- the real part of the exceptional divisor $\widetilde{L}$ is connected and has four irreducible components.

We now use Corollary 4.7.8 to show that these four irreducible components really give us four new homology classes in the real locus. In our case, $L(\mathbb{R})$ is topologically a chain of four circles, which gives us

$$\chi_{\text{top}}(\widetilde{X}(\mathbb{R})) = \chi_{\text{top}}(X(\mathbb{R})) - 4 \; .$$

Moreover the new surface is connected, as we have simply replaced a point by a connected curve. As $\mathcal{S}$ has 10 such singularities, we get that

$$\dim H_1(\widetilde{\mathcal{S}}(\mathbb{R}); \mathbb{Z}_2) = 41 \; .$$

Any finally as each new cohomology class was obtained as the real part of an exceptional curve we get that

$$\dim H_1^{\text{alg}}(\widetilde{\mathcal{S}}(\mathbb{R}); \mathbb{Z}_2) = \dim H_1(\widetilde{\mathcal{S}}(\mathbb{R}); \mathbb{Z}_2) = 41 \; .$$

Example 4.7.14 (*A real numerical quintic such that* $b_1^{\text{alg}} < b_1$) We now discuss an example of a numerical quintic constructed by Ulf Persson [Per82, p. 309] and calculate the invariants of its real part and a lower bound for the number b^1_{alg}.

We start by constructing a real curve in $W = \mathbb{P}^2$. Consider a quadrilateral defined by two pairs of lines meeting in points p and q. Each side of Σ meets two others in vertices of Σ. Each side of Σ therefore has three canonical points, namely the intersections with the three other sides. The two diagonals of Σ meet in a point r, the centre of Σ. Linking r to p and q respectively each side of Σ is now cut in a fourth point. These four points on any given line form a harmonic set, by which we mean that if we normalise the coordinates such that the point p or q is ∞ and the vertices of Σ are ± 1 then the fourth point is 0.

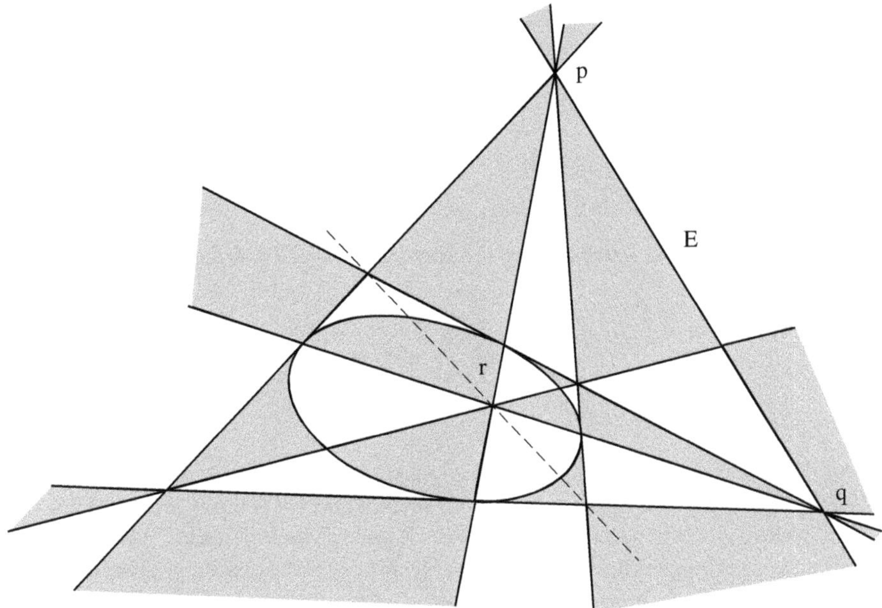

Fig. 4.14 Projection of $\mathcal{P}(\mathbb{R})$ onto $\mathbb{P}^2$

This is in fact the classical construction of the fourth point in a harmonic set given three of them. There is now a unique conic which is tangent to all the sides of Σ at the “0-points”. Let C be the reducible curve containing the six lines above, the conic and one of the diagonals (Figure 4.14), plus the curve E passing through p and q. This yields a curve of degree 10 with four ordinary quadruple points, five A_1 points, four A_6 points and three D_4 points.

Let $(\widetilde{W}, \widetilde{C})$ be the canonical resolution of the pair (W, C) as in Definition 4.7.10. Let $\mathcal{P}$ be the double cover defined by the pair (W, C) and let $\widetilde{\mathcal{P}}$ be the double cover defined by the pair $(\widetilde{W}, \widetilde{C})$.

The self intersection E^2 of E decreases by 1 each time we blow up one of its three singular points, so the line E in $\mathcal{P}$ satisfies $E^2 = -2$. By [Per81, Proposition 1.3] the curve $\widetilde{E}$ obtained as a double cover of E is a (-1)-curve in $\widetilde{\mathcal{P}}$. Using [Per81, Proposition 1.3] once more we can check that $\widetilde{E}$ is the only (-1)-curve in $\widetilde{\mathcal{P}}$.

We contract $\widetilde{E}$ and we denote by $\mathcal{P}'$ the minimal surface thus obtained.

Proposition 4.7.15 *Let $\widetilde{X}$ be the double cover obtained by canonical resolution of a pair $(\mathbb{P}^2(\mathbb{C}), D)$ where D is a curve in $\mathbb{P}^2(\mathbb{C})$ of degree $2d$. If every singular point P_k is of multiplicity $m_k = 2d_k$ or $m_k = 2d_k + 1$ we have that*

$$c_1^2(\widetilde{X}) = 2(d-3)^2 - 2\sum_{P_k}(d_k-1)^2 ,$$

$$h^{0,2}(\widetilde{X}) = 1 + \frac{1}{2}(d(d-3)) - \sum_{P_k}\frac{1}{2}d_k(d_k-1) .$$

Proof See [BHPVdV04, p. 183]. □

Corollary 4.7.16 *The invariants of the surfaces* $\widetilde{\mathcal{P}}$ *and* $\mathcal{P}'$ *defined above are*

$$\begin{aligned} c_1^2(\widetilde{\mathcal{P}}) &= 4, & c_1^2(\mathcal{P}') &= 5 , \\ h^{0,2}(\widetilde{\mathcal{P}}) &= 4, & h^{0,2}(\mathcal{P}') &= 4 , \\ \rho(\widetilde{\mathcal{P}}) &\geqslant 44, & \rho(\mathcal{P}') &\geqslant 43 . \end{aligned}$$

Proof For the first two equations we recall that as in Proposition 4.1.30 if $\pi\colon \widetilde{X} \to X'$ is the blow up of a point P of a non singular surface and E_P is the exceptional line then $K_{\widetilde{X}} = \pi^* K_{X'} + E_P$ and hence $c_1^2(X') = c_1^2(\widetilde{X}) + 1$. Moreover, as $h^{0,2}$ is a birational invariant we get two additional equations. To prove the last two equation, recall that as well as the cycles arising from resolution of singularities of type A_n, D_n or E_n which each generate n independent algebraic cycles, there is a cycle arising from the hyperplane section and two arising from the resolution of quadruple points. □

The non singular surface $\mathcal{P}'$ therefore has the same numerical invariants as a quintic in $\mathbb{P}^3(\mathbb{C})$ (see Remark 4.7.2) and its Picard number is bounded below by 43. We now calculate the real locus of $\mathcal{P}'$. Referring once more to Figure 4.14, we can calculate the Euler characteristic of the initial singular surface $\mathcal{P}(\mathbb{R})$ which is homeomorphic to a finite number of spheres glued together at singular points. In particular, $\mathcal{P}(\mathbb{R})$ is connected and

$$\chi_{\text{top}}(\mathcal{P}(\mathbb{R})) = 2(^{\#}\{\text{spheres}\}) - \sum_{P_k}(m_k - 1)$$

where m_k is the multiplicity of the point P_k. With its 13 spheres, 5 double points, 7 triple points and 2 quadruple points we therefore have that

$$\chi_{\text{top}}(\mathcal{P}(\mathbb{R})) = 1, \qquad b^1(\mathcal{P}(\mathbb{R}); \mathbb{Z}_2) = 13 .$$

Proposition 4.7.17 *The numerical quintic* $\mathcal{P}'$ *has a connected real locus.*

After blowing up the two quadruple points the branching locus is connected (see Figure 4.15), and there are only simple singularities left in the real locus. By Lemma 4.7.11 the resolution of these singularities yields a connected surface and we complete the construction by contracting the curve E, which obviously preserves connectedness.

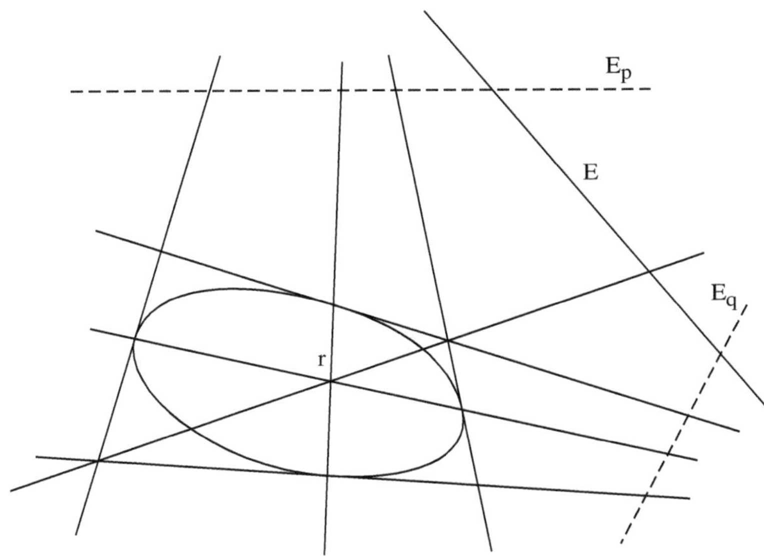

Fig. 4.15 The curve $\widetilde{C}$ on $\widetilde{W}$

Proposition 4.7.18 *For the surfaces $\widetilde{\mathcal{P}}(\mathbb{R})$ and $\mathcal{P}'(\mathbb{R})$ defined above we have the following equations:*

$$\chi_{top}(\widetilde{\mathcal{P}}(\mathbb{R})) = \chi_{top}(\mathcal{P}(\mathbb{R})) - \sum_{P_k} n_k, \qquad \chi_{top}(\mathcal{P}'(\mathbb{R})) = \chi_{top}(\widetilde{\mathcal{P}}(\mathbb{R})) + 1$$

$$b^1(\widetilde{\mathcal{P}}(\mathbb{R}); \mathbb{Z}_2) = \chi_{top}(\widetilde{\mathcal{P}}(\mathbb{R})) + 2, \qquad b^1(\mathcal{P}'(\mathbb{R}); \mathbb{Z}_2) = b^1(\widetilde{\mathcal{P}}(\mathbb{R}); \mathbb{Z}_2) - 1$$

$$b^1_{\text{alg}}(\mathcal{P}'(\mathbb{R}); \mathbb{Z}_2) = b^1_{\text{alg}}(\mathcal{P}(\mathbb{R}); \mathbb{Z}_2) + (b^1(\mathcal{P}'(\mathbb{R}); \mathbb{Z}_2) - b^1(\mathcal{P}(\mathbb{R}); \mathbb{Z}_2))$$

where $n_k = n$ if P_k is a point of type A_n, n odd or D_n, n even, and $n_k = 1$ if P_k is a quadruple ordinary point.

Proof For simple singularities this follows from the second part of Lemma 4.7.11. Moreover, let P be a quadruple ordinary point of C: we can then find a neighbourhood of P such that $\mathcal{P}$ is locally defined by the equation:

$$z^2 = xy(x^2 - y^2) .$$

The canonical resolution turns this equation into

$$z^2 = v - v^3 ,$$

so in the real world $E_P(\mathbb{R})$ is the union of two disjoint circles whence it follows that $\chi_{\text{top}}(E_P(\mathbb{R})) = 0$. We now simply apply Lemma 4.7.7. □

Corollary 4.7.19 *For the surfaces* $\widetilde{\mathcal{P}}(\mathbb{R})$ *and* $\mathcal{P}'(\mathbb{R})$ *defined above we have the equations:*

$$\chi_{top}(\mathcal{P}') = -41, \qquad b^1(\mathcal{P}'(\mathbb{R}); \mathbb{Z}_2) = 43, \qquad b^1_{\text{alg}}(\mathcal{P}'(\mathbb{R}); \mathbb{Z}_2) \geqslant 31 \ .$$

Indeed, we have that $\sum_{P_k} n_k = 43$ and $b^1_{\text{alg}}(\mathcal{P}(\mathbb{R}); \mathbb{Z}_2) \geqslant 1$ because of the hyperplane section class.

4.8 Solution to exercises of Chapter 4

4.1.45 If C' is a non singular quartic then $g(C') = 3$ by the genus formula, see Theorem 1.6.17. Since the genus of C is zero by hypothesis, we have that $\delta(C) = 3$. The multiplicities of the singular points (including the infinitely close points) are therefore $(2, 2, 2)$ or (3). Since the set of singular points is invariant under $\sigma_{\mathbb{P}}$, the only possibilities are: a unique triple point, three double ordinary points or a unique singular point whose sequence of multiplicities is $(2, 2, 2)$ (which implies that it is either a ramphoid cusp or a double point whose first blow up contains two double ordinary points). Since the number of singular points is always odd, at least one of them must be real.
4.1.48 The Euler characteristic is additive so $\chi_{top}(C) = \chi_{top}(C_1) + \chi_{top}(C_2) = 4$. Moreover $p_a(C_i) = g(C_i) = 0$ for $i = 1, 2$ and the result follows from formula (4.4), Sect. 4.1.4.
4.2.11 2. Calculating the derivative at any point of a reduced fibre gives us the result.
4. Consider the conic bundle given by the equation

$$x^2 + y^2 = (t-1)(t-2)(t-3)(t-4) \ .$$

4.3.3 By definition any (-1)-curve is rational and non singular so $p_a(C) = g(C) = 0$ and $C^2 = -1$. The result follows by the adjunction formula (4.1), Sect. 4.1.4.
4.3.14 Let E_1 and E_2 be the lines that are irreducible components of the fibre F in question. We then have that $0 = F^2 = E_1^2 + E_2^2 + 2(E_1 \cdot E_2)$ so $E_1^2 = E_2^2 = -1$.
4.3.15 Every singular fibre consist of two (-1)-curves meeting in a point. Let $X \to \mathbb{F}_n$ be the birational morphism obtained by contracting one (-1)-curve in each singular fibre. Let $E_1, \dots, E_r$ be the contracted curves. We then have that

$$\pi^*(K_{\mathbb{F}_n}) = K_X + E_1 + \cdots + E_r \ ,$$

whence $K_X^2 = \pi^*(K_{\mathbb{F}_n})^2 - r$. We now simply calculate $K_{\mathbb{F}_n}^2$ given that $K_{\mathbb{F}_n} = -2\Sigma_0 + (n-2)F$ for any fibre F and a general section Σ_0 as in [Bea78, III.18].

4.3.18 Recall that by the adjunction formula any (-1)-curve C satisfies $K_X \cdot C = -1$ as in Exercise 4.3.3. Let C_1 and C_2 be (-1)-curves such that $C_1 \cdot C_2 > 0$. We then have that $(C_1 + C_2)^2 = -2 + 2C_1 \cdot C_2 \geqslant 0$ and $(C_1 + C_2)^2 = 0$ if and only if $C_1 \cdot C_2 = 1$. By Zariski's lemma, (see [BHPVdV04, Lemma III.8.2]) the fibre of π containing C_1 and C_2 is therefore necessarily of the form $n(C_1 + C_2)$. Let F be a general (connected non singular) fibre. We then have that

$$K_X \cdot F = n(K_X \cdot (C_1 + C_2)) = -2n \ .$$

Moreover $F^2 = 0$ since it is a fibre and $p_a(F) = g(F) \geqslant 0$ because F is connected and non singular. The adjunction formula (4.1) (Sect. 4.1.4) then gives us $n = 1$ and $g(F) = 0$.

4.4.12 If the conic bundle is minimal the only singular fibres are of the form $x^2 + y^2 = 0$ and their number is twice the number of connected components of $X(\mathbb{R})$. By Exercise 4.3.15, the number of singular fibres of the conic bundle is equal to $8 - K_X^2$.

Chapter 5
Algebraic Approximation

5.1 Rational Models

In Chapter 4—more precisely in Section 4.4—we started with a given $\mathbb{R}$-surface and tried to determine the topology of its real locus. We now consider to the inverse problem mentioned in the Introduction.

Definition 5.1.1 (*Real rational and algebraic models*) Let M be a differentiable manifold of class $\mathcal{C}^\infty$. We say that a non singular quasi-projective algebraic $\mathbb{R}$-variety (X, σ) is a *real algebraic model* of M if and only if $X(\mathbb{R})$ is diffeomorphic to M. By abuse of notation we will sometimes also say that the non singular real affine algebraic variety $V := X(\mathbb{R})$, (Definition 1.3.9 and 2.2.17) is a *real algebraic model* of M. If additionally the real algebraic $\mathbb{R}$-variety (X, σ) is $\mathbb{R}$-rational we say that (X, σ) (resp. V) is a *real rational model* of the differentiable manifold M.

Of course, any differentiable manifold which has a real rational model also has a real algebraic model but the converse is false. For example, by Comessatti's theorem 4.4.16, any orientable surface of genus $g \geqslant 2$ does not have a real rational model. On the other hand, any orientable surface of genus g has a real algebraic model, as the following example shows.

Example 5.1.2 (*Algebraic models of orientable surfaces*) Consider one of the two real algebraic surfaces given by the affine equations $z^2 = \pm f(x, y)$, where f is the product of the equations of $g + 1$ suitable circles. Any such surface is singular because it contains non real singular points but we can easily create a non singular variety by resolution of singularities as in Theorem 1.5.54 or by a small perturbation $f_\varepsilon(x, y) = 0$ of the plane curve $f(x, y) = 0$. In the latter case the curve becomes irreducible under general deformation and the real locus of the curve $z^2 = \pm f_\epsilon(x, y)$ is diffeomorphic to the real locus of the surface $z^2 = \pm f(x, y)$ by Ehresmann's fibration theorem 4.3.28.

F. Mangolte, *Real Algebraic Varieties*, Springer Monographs in Mathematics,
https://doi.org/10.1007/978-3-030-43104-4_5

We begin this chapter with a discussion of differentiable maps that can be approximated by regular maps, concentrating initially on the case where the target variety is a sphere of small dimension or a more general rational variety. We continue with a study of a special class of diffeomorphisms, *birational diffeomorphisms*, which by Exercise 1.2.56(2) and Proposition 2.2.27 are simply isomorphisms of real non singular algebraic varieties. These birational diffeomorphisms enable us to classify the *real algebraic models*[1] of a given differentiable manifold. These "isomorphisms" of real algebraic models are central to a series of results based on special subgroups of the famous *Cremona group*. We finish this chapter with a discussion of some recent results on *fake real planes*.

In particular, this chapter contains a discussion of three important results from the end of the 00s.

- Up to birational diffeomorphism there is a unique real rational model of each non orientable surface. See Theorem 5.4.1.
- The group of birational diffeomorphisms of a rational real surface is infinitely transitive. See Theorem 5.4.3 for more details.
- The group of birational diffeomorphisms of a real rational surface V is dense in the group $\mathrm{Diff}(V)$ of $\mathcal{C}^\infty$ diffeomorphisms. See Theorem 5.4.16 for more details.

The Cremona group of birational transformations of the projective plane plays a central role in this chapter. See Section 5.4 for more details.

5.2 Smooth and Regular Maps

The Weierstrass approximation theorem we teach to undergraduates states that any continuous function $[a, b] \subset \mathbb{R} \to \mathbb{R}$ can be uniformly approximated by polynomial functions. In this chapter we study various generalisations of this theorem.

Definition 5.2.1 Let $U \subset \mathbb{R}^n$ be an open subset. We say that a real function $U \to \mathbb{R}$ is *smooth* if and only if it is $\mathcal{C}^\infty$. More generally, for any two differentiable manifolds V and W of class $\mathcal{C}^\infty$ a map $V \to W$ is said to be *smooth* if and only if it is $\mathcal{C}^\infty$.

For any two differentiable manifolds V and W of class $\mathcal{C}^\infty$ we denote by $\mathcal{C}^\infty(V, W)$ the set of smooth maps from V to W. We refer to the appendices for the Definition (B.5.21) of the *weak topology* and the Definition (B.5.22) of the *strong topology* on $\mathcal{C}^\infty(V, W)$.

Remark 5.2.2 Recall that if V is compact then the weak (or $\mathcal{C}^\infty$ compact-open) topology on $\mathcal{C}^\infty(V, W)$ is equivalent to the strong topology. See [Hir76, Chapitre 2] for more details. In this case the topology is simply called the $\mathcal{C}^\infty$ *topology*.

---

[1] Reread Proposition 2.2.22 if necessary.

The Stone–Weierstrass theorem generalises the Weierstrass approximation theorem. We use the following version of it (see [BCR98, Theorem 8.8.5] for more details).

Theorem 5.2.3 (Stone–Weierstrass theorem) *Any smooth real-valued function defined on an open neighbourhood of a compact subset C in $\mathbb{R}^n$ can be approximated on C by polynomial functions in the $\mathcal{C}^\infty$ topology.*

Given any two real affine algebraic varieties V and W we denote by $\mathcal{R}(V, W)$ the space of regular maps from V to W (see Definitions 1.2.54 and 1.3.4). If $V \subset \mathbb{R}^n$ and $W \subset \mathbb{R}^m$ a regular map is simply a rational map $\mathbb{R}^n \dashrightarrow W$ without poles on V by Exercise 1.2.56(2). If V and W are non singular then each of them has a $\mathcal{C}^\infty$ differentiable manifold structure by Remark 1.5.28 and the set $\mathcal{R}(V, W)$ is then a subset of $\mathcal{C}^\infty(V, W)$.

Definition 5.2.4 We will say that a map f in $\mathcal{C}^\infty(V, W)$ *has algebraic approximation* or *can be approximated by regular maps* if and only if f belongs to the closure of $\overline{\mathcal{R}(V, W)}$ in the $\mathcal{C}^\infty$ compact-open topology.

Remark 5.2.5 If V is compact then f has algebraic approximation if and only if f belongs to $\overline{\mathcal{R}(V, W)}$ in the strong topology.

The Stone–Weierstrass theorem quoted above implies that if $V = C$ is compact and $W = \mathbb{R}^m$ then we have density: $\overline{\mathcal{R}(C, \mathbb{R}^m)} = \mathcal{C}^\infty(C, \mathbb{R}^m)$. In general, the space $\mathcal{R}(V, W)$ is not dense in $\mathcal{C}^\infty(V, W)$: as the example below illustrates, regular maps can be rare. (We refer to [BKS97] for other interesting examples).

Example 5.2.6 (*Rareness of regular maps*) Let (X, σ) and (Y, τ) be irreducible projective $\mathbb{R}$-curves of genuses $g(X)$ and $g(Y)$ respectively. Suppose that their real loci $V = X(\mathbb{R})$ and $W = Y(\mathbb{R})$ are non empty: they are then compact real affine algebraic varieties by Theorem 2.2.17. Any regular map from V to W has a unique extension to a complex regular map X to Y: on the one hand by Proposition 2.2.22 any regular map can be extended to a $\mathbb{R}$-regular rational map, and on the other hand by Proposition 1.3.26 any rational map from a curve can be extended to a regular map. It follows that if $g(X) < g(Y)$ then any regular map from V to W is constant by the Riemann–Hurwitz theorem E.2.18. Moreover, if $g(Y) \geqslant 2$ (or in other words if Y is of general type) then there are only a finite number of non constant regular maps from V to W by a theorem due to de Franchis [Maz86, page 227].

We recall that a real algebraic variety is affine if and only if it is quasi-projective. In particular, any real quasi-affine algebraic variety is affine and any real projective algebraic variety is affine. See Proposition 1.3.11 for more details.

Exercise 5.2.7 (*Affine model of* $\mathbb{P}^n(\mathbb{R})$) We used one affine model of $\mathbb{P}^n(\mathbb{R})$ in the proof of Proposition 1.2.63. Here is another model which is often useful in practice. Let $\mathcal{M}_{n+1}(\mathbb{R})$ be the set of real square matrices $(n+1) \times (n+1)$ and set

$$P_n := \left\{A \in \mathcal{M}_{n+1}(\mathbb{R}) \mid {}^tA = A,\, A^2 = A, \operatorname{trace}(A) = 1\right\} .$$

1. Check that P_n is an algebraic set.
2. Prove that the map $\mathbb{P}^n(\mathbb{R}) \to P_n$,

$$(x_0 : x_1 : \cdots : x_n) \mapsto \left(\frac{x_{i-1}x_{j-1}}{x_0^2+x_1^2+\cdots+x_n^2}\right)_{\substack{1\leq i\leq n+1\\ 1\leq j\leq n+1}}$$

is an isomorphism of real algebraic varieties.

Projective space $\mathbb{P}^n(\mathbb{R})$ is a special type of Grassmannian and it can be proved that in general the Grassmannian $\mathbb{G}_{n,k}(\mathbb{R})$, (Definition 3.7.8) is isomorphic to the algebraic set

$$H_{n,k} := \left\{A \in \mathcal{M}_n(\mathbb{R}) \mid {}^tA = A,\, A^2 = A, \text{trace}(A) = k\right\}$$

as a real algebraic variety. (Note that $P_n = H_{n+1,1}$). (See [BCR98, Section 3.4.2] for more details). Similarly, considering the complex Grassmannian $\mathbb{G}_{n,k}(\mathbb{C})$ as a real algebraic variety we can prove it is isomorphic to the real algebraic variety

$$H'_{n,k} := \left\{A \in \mathcal{M}_n(\mathbb{C}) \mid {}^t\overline{A} = A,\, A^2 = A, \text{trace}(A) = k\right\}$$

which is a real (but not complex!) algebraic subvariety of $\mathcal{M}_n(\mathbb{C}) \simeq \mathbb{R}^{2n^2}$. See [BCR98, Section 3.4.2] for more details. This gives us a result similar to Proposition 3.7.10.

Proposition 5.2.8 *Let $n \geqslant k$ be natural numbers. Projective space $\mathbb{P}^n(\mathbb{C})$ and the Grassmannian $\mathbb{G}_{n,k}(\mathbb{C})$ are non singular compact real affine algebraic varieties.*

Proof See [BCR98, Propositions 3.4.6 and 3.4.11]. □

Remark 5.2.9 Let $\mathbb{H}$ be the field of quaternions. We can prove in a similar way[2] that $\mathbb{P}^n(\mathbb{H})$ and $\mathbb{G}_{n,k}(\mathbb{H})$ are non singular compact real affine algebraic varieties.

If $K = \mathbb{R}$, $\mathbb{C}$ or $\mathbb{H}$, then any quasi-projective algebraic variety over K has a natural real affine algebraic variety structure[3] and this enables us to generalise Definition 2.5.10 of algebraic vector bundles.

Definition 5.2.10 Let K be $\mathbb{R}$, $\mathbb{C}$ or $\mathbb{H}$ and let V be a real affine algebraic variety. An *algebraic K-vector bundle of rank r* over V is a K-vector bundle (E, π) (Definition C.3.5) such that the following hold.

1. The total space E is a real algebraic variety;
2. The projection $\pi\colon E \to V$ is a regular map of real algebraic varieties;

---

[2]The notation is somewhat awkward because of the non-commutativity of the field.

[3]Simply separate each complex (resp. quaternionic) equation into two (resp. four) equations with real coefficients.

3. The homeomorphisms $\psi_i : \pi^{-1}(U_i) \xrightarrow{\simeq} U_i \times K^r$ are biregular isomorphisms of real algebraic varieties[4];
4. The bundle (E, π) is isomorphic to a subbundle of a trivial bundle.

Definition 5.2.11 (*Universal bundle*) Let $n \geqslant k$ be natural numbers and let K be a field. We set

$$E_{n,k} := \{(A, v) \in \mathbb{G}_{n,k}(K) \times K^n \mid Av = v\}$$

and let $p_{n,k} : E_{n,k} \to \mathbb{G}_{n,k}(K)$ be the canonical projection. The K-vector bundle

$$\gamma_{n,k} := (E_{n,k}, p_{n,k})$$

of rank k over $\mathbb{G}_{n,k}(K)$ is called the *universal bundle* over $\mathbb{G}_{n,k}(K)$.

Proposition 5.2.12 *Let $n \geqslant k$ be natural numbers and let K be $\mathbb{R}, \mathbb{C}$ or $\mathbb{H}$. The universal bundle $\gamma_{n,k}$ is then an algebraic K-vector bundle of rank k over $\mathbb{G}_{n,k}(K)$.*

Proof See [BCR98, Proposition 12.1.8]. □

5.2.1 Homotopy, Approximations and Algebraic Bundles

If a smooth map between real affine algebraic varieties can be approximated by regular maps then it is homotopic to a regular map (see [BCR98, Corollaire 9.3.7]). The converse is false as the example below proves.

Example 5.2.13 (*Homotopic $\neq$ approximable*) Let F_n be the *Fermat curve* of degree n,

$$F_n := \{(x : y : z) \in \mathbb{P}^2(\mathbb{R}) \mid x^n + y^n - z^n = 0\} .$$

By Example 5.2.6, if $k > n \geqslant 2$, any regular map from F_n to F_k is constant. By the genus formula (Theorem 1.6.17) the genus of a complexification of F_n is strictly smaller than the genus of the complexification of F_k. For the same reason, if $k \geqslant 4$ then for any $n \geqslant 1$ there is only a finite number of non constant regular maps from F_n to F_k. It also follows that if $n \geqslant 4$ then any smooth map $F_n \to F_n$ of topological degree 1 is homotopic to a regular map (the identity) but only a finite number of them can be approximated by regular maps (those that are already regular).

When the target variety is a Grassmannian we can transform the approximation problem into a problem about algebraic bundles.

Theorem 5.2.14 *Let K be $\mathbb{R}, \mathbb{C}$ or $\mathbb{H}$, let V be a non singular compact real affine algebraic variety and let*

$$f : V \to \mathbb{G}_{n,k}(K)$$

[4]Note that K^r is a real algebraic variety of dimension $2r$ if $K = \mathbb{C}$ and $4r$ if $K = \mathbb{H}$.

be a smooth map. The following are then equivalent.

1. *The map f can be approximated in the $\mathcal{C}^\infty$ topology by regular maps $V \to \mathbb{G}_{n,k}(K)$.*
2. *The map f is homotopic to a regular map $V \to \mathbb{G}_{n,k}(K)$.*
3. *The pull back $f^*(\gamma_{n,k})$ of the universal bundle $\mathbb{G}_{n,k}(K)$ is topologically isomorphic to an algebraic K-vector bundle.*

Proof See [BCR98, Theorem 13.3.1]. □

Corollary 5.2.15 *Let K be $\mathbb{R}, \mathbb{C}$ or $\mathbb{H}$ and let V be a non singular compact real affine algebraic variety. If every topological K-vector bundle of rank k over V is topologically isomorphic to an algebraic K-vector bundle then $\mathcal{R}(V, \mathbb{G}_{n,k}(K))$ is dense in $\mathcal{C}^\infty(V, \mathbb{G}_{n,k}(K))$.*

Proof Immediate. □

Let V be a non singular compact real affine algebraic variety. Let $VB^1(V)$ be the group of isomorphism classes of (topological) vector bundles of rank 1. The morphism $w_1 : VB^1(V) \to H^1(V; \mathbb{Z}_2)$ which associates to an isomorphism class of rank 1 vector bundles its first Stiefel–Whitney class[5] is an isomorphism. See [BCR98, Section 12.4] for more details. If the dimension of V is n then we can compose with Poincaré duality to get an isomorphism

$$D_V \circ w_1 : VB^1(V) \overset{\simeq}{\to} H_{n-1}(V; \mathbb{Z}_2) .$$

Theorem 5.2.16 *Let V be a non singular compact real affine algebraic variety of dimension n and consider an element $\alpha \in H_{n-1}(V; \mathbb{Z}_2)$. The following are then equivalent.*

1. *The class α is the image under $D_V \circ w_1 : VB^1(V) \to H_{n-1}(V; \mathbb{Z}_2)$ of the class of an* algebraic *vector bundle of rank* 1.
2. *There is a non singular algebraic subset $W \subset V$ of dimension $n-1$ whose fundamental class is α (see Definition 3.7.1).*
3. *The class α belongs to the subgroup of algebraic cycles (Definition 3.7.2) $H^{\mathrm{alg}}_{n-1}(V; \mathbb{Z}_2)$.*

Proof See [BCR98, Theoreme 12.4.6]. □

Corollary 5.2.17 *The isomorphism $w_1 : VB^1(V) \overset{\simeq}{\to} H^1(V; \mathbb{Z}_2)$ induces an isomorphism*

$$w_1 : VB^1_{\mathrm{alg}}(V) \overset{\simeq}{\to} H^1_{\mathrm{alg}}(V; \mathbb{Z}_2) ;$$

and $D_V \circ w_1 : VB^1(V) \overset{\simeq}{\to} H_{n-1}(V; \mathbb{Z}_2)$ induces an isomorphism

$$D_V \circ w_1 : VB^1_{\mathrm{alg}}(V) \overset{\simeq}{\to} H^{\mathrm{alg}}_{n-1}(V; \mathbb{Z}_2) .$$

[5] $w_1(E)$ is the first Stiefel–Whitney class of a bundle E: see [MS74, Section 4] for more details.

The following theorem enables us to turn an approximation problem into a problem on algebraic cycles. See Section 3.7 for more details.

Theorem 5.2.18 *Let n be a non zero natural number, let u_n be the unique generator of the group $H^1(\mathbb{P}^n(\mathbb{R}); \mathbb{Z}_2) \simeq \mathbb{Z}_2$ and let V be non singular compact real affine algebraic variety. For any smooth map*

$$f: V \to \mathbb{P}^n(\mathbb{R}) ,$$

the following are equivalent.

1. *The map f is approximable in the $\mathcal{C}^\infty$ topology by regular maps $V \to \mathbb{P}^n(\mathbb{R})$.*
2. *The map f is homotopic to a regular map $V \to \mathbb{P}^n(\mathbb{R})$.*
3. *The pull back $f^*(u_n)$ where f^* is the induced map $f^*\colon H^1(\mathbb{P}^n(\mathbb{R}); \mathbb{Z}_2) \to H^1(V; \mathbb{Z}_2)$ belongs to the subgroup of algebraic cycles $H^1_{\text{alg}}(V; \mathbb{Z}_2)$.*

Proof By Corollary 5.2.17, the first two conditions are equivalent by Theorem 5.2.14 applied to $\mathbb{P}^n(\mathbb{R}) = \mathbb{G}_{n+1,1}(\mathbb{R})$. The third condition is then equivalent because $f^*(u_n) = w_1(f^*(\gamma_{n+1,1}))$ and apply Theorem 5.2.16. □

5.3 Maps to Spheres

Let P be a point on the sphere $\mathbb{S}^n \subset \mathbb{R}^{n+1}$. *Stereographic projection from pole P* is the map $\mathbb{S}^n \setminus \{P\} \to \mathbb{R}^n$ obtained by associating to any point Q in $\mathbb{S}^n$ different from P the intersection point of the unique line in $\mathbb{R}^{n+1}$ passing through P and Q an the affine subspace (of dimension n) tangent to $\mathbb{S}^n$ at the antipode of P.

Proposition 5.3.1 *Let n be a natural number, let*

$$\mathbb{S}^n := \{(x_1, \ldots, x_{n+1}) \in \mathbb{R}^{n+1} \mid x_1^2 + \cdots + x_{n+1}^2 = 1\}$$

be the quadric sphere of dimension n and let P be a point in $\mathbb{S}^n$. Stereographic projection $\mathbb{S}^n \setminus \{P\} \to \mathbb{R}^n$ is then an isomorphism of real algebraic varieties.

Remark 5.3.2 We invite the reader to compare this proof with the proof of Proposition 4.4.10(1b) for the sphere of dimension 2.

Proof Fix two points on the sphere: the north pole $P_N := (0, \ldots, 0, 1) \in \mathbb{S}^n$ and the south pole $P_S := (0, \ldots, 0, -1) \in \mathbb{S}^n$. In coordinates, stereographic projection from the north and south poles are given by:

$$\begin{array}{rccc} \varphi_N\colon & \mathbb{S}^n \setminus \{P_N\} & \longrightarrow & \mathbb{R}^n \\ & (x_1, \ldots, x_{n+1}) & \longmapsto & \left(\frac{x_1}{1-x_{n+1}}, \ldots, \frac{x_n}{1-x_{n+1}}\right) \end{array}$$

and

$$\begin{array}{rcl}\varphi_S\colon\quad \mathbb{S}^n\setminus\{P_S\} & \longrightarrow & \mathbb{R}^n\\ (x_1,\dots,x_{n+1}) & \longmapsto & \left(\frac{x_1}{1+x_{n+1}},\dots,\frac{x_n}{1+x_{n+1}}\right)\end{array}$$

and their inverses are given by

$$\begin{array}{rcl}\mathbb{R}^n & \longrightarrow & \mathbb{S}^n\setminus\{P_N\}\\ (y_1,\dots,y_n) & \longmapsto & \left(\frac{2y_1}{y_1^2+\dots+y_n^2+1},\dots,\frac{2y_n}{y_1^2+\dots+y_n^2+1},\frac{y_1^2+\dots+y_n^2-1}{y_1^2+\dots+y_n^2+1}\right)\end{array}$$

and

$$\begin{array}{rcl}\mathbb{R}^n & \longrightarrow & \mathbb{S}^n\setminus\{P_S\}\\ (y_1,\dots,y_n) & \longmapsto & \left(\frac{2y_1}{y_1^2+\dots+y_n^2+1},\dots,\frac{2y_n}{y_1^2+\dots+y_n^2+1},\frac{-y_1^2-\dots-y_n^2+1}{y_1^2+\dots+y_n^2+1}\right).\end{array}$$

□

Corollary 5.3.3 *As real algebraic varieties,*

1. $\mathbb{S}^1$ *is isomorphic to* $\mathbb{P}^1(\mathbb{R})$.
2. $\mathbb{S}^2$ *is isomorphic to* $\mathbb{P}^1(\mathbb{C})$.
3. $\mathbb{S}^4$ *is isomorphic to* $\mathbb{P}^1(\mathbb{H})$.

Proof Consider stereographic projection from $\mathbb{S}^n$ to $\mathbb{R}^n$. For $n=1$ for any $x\neq 0$ we have that $\varphi_N\circ\varphi_S^{-1}(x)=\frac{1}{x}$ and if $n=2$ for any $(x,y)\neq(0,0)$ we have that $\varphi_N\circ\varphi_S^{-1}(x,y)=(\frac{x}{x^2+y^2},\frac{y}{x^2+y^2})$ and hence in complex coordinates $z=x+iy$, $\varphi_N\circ\varphi_S^{-1}(z)=\frac{1}{\bar z}$. Similarly, we can show that $\mathbb{S}^4$ is isomorphic to $\mathbb{P}^1(\mathbb{H})$ using the map

$$(u:v)\mapsto\left(\frac{2u\overline{v}}{|u|^2+|v|^2},\frac{|u|^2-|v|^2}{|u|^2+|v|^2}\right).$$

□

Proposition 5.3.4 *Let V be a non singular compact real affine algebraic variety, let n be* $1, 2$ *or* 4 *and let*

$$f\colon V\to\mathbb{S}^n$$

be a smooth map. The following are then equivalent.

1. *The map f is approximable in the C^∞ topology by regular maps $V\to\mathbb{S}^n$.*
2. *The map f is homotopic to a regular map $V\to\mathbb{S}^n$.*

Proof This equivalence follows immediately from Theorem 5.2.14 because of Corollary 5.3.3 which states that for $n=1, 2$ or 4 the sphere $\mathbb{S}^n$ is isomorphic to $\mathbb{P}^1(K)=\mathbb{G}_{2,1}(K)$ for $K=\mathbb{R},\mathbb{C}$ or $\mathbb{H}$. □

5.3.1 Maps to $\mathbb{S}^1$

The isomorphism of real algebraic varieties between $\mathbb{S}^1$ and $\mathbb{P}^1(\mathbb{R})$ gives us a homological characterisation of density of regular maps to $\mathbb{S}^1$.

Theorem 5.3.5 *Let V be a non singular compact real affine algebraic variety of dimension n. The following are then equivalent.*

1. *The subspace $\mathcal{R}(V, \mathbb{S}^1)$ is dense in $\mathcal{C}^\infty(V, \mathbb{S}^1)$.*
2. *Any $\mathcal{C}^\infty$ map from V to $\mathbb{S}^1$ is homotopic to a regular map.*
3. *$H^{nt}_{n-1}(V; \mathbb{Z}_2) \subset H^{alg}_{n-1}(V; \mathbb{Z}_2)$, where $H^{nt}_{n-1}(V; \mathbb{Z}_2)$ is the subset of $H_{n-1}(V; \mathbb{Z}_2)$ of homology classes represented by compact $\mathcal{C}^\infty$ hypersurfaces in V whose normal bundle is trivial.*

Proof See [BCR98, Theorem 13.3.5]. □

Example 5.3.6 Here are two examples where purely topological arguments enable us to refine the above criteria.

1. If V is diffeomorphic to the Klein bottle $\mathbb{K}^2$ then $\mathcal{R}(V, \mathbb{S}^1)$ is dense in $\mathcal{C}^\infty(V, \mathbb{S}^1)$. In this case $H^{\mathrm{nt}}_1(V; \mathbb{Z}_2)$ is generated by the dual Poincaré class of the *algebraic* bundle $\bigwedge^2 T_V$. See Theorem 5.2.16 for more details.
2. If V is an orientable surface then $\overline{\mathcal{R}(V, \mathbb{S}^1)} = \mathcal{C}^\infty(V, \mathbb{S}^1)$ if and only if $H_1(V; \mathbb{Z}_2) = H^{\mathrm{alg}}_1(V; \mathbb{Z}_2)$. In this case, any $\mathcal{C}^\infty$ curve in V has trivial normal bundle in V and hence $H^{\mathrm{nt}}_1(V; \mathbb{Z}_2) = H_1(V; \mathbb{Z}_2)$.

Corollary 5.3.7 *Let V be a non singular compact real affine algebraic variety of dimension n. If $H_{n-1}(V; \mathbb{Z}_2) = H^{\mathrm{alg}}_{n-1}(V; \mathbb{Z}_2)$ then $\mathcal{R}(V, \mathbb{S}^1)$ is dense in $\mathcal{C}^\infty(V, \mathbb{S}^1)$.*

The converse is false, even in dimension 2, as the example below shows.

Example 5.3.8 (*Not totally algebraic Klein bottle*) Let Y be the quotient of the non singular quartic hypersurface[6] $X := \mathcal{Z}(x^4 + y^4 + z^4 - t^4)$ in $\mathbb{P}^3(\mathbb{C})_{t:x:y:z}$ by the fixed point free involution $(t : x : y : z) \mapsto (-x : t : -z : y)$. Let τ be the real structure induced on Y by the restriction of $\sigma_{\mathbb{P}}$ to X. By construction, $(X, \sigma_{\mathbb{P}}|_X)$ is a real K3 surface (Section 4.5) whose real locus is diffeomorphic to $\mathbb{S}^2$ and (Y, τ) is a real Enriques surface whose real locus is diffeomorphic to the Klein bottle $\mathbb{K}^2$. Moreover, by Theorem 4.5.17 we have that $H_1(Y(\mathbb{R}); \mathbb{Z}_2) \neq H^{\mathrm{alg}}_1(Y(\mathbb{R}); \mathbb{Z}_2)$, but Example 5.3.6(1) shows that $\overline{\mathcal{R}(Y(\mathbb{R}), \mathbb{S}^1)} = \mathcal{C}^\infty(Y(\mathbb{R}), \mathbb{S}^1)$.

These examples show that different real algebraic models (Definition 5.1.1) of the same differentiable manifold generally have non isomorphic algebraic cycle groups. This holds for the Klein bottle, but for any algebraic model V of the Klein bottle $\mathcal{R}(V, \mathbb{S}^1)$ is nevertheless dense in $\mathcal{C}^\infty(V, \mathbb{S}^1)$. This property holds for a very small number of topological surfaces.

[6] Often called the Fermat hypersurface in the literature.

Theorem 5.3.9 *The following conditions on a compact connected $\mathcal{C}^\infty$ surface M are equivalent.*

1. *The subspace $\mathcal{R}(V, \mathbb{S}^1)$ is dense in $\mathcal{C}^\infty(V, \mathbb{S}^1)$ for any real algebraic model V of M.*
2. *The topological manifold M is diffeomorphic to the sphere $\mathbb{S}^2$, the real projective plane $\mathbb{RP}^2$ or the Klein bottle $\mathbb{K}^2$.*

Proof See [BK10, Thm. 3.4]. □

5.3.2 *Maps to $\mathbb{S}^2$*

In this section we consider a non singular real algebraic variety V and a non singular complexification (Definition 2.3.2) $V_\mathbb{C}$ of V. The results presented below do not depend on the choice of the complexification $V_\mathbb{C}$ by Proposition 2.3.3. We will concentrate on approximation of smooth maps from a compact surface V of negative Kodaira dimension to the sphere. When we say that $\kappa(V) = -\infty$ we mean that some non singular projective complexification $V_\mathbb{C}$ has $\kappa(V_\mathbb{C}) = -\infty$. The number κ is a birational invariant of complete varieties and we can therefore set $\kappa(V) := \kappa(V_\mathbb{C})$ whenever V is compact and $V_\mathbb{C}$ is projective.

We start by giving a historical overview of known results, followed by a sketch of their proofs. By Theorem 4.4.8, real algebraic surfaces of negative Kodaira dimension can be classified in three categories, each of which includes the previous one: rational surfaces, geometrically rational surfaces and surfaces which are birationally equivalent to the total space of a conic bundle.

Rational Surfaces

When V is a compact non singular rational surface (Definition 1.3.37) Bochnak and Kucharz proved in the nineties that $\overline{\mathcal{R}(V, \mathbb{S}^2)} = \mathcal{C}^\infty(V, \mathbb{S}^2)$ if and only if V is not diffeomorphic to a torus, and if V is a torus only the homotopically trivial maps can be approximated algebraically. It turns out that we can do better (see [Kuc99, Th. 1.2]):

Theorem 5.3.10 *Let V and W be non singular compact real rational algebraic surfaces. The space $\mathcal{R}(V, W)$ is then dense in $\mathcal{C}^\infty(V, W)$ except when V is diffeomorphic to $\mathbb{T}^2$ and W is diffeomorphic to the sphere $\mathbb{S}^2$. In this case the topological closure of $\mathcal{R}(V, W)$ in $\mathcal{C}^\infty(V, W)$ is the set of homotopically trivial maps.*

Proof Since V is connected by Theorem 4.4.8 the theorem for $W = \mathbb{S}^2$ follows from Theorems 5.3.18 and 5.3.34 below. We refer to [Kuc99, Th. 1.2] for a proof when W is another rational surface. □

Geometrically Rational Surfaces, del Pezzo Surfaces

Rational surfaces are a special case of geometrically rational surfaces (Definition 1.3.37). By Theorem 4.4.8, real rational algebraic surfaces are exactly *connected* geometrically rational real surfaces. In 2004, surfaces with four connected components which have algebraic approximation for maps of even degree only were discovered. See [JPM04, Theorem 0.3] for more details. The surfaces in question are the real algebraic surfaces which are topologically the disjoint union of four spheres and whose complexifications are del Pezzo surfaces of degree 2 (Definition 4.2.12). Recall that a del Pezzo surface X is a complex algebraic surface whose anti-canonical divisor $-K_X$ is ample. The degree d of a del Pezzo surface is its first Chern number K_X^2. Apart from $\mathbb{P}^1(\mathbb{C}) \times \mathbb{P}^1(\mathbb{C})$, all del Pezzo surfaces are blow ups of the projective plane $\mathbb{P}^2(\mathbb{C})$ in $9 - d$ points. If the set Σ of 7 blow up centres is stable under $\sigma_{\mathbb{P}}$, the degree 2 del Pezzo surface given by Σ has two real structures, one of which has a connected real part, namely the one that arising from complex conjugaison on $\mathbb{P}^2(\mathbb{C})$ via blow up.

Theorem 5.3.11 *Let V be a non singular real affine algebraic surface diffeomorphic to the disjoint union of 4 spheres which has a complexification $V_{\mathbb{C}}$ which is a del Pezzo surface of degree 2. Let*

$$f : V \to \mathbb{S}^2$$

be a smooth map. The map f can then be approximated by regular maps if and only if it is of even topological degree.

In particular, $\overline{\mathcal{R}(V, \mathbb{S}^2)} \neq \mathcal{C}^\infty(V, \mathbb{S}^2)$.

Proof This result follows on combining Theorems 5.3.18 and 5.3.35 below. □

Remark 5.3.12 There are real geometrically rational surfaces which are topologically a disjoint union of four spheres on which all smooth maps can be approximated by regular maps. There surfaces have a conic bundle structure. See Example 4.2.8 and Theorem 5.3.13 below for more details.

It turns out that the torus and the del Pezzo surface of Theorem 5.3.11 are the only geometrically rational surfaces for which density does not hold: $\overline{\mathcal{R}(V, \mathbb{S}^2)} \neq \mathcal{C}^\infty(V, \mathbb{S}^2)$.

Theorem 5.3.13 *Let V be a non singular compact geometrically rational real algebraic surface. The space of regular maps $\mathcal{R}(V, \mathbb{S}^2)$ is dense in the space $\mathcal{C}^\infty(V, \mathbb{S}^2)$ of maps $\mathcal{C}^\infty$ unless V is diffeomorphic to the torus $\mathbb{T}^2$ or V is diffeomorphic to a disjoint union of 4 spheres and has a complexification $V_{\mathbb{C}}$ which is a real del Pezzo surface of degree 2 as in Theorem 5.3.11.*

Proof This follows from Corollary 5.3.17 and Theorems 5.3.18 and 5.3.34 below. See [JPM04, Theorem 0.4] for more details. □

Uniruled Surfaces and Conic Bundles

All that is now left to complete the classification for surfaces of negative Kodaira dimension is to deal with conic bundles whose base has non zero genus and their blow ups. In particular, if $V_{\mathbb{C}}$ has a conic bundle structure with a non rational base, $V_{\mathbb{C}}$ is not simply connected and V can have several orientable non spherical components.

Let V be a non singular compact real affine algebraic surface of negative Kodaira dimension which is not geometrically rational. By Theorem 4.3.23(7), V then has a real ruling $\rho\colon V \longrightarrow B$, since the blow-up of a conic bundle is a ruling, i.e. a genus 0 bundle. We recall that a connected component of V can be diffeomorphic to a sphere, a torus or an arbitrary non orientable surface (Theorem 4.4.14). Let K' be the set of components of V which are diffeomorphic to the Klein bottle and whose image under ρ is a connected component of B.

Theorem 5.3.14 *Let V be a non singular compact real affine algebraic surface which is of negative Kodaira dimension but is not geometrically rational. For any smooth map $f : V \longrightarrow \mathbb{S}^2$, the following are equivalent.*

1. *The map f can be approximated by regular maps.*
2. *The map f is homotopic to a regular map.*
3. *For any connected component M of V diffeomorphic to a torus we have that $\deg(f|_M) = 0$ and for any pair of components N and L belonging to K' we have that $\deg_{\mathbb{Z}_2}(f|_N) = \deg_{\mathbb{Z}_2}(f|_L)$.*

Proof See [Man06, Theorem 1.1]. □

Maps to Rational Surfaces

We end this section with a generalisation of Theorem 5.3.10.

Theorem 5.3.15 *Let V and W be non singular compact connected real affine algebraic surfaces such that $\kappa(V) = -\infty$ and W is rational. The space $\mathcal{R}(V, W)$ is then dense in $\mathcal{C}^\infty(V, W)$ unless V is diffeomorphic to the torus $\mathbb{T}^2$ and W is diffeomorphic to the sphere $\mathbb{S}^2$. In this last case, the closure of $\mathcal{R}(V, W)$ in $\mathcal{C}^\infty(V, W)$ is the set of homotopically trivial maps.*

Proof See [Man06, Theorem 1.2]. □

Algebraic $\mathbb{C}$-Vector Bundles

We now sketch the proofs of the results stated above. Let V be a real affine algebraic variety, let $V_{\mathbb{C}}$ be a non singular complexification of V and let $i : V \hookrightarrow V_{\mathbb{C}}$ be the inclusion map. By Corollary 2.5.16, a *topological* $\mathbb{R}$-vector bundle $\mathcal{L}$ on V (Definition C.3.5) is *algebraic* (Definition 5.2.10) if and only if its tensorisation by $\mathbb{C}$ is the restriction to V of an algebraic $\mathbb{C}$-vector bundle $\mathcal{E}$ over $V_{\mathbb{C}}$ with a real structure $\mathcal{L} \otimes \mathbb{C} \simeq \mathcal{E}|_V$.

We denote by $VB^1_{\mathbb{C}}(V_{\mathbb{C}})$ the group of isomorphism classes of (topological) rank 1 $\mathbb{C}$-vector bundles over $V_{\mathbb{C}}$. We denote by $H^2_{\mathbb{C}-\mathrm{alg}}(V;\mathbb{Z})$ the subgroup of $H^2(V;\mathbb{Z})$ of restrictions to V of *algebraic* rank 1 $\mathbb{C}$-vector bundles on $V_{\mathbb{C}}$. In other words, $H^2_{\mathbb{C}-\mathrm{alg}}(V;\mathbb{Z})$ is the image of classes of algebraic bundles under the restriction map $VB^1_{\mathbb{C}}(V_{\mathbb{C}}) \overset{c_1}{\simeq} H^2(V_{\mathbb{C}};\mathbb{Z}) \overset{i^*}{\to} H^2(V;\mathbb{Z})$. Note that if a bundle has, like $\mathcal{E}$ above, a real structure then its restriction to V is 2-torsion (Proposition 5.3.23) and the group $H^2_{\mathbb{C}-\mathrm{alg}}(V;\mathbb{Z})$ also contains classes of restrictions of algebraic rank 1 $\mathbb{C}$-vector bundles on V without real structure. We denote by

$$\Gamma(V) = H^2(V;\mathbb{Z})/H^2_{\mathbb{C}-\mathrm{alg}}(V;\mathbb{Z})$$

the quotient group. The group $H^2_{\mathbb{C}-\mathrm{alg}}(V;\mathbb{Z})$ does not depend on the choice of non singular complexification $V_{\mathbb{C}}$. (This follows from Proposition 2.3.3: we also refer the interested reader to [BBK89]). The group $\Gamma(V)$ is therefore also independent of this choice. Our definition of $H^2_{\mathbb{C}-\mathrm{alg}}(V;\mathbb{Z})$ is motived by the following result, which invites comparison with Theorem 5.2.18.

Theorem 5.3.16 *Let V be a non singular compact real affine algebraic variety and let s_2 be a generator of $H^2(\mathbb{S}^2;\mathbb{Z}) \simeq \mathbb{Z}$. For any smooth map*

$$f\colon V \to \mathbb{S}^2\,,$$

the following are equivalent.

1. *The map f can be approximated in the $\mathcal{C}^\infty$ topology by regular maps $V \to \mathbb{S}^2$.*
2. *The map f is homotopic to a regular map $V \to \mathbb{S}^2$.*
3. *The pull back $f^*(s_2)$, where f^* is the induced map $f^*\colon H^2(\mathbb{S}^2;\mathbb{Z}) \to H^2(V;\mathbb{Z})$, belongs to the subgroup $H^2_{\mathbb{C}-\mathrm{alg}}(V;\mathbb{Z})$.*

Proof This follows immediately from Theorem 5.2.14 and Corollary 5.3.3. □

Corollary 5.3.17 *Let V be a non singular compact real affine algebraic variety such that any topological rank 1 $\mathbb{C}$-vector bundle on V is topologically isomorphic to an algebraic $\mathbb{C}$-vector bundle. The set $\mathcal{R}(V,\mathbb{S}^2)$ is then dense in $\mathcal{C}^\infty(V,\mathbb{S}^2)$ for the $\mathcal{C}^\infty$ topology. In other words if $\Gamma(V)=0$ then $\overline{\mathcal{R}(V,\mathbb{S}^2)} = \mathcal{C}^\infty(V,\mathbb{S}^2)$.*

If V is a connected surface then the converse holds.

Theorem 5.3.18 *Let V be a compact connected non singular affine real algebraic surface. We then have that $\mathcal{R}(V,\mathbb{S}^2)$ is dense in $\mathcal{C}^\infty(V,\mathbb{S}^2)$ for the $\mathcal{C}^\infty$ topology if and only if every rank 1 topological $\mathbb{C}$-vector bundle on V is topologically isomorphic to an algebraic $\mathbb{C}$-vector bundle.*

In other words, $\overline{\mathcal{R}(V,\mathbb{S}^2)} = \mathcal{C}^\infty(V,\mathbb{S}^2)$ if and only if $\Gamma(V)=0$.

Remark 5.3.19 Note that $\Gamma(V) = 0$, or in other words $H^2_{\mathbb{C}}(V;\mathbb{Z}) = H^2_{\mathbb{C}-\mathrm{alg}}(V;\mathbb{Z})$ if and only if $VB^1_{\mathbb{C}}(V) = VB^1_{\mathbb{C}-\mathrm{alg}}(V)$ via the first Chern class map.

Proof See [BCR98, Corollary 13.3.12]. □

Blow Ups and Regular Maps

We calculate the group $\Gamma(V)$ of a geometrically rational real algebraic surface V in two steps. We first prove Proposition 5.3.20 below that enables us to control the behaviour of the quotient group $\Gamma(V)$ under blow up $\pi\colon B_P V \to V$ centred at $P \in V$. We then calculate $\Gamma(V_0)$ for the various minimal models V_0 of a given geometrically rational real algebraic surface V. Finally, we prove that if V is both minimal and non connected then either $\Gamma(V) = 0$ or $\Gamma(V) = \mathbb{Z}_2$.

Proposition 5.3.20 *Let V be a compact non singular geometrically rational real algebraic surface such that $\Gamma(V) = 0$ or $\Gamma(V) = \mathbb{Z}_2$. We then have that $\Gamma(B_P V) = 0$ for any blow up $B_P V$ centred at a point P in V.*

Before proving Proposition 5.3.20 we need some auxiliary results. Let $i : V \hookrightarrow V_{\mathbb{C}}$ be the canonical injection of a non singular compact real affine algebraic surface into a non singular projective complexification.

Lemma 5.3.21 *If the geometrical genus of $V_{\mathbb{C}}$ is zero then $H^2_{\mathbb{C}-\mathrm{alg}}(V;\mathbb{Z}) = \operatorname{Im} i^*$. If moreover the irregularity of $V_{\mathbb{C}}$ vanishes then $H^2(V_{\mathbb{C}};\mathbb{Z}) \simeq \operatorname{Pic}(V_{\mathbb{C}})$.*

Proof By definition $H^2_{\mathbb{C}-\mathrm{alg}}(V;\mathbb{Z}) \subseteq \operatorname{Im} i^*$. Inserting the isomorphism $\operatorname{Pic}(V_{\mathbb{C}}) \simeq H^1(V_{\mathbb{C}}, \mathcal{O}^*_{V_{\mathbb{C}}})$ into the long exact sequence associated to the exponential sequence (Proposition D.6.3), we get an exact sequence

$$\cdots \to H^1(V_{\mathbb{C}}, \mathcal{O}_{V_{\mathbb{C}}}) \to \operatorname{Pic}(V_{\mathbb{C}}) \xrightarrow{c_1} H^2(V_{\mathbb{C}};\mathbb{Z}) \to H^2(V_{\mathbb{C}}, \mathcal{O}_{V_{\mathbb{C}}}) \to \cdots$$

The result follows immediately since by definition $p_g(V_{\mathbb{C}}) = \dim H^2(V_{\mathbb{C}}, \mathcal{O}_{V_{\mathbb{C}}})$ and $q(V_{\mathbb{C}}) = \dim H^1(V_{\mathbb{C}}, \mathcal{O}_{V_{\mathbb{C}}})$. □

Corollary 5.3.22 *If V is a compact non singular geometrically rational real algebraic surface then $H^2_{\mathbb{C}-\mathrm{alg}}(V;\mathbb{Z}) = \operatorname{Im} i^*$.*

In other words, for any geometrically rational compact and non singular real algebraic surface V, the quotient group $\Gamma(V)$ is trivial if and only if $i^*\colon H^2(V_{\mathbb{C}};\mathbb{Z}) \to H^2(V;\mathbb{Z})$ is surjective.

Since $B_P V$ is the real blow up of a geometrically rational real algebraic surface it is also a geometrically rational real algebraic surface and $\Gamma(B_P V) = 0$ if and only if the morphism $i^*\colon H^2(B_P V_{\mathbb{C}};\mathbb{Z}) \to H^2(B_P V;\mathbb{Z})$ is surjective by Corollary 5.3.22.

We denote by $\beta\colon H^1(V;\mathbb{Z}_2) \to H^2(V;\mathbb{Z})$ the *Bockstein* morphism induced in cohomology by the exact sequence

$$0 \to \mathbb{Z} \xrightarrow{\times 2} \mathbb{Z} \to \mathbb{Z}_2 \to 0\,. \tag{5.1}$$

Proposition 5.3.23 *Let V be a compact non singular geometrically rational real algebraic surface. We then have that*

$$\beta(H^1(V;\mathbb{Z}_2)) \subseteq H^2_{\mathbb{C}-\mathrm{alg}}(V;\mathbb{Z}) .$$

Proof Recall that the group of isomorphism classes of rank 1 topological $\mathbb{C}$-vector bundles on V is denoted $VB^1_{\mathbb{C}}(V)$. There is a commutative diagram (see [JP00] for example):

$$\begin{array}{ccccc} \mathrm{Pic}(V_{\mathbb{C}}) & \xrightarrow{\mathcal{L}\mapsto\mathcal{L}|_V} & VB^1_{\mathbb{C}}(V) & \xleftarrow{\mathcal{L}\mapsto\mathcal{L}\otimes\mathbb{C}} & VB^1_{\mathrm{alg}}(V) \\ {\scriptstyle c_1}\downarrow & & {\scriptstyle c_1}\downarrow & & {\scriptstyle w_1}\downarrow \\ H^2(V_{\mathbb{C}};\mathbb{Z}) & \xrightarrow{i^*} & H^2(V;\mathbb{Z}) & \xleftarrow{\beta} & H^1(V;\mathbb{Z}_2) . \end{array}$$

Here w_1 is the morphism induced by the first *Stiefel–Whitney class* and c_1 is the morphism induced by the first *Chern class*.

Note that the restriction morphism $\mathcal{L} \mapsto \mathcal{L}|_V$ in the first line is surjective and in this case the morphisms c_1 and w_1 are also surjective (see [JP00, Proof of Claim 2]). Recall also that by Corollary 5.3.22 we have that $H^2_{\mathbb{C}-\mathrm{alg}}(V;\mathbb{Z}) = i^*(H^2(V_{\mathbb{C}};\mathbb{Z}))$. □

Remark 5.3.24 In the above proof we have used the fact that the morphisms $w_1\colon VB^1_{\mathrm{alg}}(V) \to H^1(V;\mathbb{Z}_2)$ and $c_1\colon \mathrm{Pic}(V_{\mathbb{C}}) \to H^2(V_{\mathbb{C}};\mathbb{Z}))$ are surjective. This does not hold for a general non singular projective surface because the surjectivity of c_1 fails as soon as $p_g(V_{\mathbb{C}}) > 0$. Indeed, $\dim H^{2,0}(V_{\mathbb{C}}) = p_g(V_{\mathbb{C}})$ and the image of c_1 is contained in $H^{1,1}(V_{\mathbb{C}})$ (see Theorem D.9.3). All K3 surfaces, abelian surfaces, bi-elliptic surfaces, as well as most properly elliptic surfaces and surfaces of general type, satisfy $p_g > 0$.

Let $V_1, V_2, \ldots, V_s$ be the connected components of V and assume for simplicity that the centre P of the blow up belongs to V_1. The blown up surface $B_P V$ is diffeomorphic to the disjoint union of $B_P V_1 = V_1\#\mathbb{RP}^2$ and the other connected components of V.

Proposition 5.3.25 *Via the identification*

$$H^2(B_P V;\mathbb{Z}) \simeq H^2(B_P V_1;\mathbb{Z}) \oplus H^2(V_2;\mathbb{Z}) \oplus \cdots \oplus H^2(V_s;\mathbb{Z}),$$

we get an inclusion $H^2(B_P V_1;\mathbb{Z}) \subseteq H^2_{\mathbb{C}-\mathrm{alg}}(B_P V;\mathbb{Z})$.

Proof Consider the cohomology exact sequence induced by multiplication by 2 (see the exact sequence (5.1) preceding the proof of Proposition 5.3.23)

$$\cdots \to H^1(B_P V;\mathbb{Z}_2) \xrightarrow{\beta} H^2(B_P V;\mathbb{Z}) \xrightarrow{\times 2} H^2(B_P V;\mathbb{Z}) \to \cdots$$

As $B_P V_1$ is connected and non orientable the group $H^2(B_P V_1;\mathbb{Z})$ is isomorphic to $\mathbb{Z}_2$ by Theorem B.5.7 and Corollary B.4.2. It is therefore contained in the kernel of the morphism $\times 2$. It follows that the subgroup $H^2(B_P V_1; \mathbb{Z})$ of $H^2(B_P V; \mathbb{Z})$ is contained in the image of the Bockstein morphism $\beta\colon H^1(B_P V; \mathbb{Z}_2) \to H^2(B_P V; \mathbb{Z})$. The result follows from Proposition 5.3.23. □

Remark 5.3.26 Let V be a non singular compact real affine algebraic surface. If the homology of some non singular complexification $V_\mathbb{C}$ has no 2-torsion then $H_*(V_\mathbb{C}; \mathbb{Z}) \otimes \mathbb{Z}_2 \simeq H_*(V_\mathbb{C}; \mathbb{Z}_2)$ (see Section B.4) and it follows that $H^*(V_\mathbb{C}; \mathbb{Z}) \otimes \mathbb{Z}_2 \simeq H^*(V_\mathbb{C}; \mathbb{Z}_2)$. Moreover we have that $H^2(V; \mathbb{Z}) \otimes \mathbb{Z}_2 \simeq H^2(V; \mathbb{Z}_2) \simeq (\mathbb{Z}_2)^s$ where s is the number of connected components of the compact topological surface V.

As above, we denote by i the inclusion of V in its complexification $V_\mathbb{C}$ (resp. $B_P V$ in $B_P V_\mathbb{C}$) and set $H^2_{\mathbb{C}-\mathrm{alg}}(V; \mathbb{Z}_2) = i^*(H^2_{\mathrm{alg}}(V_\mathbb{C}; \mathbb{Z}_2))$. Note that $H^2_{\mathbb{C}-\mathrm{alg}}(V; \mathbb{Z}_2) = i^*(H^2(V_\mathbb{C}; \mathbb{Z}_2)) \simeq H^2_{\mathbb{C}-\mathrm{alg}}(V; \mathbb{Z}) \otimes \mathbb{Z}_2$. By the above discussion, Proposition 5.3.25 then implies the following result:

Corollary 5.3.27 *Let V_1 be a connected component of V and let Q be a point on $B_P V_1$. The Poincaré dual class $D^{-1}_{B_P V}([Q])$ then belongs to $H^2_{\mathbb{C}-\mathrm{alg}}(B_P V; \mathbb{Z}_2)$.*

Remark 5.3.28 Corollary 5.3.27 turns out to hold for any non singular real projective algebraic surface V. Let P be a real point of V, let $E = \pi^{-1}(P)$ be the exceptional curve and let $\mathcal{E}$ be the complex line bundle associated to $E_\mathbb{C}$: we then have that $i^*(c_1(\mathcal{E})) = c_1(\mathcal{L} \otimes \mathbb{C})$, where $\mathcal{L}$ is the real line bundle associated to E.

By characteristic class theorem (see [MS74, Exercise 15-D]) we have that $c_1(\mathcal{L} \otimes \mathbb{C}) = \beta(w_1(\mathcal{L}))$. Reducing $\beta(w_1(\mathcal{L}))$ modulo 2 we obtain $D^{-1}_{B_P V}([Q])$, the element of $H^2(B_P V; \mathbb{Z}_2)$ generating the subspace $H^2(B_P V_1; \mathbb{Z}_2)$, and hence $i^*([c_1(\mathcal{E})]_2) = [\beta(w_1(\mathcal{L}))]_2 = D^{-1}_{B_P V}([Q])$.

The following corollary remains valid, with the same proof, for an arbitrary non singular real projective algebraic surface.

Corollary 5.3.29 *If $D_V^{-1}([P])$ belongs to $H^2_{\mathbb{C}-\mathrm{alg}}(V; \mathbb{Z}_2)$ then $H^2_{\mathbb{C}-\mathrm{alg}}(B_P V; \mathbb{Z}_2) = \pi^*(H^2_{\mathbb{C}-\mathrm{alg}}(V; \mathbb{Z}_2))$. Otherwise, there is a canonical decomposition*

$$H^2_{\mathbb{C}-\mathrm{alg}}(B_P V; \mathbb{Z}_2) = \pi^*(H^2_{\mathbb{C}-\mathrm{alg}}(V; \mathbb{Z}_2)) \oplus D^{-1}_{B_P V}([Q]).$$

Proof Since $\pi\colon B_P V \to V$ is a blow up centred at P, π is a morphism inducing an isomorphism from $B_P V \setminus E$ to $V \setminus \{P\}$. As V and $B_P V$ are surfaces, π induces an isomorphism $\pi^*\colon H^2(V; \mathbb{Z}_2) \to H^2(B_P V; \mathbb{Z}_2)$. As π is a morphism, the isomorphism π^* satisfies

$$\pi^*(H^2_{\mathbb{C}-\mathrm{alg}}(V; \mathbb{Z}_2)) \subseteq H^2_{\mathbb{C}-\mathrm{alg}}(B_P V; \mathbb{Z}_2)\ .$$

We complete the proof by recalling that we can identify

$$\pi^*(H^2_{\mathbb{C}-\mathrm{alg}}(V; \mathbb{Z}_2) \cap (H^2(V_2; \mathbb{Z}_2) \oplus \cdots \oplus H^2(V_s; \mathbb{Z}_2)))$$

with

$$H^2_{\mathbb{C}-\text{alg}}(B_P V; \mathbb{Z}_2) \cap (H^2(V_2; \mathbb{Z}_2) \oplus \cdots \oplus H^2(V_s; \mathbb{Z}_2))$$

and $\pi^*(D_V^{-1}([P])) = D_{B_P V}^{-1}([Q])$. □

Proposition 5.3.30 *Let V be a non singular compact real affine algebraic surface. Let $i\colon V \hookrightarrow V_{\mathbb{C}}$ be its inclusion in a non singular projective complexification. The morphism $i^*\colon H^2(V_{\mathbb{C}}; \mathbb{Z}) \to H^2(V; \mathbb{Z})$ is then surjective if and only if $i^*\colon H^2(V_{\mathbb{C}}; \mathbb{Z}_2) \to H^2(V; \mathbb{Z}_2)$ is surjective.*

Proof A non singular compact real affine algebraic surface V satisfies $2H^2(V; \mathbb{Z}) \subseteq \operatorname{Im}(i^*)$ by [JP00, Claim 1, proof of Theorem 1.1]. The result follows by Remark 5.3.26 and the linear algebra lemma below.

Lemma 5.3.31 *Let A and B be abelian groups and let $h\colon A \to B$ be a group morphism. Let $h_2\colon A_2 \to B_2$ be the morphism obtained by reduction modulo 2 and assume that $2B \subseteq \operatorname{Im}(h)$. If the morphism h_2 is surjective then so is h.* □

Proposition 5.3.32 *Let V be a non singular compact real affine algebraic surface and let $V_{\mathbb{C}}$ be a non singular projective complexification. The blow up $B_P V_{\mathbb{C}}$ of $V_{\mathbb{C}}$ at a point P in $V \subset V_{\mathbb{C}}$ is then a non singular projective complexification of $B_P V$. Assume that the restriction morphism*

$$i^*\colon H^2(V_{\mathbb{C}}; \mathbb{Z}_2) \to H^2(V; \mathbb{Z}_2)$$

is surjective. The morphism

$$i^*\colon H^2(B_P V_{\mathbb{C}}; \mathbb{Z}_2) \to H^2(B_P V; \mathbb{Z}_2)$$

is then also surjective.

Proof Let $P_j \in V_j,\ j = 1, \ldots, s$ be points such that

$$D_V^{-1}([P_1]), D_V^{-1}([P_2]), \ldots, D_V^{-1}([P_s])$$

generate $H^2(V; \mathbb{Z}_2)$. Since P belongs to V_1 and $B_P V$ is the blow up of V centred at P, $D_{B_P V}^{-1}([P_2]), \ldots, D_{B_P V}^{-1}([P_s])$ belongs to $H^2(B_P V; \mathbb{Z}_2)$. Moreover, since the restriction morphism i^* is supposed surjective, the classes $D_{B_P V}^{-1}([P_2]), \ldots, D_{B_P V}^{-1}([P_s])$ are also contained in the image $i^*(H^2(B_P V_{\mathbb{C}}; \mathbb{Z}_2))$. The result follows from Corollary 5.3.29. □

Proof of Proposition 5.3.20 Case 1. If $\Gamma(V) = 0$ then $i^*\colon H^2(V_{\mathbb{C}}; \mathbb{Z}) \to H^2(V; \mathbb{Z})$ is surjective. By Proposition 5.3.30, $i^*\colon H^2(V_{\mathbb{C}}; \mathbb{Z}_2) \to H^2(V; \mathbb{Z}_2)$ is also surjective, as is $i^*\colon H^2(B_P V_{\mathbb{C}}; \mathbb{Z}_2) \to H^2(B_P V; \mathbb{Z}_2)$ (Proposition 5.3.32). It follows that $i^*\colon H^2(B_P V_{\mathbb{C}}; \mathbb{Z}) \to H^2(B_P V; \mathbb{Z})$ is surjective by Proposition 5.3.30 and finally we get that $\Gamma(B_P V) = 0$.

Case 2. If $\Gamma(V) = \mathbb{Z}_2$, it is easy to check that $\Gamma_2(V) = \mathbb{Z}_2$ (here we have set $\Gamma_2(V) = H^2(V; \mathbb{Z}_2)/H^2_{\mathbb{C}-\mathrm{alg}}(V; \mathbb{Z}_2)$). Considering the two possible cases in Corollary 5.3.29, it is easy to see that $\Gamma_2(B_P V) = 0$ and hence $i^*\colon H^2(B_P V_{\mathbb{C}}; \mathbb{Z}_2) \to H^2(B_P V; \mathbb{Z}_2)$ is surjective. We complete the proof as in the previous case. □

Corollary 5.3.33 *Let V be a non singular compact geometrically rational real algebraic surface. If V is not connected then the space $\mathcal{R}(B_P V, \mathbb{S}^2)$ is dense in $\mathcal{C}^\infty(B_P V, \mathbb{S}^2)$ for any blow up $B_P V$ centred at a point P contained in V.*

Calculating $\Gamma(V)$

Given Theorem 5.3.18, Theorem 5.3.13 is a consequence of the following theorem.

Theorem 5.3.34 *Let V be a non singular compact geometrically rational real algebraic surface. We then have that*

$$\Gamma(V) = \begin{cases} \mathbb{Z} & \text{if } V \approx \mathbb{S}^1 \times \mathbb{S}^1 \\ \mathbb{Z}_2 & \text{if } V \text{ satisfies the hypotheses of Theorem 5.3.35} \\ 0 & \text{otherwise.} \end{cases}$$

Sketch Proof As we saw in Proposition 5.3.20, the quotient $\Gamma(V)$ is not a birational invariant.

We prove this theorem using the classification of relatively minimal models over $\mathbb{R}$ presented in Theorem 4.3.23. The case where V is connected has already been proved (Theorems 5.3.10 and 5.3.18): the remaining possible models are conic bundles and certain real del Pezzo surfaces of degrees 2 and 1. In each case, the group $\Gamma(V)$ can be determined using Theorem 5.3.35, Propositions 5.3.42 and 5.3.43. □

Theorem 5.3.11 is then a consequence of the following result.

Theorem 5.3.35 *Let V be a non singular real affine algebraic surface. If V is diffeomorphic to a disjoint union of four spheres and has a complexification $V_{\mathbb{C}}$ which is a degree 2 del Pezzo surface then $\Gamma(V) = \mathbb{Z}_2$.*

Remark 5.3.36 The following statement is equivalent to Theorem 5.3.35. Let (X, σ) be a non singular projective $\mathbb{R}$-surface such that $X(\mathbb{R}) \approx \sqcup^4 \mathbb{S}^2$ and which has an $\mathbb{R}$-minimal model (X^0, σ^0) which is a degree 2 del Pezzo surface. We then have that $\Gamma(X(\mathbb{R})) = \mathbb{Z}_2$ (Figure 5.1).

We need some auxiliary results before giving the proof of the theorem. Let $i\colon V \hookrightarrow V_{\mathbb{C}}$ be the canonical injection of a non singular compact real affine algebraic surface into a non singular projective complexification. The differentiable manifolds $V_{\mathbb{C}}$ and V are compact and $V_{\mathbb{C}}$ is orientable. If V is also orientable we can define the Gysin morphism $i_!$ via the commutative diagram:

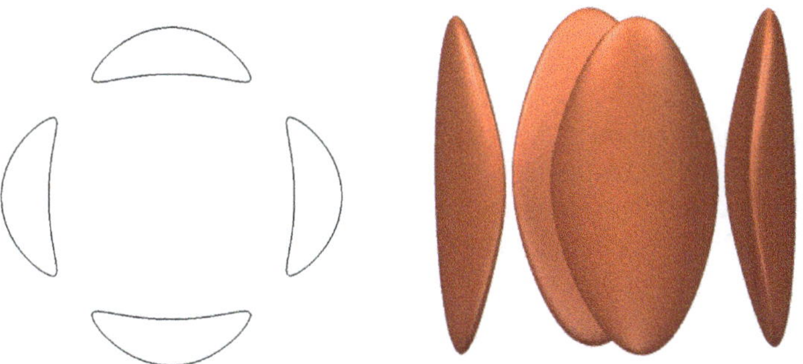

Fig. 5.1 Degree 2 del Pezzo surface defined by Equation (4.6), Section 4.2. See also Figure 4.4

$$\begin{array}{ccc} H^2(V_{\mathbb{C}};\mathbb{Z}) & \xrightarrow{i^*} & H^2(V;\mathbb{Z}) \\ {\scriptstyle D_{V_{\mathbb{C}}}^{-1}}\big\uparrow\simeq & & {\scriptstyle D_V^{-1}}\big\uparrow\simeq \\ H_2(V_{\mathbb{C}};\mathbb{Z}) & \xrightarrow{i_!} & H_0(V;\mathbb{Z}) \end{array} \tag{5.2}$$

where $D_{V_{\mathbb{C}}}$ (resp. D_V) is the Poincaré duality isomorphism (Proposition 3.1.8) of the compact oriented topological manifold $V_{\mathbb{C}}$ (resp. V).

Remark 5.3.37 If V is non orientable we can use a similar argument with coefficient group $\mathbb{Z}_2$.

Lemma 5.3.38 *Let S be a closed differentiable submanifold of dimension 2 in $V_{\mathbb{C}}$ which is transverse to V. We then have that*

$$i_!([S]) = [S \pitchfork V] .$$

Proof See Section B.7. □

We can therefore describe the image of the fundamental class of certain complex curves under the Gysin morphism $i_!\colon H_2(V_{\mathbb{C}};\mathbb{Z}) \longrightarrow H_0(V;\mathbb{Z})$ in the following way. In the underlying differentiable manifold structure $V_{\mathbb{C}}$ is of dimension 4 and V is of dimension 2. Let $L \subset V_{\mathbb{C}}$ be a complex algebraic curve. Outside of a finite number of points L is also a topological submanifold of dimension 2 in $V_{\mathbb{C}}$ and if L is transverse to V in $V_{\mathbb{C}}$ then we have that

$$i_!([L]) = [L \pitchfork V] .$$

Proposition 5.3.39 *Let V be a non singular compact geometrically rational real algebraic surface. If V is orientable then*

$$\Gamma(V) \simeq H_0(V;\mathbb{Z})/\operatorname{Im} i_! .$$

Proof By Diagram (5.2), we have that $\operatorname{Im} i^* \simeq \operatorname{Im} i_!$ and the isomorphism follows from Corollary 5.3.22. □

Sketch Proof of Theorem 5.3.35 We want to calculate the image of $i_!$. If (X, σ) is a degree 2 real del Pezzo surface then the homology group $H_2(X; \mathbb{Z})$ is generated by classes of exceptional curves and the hyperplane section. The anti-canonical map $\varphi_{-K_X} : X \to \mathbb{P}^2(\mathbb{C})$ is a double cover of the plane branched along a quartic $\mathbb{R}$-curve Δ. The (-1)-curves of X are sent by φ_{-K_X} to the bitangents of Δ. When $X(\mathbb{R}) \approx \bigsqcup 4\mathbb{S}^2$, the $\mathbb{R}$-curve $(\Delta, \sigma_{\mathbb{P}}|_\Delta)$ is maximal, by which we mean that $\Delta(\mathbb{R}) \approx \bigsqcup 4\mathbb{S}^1$, and all the bitangents are real.

We complete the proof by showing that the (-1)-curves in X are non real and are transverse to $X(\mathbb{R})$ in X. The theorem follows from our detailed knowledge of the possible configurations of bitangents of a plane quartic—see [Zeu74] for more details. The interested reader will find the details of this proof in [JPM04, Section 5]. □

Example 5.3.40 Let $(\Delta, \sigma_{\mathbb{P}}|_\Delta) \subset (\mathbb{P}^2(\mathbb{C}), \sigma_{\mathbb{P}})$ be a non singular plane $\mathbb{R}$-curve of even degree and let $\varphi\colon (X, \sigma) \to (\mathbb{P}^2(\mathbb{C}), \sigma_{\mathbb{P}})$ be one of the two real double covers branched along Δ (see the proof of Theorem 3.3.14). Let L be an $\mathbb{R}$-line tangent to Δ and let $P \in \Delta \cap L$ be a simple tangency point. In an analytic neighbourhood of P the surface X is defined by one of the two equations $z^2 = \pm(y - x^2)$ and in the plane $z = 0$ the curves Δ and L are given by equations $y = x^2$ and $y = 0$.

Let $C := \varphi^{-1}(L)$ be the inverse image of L in X. Locally analytically in a neighbourhood of P in X the complex curve C has equation $\{y = 0,\ (z - x)(z + x) = 0\}$ or $\{y = 0,\ (z - ix)(z + ix) = 0\}$. Globally, C can be either reducible or irreducible. Suppose that C is reducible. In this case C decomposes as $C = E + \tau E$ where τ is the involution of the double cover. For one of the two real structures, E and τE are defined over $\mathbb{R}$ and for the other they are conjugate complex curves. In the latter case, since the complex surface X is non singular, the local ring of regular functions is factorial and E (resp. τE) has a local equation $\{y = 0,\ z - ix = 0\}$ (resp. $\{y = 0,\ z + ix = 0\}$). The tangent plane at P of the topological surface $X(\mathbb{R})$ in the manifold X of real dimension 4 is generated by $\frac{\partial}{\partial x_1}$ and $\frac{\partial}{\partial z_1}$ where $x = x_1 + ix_2$, $y = y_1 + iy_2$ and $z = z_1 + iz_2$. It is easy to check that the tangent plane at P of the topological surface E in X is generated by $\frac{\partial}{\partial x_1} + i\frac{\partial}{\partial x_2}$ and $i\frac{\partial}{\partial z_1} - \frac{\partial}{\partial z_2}$. It follows that E is transverse to $X(\mathbb{R})$ at the point P in X.

Remark 5.3.41 The key point here is the reducibility of $i^{-1}(L)$, which is global information.

Proposition 5.3.42 *Let V be a non singular real affine algebraic surface which is diffeomorphic to the disjoint union of four spheres and a real projective plane and which has a complexification $V_{\mathbb{C}}$ which is a degree 1 del Pezzo surface. The quotient group $\Gamma(V)$ is then trivial.*

Proof See [JPM04, Theorem 6.1]. □

Proposition 5.3.43 *Let V be a non singular compact geometrically rational real algebraic surface. Assume that $V_{\mathbb{C}}$ has a conic bundle structure. If V is not connected, or in other words if V is not rational over $\mathbb{R}$, then $\Gamma(V) = 0$.*

Proof See [JPM04, Proposition 4.2]. □

Sketch Proof of Theorem 5.3.14 (*See* [Man06, Theorem 1.1] *for a full proof*) Let V be a non singular real algebraic surface with a complexification that has a conic bundle structure over a non rational base. We generalise Proposition 5.3.39 to the non geometrically rational and non orientable case. This reduces the problem to an examination of the incidence relations between V and real or complex curves in $V_{\mathbb{C}}$. The Néron Severi group $NS(V_{\mathbb{C}})$ is generated by the class of a fibre, the class of a (not necessarily real) section and by the classes of those complex (-1)-curves that meet their conjugates. It is easy to deal with the (real) fibre class. We deal with the (-1)-curves using a transversality argument similar to the one used for del Pezzo surfaces. The tricky part of the proof is dealing with the class of a section. □

5.3.3 Regulous Functions

This subsection is based on Section 5 of the survey article [Man17a].

Approximation of smooth maps by regular maps is still an open problem in general. For example, the question of the existence of algebraic representants in each homotopy class of continuous maps between spheres is not completely solved at the time of writing—see [BCR98, Chapter 13] for more details. Here is an example of the kind of theorem that can be proved: if n is a power of two and $p < n$ then any polynomial map from $\mathbb{S}^n$ is $\mathbb{S}^p$ is constant. See [BCR98, Theorem 13.1.9] for more details.

In [Kuc09], Kucharz introduced the notion of *continuous rational maps* generalising regular maps between real algebraic varieties. The special case of continuous rational functions was also studied by Kollár and Nowak—see [KN15] for more details. Continuous rational maps between non singular real algebraic varieties are now often called *regulous maps* as in [FHMM16]. If the variety is singular, these two notions can differ (see [KN15] or [Mon18]).

Definition 5.3.44 Let V and W be non singular geometrically irreducible real affine algebraic varieties. A *regulous map* from V to W is a rational map $f : V \dashrightarrow W$ satisfying the following property: if $U \subset V$ is the domain of f then the restriction of f to U can be extended to a map from V to W which is continuous in the Euclidean topology.

Kucharz proved that all homotopy classes of continuous maps between spheres contain regulous maps.

Theorem 5.3.45 *Let n and p be two strictly positive integers. Any continuous map from $\mathbb{S}^n$ to $\mathbb{S}^p$ is homotopic to a regulous map.*

Proof See [Kuc09, Theorem 1.1]. □

The article [FHMM16] laid the foundations for *regulous geometry*, namely the definitions and basic properties of the algebra of regulous functions and the associated topology, which we now summarise. Recall that a rational function f on $\mathbb{R}^n$ is said to be *regular* on $\mathbb{R}^n$ if and only if f has no poles in $\mathbb{R}^n$ (Definition 1.2.35): for example, the rational function $f(x) = 1/(x^2+1)$ is regular on $\mathbb{R}$. The set of regular functions on $\mathbb{R}^n$ is a subring of the field $\mathbb{R}(x_1, \dots, x_n)$ of rational functions on $\mathbb{R}^n$. A *regulous function* on $\mathbb{R}^n$ is a real valued function on $\mathbb{R}^n$ which is continuous in the Euclidean topology and whose restriction to some non empty Zariski open subset is regular. A typical example is the function defined by

$$f(x, y) = \frac{x^3}{x^2+y^2}, \quad f(0,0) = 0$$

which is regular on $\mathbb{R}^2 \setminus \{0\}$ and regulous on $\mathbb{R}^2$. Its graph is the head of the famous *Cartan umbrella* met in Chapter 1, Figure 1.5.

The set of regulous functions on $\mathbb{R}^n$ is a subring $\mathcal{R}^0(\mathbb{R}^n)$ of the field $\mathbb{R}(x_1, \dots, x_n)$. More generally, a real valued function defined on $\mathbb{R}^n$ is said to be *k-regulous* if and only if it is regular on a non empty Zariski open set and $\mathcal{C}^k$ on $\mathbb{R}^n$. Here, $k \in \mathbb{N} \cup \{\infty\}$. For example, the function defined by

$$f(x, y) = \frac{x^{3+k}}{x^2+y^2}$$

is k-regulous on $\mathbb{R}^2$ for any natural number $k' \leq k$. By [FHMM16, Théorème 3.3], any ∞-regulous function on $\mathbb{R}^n$ is in fact regular (the converse is immediate) and this gives us an infinite chain of subrings

$$\mathcal{R}^\infty(\mathbb{R}^n) \subseteq \cdots \subseteq \mathcal{R}^2(\mathbb{R}^n) \subseteq \mathcal{R}^1(\mathbb{R}^n) \subseteq \mathcal{R}^0(\mathbb{R}^n) \subseteq \mathbb{R}(x_1, \dots, x_n).$$

where $\mathcal{R}^k(\mathbb{R}^n)$ is the subring of $\mathbb{R}(x_1, \dots, x_n)$ of k-regulous functions.

The *k-regulous topology* is the topology whose closed sets are the vanishing loci of k-regulous functions. Figure 5.2 below, which reproduces Figure 1.8 of Chapter 1, is the algebraic subset of $\mathbb{R}^3$ defined by the equation $x^2 + y^2\big((y-z^2)^2 + yz^3\big) = 0$.

This set is irreducible in the ∞-regulous topology but reducible in the k-regulous topology for any natural number k. The "horn" of the umbrella is closed in the 0-regulous topology because it is the vanishing locus of the regulous function defined by

$$(x, y, z) \mapsto z^2 \frac{x^2 + y^2\big((y-z^2)^2 + yz^3\big)}{x^2 + y^4 + y^2 z^4}.$$

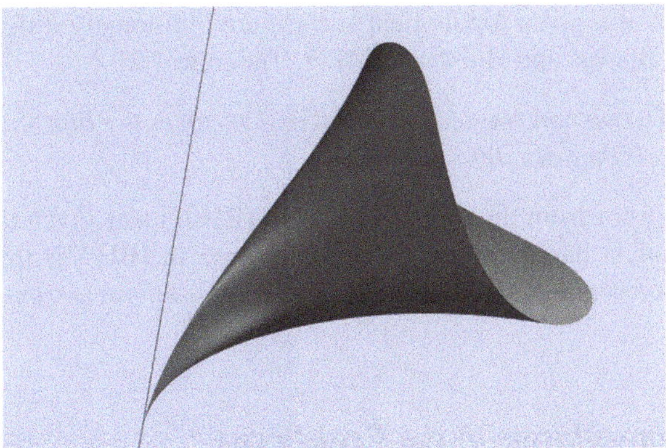

Fig. 5.2 Horned umbrella

The "handle" of the umbrella is also closed: it is the set of zeros of the function $(x, y, z) \mapsto x^2 + y^2$. The horned umbrella is therefore reducible in the regulous topology. See [FHMM16, Exemple 6.12] for more details.

Several properties of the ring of regulous functions are established in the article [FHMM16], notably a strong Nullstellensatz. Their scheme theoretic properties are analysed and regulous versions of Cartan's A and B theorems are proved. A geometric characterisation of prime ideals in $\mathcal{R}^k(\mathbb{R}^n)$ via vanishing loci of regulous functions and a relationship between the k-regulous topology and the topology generated by Euclidean closed Zariski constructible sets are proved. There are many articles linked to this new field of research and we particularly recommend the following: [Kuc13, BKVV13, Kuc14a, Kuc14b, KK16, Kuc16a, Kuc16b, FMQ17, Now17, PP17, KK18, KKK18, Mon18].

5.4 Diffeomorphisms and Biregular Maps

5.4.1 Rational Models

Let M be a $\mathcal{C}^\infty$ differentiable manifold. Recall that a real algebraic model of M (Definition 5.1.1) is a non singular real affine algebraic variety V diffeomorphic to M.

By Comessatti's theorem 4.4.16, if a topological surface M has a *rational* real algebraic model then M is diffeomorphic to $\mathbb{S}^2$, $\mathbb{T}^2$ or a non orientable connected surface. It has been known for a long time that any rational real algebraic model of $\mathbb{S}^2$, $\mathbb{T}^2$ or $\mathbb{RP}^2$ is birationally diffeomorphic to $Q_{3,1}(\mathbb{R})$, $Q_{2,2}(\mathbb{R})$ or $\mathbb{P}^2(\mathbb{R})$ respectively. (See Example 4.2.19 for the notations). Answering a questions of Bochnak, it was proved in [Man06, Theorem 1.3] that the Klein bottle also has a unique rational algebraic model up to birational diffeomorphism. Surprisingly, it turns out that all rational models of a given topological surface are birationally diffeomorphic. This was proved by Biswas and Huisman [BH07, Theorem 1.2].

Theorem 5.4.1 *Two non singular rational real surfaces are birationally diffeomorphic if and only if they are diffeomorphic.*

A different proof from the original proof of [BH07] was given in [HM09]. This alternative proof is based on the fact (conjectured in [BH07]) that the group of birational diffeomorphisms from the sphere to itself is *infinitely transitive* (definition below).

5.4.2 Automorphisms of the Real Locus

The automorphism group of a complex projective algebraic variety is "small"—it is always finite dimensional and for most varieties it is finite. On the other hand, the group $\mathrm{Aut}(V)$ of birational diffeomorphisms of a real rational algebraic surface V (which are also called automorphisms of V) is rather "big", as the following result shows.

Definition 5.4.2 Let G be a group acting on a set M and let $n > 0$ be an integer. We say that G acts *n-transitively* on M if for any pair of n-uplets $(P_1, \dots, P_n)$ and $(Q_1, \dots, Q_n)$ of distinct elements of M there is an element g in G such that $g \cdot P_j = Q_j$ for every j. The group G is said to act *infinitely transitively*[7] on M if and only if for any strictly positive integer n its action is n-transitive on M.

Theorem 5.4.3 *Let V be a compact connected non singular rational real algebraic surface. For any natural number n the group* $\mathrm{Aut}(V)$ *acts n-transitively on V.*

Proof We refer to [HM09] for the full proof, which we now illustrate by showing how to construct many birational diffeomorphisms when V is the sphere $Q_{3,1}(\mathbb{R}) \simeq \mathbb{S}^2$. Let I be the interval $[-1, 1]$ in $\mathbb{R}$ and let $\mathbb{S}^1 \subset \mathbb{R}^2$ be the unit circle. Consider a regular map $f : I \to \mathbb{S}^1$: the two components of f are then rational functions of one variable without poles in I. We define a map, called the *twisting map* associated to f, given by $\phi_f : \mathbb{S}^2 \to \mathbb{S}^2$

$$\phi_f(x, y, z) = (f(z) \cdot (x, y), z)$$

where $\cdot$ is complex multiplication in $\mathbb{R}^2 = \mathbb{C}$. In other words, $(x, y) \mapsto f(z) \cdot (x, y)$ is a rotation in the $\mathbb{R}_{x,y}$ plane depending algebraically on z. The map ϕ_f is a birational diffeomorphism from $\mathbb{S}^2$ to itself. Its inverse is ϕ_g where $g : I \to \mathbb{S}^1$ sends z to the multiplicative inverse $(f(z))^{-1}$ of $f(z)$. Now, consider n distinct points $z_1, \dots, z_n$ in I and let $\rho_1, \dots, \rho_n$ be elements in $\mathbb{S}^1$. By Lagrange's interpolation theorem, there is a regular map $f : I \to \mathbb{S}^1$ such that $f(z_j) = \rho_j$ for any $j = 1, \dots, n$. Multiplication

[7] In the literature an *infinitely* transitive group action is sometimes said to be *very* transitive.

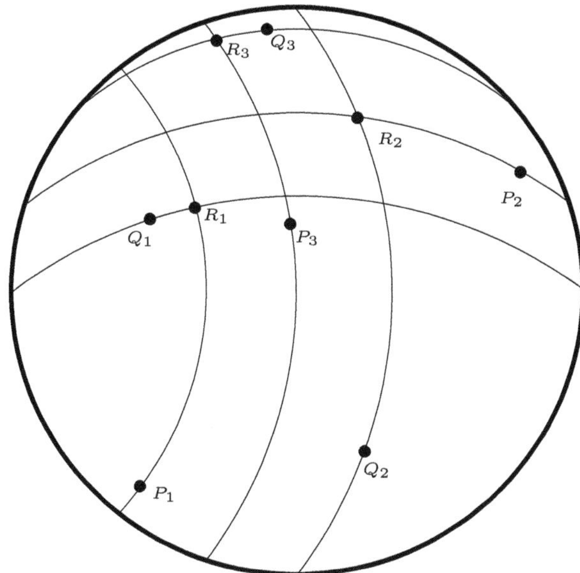

Fig. 5.3 The sphere $\mathbb{S}^2$ with two parallel families of lines

by ρ_j is then a rotation in the plane $z = z_j$: there is therefore a twisting map ϕ_f which *moves* n distinct points $P_1, \ldots, P_n$ on the sphere to n given points $R_1, \ldots, R_n$ provided that any pair of points P_j, R_j with the same value of j belong to the same horizontal plane ($z = const$). Let $(P_1, \ldots, P_n)$ and $(Q_1, \ldots, Q_n)$ be n-uplets of distinct elements of $\mathbb{S}^2$. We get a birational diffeomorphism from $\mathbb{S}^2$ to itself sending P_j to Q_j by simply considering two transverse families of parallel planes to obtained n intersection points R_j as in Figure 5.3. The fact that the transverse families of parallel planes can be chosen in such a way that the lines of intersection meet the sphere is established in [HM09, Theorem 2.3].

After linear change of coordinates the above construction yields two twisting maps, one which sends P_j to R_j for $j = 1, \ldots, n$ and the second which sends R_j to Q_j for $j = 1, \ldots, n$. The composition of these two maps is the automorphism we seek. □

Remark 5.4.4 By induction on the dimension we can prove using the same construction that the group $\mathrm{Aut}(\mathbb{S}^d)$ where $\mathbb{S}^d$ is the quadric hypersurface $Q_{d+1,1}(\mathbb{R}) := \mathcal{Z}(x_1^2 + x_2^2 + \cdots + x_{d+1}^2 - x_0^2) \subset \mathbb{P}^{d+1}_{x_0:x_1:\cdots:x_{d+1}}(\mathbb{R})$ acts infinitely transitively on $\mathbb{S}^d$ for any $d > 1$.

The above theorem can be generalised to any real algebraic surface in the following form. The action of the group $\mathrm{Aut}(V)$ on V is said to be infinitely transitive *on each connected component* if and only if for any $n \geq 1$ and any pair of n-tuplets $(P_1, \ldots, P_n)$ and $(Q_1, \ldots, Q_n)$ of distinct elements of V such that for any j, P_j and Q_j belong to the same connected component of V, there is a birational diffeomorphism $f : V \to V$ such that $f(P_j) = Q_j$.

Theorem 5.4.5 *Let V be a non singular compact real affine algebraic surface. The group* $\mathrm{Aut}(V)$ *of birational diffeomorphisms from V to V is infinitely transitive on each connected component of V if and only if $\#\pi_0(V) \leqslant 3$ and V is geometrically rational.*

Proof See [BM11, Proposition 41]. □

Remark 5.4.6 We mention two other transitivity results. By [HM10], if V is a compact connected real rational algebraic surface with certain types of singularities[8] then $\mathrm{Aut}(V)$ acts n-transitively on V for any n. By [KM12], if V is an affine suspension then the special linear group $\mathrm{SAut}(V)$ acts n-transitively on V for all n.

5.4.3 Cremona Groups of Real Surfaces

Recently, progress has been made in the study of sets of generators of the group $\mathrm{Aut}(V)$ for several specific rational surfaces V. These groups are conjugate to subgroups of the real Cremona group $\mathrm{Bir}_{\mathbb{R}}(\mathbb{P}^2)$ of birational transformations with real coefficients.

The Noether–Castelnuovo theorem [Cas01] (see [AC02, Chapter 8] for a modern presentation of the proof) describes a set of generators of the group $\mathrm{Bir}_{\mathbb{C}}(\mathbb{P}^2)$ of birational transformations of the complex projective plane. This group is generated by the biregular maps $\mathrm{Aut}_{\mathbb{C}}(\mathbb{P}^2) = \mathbf{PGL}(3, \mathbb{C})$ and the standard quadratic transformation

$$\sigma_0\colon (x : y : z) \dashrightarrow (yz : xz : xy).$$

This result does not hold over the real numbers. We recall that a *base point* of a birational transformation is a (possibly infinitely near) point at which the transformation is not defined, and we note that two of the base points of the quadratic involution

$$\sigma_1\colon (x : y : z) \dashrightarrow (y^2 + z^2 : xy : xz)$$

are not real. The involution σ_1 therefore cannot be constructed using real projective transformations and σ_0. More generally, we cannot obtain any transformation with non real base points in this way and it follows that the group $\mathrm{Bir}_{\mathbb{R}}(\mathbb{P}^2)$ of birational transformations of the real projective plane is not generated by $\mathrm{Aut}_{\mathbb{R}}(\mathbb{P}^2) = \mathbf{PGL}(3, \mathbb{R})$ and σ_0.

The main result of [BM14] is that $\mathrm{Bir}_{\mathbb{R}}(\mathbb{P}^2)$ is generated by $\mathrm{Aut}_{\mathbb{R}}(\mathbb{P}^2)$, σ_0, σ_1 and a family of birational transformations of degree 5 whose base points are all non real.

Example 5.4.7 (*Standard quintic transformation*) Let $p_1, \overline{p_1}, p_2, \overline{p_2}, p_3, \overline{p_3} \in \mathbb{P}^2(\mathbb{C})$ be three pairs of non real points on $\mathbb{P}^2(\mathbb{C})$ which do not lie on the same

[8] More precisely, V is assumed *dantesque*, by which we mean that V is a singular surface obtained from a non singular surface by weighted blow ups. See [HM10] for more details.

conic. Let $\pi\colon X \to \mathbb{P}^2(\mathbb{C})$ be the blow up of these three pairs of points: π induces a birational diffeomorphism $X(\mathbb{R}) \to \mathbb{P}^2(\mathbb{R})$. We note that the complex surface X thus obtained is isomorphic to a non singular conic in $\mathbb{P}^3(\mathbb{C})$. The set of strict transforms of conics passing through five of the six points is a family of three pairs of non real (-1)-curves (which are lines on the cubic), and these six curves are pairwise disjoint. The contraction of these six curves gives us a birational map $\eta\colon X \to \mathbb{P}^2(\mathbb{C})$ inducing a birational diffeomorphism $X(\mathbb{R}) \to \mathbb{P}^2(\mathbb{R})$ contracting the (-1)-curves to three pairs of non real points $q_1, \overline{q_1}, q_2, \overline{q_2}, q_3, \overline{q_3} \in \mathbb{P}^2(\mathbb{C})$. Permuting if necessary, we can assume that q_i is not in the image of the conic which avoids p_i. The birational map $\psi = \eta\pi^{-1}\colon \mathbb{P}^2(\mathbb{C}) \dashrightarrow \mathbb{P}^2(\mathbb{C})$ induces a birational diffeomorphism $\mathbb{P}^2(\mathbb{R}) \to \mathbb{P}^2(\mathbb{R})$.

Let $L \subset \mathbb{P}^2(\mathbb{C})$ be a general line in $\mathbb{P}^2(\mathbb{C})$. The strict transform of L on X under π^{-1} has self intersection 1 and meets each of the six curves contracted by η in two points because they derive from conics. The image $\psi(L)$ has six singular points of multiplicity 2 and self intersection 25: it is therefore a quintic with a double ordinary point at each of the points q_i. As the constructions of ψ^{-1} and ψ are symmetric, the linear system associated to ψ is formed of quintics in $\mathbb{P}^2(\mathbb{C})$ with an ordinary double point at each of the points $p_1, \overline{p_1}, p_2, \overline{p_2}, p_3, \overline{p_3}$.

We can moreover check that ψ sends the pencil of conics passing through $p_1, \overline{p_1}, p_2, \overline{p_2}$ to the pencil of conics passing through $q_1, \overline{q_1}, q_2, \overline{q_2}$, and similarly for the two other pencils of real conics, those passing through $p_1, \overline{p_1}, p_3, \overline{p_3}$ and those passing through $p_2, \overline{p_2}, p_3, \overline{p_3}$.

Definition 5.4.8 The degree 5 birational maps of $\mathbb{P}^2$ constructed in Example 5.4.7 are called the *standard quintic transformations* of $\mathbb{P}^2$.

Theorem 5.4.9 *The group* $\mathrm{Bir}_{\mathbb{R}}(\mathbb{P}^2)$ *is generated by* $\mathrm{Aut}_{\mathbb{R}}(\mathbb{P}^2)$, σ_0, σ_1 *and the standard quintic transformations of* $\mathbb{P}^2$.

Proof See [BM14, Theorem 1.1]. □

Remark 5.4.10 It has since been proved that the set of generators given above is essentially minimal: in particular, $\mathrm{Bir}_{\mathbb{R}}(\mathbb{P}^2)$ cannot be generated by $\mathrm{Aut}_{\mathbb{R}}(\mathbb{P}^2)$ and a countable set of elements. See [Zim18, Theorem 1.1].

The strategy used to prove Theorem 5.4.9 is based on a detailed study of *Sarkisov links*. This methods enables [BM14] to study several natural subgroups of $\mathrm{Bir}_{\mathbb{R}}(\mathbb{P}^2)$ in a similar way: in particular, they recover in this article the system of generators of $\mathrm{Aut}(\mathbb{P}^2(\mathbb{R}))$ given in [RV05, Teorema II] and the system of generators of $\mathrm{Aut}(Q_{3,1}(\mathbb{R}))$ given in [KM09, Thm. 1].

Theorem 5.4.11 *For a given* $\mathbb{R}$*-variety* (X, σ), $\mathrm{Aut}(X(\mathbb{R}))$ *denotes the group of birational diffeomorphisms of the real locus to itself.*

1. *The group* $\mathrm{Aut}(\mathbb{P}^2(\mathbb{R}))$ *is generated by* $\mathrm{Aut}_{\mathbb{R}}(\mathbb{P}^2) = \mathbf{PGL}(3, \mathbb{R})$ *and the standard quintic transformations (see Example 5.4.7 above).*
2. *The group* $\mathrm{Aut}(Q_{3,1}(\mathbb{R}))$ *is generated by* $\mathrm{Aut}_{\mathbb{R}}(Q_{3,1}) = \mathbf{PO}(3, 1)$ *and the standard cubic transformations (see Example 5.4.12 below).*

3. *The group* $\mathrm{Aut}(\mathbb{F}_0(\mathbb{R}))$ *is generated by* $\mathrm{Aut}_{\mathbb{R}}(\mathbb{F}_0) \simeq \mathbf{PGL}(2,\mathbb{R})^2 \rtimes \mathbb{Z}_2$ *and the involution*

$$\tau_0\colon ((x_0:x_1),(y_0:y_1)) \dashrightarrow ((x_0:x_1),(x_0y_0+x_1y_1:x_1y_0-x_0y_1))\ .$$

Note that $\mathrm{Aut}(Q_{3,1}(\mathbb{R}))$ and $\mathrm{Aut}(\mathbb{F}_0(\mathbb{R}))$ are not really subgroups of $\mathrm{Bir}_{\mathbb{R}}(\mathbb{P}^2)$, but each of them is isomorphic to a subgroup which is determined up to conjugaison. For any choice of birational map $\psi\colon \mathbb{P}^2 \dashrightarrow X$ ($X = Q_{3,1}$ or $\mathbb{F}_0$) we have that

$$\psi^{-1}\,\mathrm{Aut}(X(\mathbb{R}))\psi \subset \mathrm{Bir}_{\mathbb{R}}(\mathbb{P}^2)\ .$$

Example 5.4.12 (*Standard cubic transformations*) Let $p_1, \overline{p_1}, p_2, \overline{p_2} \in Q_{3,1} \subset \mathbb{P}^3$ be two pairs of non real points which are conjugate and not coplanar. Let $\pi\colon X \to Q_{3,1}$ be the blow up of these four points. The non real plane in $\mathbb{P}^3$ passing through $p_1, \overline{p_1}, \overline{p_2}$ meets $Q_{3,1}$ in a conic C of self intersection 2, since two general conics on $Q_{3,1}$ are hyperplane intersections and therefore meet in two points lying on the intersection line of the hyperlanes. The strict transform of this conic C on X is therefore a (-1)-curve. Proceeding similarly with all the conics passing through three points out of $p_1, \overline{p_1}, p_2, \overline{p_2}$, we obtain four disjoint (-1)-curves on X, which can be contracted to get a birational morphism $\eta\colon X \to Q_{3,1}$. The surface thus obtained is indeed $Q_{3,1}$ because it is a non singular rational projective surface with real Picard number 1. This gives us a birational map $\psi = \eta\pi^{-1}\colon Q_{3,1} \dashrightarrow Q_{3,1}$ which induces a birational diffeomorphism $Q_{3,1}(\mathbb{R}) \to Q_{3,1}(\mathbb{R})$.

Let $H \subset Q_{3,1}$ be a general hyperplane section: the strict transform of H on X under π^{-1} has an intersection 2 with each of the four (-1)-curves. The image $\psi(H)$ has four points of multiplicity 2 and self intersection 18: it is therefore a cubic section. As the constructions of ψ and ψ^{-1} are symmetric, the linear system of ψ is formed of cubic sections of multiplicity 2 at $p_1, \overline{p_1}, p_2, \overline{p_2}$.

Definition 5.4.13 The degree three birational maps of $\mathbb{P}^2$ constructed in Example 5.4.12 are called the *standard cubic transformations* of $\mathbb{P}^2$.

We refer to [Rob16], [Yas16], [RZ18] and [Zim18] for recent results on the Cremona group $\mathrm{Bir}_{\mathbb{R}}(\mathbb{P}^2)$. In particular, the abelianisation of $\mathrm{Bir}_{\mathbb{R}}(\mathbb{P}^2)$ has been calculated: the result is particularly surprising as the complex Cremona group $\mathrm{Bir}_{\mathbb{C}}(\mathbb{P}^2)$ is a perfect group. See [CD13] for more details. We recall that a group is perfect if and only if it is equal to its derived subgroup or in other words if its abelianisation (Definition B.3.8) is trivial.

$$\mathrm{Bir}_{\mathbb{C}}(\mathbb{P}^2)/[\mathrm{Bir}_{\mathbb{C}}(\mathbb{P}^2),\mathrm{Bir}_{\mathbb{C}}(\mathbb{P}^2)] \simeq \{\mathrm{id}\}\ .$$

Theorem 5.4.14 *The abelianisation of* $\mathrm{Bir}_{\mathbb{R}}(\mathbb{P}^2)$ *is isomorphic to*

$$\mathrm{Bir}_{\mathbb{R}}(\mathbb{P}^2)/[\mathrm{Bir}_{\mathbb{R}}(\mathbb{P}^2),\mathrm{Bir}_{\mathbb{R}}(\mathbb{P}^2)] \simeq \bigoplus_{\mathbb{R}} \mathbb{Z}_2\ .$$

Proof See [Zim18, Theorem 1.2]. □

5.4.4 Density of Birational Diffeomorphisms

One of the most famous applications of Nash's theorem (Introduction) is the Artin–Mazur theorem below. For any endomorphism $f\colon M \to M$ of a compact differentiable manifold M of class $\mathcal{C}^\infty$ without boundary we denote by $N_\nu(f)$ the number of *isolated* periodic points of f of period ν (i.e. the number of isolated fixed points of f^ν).

Theorem 5.4.15 *Let M be a compact $\mathcal{C}^\infty$ manifold without boundary*[9] *and let $\mathcal{C}^\infty(M) := \mathcal{C}^\infty(M, M)$ be the space of $\mathcal{C}^\infty$ endomorphisms equipped with the $\mathcal{C}^\infty$ topology. There is a dense subspace $A \subset \mathcal{C}^\infty(M)$ such that if $f \in A$ then $N_\nu(f)$ grows at most exponentially with ν, or in other words there is a constant $c := c(f) < +\infty$ such that*

$$N_\nu(f) \leqslant c^\nu \quad \text{for any } \nu \geqslant 1.$$

Proof See [AM65]. □

The proof of the Artin–Mazur theorem uses the fact that any $\mathcal{C}^\infty$ endomorphism of M can be approximated by *Nash diffeomorphisms* (Definition B.2.4). The reader should be aware of an important difference between Nash diffeomorphisms and birational diffeomorphisms. A diffeomorphism which is a rational map without real poles is a Nash diffeomorphism but is not necessarily a birational diffeomorphism because there is no guarantee the inverse map is rational. For example, the map $x \mapsto x + x^3$ is a Nash diffeomorphism from $\mathbb{R}$ to itself but it is not birational because the inverse map is written using radicals. This phenomenon arises because the inverse function theorem holds in the analytic category but not in the algebraic category. It is then natural to ask the following question: for a given non singular real algebraic variety V, is the group $\mathrm{Aut}(V)$ of birational diffeomorphisms dense in the group $\mathrm{Diff}(V)$ of diffeomorphisms from V to itself? The answer is yes for rational surfaces.

Theorem 5.4.16 *Let S be a connected compact topological surface without boundary and let* $\mathrm{Diff}(S)$ *be its group of diffeomorphisms with the C^∞ topology.*

If S is non orientable or of genus $g(S) \leq 1$ then there is a real rational model V of S such that

$$\overline{\mathrm{Aut}\big(V\big)} = \mathrm{Diff}\big(V\big) \simeq \mathrm{Diff}\big(S\big)$$

or in other words $\mathrm{Aut}\big(V\big)$ *is a dense subgroup* $\mathrm{Diff}\big(V\big)$ *in the C^∞ topology.*

Proof See [KM09, Theorem 4]. □

[9] In fact, this theorem holds for $\mathcal{C}^k$ manifolds for any $k = 1, \ldots, \infty$.

Remark 5.4.17 If S is orientable of genus $g(S) \geq 2$ then for any real algebraic model V of S we have that $\overline{\mathrm{Aut}(V)} \neq \mathrm{Diff}(V)$. Let V be a compact connected orientable non singular real affine algebraic surface and let $V_{\mathbb{C}}$ be a *minimal* non singular complex projectivisation of V. Such a complexification exists because V is orientable. By the classification of $\mathbb{R}$-surfaces (Chapter 4), we are in one of the following situations:

1. If $\kappa(V) = -\infty$ then $V \approx \mathbb{S}^2$ or $V \approx \mathbb{S}^1 \times \mathbb{S}^1$;
2. If $V_{\mathbb{C}}$ is a K3 surface or an abelian surface, $\kappa(V) = 0$, then $\mathrm{Aut}(V)$ preserves a volume form;
3. If $V_{\mathbb{C}}$ is an Enriques surface or a bi-elliptic surface, $\kappa(V) = 0$, then it has a finite covering by a K3 surface or an elliptic surface;
4. If $V_{\mathbb{C}}$ is a properly elliptic surface, $\kappa(V) = 1$, then $\mathrm{Aut}(V)$ preserves the canonical elliptic fibration;
5. If V is of general type, $\kappa(V) = 2$, then $\mathrm{Aut}(V)$ is finite. See [Uen75], for a proof.

In short, if $g(S) > 1$ then for any real algebraic model V of S density fails.

The above theorem has been generalised to geometrically rational surfaces.

Theorem 5.4.18 *Let V be a geometrically rational compact surface. We then have that*

1. *If* $\#\pi_0(V) > 4$ *then* $\mathrm{Aut}(V)$ *is not dense in* $\mathrm{Diff}(V)$.
2. *If* $\#\pi_0(V) = 3$ *or* 4 *then* $\mathrm{Aut}(V)$ *may not be dense in* $\mathrm{Diff}(V)$.

Proof Voir [BM11, Proposition 41]. □

At the time of writing there are still cases of varieties for which $\#\pi_0(V) = 3$ or $\#\pi_0(V) = 4$ and we do not know whether $\mathrm{Aut}(V)$ is dense in $\mathrm{Diff}(V)$ or not. Similarly, the case where there are two connected components is still open.

5.4.5 Approximation by Rational Curves

We end this section with an important application of the density Theorem 5.4.16. We start with an example.

Example 5.4.19 Consider the rational curve parameterised by

$$\begin{aligned} f\colon \mathbb{R} &\longrightarrow \mathbb{R}^2 \\ t &\longmapsto \big(t^2+1, t(t^2+1)\big) \end{aligned}$$

whose image set is represented in Figure 5.4.

Prolonging f to the compactification $\mathbb{P}^1(\mathbb{R})$ of $\mathbb{R}$ and composing with a birational map $\mathbb{R}^2 \dashrightarrow V$ to a rational surface V, we get a regular map $\mathbb{P}^1(\mathbb{R}) \to V$, or in other words a rational curve in V.

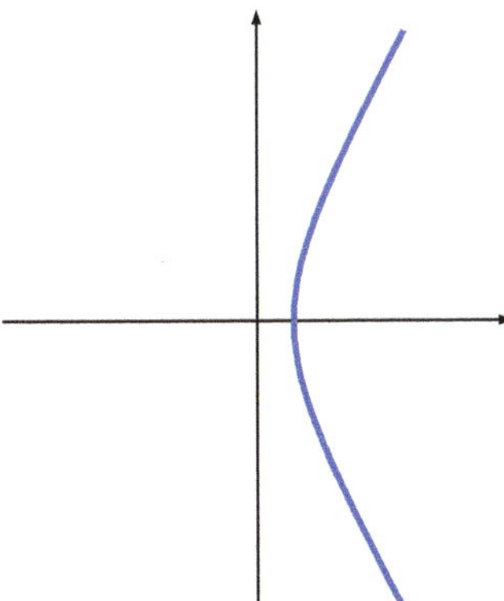

Fig. 5.4 Image of $\mathbb{R}$ under $f: t \longmapsto \left(t^2+1, t(t^2+1)\right)$

J. Bochnak and W. Kucharz proved that all differentiable maps $\mathbb{S}^1 \to V$ from the circle $\mathbb{S}^1 \approx \mathbb{P}^1(\mathbb{R})$ to a rational variety by rational curves can be approximated by rational curves.

Definition 5.4.20 Let $f: \mathbb{S}^1 \to V$ be a $\mathcal{C}^\infty$ map. We say that f can be *approximated by rational curves* if every neighbourhood of f in $\mathcal{C}^\infty\left(\mathbb{S}^1, V\right)$ contains a regular map $\mathbb{P}^1(\mathbb{R}) \to V$.

Theorem 5.4.21 *Let V be a non singular real rational algebraic variety. Every $\mathcal{C}^\infty$ map $f: \mathbb{S}^1 \to V$ can then be approximated by rational curves.*

Proof See [BK99, Theorem 1.1]. □

Note that the rational curves in this theorem are parameterisations: the Zariski closure of the image curve may contain extra isolated points.

Example 5.4.22 (*Continuation of*5.4.19) The map f can be naturally extended to the complexifications $\mathbb{C}$ of $\mathbb{R}$ and $\mathbb{C}^2$ of $\mathbb{R}^2$.

$$\begin{aligned} f: \mathbb{C} &\longrightarrow \mathbb{C}^2 \\ t &\longmapsto \left(t^2+1, t(t^2+1)\right) . \end{aligned}$$

The image set then contains an extra real point (Figure 5.5) and in particular

$$f(\mathbb{R}) \subsetneq f(\mathbb{C}) \cap \mathbb{R}^2 .$$

Extending the map f in the above example to $\mathbb{P}^1$ we get a map $f: \mathbb{P}^1 \to \mathbb{P}^2$, $(u, v) \longrightarrow \left(v(u^2+v^2), u(u^2+v^2), v^3\right)$. The image $f(\mathbb{P}^1(\mathbb{R}))$ is a simple closed curve in $\mathbb{P}^2(\mathbb{R})$ but its Zariski closure, the nodal cubic $\mathcal{Z}\left((x^2+y^2)z - x^3\right) \subset \mathbb{P}^2(\mathbb{R})$, has an isolated real point at $(0, 0, 1)$. We can remove this point by deforming the equation to $\mathcal{Z}\left(z(x^2+y^2+\varepsilon^2 z^2) - x^3\right)$ or blowing up the point $(0, 0, 1)$, but the first modification would render the curve elliptic and the second would change the

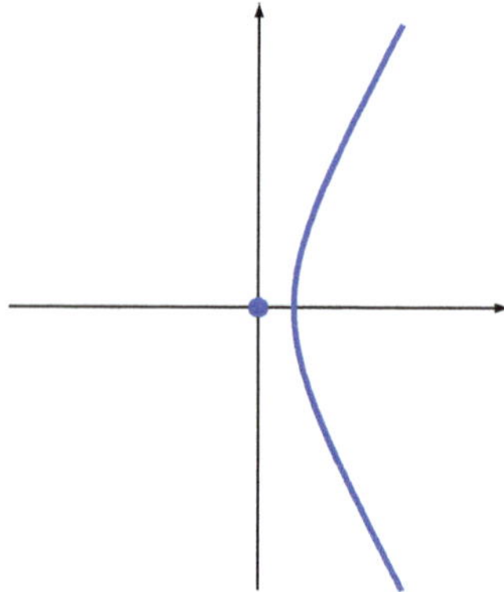

Fig. 5.5 Real locus of the image of $\mathbb{C}$ under $f : t \longmapsto \big(t^2+1, t(t^2+1)\big)$

topology of the surface. We show below how to get rid of these isolated points using Theorem 5.4.16.

Let (X, σ) be a projective non singular algebraic $\mathbb{R}$-variety and let $(C, \sigma|_C) \subset (X, \sigma)$ be a rational $\mathbb{R}$-curve. Choosing an isomorphism between the normalisation $\widetilde{C}$ of C and the plane conic $\mathcal{Z}(x^2+y^2-z^2) \subset \mathbb{P}^2$, we get a $\mathcal{C}^\infty$ map $\mathbb{S}^1 \to X(\mathbb{R})$ whose image coincides with $C(\mathbb{R})$ away from its singular real points. We call such curves *real-smooth*.

Definition 5.4.23 Let $f : L \hookrightarrow X(\mathbb{R})$ be an embedded circle. We say that L has $\mathcal{C}^\infty$ *approximation by real-smooth rational curves* if and only if every neighbourhood of f in $C^\infty\big(\mathbb{S}^1, X(\mathbb{R})\big)$ contains a map $\mathbb{S}^1 \to X(\mathbb{R})$ defined as above using a rational curve C without isolated singular real points.

It is not always possible to approximate an embedded circle by rational curves that are non singular at their real points. For example, this is impossible for a homotopically[10] trivial circle on the torus $\mathbb{T}^2 = \mathbb{P}^1(\mathbb{R}) \times \mathbb{P}^1(\mathbb{R})$.

Proposition 5.4.24 *Let $C \subset \mathbb{P}^1(\mathbb{C}) \times \mathbb{P}^1(\mathbb{C})$ be an algebraic $\mathbb{R}$-curve which is non singular at its real points. The fundamental class of its real locus*

$$[C(\mathbb{R})] \in H_1(\mathbb{T}^2; \mathbb{Z}_2)$$

is then non zero.

Proof Let E_1 (resp. E_2) be a vertical (resp. horizontal) complex line in $\mathbb{P}^1 \times \mathbb{P}^1$. The Picard group of $\mathbb{P}^1 \times \mathbb{P}^1$ is generated by the classes of E_1 and E_2. Any complex algebraic curve $D \subset \mathbb{P}^1 \times \mathbb{P}^1$ is therefore linearly equivalent to a linear combination $a_1E_1 + a_2E_2$ with $a_1, a_2 \geq 0$.

If D is an $\mathbb{R}$-curve in $\mathbb{P}^1 \times \mathbb{P}^1$ then

$$a_i = (D \cdot E_{3-i}) \equiv \big(D(\mathbb{R}) \cdot E_{3-i}(\mathbb{R})\big) \mod 2.$$

[10]Recall that $\pi_1(\mathbb{T}^2) \simeq \mathbb{Z} \oplus \mathbb{Z} \simeq H_1(\mathbb{T}^2; \mathbb{Z})$ and $H_1(\mathbb{T}^2; \mathbb{Z}_2) = H_1(\mathbb{T}^2; \mathbb{Z}) \otimes \mathbb{Z}_2$.

and if $[D(\mathbb{R})] = 0$ in $H_1(\mathbb{T}^2, \mathbb{Z}/2)$ then a_1 and a_2 are even. By the adjunction formula (Theorem 4.1.44) we have that

$$2p_a(D) - 2 = \big(a_1E_1 + a_2E_2\big) \cdot \big((a_1 - 2)E_1 + (a_2 - 2)E_2\big)$$
$$= a_1(a_2 - 2) + a_2(a_1 - 2)$$

and hence $p_a(D) = (a_1 - 1)(a_2 - 1)$. It follows that if a_1 and a_2 are even then the arithmetic genus $p_a(D)$ is odd. If D is rational then by Proposition 4.1.43 it has an odd number of singular points and at least one of them must be real. □

Surprisingly, this is the only example in which approximation does not hold.

Theorem 5.4.25 *An embedded circle L in a compact non singular real rational algebraic variety V has $\mathcal{C}^\infty$ approximation by real-smooth rational curves if and only if the pair (V, L) is not diffeomorphic to the pair $(\mathbb{T}^2, L_0)$ where L_0 is a contractible circle on the torus $\mathbb{T}^2$.*

Proof If $\dim V \geqslant 3$ the result can be deduced from Theorem 5.4.21 as follows. Let $f : \mathbb{S}^1 \to V$ be a $\mathcal{C}^\infty$ embedding of image L and let $V_\mathbb{C}$ be a non singular projective complexification of V. The proof of Theorem 5.4.21 (see [BK99] for more details) produces approximations of f by restriction to $\mathbb{P}^1(\mathbb{R})$ of morphisms $g : \mathbb{P}^1(\mathbb{C}) \to V_\mathbb{C}$ such that $g^*T_{V_\mathbb{C}}$ is ample. Now, if $\dim V \geqslant 3$, any general small perturbation of a morphism $g : \mathbb{P}^1(\mathbb{C}) \to V_\mathbb{C}$ such that $g^*T_{V_\mathbb{C}}$ is ample is an embedding ([Kol96, II.3.4]).

Suppose now that V is a surface: we give a sketch of the proof in this case and refer to [KM16, Theorem 3] for the full proof. First of all, if V_1, V_2 are diffeomorphic compact non singular real rational algebraic surfaces there is a birational diffeomorphism $g : V_1 \to V_2$ by Theorem 5.4.1. Suppose there is a rational curve $C \subset V$ which is non singular at real points and a diffeomorphism

$$\phi : \big(V, L\big) \overset{\approx}{\to} \big(V, C\big).$$

By Theorem 5.4.16, the diffeomorphism ϕ^{-1} can be approximated in the $\mathcal{C}^\infty$ topology by birational diffeomorphisms $\psi_n : V \to V$. It follows that

$$C_n := \psi_n(C) \subset V$$

is a sequence of rational curves and $(C_n)_n$ tends to L in the $\mathcal{C}^\infty$ topology. We can resolve the non real points of $C_\mathbb{C}$ to get an approximation of L by non singular rational curves $C'_n \subset V_n$. Here, the surfaces V_n are birationally diffeomorphic to V.

We complete the proof by giving a list of possible topological pairs and constructing for every pair except $(\mathbb{T}^2, L_0)$ a real rational model, by which we mean a non singular real rational algebraic surface V and a real-smooth rational curve $C \subset V$. As an illustration, we present below the construction of a real rational algebraic model of the pair $(\#^g\mathbb{T}^2\#\mathbb{RP}^2, L)$ where L is a separating curve concretising the connected sum of the orientable surface $\#^g\mathbb{T}^2$ with $\mathbb{RP}^2$. (A connected sum (Definition B.5.12) is constructed by gluing two surfaces from which a disk has been removed).

Example 5.4.26 Let $L_1, \dots, L_{g+1}$ be distinct lines passing through the origin in $\mathbb{R}^2$ and let $H(x, y) = 0$ be the equation of their disjoint union. For suitable $0 < \varepsilon \ll 1$ let $\overline{C^\pm} \subset \mathbb{P}^2$ be the Zariski closure of the image of the unit circle $\mathcal{Z}(x^2 + y^2 = 1) \subset \mathbb{R}^2$ under the map

$$(x, y) \mapsto \big(1 \pm \varepsilon H(x, y)\big)(x, y).$$

The curves $\overline{C^\pm}$ are rational and meet each other in the $2g + 2$ points where the unit circle meets one of the lines L_i. The curves $\overline{C^+}$ and $\overline{C^-}$ also meet in a pair of conjugate points $(1 : \pm i : 0)$. Note that $(1 : \pm i : 0)$ are the only points of $\overline{C^\pm}$ at infinity.

We now use the inverse of stereographic projection centred at the south pole (see the proof of Proposition 5.3.1) to compactify $\mathbb{R}^2$ as the quadric $Q_{3,1} := \mathcal{Z}(z_1^2 + z_2^2 + z_3^2 - z_0^2) \subset \mathbb{P}^3$. Starting with $\mathbb{P}^2$ we obtain this projection by blowing up the pair of points $(1 : \pm i : 0)$ and then contracting the strict transform of the line at infinity.

We think of the equator as the image of the unit circle, giving us rational curves $C^\pm \subset Q_{3,1}$. Since $(1 : \pm i : 0)$ are the only points of $\overline{C^\pm}$ at infinity, the south pole does not belong to either of the curves $C^\pm$ so the real points of $C^\pm$ are all non singular. Moreover, the curves C^+ and C^- meet at $2g + 2$ points on the equator.

Choose one of these points p and consider $C_0 := C^+ \cup C^-$ as the image under a map ϕ_0 of the reducible curve $B_0 := \mathcal{Z}(uv) \subset \mathbb{P}^2_{uvw}$ to $Q_{3,1}$ sending the point $(0 : 0 : 1)$ to p. By [AK03, Appl. 17] or [Kol96, II.7.6.1], ϕ_0 can be deformed to a morphism

$$\phi_t \colon B_t := \mathcal{Z}(uv - tw^2) \to Q_{3,1}.$$

Let $C_t \subset Q_{3,1}$ be the image of B_t. For t close to zero with the appropriate sign, $C_t(\mathbb{R}) \subset \mathbb{S}^2 = Q_{3,1}(\mathbb{R})$ goes twice around a neighbourhood of the equator and has $2g + 1$ self-intersection points. See Figure 5.6.

We complete the construction by blowing up the $2g + 1$ real singular points of C_t to get a rational surface $X_g \to Q_{3,1}$. The strict transform of C_t is a rational curve $C_g \subset X_g$ which is non singular at its real points.

The union of the $2g + 1$ regions of $\mathbb{S}^2 \setminus C_t(\mathbb{R})$ close to the equator is a Möbius band on $X_g(\mathbb{R}) \setminus C_g(\mathbb{R})$ and the union of the north and south hemispheres is a copy of a punctured $\#^g\mathbb{T}^2$. It follows that

$$\big(X_g(\mathbb{R}), C_g(\mathbb{R})\big) \approx \mathbb{RP}^2\#(\mathbb{S}^2, \mathbf{L})\#^g\mathbb{T}^2.$$

□

The constructions used in the proof of Theorem 5.4.25 can also be used to give a purely topological characterisation of the simple connected closed curves L on S which can be approximated by (-1)-curves, which are rigid objects.

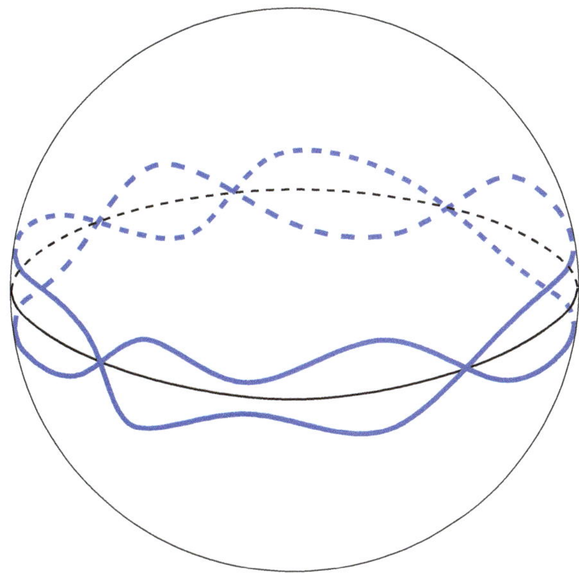

Fig. 5.6 The case $g = 2$

Definition 5.4.27 Let $C \subset V$ be a real algebraic curve on a real algebraic surface. We say that C is a (-1)-curve if and only if there is a birational morphism $\pi \colon V \to W$ such that $\pi(C)$ is a non singular point of W and π restricted to $V \setminus C \to W \setminus \pi(C)$ is an isomorphism.

This definition is motivated by Castelnuovo's criterion (Theorem 4.3.4) which states that there exists such a birational map $\pi \colon V \to W$ if and only if there is a complexification $V_{\mathbb{C}}$ of V such that $(C_{\mathbb{C}} \cdot C_{\mathbb{C}}) = -1$ (complex intersection in $V_{\mathbb{C}}$).

Theorem 5.4.28 *Let V be a non singular compact real rational algebraic surface and let $L \subset V$ be a $\mathcal{C}^\infty$ curve. The following are equivalent.*

1. *The surface V is non orientable in a neighbourhood of L and one of the following conditions is satisfied:*
   - *$V \setminus L$ is a punctured sphere,*
   - *$V \setminus L$ is a punctured torus,*
   - *$V \setminus L$ is non orientable;*
2. *The curve L is homotopic to a (-1)-curve;*
3. *The curve L has $\mathcal{C}^\infty$ approximation by (-1)-curves.*

Proof This result can be deduced from [KM16, Theorem 6] or proved directly as follows. If L is homotopic to a (-1)-curve then the open surface $V \setminus L$ is homeomorphic to the real locus of a punctured non singular rational surface. By Comessatti's theorem on rational surfaces, we know that the open surface $V \setminus L$ is non orientable, homeomorphic to a punctured sphere or homeomorphic to a punctured torus. Theorem 5.4.25, or [KM16, Theorem 3] implies that these necessary conditions are in fact sufficient. □

We end this section with the statement of two conjectures that are still open at the time of writing.

Conjecture 5.4.29 ([KM16, Conjecture 26.3]) *Let V be a non singular rationally connected real affine algebraic variety. A differentiable map $\mathbb{S}^1 \to V$ of class $\mathcal{C}^\infty$ can be approximated by rational curves if and only if it is homotopic to a rational curve $\mathbb{P}^1(\mathbb{R}) \to V$.*

Conjecture 5.4.30 ([BK10, Conjecture 1.12]) *Given two non singular real affine algebraic varieties V and W where V is compact and W is rational any smooth map $V \to W$ can be approximated by regular maps if and only if it is homotopic to a regular map.*

We refer the interested reader to [BW18a, BW18b] for recent progress on these conjectures.

5.5 Fake Real Planes

We saw in Theorem 5.4.1 that two compact non singular real rational algebraic surfaces are birationally diffeomorphic if and only if they are diffeomorphic. The situation is more complicated for non compact surfaces: in this section, we present a series of recent results on *fake real planes*, [DM17, DM16, BCM+16].

Fake Projective Planes

A *fake projective plane* was defined by Mumford [Mum79] to be a non singular complex projective surface X which has the same Betti numbers as the projective plane $\mathbb{P}^2(\mathbb{C})$ but is not biregularly isomorphic to the projective plane.

It would be tempting to define a *real fake projective plane* to be a complex fake projective plane with a real structure whose real locus is diffeomorphic to $\mathbb{P}^2(\mathbb{R})$ but is not isomorphic to the $\mathbb{R}$-variety $(\mathbb{P}^2(\mathbb{C}), \sigma_{\mathbb{P}})$. Despite the fact that 100 different fake projective planes up to biregular isomorphism are known to exist [PY07, PY10, CS10], it was proved in [KK02, Section 5] that none of them have a real structure, so there is no real fake projective plane.

Proposition 5.5.1 *Let (X, σ) be a non singular projective $\mathbb{R}$-surface such that the Betti numbers of the complex surface X are*

$$(b_0(X), b_1(X), b_2(X), b_3(X), b_4(X)) = (1, 0, 1, 0, 1) ,$$

The $\mathbb{R}$-variety (X, σ) is then isomorphic to the $\mathbb{R}$-variety $(\mathbb{P}^2(\mathbb{C}), \sigma_{\mathbb{P}})$.

Fake Real Planes

Analogy with fake projective planes motivates the following definition.

Definition 5.5.2 A *fake real plane* is a non singular quasi-projective $\mathbb{R}$-surface (X, σ) which has the same Betti numbers as the affine plane $\mathbb{A}^2(\mathbb{C})$

$$(b_0(X), b_1(X), b_2(X), b_3(X), b_4(X)) = (1, 0, 0, 0, 0) ,$$

and whose real locus $X(\mathbb{R})$ is diffeomorphic to $\mathbb{R}^2$ but is not isomorphic to the $\mathbb{R}$-variety $(\mathbb{A}^2(\mathbb{C}), \sigma_{\mathbb{A}})$.

Complex surfaces with the same Betti numbers as the affine plane have been much studied. They are often called $\mathbb{Q}$-*planes* or $\mathbb{Q}$-*acyclic surfaces*.

Definition 5.5.3 A complex surface whose higher Betti numbers are all zero is said to be $\mathbb{Q}$-*acyclique*.

Proposition 5.5.4 *Any complex algebraic* $\mathbb{Q}$-*acyclic surface is affine and rational.*

Proof See [Fuj82] for a proof of the fact that a $\mathbb{Q}$-acyclic surface is affine and [GPS97, GP99] for the proof that such a surface is rational. □

Corollary 5.5.5 *Let* (X, σ) *be a fake real plane. The complex algebraic surface* X *is then affine and rational.*

Given that there are no fake real projective planes it does not seem obvious that fake real planes exist. We give an example below: it turns out that there are an infinity of such surfaces and we refer the interested reader to [DM17, DM16] for more details.

Before presenting our example, we discuss for the sake of completeness the existence of *exotic* complex planes.

Theorem 5.5.6 *There are differentiable manifolds not diffeomorphic to* $\mathbb{R}^4$ *whose underlying topological space is homeomorphic to* $\mathbb{R}^4$.

Proof See [FQ90]. □

Definition 5.5.7 Such differentiable manifolds are called *exotic* $\mathbb{R}^4$*s*.

In the statement below we use *simply connectedness at infinity* whose precise definition is rather technical. For a detailed study of homotopy at infinity we refer to the proof of the main theorem in [Ram71] or [HR96, Chapter 9]. To summarise, simply connectedness at infinity roughly means that the space has an exhaustion by nested compact sets whose complements are simply connected from a certain point onwards.

Theorem 5.5.8 (Myanishi, Ramanujam) *Any non singular complex algebraic surface whose underlying topological space in the strong topology is contractible and simply connected at infinity is isomorphic to* $\mathbb{A}^2(\mathbb{C})$.

Proof See [Ram71]. □

Corollary 5.5.9 *There is no non singular complex algebraic surface X such that the underlying four dimensional real differentiable manifold is an exotic* $\mathbb{R}^4$.

Remark 5.5.10 For any $n \neq 4$, any differentiable manifold whose underlying topological space is homeomorphic to $\mathbb{R}^n$ is diffeomorphic to $\mathbb{R}^n$. See [Sta62] for more details.

We now construct our fake real plane. There are many other examples described in detail in [DM17, DM16].

Example 5.5.11 (*Ramanujan surface*) Consider the $\mathbb{R}$-curve

$$D = C \cup Q \subset \mathbb{P}^2$$

which is the union of a cuspidal cubic $C = \mathcal{Z}(x^2z + y^3)$ and its osculating conic Q at a general real point $q \in C(\mathbb{R})$. The conic Q is therefore a non singular $\mathbb{R}$-conic meeting C at the real point q with multiplicity 5; Q therefore meets C transversally at another real point p. We consider the projective surface $Y = B_p\mathbb{P}^2$ obtained by blowing up $\mathbb{P}^2$ at p. The complement in Y of the strict transform $\widetilde{D}$ of D, the surface $X := Y \setminus \widetilde{D}$, is a *Ramanujam surface*, see [Ram71]. In particular, as explained in [DM17, Example 3.8], the complex surface X is a contractible surface which is not isomorphic to the affine plane $\mathbb{A}^2(\mathbb{C})$. Moreover, by construction, X has a real structure σ induced by $\sigma_{\mathbb{P}}$ on $\mathbb{P}^2(\mathbb{C})$ and it is easy to check that $X(\mathbb{R})$ is connected and diffeomorphic to $\mathbb{R}^2$. (See the left hand side of Figure 5.7 where E_p is the exceptional

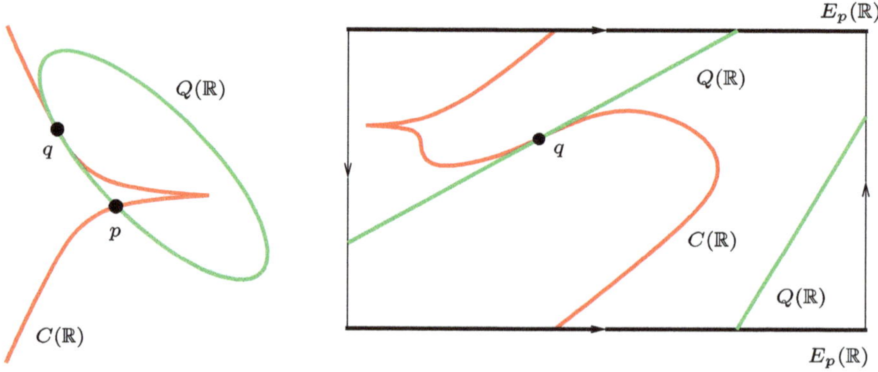

Fig. 5.7 Construction of a Ramanujam surface and connectedness of the complement

curve of the blow up $B_p\mathbb{P}^2 \to \mathbb{P}^2$. We recall that the blow up of $\mathbb{P}^2(\mathbb{R})$ at a point is a Klein bottle $B_p\mathbb{P}^2(\mathbb{R}) \approx \mathbb{K}^2$, see Corollary F.3.2. It follows that (X, σ) is a fake real plane.

In fact it is possible to construct a series of such *Ramanujam surfaces* by blowing up various configurations of points on E_p and only "keeping" the last exceptional line in the final affine surface. See [DM17, Example 3.8] for more details.

Chapter 6
Three Dimensional Varieties

Most of this chapter was previously published in [Man14] in the “Gazette de la SMF”.

6.1 The Nash Conjecture from 1952 to 2000 via 1914

6.1.1 Rational Varieties

We recall the Nash conjecture, previously discussed in the Introduction.

Conjecture Any compact connected $\mathcal{C}^\infty$ manifold of positive dimension without boundary has a real rational model (Definition 5.1.1).

This conjecture was disproved for non singular projective surfaces in 1914-before it had even been stated-proved for singular projective threefolds in the early 90s, disproved in the late 90s for higher dimensional non singular projective varieties and finally proved for non singular compact non projective threefolds! We will discuss all these results on properties of different types of algebraic models in more detail in the rest of this section.

6.1.2 The Nash Conjecture for Surfaces

When it was first stated, the Nash conjecture had already been disproved in dimension 2 by a theorem of Comessatti's proved in Chapter 4 (Corollary 4.4.16), originally published in 1914 in [Com14]. It seems likely this article had been forgotten by the time Nash was active.

Theorem (Comessatti's theorem) *Let X be a non singular projective $\mathbb{R}$-surface. If X is rational then its real locus $X(\mathbb{R})$ is diffeomorphic to $\mathbb{S}^2$, $\mathbb{S}^1 \times \mathbb{S}^1$ or a non orientable surface.*

F. Mangolte, *Real Algebraic Varieties*, Springer Monographs in Mathematics,
https://doi.org/10.1007/978-3-030-43104-4_6

Note that things change if we replace "non singular" by "possibly singular". In particular, the real locus of a singular $\mathbb{R}$-surface is not necessarily a topological manifold (ie. something that is locally homeomorphic to $\mathbb{R}^2$) and hence the notion of diffeomorphism between such surfaces does not always make sense. This leads us to replace "diffeomorphic" with "homeomorphic" in the above theorem.

Example 6.1.1 (*Real rational models of compact surfaces*) We construct a rational model for any topological surface using [BM92]. Comessatti's theorem implies that any such model is necessarily singular if the surface is orientable and of genus greater than 2. We start by constructing a non singular rational model of any non orientable surface. This can be done by blowing up k points on the projective plane $\mathbb{P}^2(\mathbb{R})$ which yields an algebraic surface X_k such that $X_k(\mathbb{R})$ is non orientable and has Euler characteristic $1-k$. To construct a model of an orientable surface of genus $g > 0$ we set $k = 2g$ and choose $k = 2g > 0$ points lying on a line H. After blowing up these k points, the strict transform $\widetilde{H} \subset X_{2g}$ has negative self-intersection $1-2g$ and can be contracted to yield an algebraic surface Y_g. If $g = 1$ the surface Y_1 is non singular and $Y_1(\mathbb{R})$ is diffeomorphic to a torus but if $g > 1$ then Y_g is singular at the point P which is the image of $\widetilde{H}$ under contraction. As the self-intersection of $\widetilde{H}$ is odd we can check that a small neighbourhood of P in $Y_g(\mathbb{R})$ is homeomorphic to a disc. The surface $Y_g(\mathbb{R})$ is therefore homeomorphic to an orientable surface of genus g. Indeed, as explained in Appendix F, the topological blow up of an orientable surface S_g of genus g at a point $Q \in S_g$ is diffeomorphic to a connected sum of $2g+1$ projective planes $B_Q S_g \approx S_g \# \mathbb{RP}^2 \approx \mathbb{RP}^2 \# \ldots \# \mathbb{RP}^2$ and in particular the fact that $S_g \setminus \{Q\} \approx X_{2g}(\mathbb{R}) \setminus H \approx Y_g(\mathbb{R}) \setminus \{P\}$ implies that the surface $Y_g(\mathbb{R})$ is homeomorphic to S_g.

Given these subtle distinctions it seems reasonable to study several different forms of the Nash conjecture. We summarise in the next section the various versions of this conjecture which have been proposed in the litterature.

6.1.3 *The Topological Nash Conjecture Holds*

We refer to Appendix F for the definitions of differentiable blow ups and contractions. The following result, which can be thought of as a topological analogue of the Nash conjecture, was proved in dimension 3 by Akbulut and King [AK91] and Benedetti and Marin [BM92] and later proved in all dimensions by Mikhalkin [Mik97].

Theorem 6.1.2 ([Mik97]) *Any compact connected $\mathcal{C}^\infty$ manifold without boundary is diffeomorphic to a $\mathcal{C}^\infty$ manifold obtained from $\mathbb{RP}^n$ by a sequence of differentiable blow ups and contractions.*

6.1.4 *The Projective Singular Nash Conjecture Holds if* $\mathbf{n \leqslant 3}$

The proof of this theorem in dimension 2 was presented above. In dimension 3, not every manifold can be produced by blowing up points and contracting divisors- we also need certain knot surgeries, where a knot is a circle embedded in a manifold. For simplicity we restrict ourselves to knots that have an orientable tubular neighbourhood: in this case the closed tubular neighbourhood is necessarily diffeomorphic to $\mathbb{S}^1 \times \mathbb{D}^2$. Topologically, any compact 3-manifold without boundary can be obtained from the sphere $\mathbb{S}^3$ by knot surgery. A *surgery* along a *knot* L in a manifold M is the operation of gluing a solid torus $T := \mathbb{S}^1 \times \mathbb{D}^2$ to the boundary of the complement of an open tubular neighbourhood U_L of L. This gluing is realised via a diffeomorphism $\varphi \in \mathrm{Diff}(\mathbb{S}^1 \times \mathbb{S}^1)$ from the torus $\mathbb{S}^1 \times \mathbb{S}^1 = \partial(M \setminus U_L) = \partial T$ to itself. The operation which produces $M_\varphi = M \setminus U_L \cup_\varphi T$ from M is called a *surgery* along L. Benedetti and Marin proved that apart from a handful of examples which can be dealt with on a case-by-case basis, most 3-manifolds can be produced from $\mathbb{S}^3$ by blowing up points and performing certain surgeries called *déchirures*. Their description of these topological transformations enables them to prove the topological Nash conjecture in dimension 3: they realise déchirures as algebraic operations and this gives them a possibly singular $\mathbb{R}$-variety X and a resolution of singularities $Y \to X$ such that Y is birationally equivalent to $Q_{4,1} = \mathcal{Z}(x_1^2 + x_2^2 + x_3^2 + x_4^2 - x_0^2) \subset \mathbb{P}^4$, $Y(\mathbb{R})$ is diffeomorphic to $\mathbb{S}^3$ and $X(\mathbb{R})$ is homeomorphic to M.

Theorem 6.1.3 ([BM92]) *Let M be a compact connected $\mathcal{C}^\infty$ 3-manifold without boundary. There is then a (possibly singular) rational projective algebraic $\mathbb{R}$-variety X such that $X(\mathbb{R})$ is homeomorphic to M.*

6.1.5 *Non Projective Non Singular Nash Holds for* $\mathbf{n = 3}$

Any non singular complex projective algebraic variety is also a compact complex analytic variety if we equip it with the Euclidean topology. (See Appendix D for more details). Conversely, if the field of meromorphic functions of a compact complex analytic variety is of maximal transcendence degree (which is equal to the dimension of the variety by Siegel's theorem [Sie55]) the variety is very close to being projective. (See [Moi67] for more details). Despite this, the Nash conjecture holds in dimension 3 for such varieties even though, as we will see in Theorem 6.1.9, it fails for projective varieties.

Definition 6.1.4 A non singular compact complex analytic variety of dimension n is said to be *Moishezon* if and only if it has a family of n algebraically independent meromorphic functions. (Compare this definition with the discussion preceding Definition 1.3.37). An *real Moishezon variety* is a Moishezon variety with a global anti-holomorphic involution $\sigma\colon X \to X$.

Remark 6.1.5 By [Moi67] (an English translation of [Moĭ66a, Moĭ66b, Moĭ66c]), any compact non singular complex analytic variety is Moishezon if and only if it is bimeromorphic to a projective variety.

Any non singular Moishezon surface is projective ([BHPVdV04, IV.5]). The first examples of 3 dimensional non singular non projective Moishezon varieties were constructed by Hironaka. See [Har77, Appendix B.3] for more details.

The following theorem implies that the non projective non singular Nash conjecture holds for $n = 3$.

Theorem 6.1.6 *Let M be a three dimensional compact connected $\mathcal{C}^\infty$ manifold without boundary. There is then a non singular $\mathbb{R}$-Moishezon variety (X, σ) and a bimeromorphic map $\pi\colon \mathbb{P}^3 \dashrightarrow X$ such that $\pi\sigma_0 = \sigma\pi$ and $X(\mathbb{R})$ is diffeomorphic to M.*

Proof See [Kol02, Theorem 1.3]. □

In fact, the following more specific theorem holds: there is a sequence of blow ups and contractions along non singular centres (see Appendix F for the definitions)

$$\mathbb{P}^3 = X_0 \overset{\pi_0}{\dashrightarrow} X_1 \overset{\pi_1}{\dashrightarrow} \cdots \overset{\pi_{n-1}}{\dashrightarrow} X_n = X$$

such that for every i the variety X_i is non singular. Moreover, this sequence is real in the following sense: each of the varieties has a global anti-holomorphic involution $\sigma_i\colon X_i \to X_i$ such that $\sigma_0 = \sigma_{\mathbb{P}}$, $\sigma_n = \sigma$ and $\pi_i\sigma_i = \sigma_{i+1}\pi_i$ for all i. See [Kol02] for more details.

Kollár proves this result using the Benedetti-Marin classification of what he calls "topological flops" which are a special case of the déchirures discussed above. He then shows how to realise these topological flops as algebraic flops. We briefly describe the special type of *algebraic flop* used by Kollár. There is a birational map $f\colon X \dashrightarrow X'$ which factors as $X \overset{\pi}{\leftarrow} X_1 \overset{\pi'}{\rightarrow} X'$ where π and π' have the same exceptional divisor $E \subset X_1$, isomorphic to $\mathbb{P}^1 \times \mathbb{P}^1$, and each of the morphisms contracts one of the two $\mathbb{P}^1$ factors. The associated transformation of the real locus $X(\mathbb{R}) \dashrightarrow X'(\mathbb{R})$ is then a topological flop. Conversely, such a transformation of the 3-dimensional variety X can only be carried out in the presence of a rational curve $C \subset X$ embedded in a particular way:

1. the exceptional divisor E of the blow up $\pi\colon X_1 \to X$ along C must be isomorphic to $\mathbb{P}^1 \times \mathbb{P}^1$,
2. $\pi_{|E}\colon E \to \mathbb{P}^1$ must be projection onto the first fibre,
3. there is a contraction $\pi'\colon X_1 \to X'$ of E whose restriction to E is a projection onto the second factor.

The first step in the proof of this theorem is the construction of a suitable algebraic approximation of certain embedded Möbius bands representing topological flops. This is done using the approximation theorem 5.4.21. The second step involves constructing algebraic flops using blow ups that do not alter the real locus. At the end of this process, the variety X is not generally projective but remains Moishezon because the function field is invariant under birational transformation. We refer the interested reader to [Kol01a, Section 4] for more details of this construction.

6.1.6 *Non Singular Projective Nash Fails for All Dimensions* $n > 1$

As we have seen, Comessatti's theorem refutes the non singular projective Nash conjecture for $n = 2$. The fact that the Nash conjecture fails for non singular projective varieties of dimension $n = 3$ follows from Theorem 6.1.9, proved in the series of articles [Kol98b, Kol99a, Kol99b, Kol00]. In particular, Kollár proved in this theorem that apart from a finite number of possible exceptions, hyperbolic manifolds (Definition B.8.6) of dimension 3 do not have a non singular rational projective model (Corollary 6.1.10) and conjectured that this result generalises. Not long afterwards, Viterbo and Eliashberg confirmed this conjecture by proving that in dimension $n > 2$ there is no hyperbolic manifold with a non singular rational projective model (Corollary 6.1.18).

The results of Kollár and Viterbo-Eliashberg apply to a class of varieties generalising rational varieties, namely the *uniruled* varieties mentioned in our discussion of surfaces (Definition 4.4.1).

Definition 6.1.7 A real or complex variety X of dimension n is said to be *uniruled* if and only if it is dominated by a cylinder of the same dimension, by which we mean there is a variety Y of dimension $n - 1$ and a rational map

$$Y \times \mathbb{P}^1 \dashrightarrow X$$

whose image is Zariski dense.

Remark 6.1.8 The definition above holds whether X is real or complex: being uniruled is invariant under change of base field. See [Deb01, Section 4.1, Remark 4.2(5)] for more details.

As the product variety $\mathbb{P}^{n-1} \times \mathbb{P}^1$ is birationally equivalent to $\mathbb{P}^n$, it is immediate that any rational variety over $\mathbb{R}$ or $\mathbb{C}$ is uniruled. To see that $\mathbb{P}^{n-1} \times \mathbb{P}^1$ is birational to $\mathbb{P}^n$, consider the rational map

$$((x_0 : \cdots : x_{n-1}), (y_0 : y_1)) \mapsto\dashrightarrow (x_0 y_0 : x_1 y_0 : \cdots : x_{n-1} y_0 : x_0 y_1) .$$

which induces an isomorphism between the open sets $\{x_0 \neq 0\} \times \{y_0 \neq 0\}$ in $\mathbb{P}^{n-1}_{x_0:\cdots:x_{n-1}} \times \mathbb{P}^1_{y_0:y_1}$ and the open set $\{z_0 \neq 0\}$ in $\mathbb{P}^n_{z_0:\cdots:z_n}$.

Kollár's Theorem

Before stating Kollár's theorem we need some results on the classification of three dimensional topological manifolds. Compact connected surfaces without boundary are classified topologically (Theorem E.1.6) by two invariants: orientability, which is a binary invariant, and the Euler characteristic, which is an integer. The theory of three dimensional manifolds is much richer: we refer the interested reader to Appendix B.8 for a full discussion of three dimensional manifold theory which is summarised here.

By Theorem B.8.17, any compact topological manifold of dimension 3 is constructed from "blocks" belonging to one of the following disjoint families:

1. Seifert manifolds (Definition B.8.1);
2. **Sol** manifolds (Definition B.8.8);
3. Hyperbolic manifolds (Definition B.8.6).

Note that the lens spaces (Definition B.8.2) appearing in the next result belong to the first class of manifolds by Proposition B.8.3. A connected sum of at least two such spaces, however, is not Seifert unless it is $\mathbb{RP}^3\#\mathbb{RP}^3$ by Propositions B.8.11 and B.8.13.

Theorem 6.1.9 (Kollár 1998) *Let X be a non singular projective algebraic $\mathbb{R}$-variety of dimension 3. Suppose that X is uniruled and $X(\mathbb{R})$ is orientable. Any connected component of $X(\mathbb{R})$ is then diffeomorphic to one of the following.*

1. *A Seifert manifold,*
2. *A connected sum of a finite collection of lens spaces,*
3. *A* **Sol** *manifold,*
4. *A manifold belonging to a finite list of possible exceptions.*
5. *A manifold obtained from one of the above manifolds by taking the connected sum with a finite number of copies of $\mathbb{RP}^3$ and a finite number of copies of $\mathbb{S}^2 \times \mathbb{S}^1$.*

Proof The above result can be deduced from the original statement ([Kol01b, Theorem 6.6]) using Proposition 6.2.3. We sketch its proof below. □

Corollary 6.1.10 *Apart from a finite number of possible exceptions, orientable hyperbolic manifolds of dimension 3 do not have a non singular uniruled projective model.*

Remark 6.1.11 In fact, subsequent results of Kollár's imply that the orientability hypothesis can be omitted in 6.1.10. See [Kol99a, Theorem 12.1, Theorem 1.8, Theorem 1.2], [Kol99b, Theorem 8.3] and [Kol00] for more details.

Proof (Proof of Corollary 6.1.10) Any hyperbolic manifold M is geometric and therefore indecomposable by Corollary B.8.12. By Theorem 6.1.9, except for a finite number of possible exceptions, any indecomposable connected component of the real locus of a non singular uniruled orientable projective variety is a Seifert manifold, a lens space (which is also a Seifert manifold) or a **Sol** manifold. None of these manifolds is hyperbolic by Corollary B.8.14. □

One of the main difficulties arising in the proof of Theorem 6.1.9 is that we need to control the modifications of the topology arising when we run the minimal model program (MMP) over $\mathbb{R}$. Theorem 6.1.13 below enables us to reduce our general topological classification to certain special manifolds.

Minimal Model Problem over $\mathbb{R}$

Starting from a non singular projective algebraic $\mathbb{R}$-variety (X, σ) de dimension 3 we can carry out a sequence of "elementary" birational transformations

$$X = X_0 \overset{\pi_0}{\dashrightarrow} X_1 \overset{\pi_1}{\dashrightarrow} \cdots \overset{\pi_{n-1}}{\dashrightarrow} X_n = X^*$$

until we obtain an $\mathbb{R}$-variety (X^*, σ^*) whose global structure is "simple". Moreover, this sequence of transformations is real in the following sense: every variety X_i has a real structure $\sigma_i : X_i \to X_i$ such that $\sigma_0 = \sigma$, $\sigma_n = \sigma^*$ and $\pi_i \sigma_i = \sigma_{i+1}\pi_i$ for every i.

The price we pay is that the varieties $X_i, i > 0$ and X^* are no longer non singular: we have to extend our class of varieties to include certain types of relatively mild singularities. The singularities in question are called *terminal singularities*- we refer to [Kol98b] for the definition of these singularities and their classification on an $\mathbb{R}$-variety of dimension 3. We note that in dimension 3 all such singularities are isolated.

When we say that the global structure is "simple" we mean that X^* satisfies one of the conditions below. See [Kol99a, Theorem 3.11] for a more precise statement.

- The canonical divisor K_{X^*} is nef (Definition 2.6.41).
- There is a real conic fibration $X^* \to Y$ over an $\mathbb{R}$-surface Y;
- There is a real del Pezzo fibration $X^* \to Y$ over an $\mathbb{R}$-curve Y;
- The variety X^* is Fano, or in other words its anti-canonical divisor $-K_{X^*}$ is ample (Definition 2.6.20).

Theorem 6.1.12 *Let (X, σ) be a non singular projective algebraic $\mathbb{R}$-variety of dimension 3 and let (X^*, σ^*) be the output of the real MMP applied to X. If X is uniruled then X^* is a fibration of one of the following types (known as* Mori fibre spaces*):*

1. *A conic bundle over a surface;*
2. *A Del Pezzo fibration over a curve;*
3. *A three dimensional Fano variety over a point.*

Proof If the variety X is uniruled K_{X^*} cannot be nef since in characteristic zero, any non singular uniruled variety contains a *free* rational curve ([Deb01, Corollary 4.11]) and by [Deb01, Example 4.7(1)], there is no free rational curve on a variety whose canonical divisor is nef. The variety X^* therefore belongs to one of the three classes listed above. □

One of Kollár's discoveries it that we can avoid the main difficulties of the MMP if the real locus of the variety is orientable.

Theorem 6.1.13 *Let (X, σ) be a non singular projective algebraic $\mathbb{R}$-variety of dimension 3 and let (X^*, σ^*) be the output of a real MMP on X. Suppose that the real locus $X(\mathbb{R})$ is orientable.*

The topological normalisation $\overline{X^*(\mathbb{R})} \to X^*(\mathbb{R})$ (Definition 4.4.35) is then a piecewise linear manifold and any connected component $L \subset X(\mathbb{R})$ can be obtained from $\overline{X^*(\mathbb{R})}$ as a connected sum of components of $\overline{X^*(\mathbb{R})}$, copies of $\mathbb{RP}^3$ and copies of $\mathbb{S}^2 \times \mathbb{S}^1$.

Proof See [Kol99a, Theorem 1.2]. □

Proof (Summary of the proof of Theorem 6.1.9) By Theorem 6.1.12, the $\mathbb{R}$-variety (X, σ) is birational to an $\mathbb{R}$-variety (X^*, σ^*) which is a Mori fibre space of one of the three types listed above. Theorem 6.1.13 enables us to conclude using known classifications of topological types of such fibrations:

1. A conic bundle over a surface. Kollár classified such bundles in [Kol99b, Theorem 1.1]. One consequence of his classification is that, up to connected sum with $\mathbb{RP}^3$s and $\mathbb{S}^2 \times \mathbb{S}^1$s, M is a Seifert manifold or a connected sum of lens spaces.
2. A fibration over a curve with rational fibres. [Kol00, Theorem 1.1] states that up to connected sum with $\mathbb{RP}^3$s and $\mathbb{S}^2 \times \mathbb{S}^1$s, M is then a Seifert manifold, a connected sum of lens spaces, a torus bundle over a circle or the $\mathbb{Z}_2$-quotient of such a bundle. In the last two cases, Proposition 6.2.3 implies that M is a lens space or a **Sol** manifold.
3. A Fano variety with terminal singularities. We know by [Kaw92] (see also [Kol98a, Section 6]) that there is only a finite number of families of such varieties: we even have an (improbable) upper bound on this number, since work by Kollár (see [Kol17, before Theorem 24] for example) implies that there are at most $10^{10^{500}}$ different topological types of $\mathbb{R}$-Fano varieties with terminal singularities.

□

Remark 6.1.14 Not much is currently known about the topology of 3 dimensional Fano varieties: we mention three articles on real Fano varieties for the interested reader. Real cubics in $\mathbb{P}^4$ were classified by Krasnov [Kra06, Kra09] and the possible real structures on the Fano variety V_{22} were classified by Kollár and Schreyer [KS04].

Viterbo's Theorem

The proof of Theorem 6.1.16 below on hyperbolic manifolds, like the proof of Theorem 6.2.4 on **Sol** manifolds, uses symplectic field theory (SFT), which would take us too far from algebraic geometry. Not having the necessary space to explain this important tool, we refer the interested reader to [EGH00, 1.7.5] (or [MW12] for **Sol** manifolds) for more details. We will however explain how symplectic geometry arises in this context.

Given any non singular complex projective algebraic variety X of complex dimension n, we may consider its underlying differentiable manifold. For any projective embedding $j : X \hookrightarrow \mathbb{P}^N(\mathbb{C})$, the restriction to $j(X)$ of the Fubini-Study metric on $\mathbb{P}^N(\mathbb{C})$ equips X with a Kähler form ω as in Example D.3.6. The pair (X, ω)- where X is the differentiable manifold of even real dimension $2n$ and ω is a Kähler form- is then a symplectic manifold. See Remark D.3.5 for more details.

Definition 6.1.15 Let (X, ω) be a symplectic manifold. A submanifold $M \subset X$ is said to be *Lagrangian* if and only if

$$\omega|_M \equiv 0 \quad \text{and} \quad \dim_{\mathbb{R}} M = \frac{1}{2} \dim_{\mathbb{R}} X \ ,$$

or in other words if M is an *isotropic* submanifold of maximal dimension.

Theorem 6.1.16 *Let X be a non singular complex projective algebraic variety of complex dimension > 2. Let $M \subset X$ be a Lagrangian submanifold for some underlying symplectic structure on X. If X is uniruled then M does not have a Riemannian metric with strictly negative sectional curvature.*

Proof See [Vit99], [EGH00, 1.7.5]. □

Exercise 6.1.17 Prove that the real locus $X(\mathbb{R})$ of a non singular projective $\mathbb{R}$-variety (X, σ) is a Lagrangian subvariety of the underlying symplectic manifold structure on X given by an $\mathbb{R}$-embedding of (X, σ).

Corollary 6.1.18 *Let X be a non singular projective $\mathbb{R}$-variety of dimension > 2. If X is uniruled then no connected component of $X(\mathbb{R})$ has a hyperbolic metric since any such a metric has constant sectional curvature -1.*

6.2 Real Uniruled 3-varieties from 2000 to 2012

Theorem 6.1.9 implies strong constraints on the real locus of a non singular uniruled projective $\mathbb{R}$-variety of dimension 3. Following this theorem, Kollár proposed several conjectures on the topological classification of the real loci of such varieties, whose current status is summarised below.

Recall that if M is an *oriented* compact manifold without boundary of dimension 3 then there is a decomposition $M = M' \#^a \mathbb{RP}^3 \#^b (\mathbb{S}^2 \times \mathbb{S}^1)$ with maximal $a + b$ and this decomposition is unique by Milnor's theorem [Mil62]. (See Definition B.5.15 for the definition of the connected sum).

Since the algebraic properties of rationality, rational connectedness (see below) and uniruledness are invariant under birational equivalence the corresponding topological properties of M can be detected on M': this phenomenon is illustrated in the examples below. This motivates our following *ad hoc* definition.

Definition 6.2.1 Let M be an oriented compact manifold without boundary of dimension 3 and let $M = M' \#^a \mathbb{RP}^3 \#^b (\mathbb{S}^2 \times \mathbb{S}^1)$ be a decomposition with $a + b$ maximal. The manifold M' is said to be the *essential* part of M. A property of M is said to be essential if it only depends on M'.

Example 6.2.2 This example is taken from [Kol99a, Example 1.4]. Let X be a non singular $\mathbb{R}$-variety of dimension 3.

1. Let $P \in X(\mathbb{R})$ be a real point and let M be the connected component of $X(\mathbb{R})$ containing P. We then have (Proposition F.3.1)
$$B_P M \approx M\#\mathbb{RP}^3.$$
2. Let $D \subset X$ be an $\mathbb{R}$-curve with a unique real point $\{0\} = D(\mathbb{R})$. Suppose that close to 0 this curve is given by equations $\{z = x^2 + y^2 = 0\}$. Let $Y_1 = B_D X$ be the variety obtained by a blow up of X centred at D. (See Appendix F for the definition of such a blow up). This new variety is real and has a unique singular point P. Consider $Y := B_P Y_1$, the variety obtained by blowing up Y_1 at P, which is a non singular real variety. Denoting by $\pi \colon Y \to X$ the composition of these two blow ups, the component $M \subset X(\mathbb{R})$ containing P satisfies
$$\pi^{-1} M \approx M\#(\mathbb{S}^2 \times \mathbb{S}^1),$$
or in other words
$$B_P(B_D M) \approx M\#(\mathbb{S}^2 \times \mathbb{S}^1).$$

6.2.1 *Uniruled Varieties*

Kollár's theorem 6.1.9 tells us that apart from **Sol** maniolds (Definition B.8.8), and a finite number of closed manifolds of dimension 3, real uniruled orientable varieties are essentially (Definition 6.2.1) Seifert bundles or connected sums of lens spaces. The progress made since this theorem can be briefly summarised as follows. Theorem 6.1.16 tells us that no hyperbolic manifold can be contained in the real locus of a non singular uniruled projective variety. Theorem 6.2.4 tells us that only a finite number of **Sol** manifolds can be contained in the real locus of a non singular uniruled projective variety. Conversely, Theorem 6.2.7 together with Proposition B.8.13 tells us that any orientable geometric manifold (Definition B.8.4) which is neither hyperbolic nor **Sol** is diffeomorphic to a connected component of the real locus of a non singular uniruled projective variety.

Sol Manifolds and Suspensions[1] of Diffeomorphisms of the Torus

Let z be a coordinate on the circle $\mathbb{S}^1 := \{|z| = 1\} \subset \mathbb{C}$ and let (u, v) be coordinates on the torus $\mathbb{S}^1 \times \mathbb{S}^1 := \{|u| = 1,\ |v| = 1\} \subset \mathbb{C} \times \mathbb{C}$. The group $\mathbf{GL}_2(\mathbb{Z})$ then acts on $\mathbb{S}^1 \times \mathbb{S}^1$ via the map
$$\begin{pmatrix} a & b \\ c & d \end{pmatrix} \longmapsto [(u, v) \mapsto (u^a v^b, u^c v^d)].$$

[1] This construction also appears in the litterature under the name “mapping Torus”.

For any $A \in \mathbf{GL}_2(\mathbb{Z})$ we set

$$M := \left(\mathbb{S}^1 \times \mathbb{S}^1\right) \times [0, 1]/((u, v), 0) = (A \cdot (u, v), 1).$$

The map $\rho\colon M \to \mathbb{S}^1 = [0, 1]/(0 = 1)$ is then a differentiable $\mathcal{C}^\infty$ torus bundle as in Definition C.3.5. We can prove that the total space M of this bundle is geometric (Definition B.8.4) and its geometry depends on the diffeomorphism of $\mathbb{S}^1 \times \mathbb{S}^1$ given by the matrix A. See [Sco83] for more details. Let λ be a eigenvalue of A: there are three different possible cases.

1. If $|\lambda| = 1$ and A is periodic then M is a Euclidean manifold,
2. If $|\lambda| = 1$ and A is non periodic then M is a **Nil** manifold
3. If $|\lambda| \neq 1$ (or in other words A is hyperbolic) then M is a **Sol** manifold.

Note that in the first two cases where M has a Euclidean or **Nil** geometry M is also a Seifert bundle (Proposition B.8.13). It follows that a torus bundle over the circle is always either a Seifert manifold or a **Sol** manifold.

Conversely, most **Sol** manifolds are toric bundles with hyperbolic gluing, as the proposition below shows.

Proposition 6.2.3 *(Classification of closed* **Sol** *manifolds)*

Let M be a compact **Sol** *manifold without boundary: M therefore has one of the following two forms.*

1. *M is the suspension of a hyperbolic diffeomorphism.*
2. *M is a sapphire, ie., a $\mathbb{Z}_2$-quotient of the previous case.*

Proof See [MW12, Theorem 2.1]. □

Theorem 6.2.4 *Any closed orientable* **Sol** *manifold cannot be embedded as a connected component of the real locus of a non singular projective variety of dimension* 3 *which is a bundle over a curve with rational fibres.*

Proof See [MW12, Corollary 3.1]. □

Apart from a finite number of possible exceptions, this establishes the first part of Kollár's conjecture [Kol01b, Conjecture 6.7(1)].

Corollary 6.2.5 *If a connected orientable component M of the real locus of a non singular uniruled projective $\mathbb{R}$-variety (X, σ) of dimension* 3 *is a* **Sol** *manifold then X is birationally equivalent to an $\mathbb{R}$-Fano variety (Y, τ) such that $Y(\mathbb{R})$ contains a connected component homeomorphic to M.*

Remark 6.2.6 In particular, the number of such uniruled **Sol** manifolds is finite: it is even conjectured that such things do not exist. See [Kol01b, Conjecture 6.7(1)]) for more details.

Proof (Proof of Corollary 6.2.5) We use the same line of attack as in the proof of Theorem 6.1.9. By Theorem 6.1.12, the $\mathbb{R}$-variety (X, σ) is birational to an $\mathbb{R}$-variety (Y, τ) which is a Mori fibre space. Moreover, $Y(\mathbb{R})$ contains a connected component which is homeomorphic to M by Theorem 6.1.13 because any **Sol** manifold is indecomposable by Corollary B.8.12. We conclude using Theorem 6.1.13 and our classification of topological types of Mori fibre spaces.

1. Conic bundles over surfaces. Kollár classified such surfaces in [Kol99b, Theorem 1.1]: this classification implies in particular that M is a Seifert manifold (or a lens space which is a special type of Seifert manifold) and therefore M cannot be a **Sol** manifold by Corollary B.8.14.
2. A fibration with rational fibres over a curve: Theorem 6.2.4 then states that M cannot be **Sol**.
3. A Fano variety with terminal singularities. We know that there is only a finite number of such varieties.

□

The following theorem, proved in the articles [HM05b] and [HM05a], is the converse of Theorem 6.1.9 if we assume Kollár's conjecture [Kol01b, Conjecture 6.7(1)] which states that there are in fact no exceptions in Kollár's theorem. In other words, this conjecture states that any connected component of the real locus of a non singular Fano variety of dimension 3 is essentially a Seifert manifold or a connected sum of lens spaces.

Theorem 6.2.7 *Any orientable Seifert manifold and any connected sum of lens spaces* $\#_{i=1}^{k}\mathbb{L}_{p_i,q_i}$ *can be realised as a connected component of the real locus of a non singular uniruled projective* $\mathbb{R}$*-variety of dimension* 3.

Proof See [HM05b, Theorem 1.1] for the construction of Seifert manifolds and [HM05a, Corollary 1.2] for the construction of connected sums of lens spaces. □

Remark 6.2.8 This theorem confirms a conjecture of Kollár's ([Kol01b, Conjecture 6.7.(2)]).

6.2.2 Rationally Connected Varieties

We have seen that uniruled varieties (Definition 6.1.7) generalise rational varieties. We now present a class of varieties in between rational and uniruled varieties.

Definition 6.2.9 A non singular projective $\mathbb{R}$-variety (X, σ) of dimension n is said to be *rationally connected* (r. c.) if and only if it has a non empty Zariski open subset $U \subset X$ such that for any pair of points $x, y \in U$ there is a rational curve $f : \mathbb{P}^1(\mathbb{C}) \to X$ such that $x, y \in f(\mathbb{P}^1(\mathbb{C}))$. It is not required that f should be real.

Remark 6.2.10 This definition is one of five equivalent characterisations given in [Kol01c, Definition 41] of rational connectedness of non singular varieties. For more information on rational connectedness of possibly singular varieties see [Deb01, Definition 4.3]: the link with Definition 6.2.9 is explained in Remark [Deb01, Remark 4.4(3)]. The interested reader may also wish to consult [Deb01, Remark 4.4(4)] which shows that rational connectedness ([Deb01, Definition 4.3]) is, like uniruledness, invariant under change of base field.

Geometrically rational varieties are rationally connected, hypersurfaces of degree less than or equal to n in $\mathbb{P}^n$ are rationally connected and more generally all Fano varieties are rationally connected [KMM92, Cam92]. Rationally connectedness is an intermediate property between rationality and uniruledness: a variety X is uniruled if and only if there is an open subset U such that for any $x \in U$ there is a rational curve $f: \mathbb{P}^1(\mathbb{C}) \to X$ such that $x \in f(\mathbb{P}^1(\mathbb{C}))$. Paraphrasing Kollár ([Kol01c]), "rationally connected variety" has come to be seen as the "right" generalisation of "rational curve".

Before stating Theorem 6.2.12, the main result of the subsection, we prove a corollary of Theorem 4.4.39 on rational surfaces with Du Val singularities.

If $g: M \to B$ is a Seifert bundle (Definition B.8.1), let k be the number of multiple fibres of g and for every multiple fibre $g^{-1}(P_i)$, $i = 1 \dots k$, let m_i be its multiplicity. If M is a connected sum of lens space (Definition B.8.2) let k be the number of lens spaces and for each lens space $\mathbb{L}_{p_i,q_i}$, $i = 1 \dots k$, let m_i be the order of its fundamental group.

Theorem 6.2.11 *Let (X, σ) be a non singular projective $\mathbb{R}$-variety of dimension 3 with an $\mathbb{R}$-fibration $X \longrightarrow Y$ whose general fibre is $\mathbb{P}^1$. If $X(\mathbb{R})$ is orientable and Y is a geometrically rational $\mathbb{R}$-surface then for every connected component $M \subset X(\mathbb{R})$ we have that*

1. *$k(M) \leqslant 4$;*
2. *$\sum_{i=1}^{k}(1 - \frac{1}{m_i+1}) \leqslant 2$;*
3. *If $M \to B$ is a Seifert bundle with $|B| = \mathbb{S}^1 \times \mathbb{S}^1$ then*

$$k(M) = 0 .$$

The three conclusions of this theorem confirm three conjectures of Kollár [Kol99b, Remark 1.2(1–3)] (see also [Kol01b, Conjecture 6.7.(3)]).

Proof The original proof can be found in [CM09, Theorem 0.1, Section 6].

Let $f: X \longrightarrow Y$ be a non singular projective $\mathbb{R}$-variety of dimension 3 which is $\mathbb{R}$-fibred in rational curves over a geometrically rational $\mathbb{R}$-surface Y. Suppose that $X(\mathbb{R})$ is orientable and let $M \subset X(\mathbb{R})$ be a connected component. Kollár proved (see [CM08, 3.3, 3.4, and the proof of Corollary 0.2]) that the following objects exist.

1. A pair of birational contractions $c: X \to X'$, $r: Y \to Y'$ such that
   (a) X' is a projective $\mathbb{R}$-variety of dimension 3 with terminal singularities

(b) Y' is a Du Val surface.

2. An $\mathbb{R}$-fibration in rational curves $f'\colon X' \longrightarrow Y'$ such that $-K_{X'}$ is f'-ample and $f' \circ c = r \circ f$.
3. The important property of this construction is that $M'' = M'\#^{a'}\mathbb{RP}^3$ where M' is the essential part of M and M'' is the connected component of the topological normalisation $\overline{\nu}\colon \overline{X'(\mathbb{R})} \to X'(\mathbb{R})$ which has the property that $\overline{\nu}(M'') = c(M)$.

By [Kol99b, Theorem 8.1] and [CM08, Proof of Corollary 0.2, end of Section 3], there is a small perturbation $g\colon M'' \to B$ of $f'|_{\overline{\nu}(M'')}$ such that $g|_{g^{-1}(B\setminus\partial B)}$ is a Seifert bundle and is injective on the set of singular points of Y' contained in $f'(\overline{\nu}(M''))$ which are of type A^+ and are globally separating if they are locally separating. This injection shows that the multiplicity of the fibre is $m+1$ if the singular point is of type A^+_m. The inequalities $k(M) \leqslant 4$ and $\sum_{i=1}^k(1 - \frac{1}{m_i+1}) \leqslant 2$ then follow from Theorem 4.4.39. □

The next theorem summarises our current knowledge of the topological classification of three dimensional uniruled and rationally connected varieties. It brings together work originally published in [Kol98b, Kol99a, Kol99b, Kol00, Vit99, EGH00, HM05b, HM05a, CM08, CM09, MW12].

Theorem 6.2.12 (Classification) *Let X be a non singular projective $\mathbb{R}$-variety of dimension 3 with orientable real locus $X(\mathbb{R})$. Let $M \subset X(\mathbb{R})$ be a connected component. Except for a finite number of possible exceptions,*

1. *If X is uniruled then there are integers $a, b \in \mathbb{N}$ and a variety M' such that*

$$M = M'\#^a\mathbb{RP}^3\#^b(\mathbb{S}^2 \times \mathbb{S}^1)$$

and M' is either a Seifert bundle $M' \to B$ or has a decomposition $M' = \#_{i=1}^k \mathbb{L}_{p_i,q_i}$ into lens spaces.

2. *If X is rationally connected and M is a Seifert manifold $M \to B$ whose orbit space B is orientable then M has one of the following four geometries*

$$\mathbb{S}^3,\ \mathbb{E}^3,\ \mathbb{S}^2 \times \mathbb{E}^1,\ \mathbf{Nil}.$$

Conversely, let $M = M'\#^a\mathbb{RP}^3\#^b(\mathbb{S}^2 \times \mathbb{S}^1)$ be a compact manifold without boundary of dimension 3. If M' is an orientable Seifert manifold or a connected sum of lens spaces $M' = \#_{i=1}^k \mathbb{L}_{p_i,q_i}$ then there is a non singular uniruled projective $\mathbb{R}$-variety X such that M is diffeomorphic to a connected component of $X(\mathbb{R})$.

Proof By Kollár's theorem 6.1.9 result (1) will follow if we can show that the only infinite family of possible exceptions in 6.1.9.(3) does not in fact arise. In other words, it will be enough to prove that if $M \to \mathbb{S}^1$ is a locally trivial torus bundle which does not also have a Seifert bundle structure then M belongs to the finite list of exceptions 6.1.9.(4). This follows from Corollary 6.2.5. The result (2) follows from Theorem 6.2.11 using Proposition 4.4.34 and the Table 6.1 [Sco83, Table 4.1],

Table 6.1 Geometry of Seifert manifolds $f: M \to B$

	$\chi(B) > 0$	$\chi(B) = 0$	$\chi(B) < 0$
$e(f) = 0$	$\mathbb{S}^2 \times \mathbb{E}^1$	$\mathbb{E}^3$	$\mathbb{H}^2 \times \mathbb{E}^1$
$e(f) \neq 0$	$\mathbb{S}^3$	**Nil**	$\widetilde{\mathbf{SL}_2(\mathbb{R})}$

which gives the geometry of the total space M as a function of the geometry of the base orbifold B of the Seifert bundle $f: M \to B$ and the Euler number $e(f)$ of the fibration. (See [Sco83, discussion after Theorem 3.6] for the definition of $e(f)$ and [Sco83, Lemme 3.7] for its main properties.)

The converse follows from Theorem 6.2.7. □

6.3 Questions and Conjectures

1. The following conjecture from 2004 ([Man04, Page 24]) is still open.

Conjecture 6.3.1 The real *geometric* components of orientable non singular uniruled projective $\mathbb{R}$-varieties of dimension 3 are exactly the orientable Seifert manifolds.

Given Theorem 6.2.12, the number of possible counter-examples to this conjecture is finite. Moreover, any such counter-example would have to be the resolution of a Fano variety with terminal singularities and a connected **Sol** component in its real locus.

2. Let us unearth a question first asked in [MW12]. We have seen above that there are uniruled models for all orientable Seifert manifolds (Theorem 6.2.7) but the following question is still open: what is the simplest non singular real projective model of a hyperbolic manifold? of a **Sol** manifold?
3. It should be possible to prove that any non orientable Seifert manifold has a uniruled model by improving the construction in [HM05b].

Appendix A
Commutative Algebra

In this chapter we will summarise various results from commutative algebra for the reader's convenience. Some are well known, but others are not easily available despite being standard in the specialist literature. Our main reference is Eisenbud's book [Eis95].

A.1 Inductive Limits

We refer to [Eis95, Appendices 5 and 6] for an introduction to categories and limits: the reader should be aware that our inductive limits are called filtered colimits in Eisenbud's book. We recall some basic definitions.

Definition A.1.1 (*Inductive system*) Let $\mathcal{C}$ be a category and let $(J, \leqslant)$ be a partially ordered set such that $\forall (i, j) \in J^2, \exists k \in J \mid i \leqslant k$ and $j \leqslant k$. We then call $(J, \leqslant)$ a *filtered set*, a *directed set* or a *directed preorder*. An *inductive system* indexed by J is the data of a family $\{M_j\}_{j \in J}$ of objects of $\mathcal{C}$ and morphisms $\varphi_{ij} \colon M_i \to M_j$ for all pairs of indices $(i, j) \in J^2$ such that $i \leqslant j$ which satisfy

1. $\forall j \in J, \varphi_{jj} = \mathrm{id}_{M_j}$;
2. $\forall (i, j, k) \in J^3, i \leqslant j \leqslant k \implies \varphi_{jk} \circ \varphi_{ij} = \varphi_{ik}$.

Definition A.1.2 (*Inductive limit*) Let $\{M_j\}_{j \in J}$ be an inductive system in a category $\mathcal{C}$. A object M in $\mathcal{C}$ is the *inductive limit, direct limit* or *colimit of a filtered set* of the system $\{M_j\}$ if it is equipped with morphisms $\varphi_j \colon M_j \to M$ satisfying the compatibility relations $\varphi_i = \varphi_j \circ \varphi_{ij}$ for every $i \leqslant j$ and having the following universal property: if N is an object in $\mathcal{C}$ equipped with morphisms $\psi_j \colon M_j \to N$ which are compatible with the inductive system structure then there is a unique map $M \to N$ such that for all j the morphism ψ_j factors

F. Mangolte, *Real Algebraic Varieties*, Springer Monographs in Mathematics,
https://doi.org/10.1007/978-3-030-43104-4

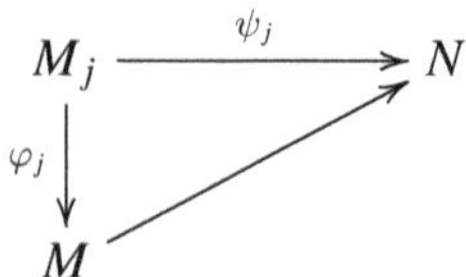

When the inductive limit of the system $\{M_j\}_{j\in J}$ exists we denote it by $\varinjlim_{j\in J} M_j$.

Example A.1.3 If the category $\mathcal{C}$ is the category of groups, rings or A-modules/ algebras for some given ring A then the inductive limit exists: it is simply the quotient of the disjoint union of the M_js modulo an equivalence relation

$$\varinjlim_{j\in J} M_j = \bigsqcup_{j\in J} M_j \Big/ \sim$$

where $x_i \in M_i$ is equivalent to $x_j \in M_j$ if and only if there is a $k \in J$ such that $\varphi_{ik}(x_i) = \varphi_{jk}(x_j)$.

Example A.1.4 Let $\mathcal{F}$ be a sheaf (see Appendix C) of elements of $\mathcal{C}$ over a topological space X. For any given $x \in X$ the set of open neighbourhoods of x ordered by inclusion ($U \leqslant V$ if and only if $U \supseteq V$) is a filtered set and $\{\mathcal{F}(U)\}_{U\ni x}$ is an inductive system. The limit of this system is called the *stalk* of $\mathcal{F}$ at x and is denoted $\mathcal{F}_x$. For every open neighbourhood of x the canonical morphism $\mathcal{F}(U) \to \mathcal{F}_x$ sends a section s of $\mathcal{F}$ over U to the *germ* $s_x \in \mathcal{F}_x$ of s at x.

A.2 Rings, Prime Ideals, Maximal Ideals and Modules

By convention all our rings are assumed to be commutative with a multiplicative unit, and ring morphisms are required to send the multiplicative unit to the multiplicative unit.

The set of invertible elements of A is denoted $U(A)$. If K is a ring then a K-algebra A is a ring equipped with a ring morphism $K \to A$. For example, $K[X_1, \ldots, X_n]$ is the K-algebra of polynomials in n variables with coefficients in K and $K(X_1, \ldots, X_n)$ is the K-algebra of rational functions in n variables with coefficients in K.

Definition A.2.1 A non zero element a in a ring A is said to be a *zero divisor in A* if and only if there is a non zero element $b \in A$ such that $ab = 0$.

A ring is said to be an *integral domain* if and only if it has at least two elements and does not contain a zero divisor.

A *field* is a ring with at least two elements such that all its non zero elements are invertible.

Definition A.2.2 Let A be a ring.

1. An ideal I in A is said to be *prime* if and only if it satisfies the following properties:
   (a) I is not equal to A
   (b) If a and b are elements of A such that $ab \in I$ then $a \in I$ or $b \in I$.
2. An ideal I in A is said to be *maximal* if and only if it is different from A and the only ideals in A containing I are I and A.

Definition A.2.3 For any ideal I in a ring A, the *radical* $\sqrt{I}$ of I in A is the ideal of roots of elements of I.

$$\sqrt{I} := \{a \in A \mid \text{ there is a natural number } n \geqslant 1, a^n \in I\} .$$

An ideal $I \subset A$ is said to be *radical* if and only if $I = \sqrt{I}$.

Exercise A.2.4 (*See Remark* 1.2.29) Let K be a field. Prove that if F is a Zariski closed subset of $\mathbb{A}^n(K)$ then $\mathcal{I}(F)$ is radical.

Definition A.2.5 An element a in a ring A is said to be *nilpotent* if and only if there is a natural number $n > 1$ such that $a^n = 0$. The *nilradical* of a ring A is the set of its nilpotent elements. A ring is said to be *reduced* if and only if its nilradical is the zero ideal, or in other words if it has no non zero nilpotent elements.

Exercise A.2.6 The nilradical of a ring A is an ideal, namely the radical ideal of the zero ideal of A.

Proposition A.2.7 *Let A be a ring and let I be an ideal of A.*

1. *The ideal I is* radical *if and only if the quotient ring A/I is* reduced.
2. *The ideal I is* prime *if and only if the quotient ring A/I is an* integral domain.
3. *The ideal I is* maximal *if and only if the quotient ring A/I is a* field.

Proof Easy exercise. □

Proposition A.2.8 (Correspondence theorem) *Let A be a ring and let $I \subset A$ be an ideal. The canonical surjection $A \to A/I$ induces a one-to-one correspondence between prime ideals of A/I and prime ideals of A containing I and a similar one-to-one correspondence between maximal ideals of A/I and maximal ideals of A containing I.*

Proof Easy exercise. □

The following lemma is extremely useful despite its simplicity.

Lemma A.2.9 *Let A be a ring and let $B \subset A$ be a sub-ring. If I is a prime ideal of A then $I \cap B$ is a prime ideal of B.*

Proof Let a and b be elements of B such that $ab \in I \cap B$ and $a \notin I \cap B$. As $a \in B$ and $a \notin I \cap B$, a does not belong to I. It follows that $b \in I$ because I is a prime ideal of A and hence b belongs to $I \cap B$. □

Example A.2.10 We calculate the dimension of the ideal $I := (x^2 + y^2)$ from Example 1.5.20. There is a unique chain (Definition 1.5.2) of prime ideals in $\mathbb{R}[x, y]$ containing $(x^2 + y^2)$ which is of maximal length

$$(x^2 + y^2) \subset (x, y) .$$

There is therefore only one chain of prime ideals of $\mathbb{R}[x, y]/(x^2 + y^2)$ of maximal length. The dimension of the ring $\mathbb{R}[x, y]/(x^2 + y^2)$ is therefore equal to 1 and according to Definition 1.5.9 $\dim I = 1$.

Lemma A.2.11 (Nakayama's Lemma) *Let A be a ring, let $\mathfrak{a} \subset A$ be an ideal and let M be a finitely generated A-module such that $M = \mathfrak{a}M$. There is then an element $a \in 1 + \mathfrak{a}$ such that $aM = 0$: in particular, if A is local and $\mathfrak{a}$ is its maximal ideal then $M = 0$.*

Proof See [Eis95, Corollary 4.8]. □

Definition A.2.12 A ring S is said to be *graded* if and only if it has a decomposition $S = \oplus_{d \geqslant 0} S_d$ as a direct sum of abelian groups S_d such that for any $d, e \geqslant 0$, $S_d \cdot S_e \subset S_{d+e}$. An ideal $I \subset S$ is said to be a *homogeneous ideal* if and only if $I = \oplus_{d \geqslant 0}(I \cap S_d)$.

A.3 Localisation

Definition A.3.1 Let A be a ring, let M be an A-module and let $S \subset A$ be a *multiplicative subset*[1] or in other words a subset stable under multiplication. The *localised module* (or *localisation*) of M in S, denoted $S^{-1}M$, is the set of equivalence classes of pairs $(m, s) \in M \times S$ for the relation $(m, s) \sim (m', s')$ if and only if there is an element $t \in S$ such that $t(s'm - sm') = 0$. This set is equipped with an obvious A-module structure. The equivalence class of (m, s) is denoted by m/s. When $M = A$, this construction yields the *localisation* $S^{-1}A$ of A at S.

If f is an element of A, the set $S = \{1, f, f^2, \ldots, f^k, \ldots, \}$ is a multiplicative subset of A and we denote these special localisations by $A_f := S^{-1}A$ and $M_f := S^{-1}M$. If f is nilpotent then A_f is the zero ring.

If $\mathfrak{p} \subset A$ is a prime ideal then $S := A \setminus \mathfrak{p}$ is a multiplicative subset and we denote these special localisations by $A_\mathfrak{p} := S^{-1}A$ and $M_\mathfrak{p} := S^{-1}M$.[2] We denote

---

[1] By convention, the product over an empty set is 1, so any multiplicative subset of a ring is assumed to contain the multiplicative unit.

[2] This notation can sometimes be confusing: if K is a field and $A = K[x]$ then $A_x = K[x, \frac{1}{x}]$ (we localise with respect to the multiplicative subset of A generated by x) but

by $\kappa(\mathfrak{p})$ the quotient ring $A_\mathfrak{p}/\mathfrak{p}_\mathfrak{p}$. There is a natural morphism $i\colon M \to S^{-1}M$ defined by $a \mapsto a/1$ and a natural localisation of morphisms: if $\varphi\colon M \to N$ is a morphism of A-modules, the localised morphism $S^{-1}\varphi\colon S^{-1}M \to S^{-1}N$ is defined by $(S^{-1}\varphi)(m/s) = \varphi(m)/s$.

Proposition A.3.2 (Universal property of localisations) *Let A and B be rings, let $S \subset A$ be a multiplicative subset and let $\varphi\colon A \to B$ be a ring morphism such that $\varphi(S) \subset U(B)$ (note that the set $U(B)$ of invertible elements of B is a multiplicative subset). There is then a unique extension $\hat{\varphi}\colon S^{-1}A \to B$ of φ such that the following diagram commutes.*

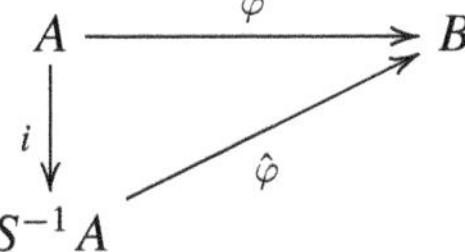

Corollary A.3.3 *The localised ring $S^{-1}A$ is a* flat *A-module (Definition A.4.5).*

Proof See [Eis95, Proposition 2.5, page 66]. □

Exercise A.3.4 Let A be a ring and let $\mathfrak{m} \subset A$ be a maximal ideal. We then have that $A/\mathfrak{m} \simeq A_\mathfrak{m}/\mathfrak{m}_\mathfrak{m}$.

Proposition A.3.5 *Let A be a ring, let $\mathfrak{p} \subset A$ be a prime ideal and let $\varphi\colon A \to A_\mathfrak{p}$, $a \mapsto a/1$ be the natural map. The map $I \mapsto \varphi^{-1}(I)$ from the set of ideals of $A_\mathfrak{p}$ to the set of ideals of A is then injective and induces a bijection between the set of prime ideals of $A_\mathfrak{p}$ and the set of prime ideals of A contained in $\mathfrak{p}$.*

Proof [Eis95, Proposition 2.2]. □

Definition A.3.6 A quotient $\kappa(\mathfrak{m}) = A/\mathfrak{m}$ of a ring A by a maximal ideal $\mathfrak{m}$ is called a *residue field* of A.

Definition A.3.7 A ring is said to be *local* if and only if it has a unique maximal ideal. *The* unique residue field of a local ring A of maximal ideal $\mathfrak{m}$ is denoted

$$\kappa = A/\mathfrak{m} .$$

Definition A.3.8 (*Total ring of fractions*) Let S be the set of non zero divisors of a ring A, which is a multiplicative subset. The *ring of fractions* (or *total ring of fractions*) of A is the localisation

$$\mathrm{Frac} A := S^{-1}A .$$

Proposition A.3.9 (Fraction field) *If A is an integral domain then* Frac$A$ *is a field called the* fraction field *of A.*

$A_{(x)} = \{\frac{p}{q},\ p, q \in K[x], q(0) \neq 0\}$ (we localise with respect to the multiplicative subset which is the complement of the prime ideal generated by x).

Remark A.3.10 If A is an integral domain then the ideal (0) is prime and A satisfies

$$\mathrm{Frac}A = A_{(0)} = \kappa((0)) \ .$$

Proposition A.3.11 *Let A be an integral domain and let $\mathfrak{m} \subset A$ be a maximal ideal. We can think of the localisation $A_{\mathfrak{m}}$ as a subring of* Frac$A$ *and we then have that*

$$A = \bigcap_{\substack{\mathfrak{m} \text{ maximal ideal} \\ \text{of } A}} A_{\mathfrak{m}}$$

Proof As A is an integral domain we may assume that the localisations $A_{\mathfrak{m}}$ are included in Frac$A$ as subrings. By definition $A_{\mathfrak{m}}$ is made up of classes of fractions $\frac{g}{h}$ where $g, h \in A$, $h \notin \mathfrak{m}$. We will now prove that

$$\bigcap_{\substack{\mathfrak{m} \text{ maximal ideal} \\ \text{of } A}} A_{\mathfrak{m}} \subset A \ ,$$

since the opposite inclusion is obvious. Let $f \in \bigcap_{\substack{\mathfrak{m} \text{ maximal ideal} \\ \text{of } A}} A_{\mathfrak{m}}$. If $f = \frac{g}{h}$ for some $g, h \in A$ then h is invertible in A, since otherwise $A.h$ would be contained in a maximal ideal so $\frac{g}{h} \in A$. □

Lemma A.3.12 (Avoidance lemma) *Let A be a ring and let $I_1, \ldots, I_n$ and I be ideals of A such that $I \subset \cup_{\ell=1}^{n} I_\ell$. If A contains an infinite field or if at most two of the I_ℓ are not prime then there is an $\ell \in \{1, \ldots, n\}$ such that $I \subset I_\ell$.*

Proof [Eis95, Lemma 3.3, page 90]. □

Definition A.3.13 (*Noetherian ring*) A ring A is said to be *Noetherian* if and only if every ideal in A is finitely generated.

Example A.3.14 If K is a field then the polynomial ring $K[X_1, \ldots, X_n]$ is Noetherian by Hilbert's famous basis theorem which states that if A is a commutative Noetherian ring then the polynomial ring $A[X]$ is Noetherian. Any ideal $I \subset K[X_1, \ldots, X_n]$ therefore has a *finite* set of generators.

A.4 Tensor Product

Proposition A.4.1 (Universal property of tensor product) *Let A be a ring and let M and N be A-modules. There is then an A-module denoted $M \otimes_A N$ equipped with a A-bilinear map $\psi \colon M \times N \to M \otimes_A N$ which satisfies the following universal property: for any A-module B and any A-bilinear map $\varphi \colon M \times N \to B$, there is a unique A-linear map*

$$\hat{\varphi} \colon M \otimes_A N \to B$$

such that the diagram

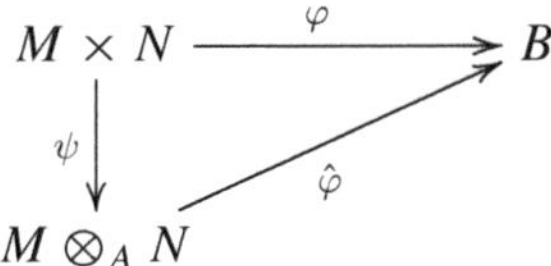

commutes.

The pair $(M \otimes_A N, \psi)$ *is unique up to isomorphism. Most of the time we omit* ψ *from the notation and call the A-module* $M \otimes_A N$ *the* tensor product *of the A-modules M and N.*

See [Eis95, Appendice 2.2] for a proof.

One possible way of constructing the tensor product $M \otimes_A N$ *of two A-modules M and N is to quotient the free abelian group generated by symbols of the form* $m \otimes n := \psi(m, n)$ *by the subgroup generated by elements of the form*

1. $m \otimes n + m' \otimes n - (m + m') \otimes n$
2. $m \otimes n + m \otimes n' - m \otimes (n + n')$
3. $(a \cdot m) \otimes n - m \otimes (a \cdot n)$

for any $m, m' \in M$, $n, n' \in N$ *and* $a \in A$.

Proposition A.4.2 *Let A be a ring and let* M, M' *and N be A-modules. We then have that*

1. $M \otimes_A N \simeq N \otimes_A M$;
2. $(M \oplus M') \otimes_A N \simeq (M \otimes_A N) \oplus (M' \otimes_A N)$.

Proposition A.4.3 *Let* $\varphi \colon A \to B$ *be a ring morphism and let I be an ideal of A. We then have that*

1. $B \otimes_A A/I = B/\varphi(I)B$;
2. $B \otimes_A A[t_1 \dots, t_n] = B[t_1 \dots, t_n]$;
3. *If S is a multiplicative subset of A then*

$$(\varphi(S))^{-1}B = S^{-1}A \otimes_A B \ .$$

Proof See [Eis95, Appendice 2.2]. □

Exercise A.4.4 Following standard notation, we denote by $\mathbb{Z}_m := \mathbb{Z}/m\mathbb{Z}$ the cyclic group of order $m \neq 1$.

1. Prove that if $m > 1$ is an odd integer then

$$\mathbb{Z}_m \otimes_{\mathbb{Z}} \mathbb{Z}_2 = 0 \ .$$

2. Prove that if m is an even integer (by convention $\mathbb{Z}_0 = \mathbb{Z}$) then

$$\mathbb{Z}_m \otimes_{\mathbb{Z}} \mathbb{Z}_2 = \mathbb{Z}_2 \ .$$

Definition A.4.5 Let A be a ring. An A-module M is said to be *flat* if and only if for any injective morphism $N' \to N$ of A-modules the induced morphism

$$M \otimes_A N' \to M \otimes_A N$$

is also injective.

Definition A.4.6 An A-module over a ring A is said to be *free* if and only if it has a basis and *projective* if and only if it is a direct summand of a free A-module.

Lemma A.4.7 *Let A be a ring.*

1. *Any free A-module is projective.*
2. *If A is a local ring then any projective A-module is free.*
3. *Any free A-module is flat.*

Proof See [Eis95, Appendice 3.2]. □

Definition A.4.8 Let A be a ring and let M be an A-module. Let $T^k(M)$ be the tensor product $M \otimes \cdots \otimes M$ of M with itself k times for any $k \geqslant 1$. By convention, we set $T^0(M) = A$. The (non commutative) A-algebra $T(M) = \bigoplus_{k \geqslant 0} T^k(M)$ is called the *tensor algebra* of M. The *symmetric algebra* $S(M) = \bigoplus_{k \geqslant 0} S^k(M)$ of M is the quotient of $T(M)$ by the two-sided ideal generated by expressions of the form $x \otimes y - y \otimes x$ with $x, y \in M$. This A-algebra is commutative. The *exterior algebra* $\bigwedge(M) = \bigoplus_{k \geqslant 0} \bigwedge^k(M)$ of M is the quotient of $T(M)$ by the two-sided ideal generated by expressions of the form $x \otimes x$ with $x \in M$. This A-algebra is anti-symmetric.

Exercise A.4.9 If M is a free A-module of rank r then $S(M) \simeq A[X_1, \ldots, X_r]$.

A.5 Rings of Integers and the Nullstellensatz

Definition A.5.1 (*Integer over A*) Let $\varphi \colon A \to B$ be a ring morphism (B is then an A-algebra) and consider an element $x \in B$. We say that x is *integral over A* if and only if it satisfies a unitary equation, or in other words if there are elements $a_0, \ldots, a_{n-1}$ of A such that:

$$x^n + \varphi(a_{n-1})x^{n-1} + \cdots + f(a_0) = 0\,.$$

If all the elements of B are integral over A, we say that B is an *integral extension of A* and that $\varphi \colon A \to B$ is an *integral morphism.*

Definition A.5.2 The *integral closure* of a subring $A \subset B$ *in* B is the ring of all the elements in B which are integral over A.

A ring A is said to be *integrally closed* if and only if it is an integral domain and it is its own integral closure in the fraction field $\operatorname{Frac} A$.

We refer to Definition 1.5.3 for the definition of the dimension dim A of a ring A.

Proposition A.5.3 *Let A and B be rings and let* $A \to B$ *be an integral morphism. We then have that* $\dim B \leqslant \dim A$. *If moreover* $A \to B$ *is injective then* $\dim A = \dim B$.

Proof See [Liu02, Proposition II.5.10]. □

Definition A.5.4 Let A and B be rings. A ring morphism $A \to B$ is said to be a *finite morphism* if and only if it is an integral morphism and B is a finitely generated A-module for the structure given by the inclusion $A \to B$. We then say that B is an *A-algebra of finite type* (or that B is a *finite* A-algebra).

It is easy to see that if B is a finitely generated A-algebra then B is integral over A (and hence finite over A) if and only if B is a finitely generated A*-module*.

Definition A.5.5 Let K be a field. A K-algebra A is said to be *affine* if and only if it is non zero and finitely generated as an algebra. A ring A is said to be *affine* if and only if there is a field K such that A is an affine K-algebra.

Lemma A.5.6 (Noether's normalisation lemma) *Let K be a field and let A be an* affine *K-algebra. There is then an integer $d \geqslant 0$ and a finite injective morphism*

$$K[X_1, \ldots, X_d] \hookrightarrow A \ .$$

Proof See [Liu02, Proposition II.1.9]. □

Definition A.5.7 Let $L|K$ be a field extension . If there exists an integer d such that L is algebraic over a subfield isomorphic to $K(X_1, \ldots, X_d)$ then we say that L is of finite transcendence degree over K. It is possible to prove that the integer d is unique: d is called the *transcendence degree* of L over K, denoted $\mathrm{trdeg}_K L$.

Definition A.5.8 Let K be a field. A *function field* over K is a field L generated over K by a finite number of elements, by which we mean that there is a finite set $f_1, \ldots, f_r \in L$ such that $L = K(f_1, \ldots, f_r)$. Such a field L is said to be a function field *in* n *variables* if the transcendence degree $\mathrm{trdeg}_K L = n$.

Theorem A.5.9 (Primitive element theorem) *Any field extension which is* finite- *i.e. of finite degree- and* separable *is generated by a single element.*

Corollary A.5.10 *Let K be a field of characteristic zero. If $L|K$ is an algebraic extension of finite type then there is an element $\alpha \in L$ such that*

$$L = K(\alpha) \ .$$

A.5.1 The Nullstellensatz over an Algebraically Closed Field

Definition A.5.11 A field K is said to be *algebraically closed* if and only if any non constant polynomial in $K[X]$ has a root in K.

There are several equivalent versions of Hilbert's Nullstellensatz.

Theorem A.5.12 (Nullstellensatz 1) *If K is an algebraically closed field then all maximal ideals of $K[X_1, \ldots, X_n]$ are of the form $\mathfrak{m}_{(a_1,\ldots,a_n)}=(X_1 - a_1, \ldots, X_l - a_n)$ for some $(a_1, \ldots, a_n) \in K^n$.*

Corollary A.5.13 (Nullstellensatz 2) *Let K be an algebraically closed field and let I be an ideal of $K[X_1, \ldots, X_n]$. We then have that*

$$\mathcal{I}(\mathcal{Z}(I)) = I \quad \text{if and only if } I \text{ is a radical ideal.}$$

A.5.2 The Nullstellensatz over a Real Closed Field

The analogue of Corollary A.5.13 over the real numbers (or more generally over any real closed field) is much less powerful.

Definition A.5.14 Let A be a commutative ring . An ideal I of A is said to be a *real ideal* if and only if for any sequence $a_1, \ldots, a_l$ of elements in A,

$$a_1^2 + \cdots + a_l^2 \in I \implies a_i \in I, \forall i = 1, \ldots, l.$$

Theorem A.5.15 (Real Nullstellensatz) *Let I be an ideal in $\mathbb{R}[X_1, \ldots, X_n]$. We then have that*

$$\mathcal{I}(\mathcal{Z}(I)) = I \quad \text{if and only if } I \text{ is a real ideal.}$$

Proof See [BCR98, 4.1.4]. □

This theorem turns out to hold over any real closed field, such as $\mathbb{R}_{\text{alg}} = \overline{\mathbb{Q}} \cap \mathbb{R}$.

Definition A.5.16 ([BCR98, 1.1.6]) A field K is said to be a *real field* if and only if for any $x_1, \ldots, x_n \in K$,

$$\sum_{k=1}^{n} x_k^2 = 0 \implies x_1 = \cdots = x_n = 0\ .$$

Remark A.5.17 Any real field is of characteristic zero.

Definition A.5.18 A *real closed* field K is a real field which does not have any non trivial real extension.

Theorem A.5.19 ([BCR98, Théorème 1.2.2]) *Let K be a field. The following are equivalent.*

1. *The field K is real closed.*
2. *The field K has a unique total order whose positive cone is exactly the squares of K and any polynomial $K[X]$ of odd degree has a root in K.*
3. *The extension $K[i] = K[X]/(X^2 + 1)$ is an algebraically closed field.*

Remark A.5.20 Only the "true" real number field can be used in problems requiring transcendental methods: we refer to [BCR98, page 2] for more details. For example, the Stone–Weierstrass Theorem 5.2.3 which helps us to compare algebraic and differentiable geometry, holds only over $\mathbb{R}$.

A.6 Quadratic $\mathbb{Z}$-Modules and Lattices

This section draws on [Ser77, Chapitre V]. We recall that any $\mathbb{Z}$-module is an abelian group and that any abelian group has a unique $\mathbb{Z}$-module structure. We further recall that any finitely generated $\mathbb{Z}$-module, M, can be decomposed as a free part and a torsion part

$$M = M_f \oplus \mathrm{Tor}(M) .$$

Definition A.6.1 Let A be a ring. A *quadratic form* on an A-module M is a map $Q \colon M \to A$ such that

1. $Q(ax) = a^2 Q(x)$ for any $a \in A$ and $x \in M$.
2. The map $(x, y) \mapsto Q(x + y) - Q(x) - Q(y)$ is a bilinear form.

Such a pair (M, Q) is called a *quadratic A-module*.

Remark A.6.2 1. The bilinear form $(x, y) \mapsto Q(x + y) - Q(x) - Q(y)$ is clearly *symmetric*.
2. If 2 is invertible in A (this holds in particular if A is a field of characteristic different from 2) then the map

$$(x, y) \mapsto \frac{1}{2}[Q(x + y) - Q(x) - Q(y)]$$

is a symmetric bilinear form sending the pair (x, x) to $Q(x)$ for any $x \in M$. This yields a one to one correspondence between bilinear symmetric forms and quadratic forms on M.

Definition A.6.3 Let K be a field of characteristic different from 2. The *discriminant* of a quadratic K-module (M, Q) is defined to be the determinant of the matrix B_Q of Q with respect to a basis of M, modulo the non zero squares of K. In other words,

$$d(Q) = \det(B_Q) \mod K^{*2} .$$

Remark A.6.4 Let (M, Q) be a free finitely generated quadratic $\mathbb{Z}$-module. As any change of basis matrix for M has determinant ± 1, the determinant of a matrix of Q is independent of the basis. The *discriminant* of the quadratic module (M, Q) is defined to be the determinant of its matrix in any basis of M

$$d(Q) = \det(Q) \ .$$

Definition A.6.5 A free $\mathbb{Z}$-module M of finite rank n (i.e. isomorphic to $\mathbb{Z}^n$ as a $\mathbb{Z}$ module) is a *lattice* if and only if it is equipped with a symmetric bilinear form $(x, y) \mapsto (x \cdot y)$. The lattice M is said to be *integral* if the form $(x \cdot y)$ is an integer for all $x, y \in M$. The *determinant* of a lattice M is the discriminant of the quadratic form $x \mapsto (x \cdot x)$. An integral lattice is said to be *unimodular* if and only if its determinant is ± 1.

Remark A.6.6 Let $(M, (x, y) \mapsto (x \cdot y))$ be an integral lattice. The map $Q\colon M \to \mathbb{Z}, x \mapsto (x \cdot x)$ is then a quadratic form on M. The pair (M, Q) is therefore a torsion free quadratic $\mathbb{Z}$-module of finite rank.

The following result is one of the main reasons for which we are interested in unimodular lattices.

Proposition A.6.7 *The degree* $2m$ *cohomology of a compact oriented simply connected topological manifold without boundary of dimension* $4m$ *with its cup product form is a unimodular lattice.*

Sketch of the Proof Proposition A.6.7 follows from Poincaré duality (Corollary B.7.7). Indeed, if M is a compact oriented topological manifold without boundary of dimension $4m$ then $H^{2m}(M; \mathbb{Z})_f$ is of finite rank and the cup product $H^{2m}(M; \mathbb{Z})_f \times H^{2m}(M; \mathbb{Z})_f \to \mathbb{Z}$ is a non degenerate symmetric bilinear form of determinant 1. If moreover M is simply connected its homology is torsion free and hence $H^{2m}(M; \mathbb{Z})$ is a free $\mathbb{Z}$-module or on other words $H^{2m}(M; \mathbb{Z})_f = H^{2m}(M; \mathbb{Z})$. □

Corollary A.6.8 *Let X be a non singular complex projective algebraic surface with its Euclidean topology. (More generally, X may be a non singular compact analytic surface). The group $H^2(X; \mathbb{Z})$ with cup product is then a unimodular lattice.*

Lemma A.6.9 *Let (M, Q) be a non degenerate quadratic $\mathbb{Z}$-module. Assume that M is free and of finite type. Let A be a submodule of M and set $B = A^{\perp}$. The absolute value of the discriminant of $Q|_A$ (resp. $Q|_B$) is then equal to the index of the subgroup $A \oplus B \subset M$:*

$$|d(Q|_A)| = |d(Q|_B)| = [M : A \oplus B] \ .$$

Proof See [Wil78, Lemma 3.14]. □

Definition A.6.10 The *signature* $(a, b) \in \mathbb{N} \times \mathbb{N}$ of a non degenerate quadratic form Q on a $\mathbb{Z}$-module M is defined to be the signature of the induced quadratic form $Q_{\mathbb{R}}$ on the $\mathbb{R}$-vector space $M \otimes_{\mathbb{Z}} \mathbb{R}$. The *index* of a non degenerate quadratic form Q of signature (a, b) is defined to be

$$\tau(Q) := a - b \ .$$

Remark A.6.11 If Q is non degenerate then so is $Q_{\mathbb{R}}$ and it follows that $a + b = \dim_{\mathbb{R}} M \otimes_{\mathbb{Z}} \mathbb{R} = \mathrm{rk}(M/\mathrm{Tor}(M))$.

Definition A.6.12 Let $(M, (x, y) \mapsto (x \cdot y))$ be a unimodular lattice. The symmetric bilinear form $(x, y) \mapsto (x \cdot y)$ is said to be *even* (or *of type* II) if and only if for any $x \in M$ the integer $(x \cdot x)$ is even. Otherwise, i.e. if there is an $x \in M$ such that $(x \cdot x)$ is odd, the form is said to be *odd* (or *of type* I).

Proposition A.6.13 *Let (M, Q) be a unimodular lattice. If Q is even then its index satisfies*

$$\tau(Q) \equiv 0 \mod 8 \ .$$

Proof See [Ser77, Corollaire 1, Section V.2]. □

A.7 Anti-linear Involutions

Definition A.7.1 Let E be a $\mathbb{C}$-vector space. A map

$$\sigma \colon E \to E$$

is an *anti-linear involution* if and only if $\sigma \circ \sigma = \mathrm{id}_E$ and for all $\lambda \in \mathbb{C}$ and all $a \in E$ we have that

$$\sigma(\lambda a) = \overline{\lambda}\sigma(a) \ .$$

Definition A.7.2 Let $G := \mathrm{Gal}(\mathbb{C}|\mathbb{R})$ be the Galois group, let E be an $\mathbb{R}$-vector space with a G action and let σ be the corresponding involution of E. We denote by $E^G := E^\sigma = \{a \in E \mid \sigma(a) = a\}$ the *subspace of invariants* and by $E^{-\sigma} = \{a \in E \mid \sigma(a) = -a\}$ the *subspace of anti-invariants*.

The following elementary lemma is often useful.

Lemma A.7.3 *Let E be $\mathbb{C}$-vector space of finite dimension with an anti-linear involution σ. Any $\mathbb{R}$-basis of E^σ is then a $\mathbb{C}$-basis of E whose elements are all σ-invariant.*

Proof The map σ is $\mathbb{R}$-linear for the $\mathbb{R}$-vector space structure on E induced by the inclusion $\mathbb{R} \subset \mathbb{C}$. As σ is an involution, its minimal polynomial is $X^2 - 1$. As this polynomial is square free, the $\mathbb{R}$-linear map σ is diagonalisable and is a direct sum of the two eigenspaces associated to the eigenvalues 1 and -1: in other words, $E = E^{\sigma} \oplus E^{-\sigma}$. Let $(a_1, \ldots, a_d)$ be an $\mathbb{R}$-basis of E^{σ}: we then have that $(ia_1, \ldots, ia_d)$ is an $\mathbb{R}$-basis of $E^{-\sigma}$ since $\sigma(ia_k) = -i\sigma(a_k) = -ia_k$ for any $k = 1, \ldots, d$. We therefore have that $\dim_{\mathbb{R}} E^{\sigma} = \dim_{\mathbb{R}} E^{-\sigma} = \frac{1}{2}\dim_{\mathbb{R}} E = \dim_{\mathbb{C}} E$. The $2d$-tuple $(a_1, \ldots, a_d, ia_1, \ldots, ia_d)$ is therefore an $\mathbb{R}$-basis of E and $(a_1, \ldots, a_d)$ is a $\mathbb{C}$-basis of E. □

A.8 Solution to Exercises of Appendix A

A.2.4 By definition, $\mathcal{I}(F) \subset \sqrt{\mathcal{I}(F)}$ and conversely for any $f \in \sqrt{\mathcal{I}(F)}$ there is an $n \in \mathbb{N}^*$ such that $f^n \in \mathcal{I}(F)$, or in other words $\forall x \in F, (f(x))^n = 0$. Since f is K-valued and the field K is reduced we have that $\forall x \in F, f(x) = 0$ and hence $f \in \mathcal{I}(F)$.

A.3.4 Consider the map $\varphi\colon A/\mathfrak{m} \to A_{\mathfrak{m}}/\mathfrak{m}_{\mathfrak{m}}, x \mapsto \frac{x}{1}$. The quotient $A/\mathfrak{m}$ is a field so φ is injective. Conversely, consider an element $\frac{x}{s} \in A_{\mathfrak{m}}$ where $x \in A$ and $s \in A \setminus \mathfrak{m}$. The class of s in $A/\mathfrak{m}$ is non zero and is therefore invertible. There is therefore an $a \in A$ and an $m_0 \in \mathfrak{m}$ such that $1 = as + m_0$: moreover $\frac{x}{s} = \frac{x(as+m_0)}{s} = \frac{(xa)s}{s} + \frac{m_0}{s} \in \frac{xa}{1} + \mathfrak{m}_{\mathfrak{m}}$ and it follows that $\varphi(xa) = \frac{xa}{1} = \frac{x}{s}$ and hence φ is an isomorphism.

A.4.4 1. If m is odd it is coprime with 2 so there is a pair $(u, v) \in \mathbb{Z}^2$ such that $2u + mv = 1$. In $\mathbb{Z}_m \otimes_{\mathbb{Z}} \mathbb{Z}_2$, we therefore have that

$$1 \otimes 1 = (2u + mv) \otimes 1 = 2u \otimes 1 = u \otimes 2 = u \otimes 0 = 0$$

and the conclusion follows because $1 \otimes 1$ generates $\mathbb{Z}_m \otimes_{\mathbb{Z}} \mathbb{Z}_2$.

2. We apply the universal property of tensor product to the map

$$\varphi\colon \mathbb{Z}_m \times_{\mathbb{Z}} \mathbb{Z}_2 \to \mathbb{Z}_2, \ (a, b) \mapsto ab$$

which is $\mathbb{Z}$-bilinear and satisfies $\varphi(1, 1) = 1 \neq 0$. There is therefore a map $\hat{\varphi}\colon \mathbb{Z}_m \otimes_{\mathbb{Z}} \mathbb{Z}_2 \to \mathbb{Z}_2$ such that $\hat{\varphi}(1 \otimes 1) \neq 0$ and in particular $1 \otimes 1$ is not zero in $\mathbb{Z}_m \otimes_{\mathbb{Z}} \mathbb{Z}_2$. Moreover, as $2(1 \otimes 1) = 1 \otimes 2 = 0$, the order of $1 \otimes 1$ in $\mathbb{Z}_m \otimes_{\mathbb{Z}} \mathbb{Z}_2$ is 2 and the result follows because $1 \otimes 1$ generates $\mathbb{Z}_m \otimes_{\mathbb{Z}} \mathbb{Z}_2$.

Appendix B
Topology

Our main reference for this appendix is Hatcher's book [Hat02].

B.1 Hausdorff Spaces

Any real or complex algebraic variety comes equipped with two natural topologies, namely the Zariski and Euclidean topologies (Definitions 1.2.3 and 1.4.1 respectively). We start this chapter by reviewing some important differences between them.

Definition B.1.1 A topological space X is said to be *Hausdorff* if and only if any two distinct points of X have disjoint neighbourhoods.

In Chapter 1 we defined the Zariski topology on $\mathbb{A}^n(K)$ and noted that it was only Hausdorff in a few special cases ($n = 0$ or $n = 1$ and $K = \mathbb{Z}_2$ for example). For algebraic varieties, there is a related notion of *separated space*, motivated by the following elementary result.

Proposition B.1.2 *Let X be a topological space: we equip the cartesian product $X \times X$ with the product topology (i.e. the topology generated by products of open spaces). X is then Hausdorff if and only if the diagonal $\Delta := \{(x, x) \in X \times X \mid x \in X\}$ is closed in $X \times X$.*

Definition B.1.3 An algebraic variety X over a field K is said to be *separated* if and only if the diagonal $\Delta := \{(x, x) \in X \times X \mid x \in X\}$ is closed in the Zariski topology on $X \times X$.

Exercise B.1.4 (*See Exercise* 1.2.2). Let K be an infinite field. The Zariski topology on the product of two affine algebraic sets $X \subset \mathbb{A}^n(K)$ and $Y \subset \mathbb{A}^m(K)$ is the topology induced on $X \times Y$ by the Zariski topology on $\mathbb{A}^{n+m}(K)$. Prove that the Zariski topology on a product $X \times Y$ is not the product of the Zariski topologies on X and Y.

F. Mangolte, *Real Algebraic Varieties*, Springer Monographs in Mathematics,
https://doi.org/10.1007/978-3-030-43104-4

Definition B.1.5 A topological space X is said to be *quasi-compact* if and only if every open cover of X has a finite sub-cover.

Proposition B.1.6 *A topological space X is Noetherian (Definition 1.2.22) if and only if every subspace of X is quasi-compact. In particular, every subspace of an algebraic set is quasi-compact in the Zariski topology.*

Definition B.1.7 A topological space is said to be *compact* if and only if it is both quasi-compact and Hausdorff.

The following proposition summarises the links between compactness and separation of algebraic varieties.

Proposition B.1.8 *Let X be a real or complex algebraic variety.*

1. *X is quasi-compact and every subspace of X is quasi-compact in the Zariski topology. If the variety X is projective, quasi-projective, affine or quasi-affine then X is a separated algebraic variety.*
2. *If we equip X with its Euclidean topology then X becomes Hausdorff. If X is projective then it is compact.*

B.2 Semi-algebraic Sets

This brief section contains several useful definitions: see [BCR98, Chapters 2 and 8] for a more detailed presentation.

Definition B.2.1 A *semi-algebraic set* in $\mathbb{R}^n$ is a finite union of sets of the form

$$\{x \in \mathbb{R}^n \mid P_1(x) = \cdots = P_l(x) = 0 \text{ and } Q_1(x) > 0, \ldots, Q_m(x) > 0\}$$

where $P_i, i = 1, \ldots, l$ and $Q_j, j = 1, \ldots, m$ are elements of $\mathbb{R}[X_1, \ldots, X_n]$.

Definition B.2.2 Let $A \subset \mathbb{R}^m$ and $B \subset \mathbb{R}^n$ be semi-algebraic sets. A map $f: A \to B$ is said to be *semi-algebraic* if and only if its graph is semi-algebraic in $\mathbb{R}^{m+n}$.

Definition B.2.3 Let $A \subset \mathbb{R}^m$ be an open semi-algebraic set. A *Nash function* $f: A \to \mathbb{R}$ is a function which is both semi-algebraic and C^∞-differentiable. Let $A \subset \mathbb{R}^m$ and $B \subset \mathbb{R}^n$ be open semi-algebraic sets. A *Nash map* $f: A \to B$ is a map which is both semi-algebraic and C^∞-differentiable.

There is an implicit functions theorem for Nash maps (see [BCR98, Proposition 2.9.7 and Corollary 2.9.8]) which justifies the following definition.

Definition B.2.4 Let $A \subset \mathbb{R}^m$ and $B \subset \mathbb{R}^n$ be open semi-algebraic sets. A map $f: A \to B$ is said to be a *Nash diffeomorphism* if and only if f is both a Nash map and a C^∞ diffeomorphism.

B.3 Simplicial Complexes and Homology

Definition B.3.1 An abstract *simplicial complex*[3] is a set K whose elements are called *vertices* equipped with a family of non empty finite subsets of K called *simplexes* such that every vertex is contained in at least one simplex and every non empty subset of a simplex is a simplex. A non empty subset of a simplex is said to be a *face* of the simplex. A *simplicial map* is a map $\varphi\colon K \to K'$ between simplicial complexes which sends simplexes to simplexes. A *simplicial pair* is a pair (K, L) of simplicial complexes such that $L \subset K$ and every simplex of L is a simplex of K.

We denote by $|K|$ the *geometric realisation* of a simplicial complex K. See [Spa66, III.1] or [Hat02, Section 2.1] for more details.

Definition B.3.2 A topological space X is said to be *triangulable* if and only if there is a simplicial complex K such that its geometric realisation $|K|$ is homeomorphic to X. A compact triangulable space—in other words, a space which is homeomorphic to the geometric realisation of a finite simplicial complex—is often called a *polyhedron*.

Remark B.3.3 (*Triangulation of (semi)-algebraic sets*) It has been known for a long time that every real or complex quasi-projective algebraic set is triangulable, as is every real or complex analytic set. We refer to [Wae30], [KB32], or [LW33] for the proof of this result. See [Hir75, Theorem, page 170] for a "modern" proof inspired by work of Łojasiewicz [Łoj64]. The interested reader may also wish to read [BCR98, Section 9.2] or [Cos02, Chapter 3]. The result proved by Hironaka is actually more general: he shows that any disjoint union of a finite number of semi-algebraic sets is triangulable. Recall that as in Definition B.2.1 a semi-algebraic set is a subset of $\mathbb{R}^N$ (or more generally of K^N for any real closed field K) for some N defined by polynomial equalities and inequalities. A real algebraic set is therefore a special case of a semi-algebraic set. To prove that Hironaka's theorem implies that any affine (for example) complex algebraic variety X is triangulable, we start by embedding X as an algebraic subset of $\mathbb{C}^n$ for some n. Separating the real and imaginary parts of the equations defining X, we obtain an embedding of the underlying Euclidean space of X in $\mathbb{R}^{2n}$: X can therefore be seen as a real algebraic set in $\mathbb{R}^{2n}$ and we now apply Hironaka's theorem.

Definition B.3.4 The *barycentric subdivision* of a simplicial complex K is a simplicial complex K' whose vertices are the simplexes of K and whose simplices are the sets $\{s_0, \ldots, s_n\}$ of simplexes of K (i.e. vertices of K') such that after permutation

$$s_0 \subset s_1 \subset \cdots \subset s_n .$$

In other words, s_i is a face of s_{i+1} for all $i = 0 \ldots n-1$.

[3] Such an object is called a *Schéma simplicial* in [God58, II.3.2].

Remark B.3.5 Simplicial complexes can be generalised to *cell complexes*, or *CW-complexes* as in [Hat02, Chapter 0 and Appendix]. In other words, every simplicial complex, and in particular every polyhedron and every graph, has a natural CW-complex structure. Moreover, every CW-complex is homotopy equivalent to a simplicial complex of the same dimension, [Hat02, Theorem 2C.5]. Every differentiable manifold and every real or complex quasi-projective algebraic variety has the homotopy type of a CW-complex (in the Euclidean topology). All these complexes are used for the same reason: calculating *singular homology*. When the space is *triangulable*—i.e. decomposable as a simplicial complex—we can calculate singular homology via simplicial homology. The modern method for this reduction is Δ-complexes [Hat02, Chapter II]: the definition of the simplicial homology of a Δ-complex given by Hatcher [*Ibid.*, Section 2.1] applies directly to the special case of a simplicial complex K provided we use its realisation as a topological space to *orient* it correctly.

Theorem B.3.6 (Homology long exact sequence of a pair) *Let G be an abelian group (in practice we generally take $G = \mathbb{Z}, \mathbb{Z}_2, \mathbb{Q}, \mathbb{C}$ or $\mathbb{R}$) and let (X, A) be a topological pair. We then have the following exact sequence*

$$\cdots \to H_k(A; G) \to H_k(X; G) \to H_k(X, A; G) \to H_{k-1}(A; G) \to \cdots \\ \to H_0(X, A; G) \to 0 \quad \text{(B.1)}$$

Proof See [Hat02, 2.13, pages 114–117]. □

Theorem B.3.7 (Cohomology long exact sequence of a pair) *Let G be an abelian group and let (X, A) be a topological pair. We then have the following exact sequence:*

$$\cdots \to H^k(X, A; G) \to H^k(X; G) \to H^k(A; G) \to H^{k+1}(X, A; G) \to \cdots \quad \text{(B.2)}$$

Proof See [Hat02, Section 3.1, page 200]. □

Definition B.3.8 Let G be a group and let $[G, G]$ be its *derived subgroup* or in other words the subgroup of G generated by its *commutators* $[x, y] := xyx^{-1}y^{-1}$. The derived subgroup is distinguished in G and the quotient group

$$G^{ab} := G/[G, G]$$

is an abelian group called the *abelianisation of G*.

Theorem B.3.9 (Hurewicz' theorem) *Let X be a path connected topological space and let x be a point in X. To any loop $\gamma\colon [0, 1] \to X$ passing through x there corresponds a 1-chain which is a cycle whose class is an element of $H_1(X; \mathbb{Z})$. This correspondence induces a functorial isomorphism between the abelianisation of $\pi_1(X, x)$ and $H_1(X; \mathbb{Z})$. In particular, if $\pi_1(X)$ is abelian then $H_1(X; \mathbb{Z}) \simeq \pi_1(X)$.*

Proof See [Hat02, Theorem 2.A1]. □

B.4 Universal Coefficients Theorem

Recall that as in Section A.6, any finitely generated $\mathbb{Z}$-module M decomposes as a free part and a torsion part

$$M = M_f \oplus \mathrm{Tor}(M)$$

where $M_f \simeq \mathbb{Z}^r$ and $\mathrm{Tor}(M) \simeq \mathbb{Z}_{d_1} \oplus \mathbb{Z}_{d_2} \oplus \cdots \oplus \mathbb{Z}_{d_l}$ and for any $i < l, d_i > 1$ and $d_i | d_{i+1}$.

Theorem B.4.1 (Universal coefficients in cohomology) *Let X be a topological space and let G be an abelian group. For any k the sequence*

$$0 \to \mathrm{Ext}(H_{k-1}(X;\mathbb{Z}), G) \to H^k(X;G) \to \mathrm{Hom}(H_k(X;\mathbb{Z}), G) \to 0$$

is then exact and split.

Proof See [Hat02, Theorem 3.2, page 195]. □

Corollary B.4.2 *If the groups $H_k(X;\mathbb{Z})$ and $H_{k-1}(X;\mathbb{Z})$ are finitely generated then*

$$H^k(X;\mathbb{Z}) \simeq (H_k(X;\mathbb{Z})/\mathrm{Tor}(H_k(X;\mathbb{Z}))) \oplus \mathrm{Tor}(H_{k-1}(X;\mathbb{Z}))\ .$$

Proof See [Hat02, Corollary 3.3]. □

Theorem B.4.3 (Universal coefficients theorem in homology) *Let X be a topological space and let G be an abelian group. For any natural number k the sequence*

$$0 \to H_k(X;\mathbb{Z}) \otimes G \to H_k(X;G) \to \mathrm{Tor}(H_{k-1}(X;\mathbb{Z}), G) \to 0$$

is exact and split.

Proof See [Hat02, Theorem 3A.3, page 264]. □

Corollary B.4.4 *Let X be a topological space. For any natural number k we then have that*

$$H_k(X;\mathbb{C}) = H_k(X;\mathbb{Q}) \otimes_{\mathbb{Q}} \mathbb{C}\ ,$$

We now apply Theorems B.4.1 and B.4.3 to cohomology with coefficients in $\mathbb{Z}_2$:

Corollary B.4.5 *Let X be a topological space. For any k the sequences*

$$0 \to \mathrm{Ext}(H_{k-1}(X;\mathbb{Z}), \mathbb{Z}_2) \to H^k(X;\mathbb{Z}_2) \to \mathrm{Hom}(H_k(X;\mathbb{Z}), \mathbb{Z}_2) \to 0$$
$$0 \to H_k(X;\mathbb{Z}) \otimes \mathbb{Z}_2 \to H_k(X;\mathbb{Z}_2) \to \mathrm{Tor}(H_{k-1}(X;\mathbb{Z}), \mathbb{Z}_2) \to 0$$

are then exact and split.

Remark B.4.6 Let $m > 1$ be a natural number. The image of the multiplication by m map

$$\mathbb{Z}_2 \xrightarrow{\times m} \mathbb{Z}_2$$

vanishes if and only if m is even. It follows (see [Hat02, page 195 and Proposition 3A.5, page 265]) that:

- If m is even then $\mathrm{Ext}(\mathbb{Z}_m, \mathbb{Z}_2) \simeq \mathbb{Z}_2$ and $\mathrm{Tor}(\mathbb{Z}_m, \mathbb{Z}_2) \simeq \mathbb{Z}_2$;
- if m is odd then $\mathrm{Ext}(\mathbb{Z}_m, \mathbb{Z}_2) \simeq 0$ and $\mathrm{Tor}(\mathbb{Z}_m, \mathbb{Z}_2) \simeq 0$.

Moreover, we have that

$$\mathrm{Ext}(M \oplus M', \mathbb{Z}_2) \simeq \mathrm{Ext}(M, \mathbb{Z}_2) \oplus \mathrm{Ext}(M', \mathbb{Z}_2) \; ;$$
$$\mathrm{Tor}(M \oplus M', \mathbb{Z}_2) \simeq \mathrm{Tor}(M, \mathbb{Z}_2) \oplus \mathrm{Tor}(M', \mathbb{Z}_2) \; .$$

If X is compact we can calculate the group $H_k(X; \mathbb{Z}_2)$ using the above remark and the invariant factors of the finitely generated $\mathbb{Z}$-module $H_{k-1}(X; \mathbb{Z})$:

$$H_{k-1}(X; \mathbb{Z}) = \mathbb{Z}_{j_1} \oplus \cdots \oplus \mathbb{Z}_{j_t} \oplus \mathbb{Z} \oplus \cdots \oplus \mathbb{Z} \; ,$$

where j_i divides j_{i+1} for every i. Let l be the number of even invariant factors j_i: we then have that $\mathrm{Ext}(H_{k-1}(X; \mathbb{Z}), \mathbb{Z}_2) \simeq \oplus^l \mathbb{Z}_2$ and $\mathrm{Tor}(H_{k-1}(X; \mathbb{Z}), \mathbb{Z}_2) \simeq \oplus^l \mathbb{Z}_2$. It follows that

$$H^k(X; \mathbb{Z}_2) \simeq \mathrm{Hom}(H_k(X; \mathbb{Z}), \mathbb{Z}_2) \oplus^l \mathbb{Z}_2 \; ;$$
$$H_k(X; \mathbb{Z}_2) \simeq H_k(X; \mathbb{Z}) \otimes \mathbb{Z}_2 \oplus^l \mathbb{Z}_2 \; .$$

Example B.4.7 (*Homology of real projective spaces*)

1. Homology of $\mathbb{RP}^2$. We have that $H_0(\mathbb{RP}^2; \mathbb{Z}) \simeq \mathbb{Z}$, $H_1(\mathbb{RP}^2; \mathbb{Z}) \simeq \mathbb{Z}_2$ and $H_2(\mathbb{RP}^2; \mathbb{Z}) = \{0\}$. It follows that $H_0(\mathbb{RP}^2; \mathbb{Z}_2) \simeq H_1(\mathbb{RP}^2; \mathbb{Z}_2) \simeq H_2(\mathbb{RP}^2; \mathbb{Z}_2) \simeq \mathbb{Z}_2$ because $\mathrm{Tor}(H_1(\mathbb{RP}^2; \mathbb{Z})) \simeq \mathbb{Z}_2$.
2. Homology of $\mathbb{RP}^3$. We have that $H_0(\mathbb{RP}^3; \mathbb{Z}) \simeq \mathbb{Z}$, $H_1(\mathbb{RP}^3; \mathbb{Z}) \simeq \mathbb{Z}_2$, $H_2(\mathbb{RP}^3; \mathbb{Z}) = \{0\}$ and $H_3(\mathbb{RP}^3; \mathbb{Z}) \simeq \mathbb{Z}$. It follows that $H_0(\mathbb{RP}^3; \mathbb{Z}_2) \simeq H_1(\mathbb{RP}^3; \mathbb{Z}_2) \simeq H_2(\mathbb{RP}^3; \mathbb{Z}_2) \simeq H_3(\mathbb{RP}^3; \mathbb{Z}_2) \simeq \mathbb{Z}_2$ because $\mathrm{Tor}(H_1(\mathbb{RP}^3; \mathbb{Z})) \simeq \mathbb{Z}_2$.

Definition B.4.8 (*Reduced homology*) The reduced homology groups of a non empty topological space X are defined as follows:

$$\widetilde{H}_k(X; \mathbb{Z}) = H_k(X; \mathbb{Z})$$

for $k > 0$ and

$$\widetilde{H}_0(X; \mathbb{Z}) \oplus \mathbb{Z} \simeq H_0(X; \mathbb{Z}) \; .$$

Example B.4.9 (*Reduced homology uniformises statements*)

1. The reduced homology of a point (and indeed of any contractible space) is trivial for all k (including $k = 0$),

$$\forall k \quad \widetilde{H}_k(\{x\}; \mathbb{Z}) = \{0\} \; .$$

2. The reduced homology of a sphere of dimension n is concentrated in dimension n:

$$\begin{cases} \widetilde{H}_n(\mathbb{S}^n; \mathbb{Z}) & \simeq \mathbb{Z} \\ \widetilde{H}_k(\mathbb{S}^n; \mathbb{Z}) & = 0 \quad \forall k \neq n \; . \end{cases}$$

   We note for completeness' sake that since $\mathbb{S}^0 = \{-1, 1\}$ we have that

$$H_0(\mathbb{S}^0; \mathbb{Z}) \simeq \mathbb{Z} \oplus \mathbb{Z} \quad \text{and} \quad \widetilde{H}_0(\mathbb{S}^0; \mathbb{Z}) \simeq \mathbb{Z} \; .$$

Proposition B.4.10 *Let (X, L) be a topological pair. If X is contractible and L is non empty then*

$$H_k(X, L; \mathbb{Z}) \simeq \widetilde{H}_{k-1}(L; \mathbb{Z})$$

for every natural number k.

Proof We deduce the following exact sequence from the exact sequence (B.1).

$$\cdots \to H_{k+1}(X, L; \mathbb{Z}) \to \widetilde{H}_k(L; \mathbb{Z}) \to \widetilde{H}_k(X; \mathbb{Z}) \to H_k(X, L; \mathbb{Z}) \to \cdots$$

The desired result follows because the reduced homology of a contractible space is trivial for all k. □

Example B.4.11 For any $n > 0$ we have that

$$\begin{cases} H_n(\mathbb{R}^n, \mathbb{R}^n \setminus \{0\}; \mathbb{Z}) & \simeq \mathbb{Z} \\ H_k(\mathbb{R}^n, \mathbb{R}^n \setminus \{0\}; \mathbb{Z}) & = 0 \quad \forall k \neq n \; . \end{cases}$$

Proof The group $H_k(\mathbb{R}^n, \mathbb{R}^n \setminus \{0\}; \mathbb{Z})$ is isomorphic to the reduced homology group $\widetilde{H}_{k-1}(\mathbb{R}^n \setminus \{0\}; \mathbb{Z})$ because $\mathbb{R}^n$ is contractible (Proposition B.4.10) which is in turn isomorphic to $\widetilde{H}_{k-1}(\mathbb{S}^{n-1}; \mathbb{Z})$ because $\mathbb{R}^n \setminus \{0\}$ is homeomorphic to $\mathbb{S}^{n-1}$. We now simply apply Example B.4.9(2). □

Note that for $n = k = 1$ we have that $H_{k-1}(\mathbb{R}^n \setminus \{0\}; \mathbb{Z}) \simeq \mathbb{Z} \oplus \mathbb{Z}$.

B.5 Topological and Differentiable Manifolds and Orientability

This section is based on [Hat02, 3.3].

Definition B.5.1 A *topological manifold* M of *dimension* n is a Hausdorff topological space such that every point has a neighbourhood which is homeomorphic to $\mathbb{R}^n$.

Proposition B.5.2 *The dimension of a topological manifold M can be characterised intrinsically using the fact that for any $x \in M$ the local homology group*

$$H_k(M, M \setminus \{x\}; \mathbb{Z})$$

is non zero only for $k = n$ and in this case it is isomorphic to $\mathbb{Z}$.

Proof Suppose that $n > 0$. Since M is locally homeomorphic to $\mathbb{R}^n$ the group $H_k(M, M \setminus \{x\}; \mathbb{Z})$ is isomorphic to $H_k(\mathbb{R}^n, \mathbb{R}^n \setminus \{0\}; \mathbb{Z})$ by excision ([Hat02, Theorem 2.20]). The result then follows from Example B.4.11.

If $n = 0$ and M is connected then $M = \{x\}$ and $H_k(M, M \setminus \{x\}; \mathbb{Z}) = H_k(\{x\}; \mathbb{Z})$. The non connected case follows from the connected case. □

Definition B.5.3 Let M be a topological manifold of dimension n. An *orientation* of M is a function $M \to \sqcup_{x \in M} H_n(M, M \setminus \{x\}; \mathbb{Z})$ associating to every $x \in M$ a generator μ_x of the cyclic group $H_n(M, M \setminus \{x\}; \mathbb{Z})$, subject to the following local constancy condition. We require that every point $x \in M$ should have a neighbourhood $B \subset \mathbb{R}^n \subset M$ which is an open ball of finite radius in $\mathbb{R}^n$ such that for every point $y \in B$ the local orientation μ_y is the image of a fixed generator $\mu_B \in H_n(M, M \setminus B; \mathbb{Z}) \simeq H_n(\mathbb{R}^n, \mathbb{R}^n \setminus B; \mathbb{Z})$ under the natural maps $H_n(M, M \setminus B; \mathbb{Z}) \to H_n(M, M \setminus \{y\}; \mathbb{Z})$.

A manifold M with an orientation is said to be *orientable*: a manifold without any orientation is said to be *non orientable*.

Remark B.5.4 It follows immediately from the above definition that M is orientable if and only if all its connected components are orientable and M is non orientable if and only if at least one of its connected components is non orientable.

Remark B.5.5 In particular, every topological manifold M of dimension 1 is orientable. Indeed, consider a connected component M_0 of M and note that an orientation at a point x determines an orientation at every point y of the same connected component via the canonical isomorphism $H_1(M_0, M_0 \setminus \{x\}; \mathbb{Z}) \simeq H_1(M_0, M_0 \setminus B; \mathbb{Z}) \simeq H_1(M_0, M_0 \setminus \{y\}; \mathbb{Z})$ where B is the image of any path in M_0 passing through x and y.

Proposition B.5.6 (Non compact manifolds*) Let M be a connected non compact manifold of dimension n. For any $k \geqslant n$ we then have that*

$$H_k(M; \mathbb{Z}) = H_k(M; \mathbb{Z}_2) = 0 .$$

See [Hat02, Proposition 3.29].

Theorem B.5.7 (Compact manifolds) *Let M be a* compact *connected manifold of dimension n.*[4] *The homology groups in degree n or more are as follows.*

1. $H_n(M; \mathbb{Z}_2) \simeq \mathbb{Z}_2$,
2. $H_n(M; \mathbb{Z}) \simeq \mathbb{Z}$ *if M is orientable,*
3. $H_n(M; \mathbb{Z}) \simeq 0$ *if M is non orientable,*
4. $H_k(M; \mathbb{Z}) = 0$ *if $k > n$.*

See [Hat02, Theorem 3.26].

Remark B.5.8 (*Fundamental class*) It follows that any compact connected manifold M without boundary of dimension n has a fundamental $\mathbb{Z}_2$-homology class $[M] \in H_n(M; \mathbb{Z}_2)$. If moreover the manifold M is *oriented* then it has a fundamental $\mathbb{Z}$-homology class $[M] \in H_n(M; \mathbb{Z})$. See [Hat02, Theorem 3.26], for example, for more details.

Corollary B.5.9 *Let M be a compact connected topological manifold of dimension n. The torsion subgroup $H_{n-1}(M; \mathbb{Z})$ then satisfies*

1. $\mathrm{Tor}(H_{n-1}(M; \mathbb{Z})) = 0$ *if M is orientable,*
2. $\mathrm{Tor}(H_{n-1}(M; \mathbb{Z})) \simeq \mathbb{Z}_2$ *if M is non orientable.*

See [Hat02, Corollary 3.28].

Proposition B.5.10 *A differentiable manifold M is orientable if and only if it has an atlas $\mathcal{A}$ such that the transition maps preserve orientation, or in other words, such that*

$$\forall (U_1, \varphi_1), (U_2, \varphi_2) \in \mathcal{A}, U_1 \cap U_2 \neq \varnothing \implies \forall x \in U_1 \cap U_2, \det d_x(\varphi_1 \circ \varphi_2^{-1}) > 0 .$$

Proof See [Hir76, Section 4.4]. □

Exercise B.5.11 Any complex analytic variety is a differentiable manifold whose transition maps are complex analytic. Prove that such a manifold is not only orientable, but *oriented* in the sense that the complex analytic manifold structure (i.e. the data of a maximal atlas whose transition maps are holomorphic) induces a canonical orientation.

B.5.1 Connected Sums

The operation 'connected sum of two topological surfaces' enables us to equip the set of classes of compact surfaces without boundary up to homeomorphism with

---

[4] We will sometimes say that M is *closed* to emphasise the fact that M is compact and has no boundary.

a semigroup structure generated by the class of the torus $\mathbb{T}^2$ and the real projective plane $\mathbb{RP}^2$. The identity element is the class of $\mathbb{S}^2$ and the only relation is $\mathbb{T}^2\#\mathbb{RP}^2 = \mathbb{RP}^2\#\mathbb{RP}^2\#\mathbb{RP}^2$.

Let M_1 and M_2 be two connected manifolds of the same dimension n. Let $B_1 \subset M_1$ and $B_2 \subset M_2$ be two open balls: the complements $F_1 := M_1 \setminus B_1$ and $F_2 := M_2 \setminus B_2$ are then manifolds whose boundary is homeomorphic to a sphere $\mathbb{S}^{n-1}$. Identifying F_1 and F_2 by a diffeomorphism along the boundary spheres ∂F_1 and ∂F_2 we obtain a manifold without boundary. See [Laf96, Laf15, Exercice II.28] or [Die70, 16.26 problèmes 12 à 15] for details of the "gluing" of the manifolds F_1 and F_2. Under certain conditions the resulting manifold is unique [*Ibid.*]:

Definition B.5.12 (*Connected sum*) If each of the manifolds M_1 and M_2 is non orientable or has an orientation-reversing differentiable automorphism then the manifold obtained from the above surgery is uniquely determined (up to homeomorphism) by M_1 and M_2 and is called the *connected sum* of M_1 and M_2, denoted $M_1\#M_2$.

Remark B.5.13 As any orientable surface S has an orientation reversing differentiable automorphism (and moreover there is always such a map without fixed points whose quotient is the non orientable surface whose Euler characteristic is half that of S) the connected sum of two arbitrary surfaces is well defined.

Exercise B.5.14 Any real projective space of even dimension is non orientable. Any real projective space of odd dimension is orientable and has an orientation-reversing differentiable automorphism. Any complex projective space is orientable and such a space has an orientation-reversing differentiable automorphism if and only if its (complex) dimension is odd. (See Proposition 2.2.28).

We will also use oriented connected sums.

Definition B.5.15 (*Oriented connected sums*) Let M_1 and M_2 be two connected oriented manifolds of the same dimension n. Let $B_1 \subset M_1$ and $B_2 \subset M_2$ be open balls. The complements $F_1 := M_1 \setminus B_1$ and $F_2 := M_2 \setminus B_2$ are manifolds with boundary whose boundary is homeomorphic to a sphere $\mathbb{S}^{n-1}$. Identifying F_1 and F_2 along the spheres ∂F_1 and ∂F_2 by a diffeomorphism which is compatible with the induced orientations we get an oriented manifold without boundary uniquely determined (up to homeomorphism) by the oriented manifolds M_1 and M_2, which we call the *connected sum* $M_1\#M_2$.

Remark B.5.16 If we denote by $-M_2$ the manifold M_2 with the inverse orientation then the connected sums $M_1\#M_2$ and $M_1\#-M_2$ are not generally homeomorphic. In dimension 2, however, $M_1\#M_2$ and $M_1\#-M_2$ are always homeomorphic: in dimension 3 we have that $M_1\#M_2$ and $M_1\#-M_2$ are homeomorphic whenever $M_2 = \mathbb{RP}^3$ or $M_2 = \mathbb{S}^2 \times \mathbb{S}^1$. See [Hem76, Chapter 3] for more details.

B.5.2 Spin Structures

Our main references for this section are [LM89, Chapter II], [Mil63b]. The group $\mathbf{SO}(n)$ of positive isometries is a topological group whose fundamental group is

$$\begin{cases} \pi_1(\mathbf{SO}(2)) & = \mathbb{Z} \\ \pi_1(\mathbf{SO}(n)) & = \mathbb{Z}_2 \quad \forall n > 2\,. \end{cases}$$

Definition B.5.17 (*The group* $\mathbf{Spin}(n)$) For all $n > 1$ we let the group $\mathbf{Spin}(n)$ be the unique double cover of $\mathbf{SO}(n)$.

$$1 \to \mathbb{Z}_2 \to \mathbf{Spin}(n) \to \mathbf{SO}(n) \to 1$$

Remark B.5.18 For any $n > 2$ the group $\mathbf{Spin}(n)$ is the universal cover of $\mathbf{SO}(n)$.

Definition B.5.19 An oriented differentiable manifold is said to be a *spin manifold* if and only if its tangent bundle has at least one spin structure. See [LM89, Chapter II, Definition 1.3, Remark 1.9] for more details.

Proposition B.5.20 *A differentiable oriented manifold V is a spin manifold if and only if its second Stiefel–Whitney class vanishes,* $\mathrm{w}_2(T_V) = 0$.

Proof See [LM89, Chapter II, Theorem 2.1] or the original paper [BH59, page 350] for more details. □

B.5.3 Topologies on a Family of Maps

Let k be an element of $\mathbb{N} \cup \{\infty\}$. For any two $\mathcal{C}^k$ differentiable manifolds V and W we denote by $\mathcal{C}^k(V, W)$ the set of differentiable $\mathcal{C}^k$ maps from V to W.

Definition B.5.21 (*Weak topology*) For any natural number k, the *weak* (or *$\mathcal{C}^k$-compact-open*) topology on the set $\mathcal{C}^k(V, W)$ is the topology generated by open sets $\Omega\left(f; (\varphi, U); (\psi, U'); K; \varepsilon\right)$ defined as follows. Consider a function $f \in \mathcal{C}^k(V, W)$, a chart (φ, U) of V, a chart (ψ, U') of W and a compact set $K \subset U$ such that $f(K) \subset U'$ and $0 < \varepsilon \leqslant \infty$. The open set

$$\Omega\left(f; (\varphi, U); (\psi, U'); K; \varepsilon\right)$$

contains those functions $g\colon V \to W$ of class $\mathcal{C}^k$ such that $g(K) \subset U'$ and

$$\left\| D^l(\psi f \varphi^{-1})(x) - D^l(\psi g \varphi^{-1})(x) \right\| < \varepsilon$$

for any $x \in \varphi(K)$ and any $l = 0, \ldots, k$.

For $k=\infty$ the *weak* topology on the set $\mathcal{C}^\infty(V,W)$ is the union of all the topologies induced by the inclusion $\mathcal{C}^k(V,W)\hookrightarrow\mathcal{C}^\infty(V,W)$ for some finite k.

Definition B.5.22 (*Strong topology*) For any natural number k the *strong* (or *Whitney*) topology on the set $\mathcal{C}^k(V,W)$ is generated by open sets $\Omega(f;\Phi;\Psi;K;\varepsilon)$ defined as follows. Let $\Phi=\{(\varphi_i,U_i)\}_{i\in\Lambda}$ be a *locally finite* family of charts on V, by which we mean that every point x in V has a neighbourhood meeting only a finite number of U_is. Let $K=\{K_i\}_{i\in\Lambda}$ be a family of compact subsets of V such that $K_i\subset U_i$ for all i. Let $\Psi=\{(\psi_i,U_i')\}_{i\in\Lambda}$ be a family of charts of W and let $\varepsilon=\{\varepsilon_i\}_{i\in\Lambda}$ be a family of strictly positive real numbers. For any map $f\in\mathcal{C}^k(V,W)$ sending K_i to U_i' the open set

$$\Omega(f;\Phi;\Psi;K;\varepsilon)$$

contains those functions $g\colon V\to W$ of class $\mathcal{C}^k$ such that for every $i\in\Lambda$, $g(K_i)\subset U_i'$ and

$$\left\|D^l(\psi_i f\varphi_i^{-1})(x)-D^l(\psi_i g\varphi_i^{-1})(x)\right\|<\varepsilon_i$$

for any $x\in\varphi_i(K_i)$ and any $l=0,\dots,k$.

For $k=\infty$ the *strong* topology on the set $\mathcal{C}^\infty(V,W)$ is the union of topologies induced by the inclusions $\mathcal{C}^k(V,W)\hookrightarrow\mathcal{C}^\infty(V,W)$ for all finite k.

Remark B.5.23 When V is compact the weak and strong topologies are equivalent on $\mathcal{C}^k(V,W)$. See [Hir76, Section 2.1] for more details.

B.6 Cohomology

Let X be a topological space and let G be a ring (in practice we generally have $G=\mathbb{Z},\mathbb{Z}_2,\mathbb{Q},\mathbb{C}$ or $\mathbb{R}$). We denote by $C^k(X;G)$ the group of cochains $\phi\colon C_k(X)\to G$ and by $H^k(X;G)$ the cohomology groups associated to the cochain complex

$$\cdots\to C^k(X;G)\to C^{k+1}(X;G)\to\dots \tag{B.3}$$

Definition B.6.1 (*Cup-product*) Let X be a topological space and let l,k be integers. There is a bilinear map called the *cup-product*:

$$\smile\colon H^l(X;\mathbb{Z})\times H^k(X;\mathbb{Z})\longrightarrow H^{l+k}(X;\mathbb{Z})$$

sending the class of an l-cochain $\psi\in C^l(X;\mathbb{Z})$ and the class of a k-cochain $\phi\in C^k(X;\mathbb{Z})$ to the class of the $(l+k)$-cochain $\psi\smile\phi\in C^{l+k}(X;\mathbb{Z})$ whose value on a singular $(l+k)$-simplex $s\colon\Delta^{l+k}\to X$ is given by

$$(\psi\smile\phi)(s):=\psi(s|_{[0,\cdots,k]})\phi(s|_{[k,\cdots,k+l]})$$

See [Hat02, Section 3.2] for the details of this construction.

Proposition B.6.2 *Let $f\colon X \to Y$ be a continuous map between topological spaces and let l, k be integers. The naturality of the cup-product manifests itself in the fact that the induced maps $f^*\colon H^{l+k}(Y;\mathbb{Z}) \to H^{l+k}(X;\mathbb{Z})$, $f^*\colon H^l(Y;\mathbb{Z}) \to H^l(X;\mathbb{Z})$ and $f^*\colon H^k(Y;\mathbb{Z}) \to H^k(X;\mathbb{Z})$ satisfy the formula*

$$f^*(\psi \smile \phi) = f^*(\psi) \smile f^*(\phi)$$

for any $\psi \in H^l(Y;\mathbb{Z})$ and any $\phi \in H^k(Y;\mathbb{Z})$.

Proof See [Hat02, Proposition 3.10]. □

The space $H^*(X; G)$ is a ring whose multiplication is given by cup-product. The following theorem can be used to calculate the cohomology ring of a product space.

Theorem B.6.3 *Let X, Y be "reasonable" topological spaces (by which we mean they should be CW-complexes) and let G be a ring. The* cross-product

$$H^*(X; G) \otimes_G H^*(Y; G) \to H^*(X \times Y; G)$$

is then an isomorphism of graded rings whenever $H^k(Y; G)$ is a free finitely generated G-module for all k.

Proof See [Hat02, Theorem 3.16]. □

Theorem B.6.4 (Künneth formula) *Let X, Y be "reasonable" topological spaces (i.e. CW complexes) and let K be a field. The homology cross product*

$$\bigoplus_l (H_l(X; K) \otimes_K H_{k-l}(Y; K)) \to H_k(X \times Y; K)$$

is then an isomorphism for any k.

Proof See [Hat02, Corollary 3.B7]. □

Example B.6.5 We can calculate the homology of tori of dimension n $\mathbb{T}^n = \mathbb{S}^1 \times \cdots \times \mathbb{S}^1$ using the Betti numbers of the circle $\mathbb{S}^1$: $b_0(\mathbb{S}^1) = b_1(\mathbb{S}^1) = 1$ and $b_k(\mathbb{S}^1) = 0$ for $k \notin \{0, 1\}$. This gives us

$$b_k(T^n) = b_{k-1}(T^{n-1}) + b_k(T^{n-1}) \, .$$

It follows that $b_0(T^n) = 1$ and $b_1(T^n) = b_1(T^{n-1}) + 1 = n$. Organising the Betti numbers of the torus into a Pascal triangle we get that

$$b_k(T^n) = \binom{n}{k} \, .$$

Remark B.6.6 See [Hat02, Theorem 3.B6] for a version of the Künneth formula valid for spaces whose homology groups contain torsion.

B.6.1 Cohomology with Compact Support

Let X be a topological space and let G be a ring (typically $G = \mathbb{Z}, \mathbb{Z}_2, \mathbb{Q}, \mathbb{C}$ or $\mathbb{R}$). We denote by $C_c^k(X; G)$ the subgroup of $C^k(X; G)$ of cochains $\phi\colon C_k(X) \to G$ such that there exists a compact set $K = K_\phi \subset X$ such that ϕ vanishes on all chains in $X \setminus K$. If $\delta\colon C^k(X; G) \to C^{k+1}(X; G)$ is the coboundary map then $\delta\phi$ vanishes on chains in $X \setminus K$. The subgroups

$$\cdots \to C_c^k(X; G) \to C_c^{k+1}(X; G) \to \dots \tag{B.4}$$

therefore form a subcomplex of the complex of singular cochains of X with values in G.

Definition B.6.7 (*Cohomology with compact support*) The cohomology groups

$$H_c^k(X; G)$$

associated to the subcomplex (B.4) are called the *compact support cohomology* groups of X with coefficients in G.

Proposition B.6.8 *We have the following exact sequence for cohomology with compact support of a compact pair* (X, L)*:*

$$\cdots \to H_c^k(X \setminus L; G) \to H^k(X; G) \to H^k(L; G) \to H_c^{k+1}(X \setminus L; G) \to \cdots \tag{B.5}$$

B.7 Poincaré Duality

In this section, we discuss several different versions of Poincaré duality, whose common hypothesis is that the topological space in question should be a *topological manifold*, or in other words every point in this space should have a neighbourhood isomorphic to $\mathbb{R}^n$ (Definition B.5.1). We refer the interested reader to [PP12] for a historical discussion of this result. Most of the proofs omitted below can be found in [Hat02, Section 3.3] or [Gre67].

Theorem B.7.1 *Let M be a topological manifold of dimension n. There is then a dualising isomorphism*

$$D_M\colon H_c^k(M; \mathbb{Z}_2) \xrightarrow{\cong} H_{n-k}(M; \mathbb{Z}_2)\ .$$

If M is orientable then there is a dualising isomorphism

$$D_M\colon H_c^k(M; \mathbb{Z}) \xrightarrow{\cong} H_{n-k}(M; \mathbb{Z})$$

which does not depend on the choice of an orientation on M.

Proof See [Hat02, Theorem 3.35]. □

Corollary B.7.2 *Let M be a topological manifold of dimension n. If M is compact and orientable then there is a dualising isomorphism*

$$D_M\colon H^k(M;\mathbb{Z}) \xrightarrow{\simeq} H_{n-k}(M;\mathbb{Z}).$$

Remark B.7.3 In this case, as the manifold M is compact and orientable, we may choose an orientation. With this orientation, M has a fundamental class $[M]$ and the isomorphism D_M is given for all k by *cap-product* (see below) with this fundamental class.

$$D_M(\phi) = [M] \frown \phi\,.$$

Definition B.7.4 (*Cap-product*) Let X be a topological space and let $l \geqslant k$ be integers. There is then a bilinear map, called the *cap-product*:

$$\frown\colon H_l(X;\mathbb{Z}) \times H^k(X;\mathbb{Z}) \longrightarrow H_{l-k}(X;\mathbb{Z})$$

which sends the class of a singular l-simplex $s\colon \Delta^l \to X$ and the class of a k-cochain $\phi \in C^k(X;\mathbb{Z})$ to the class of the $(l-k)$-simplex $s \frown \phi := \phi(s|_{[0,\cdots,k]})s|_{[k,\cdots,l]}$.

We refer to [Hat02, Section 3.3] for details of this construction.

Proposition B.7.5 *Let $f\colon X \to Y$ be a continuous map between topological spaces: the naturality of the cap product is expressed in the following diagram*

$$\begin{array}{ccc} H_l(X;\mathbb{Z})\times H^k(X;\mathbb{Z}) & \xrightarrow{\frown} & H_{l-k}(X;\mathbb{Z}) \\ f_*\downarrow \qquad f^*\uparrow & & \downarrow f_* \\ H_l(Y;\mathbb{Z})\times H^k(Y;\mathbb{Z}) & \xrightarrow{\frown} & H_{l-k}(Y;\mathbb{Z}) \end{array}$$

by

$$f_*(\alpha) \frown \phi = f_*(\alpha \frown f^*(\phi))$$

for any $\alpha \in H_l(X;\mathbb{Z})$ and $\phi \in H^k(Y;\mathbb{Z})$.

Proof See [Hat02, Section 3.3]. □

The cup-product and cap product are linked by the formula

$$\psi(\alpha \frown \phi) = (\phi \smile \psi)(\alpha)$$

for any $\alpha \in H_{k+l}(X;\mathbb{Z})$, $\phi \in H^k(X;\mathbb{Z})$ and $\psi \in H^l(X;\mathbb{Z})$.

Considering the free part of the cohomology groups we get a *perfect pairing*:

Theorem B.7.6 *Let M be a topological manifold of dimension n. If M is compact and oriented then for any $0 \leqslant k \leqslant n$ the cup-product pairing*

$$\begin{cases} P^k \colon H^k(M;\mathbb{Z})_f \times H^{n-k}(M;\mathbb{Z})_f \to \mathbb{Z} \\ (\phi,\psi) \mapsto (\phi \smile \psi)[M] \end{cases}$$

is a $\mathbb{Z}$-bilinear form such that the induced $\mathbb{Z}$-linear maps

$$H^k(M;\mathbb{Z})_f \to \mathrm{Hom}(H^{n-k}(M;\mathbb{Z})_f;\mathbb{Z})$$

and

$$H^{n-k}(M;\mathbb{Z})_f \to \mathrm{Hom}(H^k(M;\mathbb{Z})_f;\mathbb{Z})$$

are isomorphisms.

Corollary B.7.7 *Let M be a compact topological manifold of even dimension $n = 2m$.*

1. *The cup-product pairing*

$$H^m(M;\mathbb{Z}_2) \times H^m(M;\mathbb{Z}_2) \to \mathbb{Z}_2$$

is a symmetric non degenerate bilinear form of determinant 1.

2. *If M is oriented then the cup-product pairing*

$$H^m(M;\mathbb{Z})_f \times H^m(M;\mathbb{Z})_f \to \mathbb{Z}$$

is a non degenerate bilinear form of determinant ± 1, *which is symmetric if m is even and anti-symmetric if m is odd.*

Proposition A.6.7 is a key application of Poincaré duality.

Theorem B.7.8 (Coefficients in $\mathbb{Z}$) *Let M be a topological manifold of dimension n. If M is compact and oriented then for any integers $0 \leqslant k \leqslant n$ there is a bilinear form (called the "intersection form")*

$$P_k \colon H_k(M;\mathbb{Z}) \times H_{n-k}(M;\mathbb{Z}) \to \mathbb{Z}$$

which induces an identification between the free part

$$H_k(M;\mathbb{Z})_f = H_k(M;\mathbb{Z})/\mathrm{Tor}(H_k(M;\mathbb{Z}))$$

of the $\mathbb{Z}$-module $H_k(M;\mathbb{Z})$ and the dual $\mathbb{Z}$-module $\mathrm{Hom}(H_{n-k}(M;\mathbb{Z}),\mathbb{Z})$.

In particular, if $H_k(M;\mathbb{Z})$ and $H_{n-k}(M;\mathbb{Z})$ are free $\mathbb{Z}$-modules then we can associate to any basis $\{e_i\}$ of $H_k(M;\mathbb{Z})$ a *dual* basis $\{f_j\}$ of $H_{n-k}(M;\mathbb{Z})$ such that

$$P_k(e_i, f_j) = \delta_{ij} .$$

Proposition B.7.9 *If M is a compact oriented topological manifold of even dimension $n = 2m$ then the bilinear form in dimension m*

$$P_m : H_m(M; \mathbb{Z}) \times H_m(M; \mathbb{Z}) \to \mathbb{Z}$$

is symmetric if m is even and anti-symmetric if m is odd.

Theorem B.7.10 (Coefficients in $\mathbb{Z}_2$) *Let M be a topological manifold of dimension n. If M is compact then for all integers k such that $0 \leqslant k \leqslant n$ there is an intersection form*

$$P_k : H_k(M; \mathbb{Z}_2) \times H_{n-k}(M; \mathbb{Z}_2) \to \mathbb{Z}_2$$

inducing an identification between the $\mathbb{Z}_2$-vector spaces $H_k(M; \mathbb{Z}_2)$ and $\mathrm{Hom}(H_{n-k}(M; \mathbb{Z}_2), \mathbb{Z}_2)$ *which in turn is isomorphic to $H^{n-k}(M; \mathbb{Z}_2)$ because $\mathbb{Z}_2$ is a field.*

Proposition B.7.11 *If M is a compact topological manifold of even dimension $n = 2m$ then the bilinear form in dimension m*

$$P_m : H_m(M; \mathbb{Z}_2) \times H_m(M; \mathbb{Z}_2) \to \mathbb{Z}_2$$

is symmetric.

Remark B.7.12 When M is a differentiable manifold and $\alpha = [A] \in H_k(M; \mathbb{Z})$ and $\beta = [B] \in H_{n-k}(M; \mathbb{Z})$ are the fundamental classes of two oriented differentiable submanifolds A of dimension k and B of dimension $n - k$ which are *transverse* in M.[5] Since A and B have complementary dimensions the intersection $A \pitchfork B$ is then a finite collection of points. Let $P \in A \pitchfork B$ be such a point: we associate a sign $\varepsilon_P = \pm 1$ to P in the following way. If the orientation determined by the orientation on $T_P A$ and on $T_P B$ is the same as that on $T_P M$ then $\varepsilon_P := 1$ and otherwise $\varepsilon_P := -1$. We then have that

$$P_k(\alpha, \beta) = \sum_{P \in A \pitchfork B} \varepsilon_P \in \mathbb{Z} .$$

For $\mathbb{Z}_2$ cohomology, M, A and B do not need to be oriented (they are automatically $\mathbb{Z}_2$-oriented) and $P_k(\alpha, \beta)$ is equal to the number of points $P \in A \pitchfork B$ modulo 2.

Note that this interpretation of P_k only applies in certain special cases and cannot be used to define the intersection form in full generality. In particular, not all homology classes can be represented by embedded differentiable submanifolds—see [Tho54] for more details.

[5]This means that for any point P in the intersection $A \cap B$ we have that $T_P M = T_P A \oplus T_P B$: we denote this relationship by $A \pitchfork B$.

Remark B.7.13 In the situation of the above remark, it is easy to show that if M is a manifold of even dimension $n = 2m$ then the bilinear form $P_m\colon H_m(M; \mathbb{Z}) \times H_m(M; \mathbb{Z}) \to \mathbb{Z}$ is symmetric if m is even and anti-symmetric if m is odd.

Remark B.7.14 Analogues of Theorems B.7.1 and B.7.2 exist for not necessarily orientable manifolds. (To remember these generalisations, think of an arbitrary manifold as being $\mathbb{Z}_2$-orientable as in the discussion preceding [Hat02, Theorem 3.26].)

Theorem B.7.15 *Let M be a topological manifold of dimension n. We then have that*

$$H_c^k(M; \mathbb{Z}_2) \simeq H_{n-k}(M; \mathbb{Z}_2).$$

Corollary B.7.16 *Let M be a topological manifold of dimension n. If M is compact then*

$$H^k(M; \mathbb{Z}_2) \simeq H_{n-k}(M; \mathbb{Z}_2).$$

B.7.1 Application: Orientability of a Submanifold

Proposition B.7.17 *Let $S \subset \mathbb{RP}^3$ be a connected differentiable submanifold of dimension 2. The following are then equivalent.*

1. *S is orientable,*
2. *Any line in $\mathbb{RP}^3$ transverse to S meets S in an even number of points,*
3. *The homology class (Definition 3.7.1) $[S]_2 \in H_2(\mathbb{RP}^3; \mathbb{Z}_2)$ vanishes,*
4. *The complement $\mathbb{RP}^3 \setminus V$ has two connected components.*

Proof We refer to [BR90, Proposition 5.1.7] for more details.

$1 \implies 2$: if the surface S is orientable then it has a fundamental class $[S]$ in $H_2(S; \mathbb{Z})$ but as $H_2(\mathbb{RP}^3; \mathbb{Z}) = \{0\}$ the class $i_*([S])$ in $H_2(\mathbb{RP}^3; \mathbb{Z})$ must vanish. Let H be a line in $\mathbb{RP}^3$ transverse to S. As $[H]$ generates $H_1(\mathbb{RP}^3; \mathbb{Z}) \simeq \mathbb{Z}_2$ the intersection $H \pitchfork S$ has an even number of elements by Poincaré duality.

$2 \implies 3$: By Poincaré duality $[S]_2 = 0$ in $H_2(\mathbb{RP}^3; \mathbb{Z}_2)$ because $[H]_2$ generates $H_1(\mathbb{RP}^3; \mathbb{Z}_2)$.

$3 \implies 4$: Consider the homology exact sequence of the pair $(\mathbb{RP}^3, S)$,

$$0 \to H_3(\mathbb{RP}^3; \mathbb{Z}_2) \simeq \mathbb{Z}_2 \to H_3(\mathbb{RP}^3, S; \mathbb{Z}_2) \to$$
$$H_2(S; \mathbb{Z}_2) \simeq \mathbb{Z}_2 \xrightarrow{i_*} H_2(\mathbb{RP}^3; \mathbb{Z}_2) \simeq \mathbb{Z}_2$$

Since i_* is trivial by hypothesis $H_3(\mathbb{RP}^3, S; \mathbb{Z}_2) \simeq \mathbb{Z}_2 \oplus \mathbb{Z}_2$. By Alexander duality ([Hat02, Theorem 3.44]) we have that $H^0(\mathbb{RP}^3 \setminus S; \mathbb{Z}_2) \simeq H_3(\mathbb{RP}^3, S; \mathbb{Z}_2)$ and $\mathbb{RP}^3 \setminus V$ therefore has two connected components.

$4 \implies 1$: Since $\mathbb{RP}^3 \setminus S$ has two connected components we can orient the normal bundle of S in $\mathbb{RP}^3$. Since $\mathbb{RP}^3$ is orientable this yields an orientation of the tangent bundle of S. See [Hir76, Lemma 4.4.1 and Theorem 4.4.5] for more details. □

B.7.2 Applications to Algebraic Varieties

In this section we study Poincaré duality on algebraic varieties that may be complex or real, projective or affine.

B.7.3 Compact ANRs

A topological space X is said to be *normal* if and only if it is Hausdorff (Definition B.1.1.) and additionally satisfies the following stronger separation axiom:

For any pair of disjoint closed sets A and B there are two disjoint open sets U and V such that A is contained in U and B is contained in V. In particular, any metrisable space is normal.

A topological space is said to be *paracompact* if and only if it is Hausdorff and any open covering has a locally finite (open) refinement. We recall that a covering (X_i) of a topological space X is said to be *locally finite* if and only if every point of X has a neighbourhood which meets only a finite number of the X_is.

Every paracompact space is normal and every compact space or CW-complex is paracompact. Every metrisable space is paracompact and every paracompact manifold is metrisable.

Definition B.7.18 A topological space X is said to be an ANR (Absolute Neighborhood Retract) if and only if it satisfies the following universal property: for any normal topological space Y, any continuous map $f\colon B \to X$ defined on a closed subset B in Y can be extended to a continuous map $U \to X$ from an open neighbourhood U of B in Y.

Proposition B.7.19 (Examples of ANRs)

1. *$\mathbb{R}^n$ is an ANR.*
2. *Any open subset of an ANR is an ANR.*
3. *Any compact topological manifold is an ANR.*
4. *[Hu59, page 30 K.4] Let X be a metrisable space and let X_1 and X_2 be two closed subspaces such that $X = X_1 \cup X_2$. If X_1, X_2 and $X_1 \cap X_2$ are ANRs then X is an ANR.*

Proposition B.7.20 *Let V be a compact topological manifold and let $B \subset V$ be a compact subspace which is an ANR. We then have that*

$$H_c^k(V \setminus B; \mathbb{Z}) \simeq H^k(V, B; \mathbb{Z}).$$

Proof See [Gre67, Cor. 27.4]. □

Remark B.7.21 This can be generalised to the case where V is a compact ANR.

Proposition B.7.22 *Let V be a non singular real projective variety of dimension n and let $V_{\mathbb{C}}$ be a non singular complexification of V. We then have that*

$$H^k(V_{\mathbb{C}}, V; \mathbb{Z}) \simeq H_{2n-k}(V_{\mathbb{C}} \setminus V; \mathbb{Z}).$$

Proof Poincaré duality applied to the topological manifold $V_{\mathbb{C}} \setminus V$ of dimension $2n$ yields that

$$H_c^k(V_{\mathbb{C}} \setminus V; \mathbb{Z}) \simeq H_{2n-k}(V_{\mathbb{C}} \setminus V; \mathbb{Z}) .$$

Since moreover V is a compact *ANR* contained in the compact manifold $V_{\mathbb{C}}$ we have that $H_c^k(V_{\mathbb{C}} \setminus V; \mathbb{Z}) \simeq H^k(V_{\mathbb{C}}, V; \mathbb{Z})$. □

Proposition B.7.23 *Let S be a non singular complex affine algebraic variety of dimension n and let (V, B) be a smooth projective completion such that B is a simple normal crossing (SNC) divisor. We then have that*

$$H^k(V, B; \mathbb{Z}) \simeq H_{2n-k}(S; \mathbb{Z}) . \tag{B.6}$$

Proof Poincaré duality applied to the topological manifold S gives us that $H_c^k(S; \mathbb{Z}) \simeq H_{2n-k}(S; \mathbb{Z})$. By Proposition B.7.19(4), B is a compact *ANR* in the compact manifold V so $H_c^k(S; \mathbb{Z}) \simeq H^k(V, B; \mathbb{Z})$. □

Example B.7.24 (*Homology of affine rational surfaces*) In the above situation, assume that V is a complex surface ($n = 2$) from which it follows that B is a complex curve. The following sequence is part of the exact cohomology sequence associated to the pair (V, B):

$$\begin{aligned} 0 \to H^1(V; \mathbb{Z}) \to H^1(B; \mathbb{Z}) \to \\ H^2(V, B; \mathbb{Z}) \to H^2(V; \mathbb{Z}) \to H^2(B; \mathbb{Z}) \to \\ H^3(V, B; \mathbb{Z}) \to H^3(V; \mathbb{Z}) \to 0 \end{aligned}$$

Indeed, $H^3(B; \mathbb{Z}) = 0$ because B is a topological manifold of dimension 2 and $H^1(V, B; \mathbb{Z}) \simeq H_3(S; \mathbb{Z}) = 0$ because S is a complex affine surface. Suppose now that V is a *rational* surface and B is a tree of rational curves. In this case we have that $H^1(B; \mathbb{Z}) \simeq H_1(B; \mathbb{Z}) = 0$, $H^1(V; \mathbb{Z}) \simeq H_3(V; \mathbb{Z}) = 0$ and $H^3(V; \mathbb{Z}) \simeq H_1(V; \mathbb{Z}) = 0$. Using Poincaré duality (B.6) applied to the topological manifold S of dimension 4 we get an exact sequence:

$$0 \to H_2(S; \mathbb{Z}) \to H^2(V; \mathbb{Z}) \to H^2(B; \mathbb{Z}) \to H_1(S; \mathbb{Z}) \to 0 .$$

B.8 Three Dimensional Manifolds

B.8.1 Seifert Manifolds

Let $\mathbb{S}^1 \times \mathbb{D}^2$ be the solid torus, where $\mathbb{S}^1$ is the unit circle $\{u \in \mathbb{C} \mid |u| = 1\}$ and $\mathbb{D}^2$ is the closed unit disc $\{z \in \mathbb{C},\ |z| \leqslant 1\}$. A *Seifert fibration* of the solid torus is a differentiable map of the form

$$f : \mathbb{S}^1 \times \mathbb{D}^2 \to \mathbb{D}^2,\ (u, z) \mapsto u^q z^p,$$

where p, q are natural numbers such that $p \neq 0$ and $(p, q) = 1$. The map f is a circle bundle which is locally trivial over the punctured disc $\mathbb{D}^2 \setminus \{0\}$. If $p > 1$ then the fibre $f^{-1}(0)$ is said to be a *multiple fibre* of multiplicity p.

Definition B.8.1 A compact manifold without boundary M of dimension 3 is said to be *Seifert* if and only if it has a *Seifert fibration*, or in other words if and only if there is a differentiable map $g \colon M \to B$ to a surface B such that every point $P \in B$ has a closed neighbourhood U such that the restriction of g to $g^{-1}(U)$ is diffeomorphic to a Seifert fibration of the solid torus.

In particular, ever fibre of g is diffeomorphic to $\mathbb{S}^1$ and g is locally trivial outside of a finite set of points $\{P_1, \ldots, P_k\} \subset B$ where the fibre $g^{-1}(P_i)$ is multiple.

B.8.2 Lens Spaces

For any natural number $n \in \mathbb{N}^*$ denote by μ_n the multiplicative subgroup of $\mathbb{C}^*$ consisting of nth roots of unity.

Definition B.8.2 Let $0 < q < p$ be coprime integers. The *lens space* $\mathbb{L}_{p,q}$ is the quotient of the sphere

$$\mathbb{S}^3 = \{(w, z) \in \mathbb{C}^2 \mid |w|^2 + |z|^2 = 1\}$$

by the action of μ_p defined by

$$\zeta \cdot (w, z) = (\zeta w, \zeta^q z),$$

for any $\zeta \in \mu_p$ and $(w, z) \in \mathbb{C}^2$.

Proposition B.8.3 *Any lens space has a Seifert fibration.*

In fact, any such space has an infinite number of Seifert fibrations. Spaces of the form $\mathbb{L}_{p,1}$ have locally trivial fibrations.

Proof Such a fibration can be constructed from the Hopf fibration of the sphere $\mathbb{S}^3$ over the sphere $\mathbb{S}^2 \approx \mathbb{C} \cup \{\infty\}$:

$$\begin{array}{rcl} \mathbb{S}^3 & \longrightarrow & \mathbb{S}^2 \\ (w, z) & \longmapsto & w/z \ . \end{array} \tag{B.7}$$

A cyclic quotient of the Hopf fibration is a Seifert fibration over an orbifold of dimension 2 (Definition 4.4.32):

$$\begin{array}{rcl} \mathbb{L}_{p,q} & \longrightarrow & \mathbb{S}^2(p, q) \\ (w, z) & \longmapsto & w^q/z \ . \end{array}$$

□

We have seen that any lens space is a Seifert manifold. On the other hand, apart from $\mathbb{L}_{2,1}\#\mathbb{L}_{2,1} = \mathbb{RP}^3\#\mathbb{RP}^3$, any connected sum of at least two lens spaces has no Seifert fibration structure (Propositions B.8.11 and B.8.13).

B.8.3 C^∞ Geometric Manifolds

A Riemannian manifold Ω is said to be *homogeneous* if and only if the isometry group $\mathrm{Isom}(\Omega)$ acts transitively on Ω. A *geometry* Ω is a simply connected homogeneous Riemannian manifold that has a quotient of finite volume. If Ω is a real Lie group then we can make it into a Riemannian manifold by equipping it with a left-invariant metric and we call the resulting object "the" Ω geometry.

Definition B.8.4 A C^∞ differentiable manifold M is said to be *geometric* if and only if M is diffeomorphic to a quotient of a geometry Ω by a discrete subgroup of isometries $\Lambda \subset \mathrm{Isom}(\Omega)$ acting without fixed points. We also say that $M = {}_\Lambda\backslash\Omega$ has a *geometric structure modelled* on Ω. Extending this definition, we will say that a manifold with boundary is *geometric* if and only if its interior (see Page 351) is geometric.

When Ω is a Lie group the above hypotheses imply the existence of a lattice of finite covolume. In other words, Ω is a *unimodular* Lie group.

Definition B.8.5 A *crystallographic group* of dimension n is a discrete group Λ of isometries of Euclidean space $\mathbb{E}^n$ such that the quotient ${}_\Lambda\backslash\mathbb{E}^n$ is compact.

Definition B.8.6 A C^∞ differentiable manifold of dimension n is said to be *spherical*, (resp. *Euclidean*, resp. *hyperbolic*) if and only if it has a geometry modelled on $\mathbb{S}^n$ (resp. $\mathbb{E}^n$, resp. $\mathbb{H}^n$).[6]

[6] $\mathbb{H}^n$ is the half-space $\{(x_1, \ldots, x_n) \in \mathbb{R}^n \mid x_n > 0\}$ with the metric $\frac{1}{x_n^2}(dx_1^2 + \cdots + dx_{n-1}^2 + dx_n^2)$.

The uniformisation theorem tells us that any compact topological surface has a spherical, Euclidean or hyperbolic geometry: see [Sti92] for more details. Remarkably, all surfaces are geometric and moreover all 2 dimensional geometries have constant sectional curvature. This no longer holds in dimension 3 where as well as the constant scalar curvature geometries $\mathbb{S}^3, \mathbb{E}^3$ and $\mathbb{H}^3$ there are five geometries without constant scalar curvature: $\mathbb{S}^2 \times \mathbb{E}^1$, $\mathbb{H}^2 \times \mathbb{E}^1$, $\widetilde{\mathbf{SL}_2(\mathbb{R})}$, **Nil** and **Sol**. Thurston proved that up to equivalence these are the only three dimensional geometries if we require the isometry group to be maximal.

Theorem B.8.7 (Thurston) *Up to equivalence there are exactly eight three dimensional geometries with maximal isometry group:*

$$\mathbb{S}^3,\ \mathbb{E}^3,\ \mathbb{H}^3,\ \mathbb{S}^2 \times \mathbb{E}^1,\ \mathbb{H}^2 \times \mathbb{E}^1,\ \widetilde{\mathbf{SL}_2(\mathbb{R})},\ \mathbf{Nil},\ \mathbf{Sol}\ .$$

Proof See [BBM+10, page 2] or [Sco83]. □

We refer to [Sco83]) for more details on these eight geometries which we will not describe in depth here. We will simply give a quick definition of some of the associated Lie groups. The group $\widetilde{\mathbf{SL}_2(\mathbb{R})}$ is the universal cover of $\mathbf{SL}_2(\mathbb{R})$. The group **Nil** is the Heisenberg group of upper triangular 3×3 matrices whose diagonal elements are all equal to 1.

The **Sol** group is the only simply connected Lie group of dimension 3 with a finite volume quotient which is resoluble but not nilpotent. The Lie group **Sol** is the set $\mathbb{R}^3$ with the semi-direct product law induced by the action

$$\mathbb{R} \times \mathbb{R}^2 \to \mathbb{R}^2,\ (z, (x, y)) \mapsto \left(e^z x, e^{-z} y\right).$$

The group law on $\mathbb{R}^3$ is

$$((\alpha, \beta, \lambda), (x, y, z)) \mapsto \left(e^\lambda x + \alpha, e^{-\lambda} y + \beta, z + \lambda\right)$$

and the metric

$$ds^2 = e^{-2z} dx^2 + e^{2z} dy^2 + dz^2$$

is left invariant. The group Isom(**Sol**) has eight connected components and the identity component is the group **Sol** itself. See [Tro98, Lemme 3.2] for more details.

Definition B.8.8 A manifold M of dimension 3 is said to be a **Sol** *manifold* if and only if there is a discrete subgroup of isometries $\Lambda \subset \mathrm{Isom}(\mathbf{Sol})$ acting without fixed points such that

$$M = {}_{\Lambda}\backslash\mathbf{Sol}\ .$$

We have seen that every compact surface is geometric: on the other hand, not every manifold M of dimension 3 has a geometric structure. We do however have the following result.

Proposition B.8.9 *If a three dimensional manifold M of finite volume has a geometric structure then this structure is unique.*

Proof See[Sco83, Section 5]. □

Definition B.8.10 (*Indecomposable manifold*) A compact $\mathcal{C}^\infty$ manifold without boundary M of dimension 3 is said to be *indecomposable* if and only if for any connected sum decomposition $M = M_1 \# M_2$ one of the terms M_1 or M_2 is homeomorphic to $\mathbb{S}^3$.

Proposition B.8.11 *Let M be a three dimensional compact $\mathcal{C}^\infty$ manifold without boundary. If M is geometric and not diffeomorphic to $\mathbb{RP}^3 \# \mathbb{RP}^3$ then M is indecomposable.*

Proof See [Sco83, page 457]. □

As $\mathbb{RP}^3 \# \mathbb{RP}^3$ is modelled on $\mathbb{S}^2 \times \mathbb{E}^1$ we have the following corollary.

Corollary B.8.12 *Let M be a three dimensional compact $\mathcal{C}^\infty$ manifold without boundary. If M is geometric and its geometry is not $\mathbb{S}^2 \times \mathbb{E}^1$ then M is indecomposable.*

An important result in this area states that all Seifert manifolds have a geometric structure. We even have a characterisation of the six "Seifert" geometries. See [Sco83, Theorem 5.3] for more details.

Proposition B.8.13 *An orientable compact $\mathcal{C}^\infty$ manifold without boundary M is a Seifert manifold if and only if M has a geometry modelled on one of the six following geometries.*

$$\mathbb{S}^3,\ \mathbb{S}^2 \times \mathbb{E}^1,\ \mathbb{E}^3,\ \mathbf{Nil},\ \mathbb{H}^2 \times \mathbb{E}^1,\ \widetilde{\mathbf{SL}_2(\mathbb{R})}.$$

This proposition is also valid for non orientable M if we extend the definition of Seifert manifolds to manifolds with a foliation of circles (which essentially means that we accept non orientable local models in addition to the models used in Definition B.8.1).

Corollary B.8.14 *If $M = {}_\Lambda\backslash\Omega$ is a compact geometric manifold without boundary then either M is a Seifert manifold, $\Omega = \mathbf{Sol}$ or $\Omega = \mathbb{H}^3$.*

Amongst classes of three dimensional topological manifolds, the class of hyperbolic manifolds—i.e. quotients of the form ${}_\Lambda\backslash\mathbb{H}^3$ where $\Lambda \subset \mathbf{PO}(3, 1)$ is a discrete subgroup—is both the geometrically richest class and the least well understood.

B.8.4 Geometrisation and Classification

Several articles have appeared in the journal 'Gazette des Mathématiciens' on the Poincaré Conjecture and Thurston's Geometrisation Conjecture ([And05], [Mil04],

[Bes05], [Bes13]). The following rapid summary of the state of the art since Perelman's work opened up the possibility of a complete classification is based on [BBM+10].

Classification of C^∞ Manifolds of Dimension 3

Throughout this paragraph, a *manifold* of dimension n can have a *boundary*, denoted ∂M, characterised by the fact that any point $p \in \partial M$ has a neighbourhood in M which is locally homeomorphic to a product $\mathbb{R}^{n-1} \times \mathbb{R}_{\geqslant 0} = \{(x_1, \dots, x_n) \in \mathbb{R}^n \mid x_n \geqslant 0\}$ where p is sent to $(0, \dots, 0)$. If the boundary of a manifold of dimension n is non empty then it is a manifold of dimension $n - 1$ whose boundary is empty: $\partial\partial M = \varnothing$.

The *interior* of a manifold with boundary is the subvariety that is the complement of the boundary $M \setminus \partial M$. Any point $p \in M \setminus \partial M$ has a neighbourhood homeomorphic to $\mathbb{R}^n$. A manifold is said to be *closed* if and only if it is compact and $\partial M = \varnothing$.

A connected closed surface S in a compact orientable manifold M of dimension 3 is said to be *essential* if and only if its fundamental group injects into $\pi_1(M)$ and S neither bounds a 3-ball nor is cobordant to a product with a connected component of ∂M.

We now state Thurston's famous geometrisation conjecture. (See Definition B.8.4 for the definition of a *geometric manifold*).

Conjecture B.8.15 (Thurston's geometrisation conjecture) *The interior of a compact orientable manifold of dimension* 3 *can be cut along a finite family of essential embedded pairwise disjoint* 2*-spheres and* 2*-tori into a canonical collection of geometric* 3*-manifolds after filling up the spherical boundaries with* 3*-balls.*

Every connected component of the complement of the family of tori and spheres has a locally homogeneous metric of finite volume. Let M be such a connected component and let $\widehat{M}$ be the compact manifold obtained by "filling up" the holes. The manifold M then has a geometric structure modelled on one of the eight geometries of Theorem B.8.7.

The famous Poincaré conjecture is a special case of this conjecture.

Conjecture B.8.16 (Poincaré conjecture) *Let M be a simply connected closed topological manifold of dimension* 3*. M is then homeomorphic to the* 3*-sphere.*

Bringing together Thurston's hyperbolisation theorem [BBM+10, 1.1.5] and Perelman's theorem [BBM+10, 1.1.6] we obtain the following classification of orientable manifolds.

Theorem B.8.17 (Geometrisation theorem) *The above geometrisation Conjecture B.8.15 holds for any compact orientable manifold of dimension* 3.

Appendix C
Sheaves and Ringed Spaces

This appendix is based on books by Godement [God58, Chapitre II], Liu [Liu02, Section 2.2.1] and Hartshorne [Har77, Section II.1]. We will not discuss sheaf cohomology, for which we refer to [Liu02, Section 5.2] or [Har77, Chapter III].

C.1 Sheaves

Technical warning: we quote [Ser55a] several times in this chapter. In this article, a sheaf over X is defined to be a sheaf space (i.e. a certain type of topological space with a continuous map to X as in Definition C.2.2) whereas elsewhere in the literature, notably in[Har77, Chapter II], a sheaf is a presheaf (i.e. a certain type of contravariant functor as in Definition C.1.1 below) satisfying certain axioms. See Definition C.1.4 for more details. Corollary C.4.3 establishes that these two notions are equivalent. (In particular, Godement freely identifies them in his book [God58, Remarque II.1.2.1]).

Definition C.1.1 Let X be a topological space. A *presheaf* (of abelian groups) $\mathcal{F}$ over X is the data of an abelian group $\mathcal{F}(U)$ for every open set $U \subset X$ and for every nested pair of open sets $V \subset U \subset X$ a group morphism called the *restriction morphism* $\rho_{UV} \colon \mathcal{F}(U) \to \mathcal{F}(V)$ satisfying the following conditions

1. $\mathcal{F}(\varnothing) = \{0\}$;
2. $\rho_{UU} = \mathrm{id}_U$;
3. If $W \subset V \subset U \subset X$ are nested open sets then $\rho_{UW} = \rho_{VW} \circ \rho_{UV}$.

We define *presheaves of sets* (resp. *presheaves of rings*) in a similar way: the $\mathcal{F}(U)$s are then sets (resp. rings) and the $\rho_{UV} \colon \mathcal{F}(U) \to \mathcal{F}(V)$ are maps (resp. ring morphisms). There are many variations on this theme: for example, given a ring A we can define *presheaves of A-modules* and *presheaves of A-algebras*.

There is a natural notion of *subpresheaf* $\mathcal{F}'$ of $\mathcal{F}$: it is a presheaf such that for any open set $\mathcal{F}'(U)$ is a subgroup (resp. subset, subring etc.) of $\mathcal{F}(U)$ and the morphism ρ'_{UV} is required to be induced by ρ_{UV}.

F. Mangolte, *Real Algebraic Varieties*, Springer Monographs in Mathematics,
https://doi.org/10.1007/978-3-030-43104-4

Example C.1.2 Let X be a topological space.

1. Let K be a ring. We define a presheaf $\mathcal{F}$ as follows: for any $U \in X, \mathcal{F}(U) := K^U$ is the ring of K-valued functions on U and the morphisms ρ_{UV} are the restriction maps of functions.
2. The presheaf $\mathcal{C}^0$ of continuous real valued functions is a subpresheaf of the presheaf of real valued functions $\mathcal{C}^0(U) \subset \mathbb{R}^U$.

Let $\mathcal{F}$ be a sheaf over X and let $U \subset X$ be an open subset. An element $s \in \mathcal{F}(U)$ is called a *section* of $\mathcal{F}$ over U. A *global section* of $\mathcal{F}$ is a section over the space X. By analogy with presheaves of functions we denote by $s|_V$ the element $\rho_{UV}(s)$ in $\mathcal{F}(V)$ and we call it the *restriction* of s to V.

Definition C.1.3 The *set of sections* $\mathcal{F}(U)$ of a presheaf $\mathcal{F}$ over an open set U is sometimes denoted $\Gamma(U, \mathcal{F})$.

Definition C.1.4 (*Sheaf*) A presheaf $\mathcal{F}$ is said to be a *sheaf* if and only if for any open subset $U \subset X$ and any open cover $\{U_i\}_{i \in I}$ of U the following two conditions hold.

1. (Uniqueness.) If $s \in \mathcal{F}(U)$ and $s|_{U_i} = 0$ for all $i \in I$ then $s = 0$;
2. (Gluing) If the collection $s_i \in \mathcal{F}(U_i)$, $i \in I$ has the property that $s_i|_{U_i \cap U_j} = s_j|_{U_i \cap U_j}$ for any pair $i, j \in I$ then there is a section $s \in \mathcal{F}(U)$ such that $s|_{U_i} = s_i$ for any $i \in I$. (This section s is then unique by (1).)

Example C.1.5 (*Sheaves of functions*) Let X be a topological space and let K be a ring ($K = \mathbb{R}$ or $\mathbb{C}$ for example). The presheaf of K-valued functions of Example C.1.2(1) is then a sheaf, called the *sheaf of K valued functions on X*. By a *sheaf of functions* we mean a subsheaf of the sheaf of K-valued functions.

1. The sheaf $\mathcal{C}^0$ of continuous real (or complex) functions.
2. The sheaves of real or complex $\mathcal{C}^k$ or $\mathcal{C}^\infty$ functions.
3. The sheaf of holomorphic functions is a subsheaf of the sheaf of complex valued functions.
4. In general, a sub-presheaf of the sheaf of functions is a sheaf whenever the sub-presheaf is defined locally. This is notably the case for the sheaf of regular functions on a quasi-algebraic set, see Definitions 1.2.33, 1.2.34 and 1.2.35).

Definition C.1.6 (*Sheaf of restrictions to a subspace*) Let K be a ring, let X be a topological space, let $\mathcal{F}$ be a sheaf of K-valued functions on X and let $Y \subset X$ be a topological subspace with the induced topology. We define a sheaf $\mathcal{F}_Y$ on Y by deciding that for any open set U in Y a function $f: U \to K$ belongs to $\mathcal{F}_Y(U)$ if and only if for any x in U there is an open neighbourhood V of x in X and a function $g \in \mathcal{F}_X(V)$ such that $g(y) = f(y)$ for any $y \in V \cap U$.

The sheaf $\mathcal{F}_Y$ is in fact the *sheafification* (Definition C.4.1) of the presheaf of restrictions of functions of $\mathcal{F}$ to open subsets of Y.

Remark C.1.7 Note that this definition $\mathcal{F}_Y$ is specific to sheaves of functions. Note in particular that the sheaf $\mathcal{F}_Y$ is a sheaf of K-valued functions on Y.

Remark C.1.8 If $U \subset X$ is an open subset then $\mathcal{F}_U = \mathcal{F}_X|_U$ where for any open set $V \subset U$ we set $\mathcal{F}_X|_U(V) := \mathcal{F}(V)$. See Definition C.4.8 and Example C.4.9.

C.2 Sheaf Spaces over X

Let X be a topological space and let (E, π) be a pair such that E is a topological space and $\pi: E \to X$ is a continuous map. Let Y be a subset of X. By a (continuous) *section* s of (E, π) over Y we mean a continuous map $s: Y \to E$ such that $\pi(s(x)) = x$ for any $x \in Y$. We define a sheaf $\widehat{E}$ over X associated to (E, π) in the following way: for any $U \subset X$, $\widehat{E}(U)$ is the set of continuous sections of (E, π) over U and whenever $U \supset V$, the restriction to V of a section over U is the restriction of the corresponding map $U \to E$.

Definition C.2.1 The sheaf $\widehat{E}$ is called the *sheaf of sections* of (E, π).

Definition C.2.2 If π is a local homeomorphism (by which we mean that every point in $p \in E$ has an open neighbourhood homeomorphic via π to an open neighbourhood of $\pi(p)$ in X), we say that (E, π) is a *sheaf space over* X.

We will see in Definition C.4.1 that conversely we can associate a sheaf space to any presheaf.

Remark C.2.3 There are several different versions of construction C.2.1. In particular, if X is a differentiable (respectively analytic) manifold there is a similar definition of the sheaf of differentiable (resp. analytic) sections of (E, π) where E is a differentiable (resp. analytic) manifold and $\pi: E \to X$ is a differentiable (resp. analytic) map.

Remark C.2.4 It can be proved that a sheaf $\mathcal{F}$ is uniquely determined by the data of $\mathcal{F}(U)$ and ρ_{UV} for all U, V in some basis of open sets $\mathcal{B}$ of the topological space X. Recall that $\mathcal{B}$ is a basis for X if and only if any open set in X is a union of elements of $\mathcal{B}$ and any finite intersection of members of $\mathcal{B}$ is a member of $\mathcal{B}$. For example, if X is a real or complex algebraic variety, the data of $\mathcal{F}(\mathcal{D}(f))$ for any open affine set $\mathcal{D}(f)$ (f is a regular function) characterises $\mathcal{F}$. (See Exercise 1.3.15(3)).

C.3 Stalks of a Sheaf

Definition C.3.1 Let $\mathcal{F}$ be a presheaf of abelian groups over a topological space X and let x be a point in X. The *stalk* of the presheaf $\mathcal{F}$ at x is the inductive limit

$$\mathcal{F}_x := \varinjlim_{U \ni x} \mathcal{F}(U)$$

(see Definition A.1.2 and Example A.1.4). This inductive limit is taken over all open sets in X and if $\mathcal{F}$ is a presheaf of rings (resp. of A-algebras or A-modules for some fixed ring A) then $\mathcal{F}_x$ is a ring (resp. an A-module or an A-algebra). Let $s \in \mathcal{F}(U)$ be a section: for any $x \in U$, we denote the image of s in $\mathcal{F}_x$ by s_x. The element s_x is called the *germ* of s at x. The map $\mathcal{F}(U) \to \mathcal{F}_x$, $s \mapsto s_x$ is then a morphism of groups (resp. of rings, resp. of A-modules, resp. of A-algebras).

Remark C.3.2 In the special case of a sheaf of functions $\mathcal{F}$ such that all the sets of sections $\mathcal{F}(U)$ are subsets of a single common set and all the restriction morphisms are inclusions $\rho_{UV} : \mathcal{F}(U) \subset \mathcal{F}(V)$, inductive limit is simply union and we have that

$$\mathcal{F}_x = \bigcup_{U \ni x} \mathcal{F}(U) .$$

Lemma C.3.3 *Let $\mathcal{F}$ be a sheaf of abelian groups on a topological space X. Let s and $t \in \mathcal{F}(X)$ be global sections such that $s_x = t_x$ for any $x \in X$. We then have that $s = t$.*

Proof We can assume that $t = 0$. For any $x \in X$, there is an open neighbourhood U_x of x such that $s|_{U_x} = 0$, since $s_x = 0$. As the open sets $\{U_x\}_{x \in X}$ cover X we have that $s = 0$ by definition of a sheaf. □

Let (E, π) be a sheaf space over X and let $\widehat{E}$ be its sheaf of sections (see Definition C.2.1). For any section s of (E, π) over an open set U in X, the image $s(U)$ is open in E, so any section of (E, π) is an open mapping. For any $p \in E$ the fact that π is a local homeomorphism implies that there is a section s of (E, π) defined in a neighbourhood of $x = \pi(p)$ such that $s(x) = p$. Sets of the form $s(U)$ are therefore a basis of open sets for E.

Proposition C.3.4 *The fibre $E_x := \pi^{-1}(x)$ of the sheaf space (E, π) can be identified with the stalk of its space of sections.*

$$E_x \simeq \widehat{E}_x = \varinjlim_{U \ni x} \widehat{E}(U) .$$

Proof It will be enough to show that if two sections s and t of (E, π) defined on open neighbourhoods U and V of a point x in X are equal at x then they are equal on some open neighbourhood $W \subset U \cap V$ of x. As sections are open maps, $s(U)$ and $t(V)$ are open subsets of E. As $s(x) = t(x)$, the intersection $s(U) \cap t(V)$ is non empty. As π is a local homeomorphism there is an open subset $W' \subset s(U) \cap t(V)$ such that $\pi|_{W'} : W' \to \pi(W')$ is a bijection. The equality $\pi \circ s = \pi \circ t$ therefore implies $s = t$ on $W := \pi(W')$. □

There is an inverse of this construction: equipped with a suitable topology, the disjoint union of the stalks $\sqcup_{x \in X} \mathcal{F}_x$ of a sheaf $\{U \mapsto \mathcal{F}(U)\}_{U \text{ open set in } X}$ over a topological space X is the *sheaf space* of $\mathcal{F}$ as in Definition C.4.1.

C.3.1 Locally Trivial Fibrations

Definition C.3.5 Let X and F be topological spaces. We recall that a *locally trivial bundle*[7] with fibre F over X is the data of a pair (E, π) where $\pi\colon E \to X$ is a continuous map which locally (on X) has the form of a product. In other words, there is an open cover $\{U_i\}_{i\in I}$ of X and a family of homeomorphisms $\{\psi_i : \pi^{-1}(U_i) \xrightarrow{\simeq} U_i \times F\}_{i\in I}$ such that for any $i \in I$ the following diagram commutes.

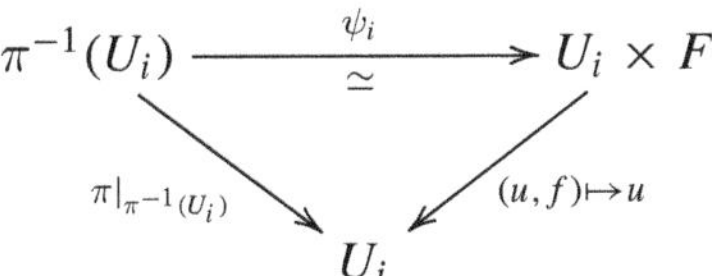

If $x \in U_i \cap U_j$ then $\psi_j|_{U_i\cap U_j} \circ \psi_i^{-1}|_{U_i\cap U_j}$ induces a continuous map ψ_{ij} from $U_i \cap U_j$ to the symmetric group of F.

A *covering space* of X is a locally trivial bundle whose fibre is a discrete topological space: the map π is then a local homeomorphism.

A *(locally trivial)* $\mathcal{C}^k$ *bundle* is defined as above, except that we require the topological spaces to be $\mathcal{C}^k$ differentiable manifolds and the continuous maps to be $\mathcal{C}^k$ differentiable maps.

A K*-vector bundle of rank* r is a locally trivial bundle of fibre $F = K^r$ such that for any $x \in X$ and any pair of open sets U_i, U_j containing x, $\psi_{ij}(x) \in \mathbf{GL}_r(K)$. More generally, a *vector bundle* is a vector bundle of constant rank on each connected component of X.

Example C.3.6 A locally trivial bundle of fibre F is not generally a sheaf space unless F is a discrete topological space (such as a finite set with the discrete topology), in which case π is a local homeomorphism. In other words, a sheaf space (E, π) is a *covering* of X.

C.3.2 Sheaf Morphisms

Definition C.3.7 Let $\mathcal{F}$ and $\mathcal{G}$ be presheaves (resp. sheaves) of abelian groups on a topological space X. A *presheaf morphism* (resp. *sheaf morphism*) $\alpha\colon \mathcal{F} \to \mathcal{G}$ is a family of group morphisms $\{\alpha(U) : \mathcal{F}(U) \to \mathcal{G}(U)\}_{U \text{ open in} X}$ which are compatible with the restriction morphisms ρ_{UV}.

[7] *Locally trivial* is often implicit in the literature, which can sometimes be confusing.

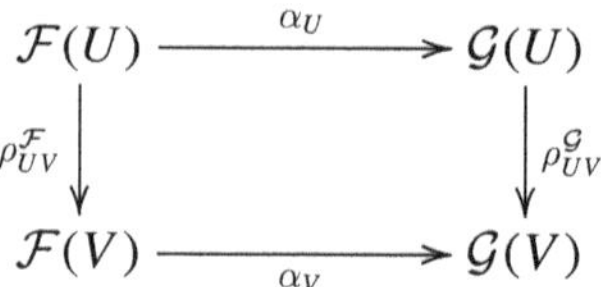

The composition of presheaf morphisms is clearly defined. An *isomorphism* can therefore be defined to be a morphism that has an inverse. In other words, a morphism of presheaves $\alpha\colon \mathcal{F} \to \mathcal{G}$ is an isomorphism if and only if $\alpha(U)\colon \mathcal{F}(U) \to \mathcal{G}(U)$ is an isomorphism of groups for any open set U in X.

Definition C.3.8 Let X be a topological space and let $\alpha\colon \mathcal{F} \to \mathcal{G}$ be a morphism of sheaves of abelian groups over X. The *kernel* of α, denoted $\ker\alpha$, is the presheaf $U \mapsto \ker(\alpha(U))$, which turns out to be a sheaf. The *image* of α, denoted $\mathrm{Im}\alpha$, is the sheaf associated to the presheaf $U \mapsto \mathrm{Im}(\alpha(U))$.

Remark C.3.9 By definition, the sheaf $\ker\alpha$ is a subsheaf of $\mathcal{F}$. By the universal property of sheafification, there is a natural map $\theta\colon \mathrm{Im}\alpha \to \mathcal{G}$ which is in fact injective. We can therefore identify $\mathrm{Im}\alpha$ with a subsheaf of $\mathcal{G}$.

Let $\alpha\colon \mathcal{F} \to \mathcal{G}$ be a morphism of presheaves on X. By the universal property of inductive limit, for every $x \in X$ the morphism α induces a canonical group morphism $\alpha_x : \mathcal{F}_x \to \mathcal{G}_x$ such that $(\alpha(U)(s))_x = \alpha_x(s_x)$ for any open neighbourhood U in x and any element $s \in \mathcal{F}(U)$.

Definition C.3.10 Let $\mathcal{F}$ and $\mathcal{G}$ be presheaves on a topological space X. A morphism of presheaves $\alpha\colon \mathcal{F} \to \mathcal{G}$ is said to be *injective* (resp. *surjective*) if and only if for any $x \in X$ the map $\alpha_x : \mathcal{F}_x \to \mathcal{G}_x$ is injective (resp. surjective).

If α is injective then for any open set U in X the map $\alpha(U)\colon \mathcal{F}(U) \to \mathcal{G}(U)$ is injective. In particular α is injective if and only if $\ker\alpha$ is trivial.

There are surjective sheaf morphisms α such that the maps $\alpha(U)\colon \mathcal{F}(U) \to \mathcal{G}(U)$ are not all surjective.

Example C.3.11 Let X be a complex analytic variety, let $\mathcal{O}_X$ be the additive sheaf of holomorphic functions and let $\mathcal{O}_X^*$ be the multiplicative sheaf of invertible holomorphic functions (i.e. everywhere non vanishing holomorphic functions). Associating to any holomorphic function $f\colon U \to \mathbb{C}$ the function $\alpha(U)(f) := \exp\circ f\colon U \to \mathbb{C}^*$ for any open set U in X, we get a sheaf morphism $\alpha\colon \mathcal{O}_X \to \mathcal{O}_X^*$. As any non vanishing holomorphic function is locally of the form $\exp\circ f$ this sheaf morphism is surjective but it is well known that if U is not simply connected then the map $f \mapsto \exp\circ f$ from $\mathcal{O}_X(U)$ to $\mathcal{O}_X^*(U)$ is not surjective: the identity map may not have a preimage.

Proposition C.3.12 *A sheaf morphism $\alpha\colon \mathcal{F} \to \mathcal{G}$ over X is an isomorphism if and only if $\alpha_x : \mathcal{F}_x \to \mathcal{G}_x$ is a group isomorphism for all $x \in X$.*

Proof See [Liu02, Proposition 2.12]. □

To summarise, if α_x is surjective for every x then α_U is not necessarily surjective for every U but if α_x is both injective and surjective for every x then α_U is both injective and surjective for every U.

Corollary C.3.13 *Let $\alpha\colon \mathcal{F} \to \mathcal{G}$ be a sheaf morphism over X. The morphism α is then an isomorphism if and only if it is both injective and surjective.*

Definition C.3.14 Let X be a topological space. A sequence of sheaves $\mathcal{F}\to\mathcal{G}\to\mathcal{H}$ is said to be *exact* if and only if $\mathcal{F}_x \to \mathcal{G}_x \to \mathcal{H}_x$ is an exact sequence of groups for all $x \in X$.

Example C.3.15 Returning to Example C.3.11, the exponential function induces an exact sequence of abelian groups

$$0 \to \mathbb{Z} \xrightarrow{\text{incl.}} \mathbb{C} \xrightarrow{z \mapsto \exp(2\pi i z)} \mathbb{C}^* \to 0 \tag{C.1}$$

where $\mathbb{C}$ has its additive structure and $\mathbb{C}^*$ has its multiplicative structure. Let X be a reduced complex analytic space: considering holomorphic functions with values in the exact sequence (C.1) we get an exact sequence of sheaves. For any reduced complex analytic space X we therefore have an exact sequence of sheaves

$$0 \to \mathbb{Z} \longrightarrow \mathcal{O}_X \longrightarrow \mathcal{O}_X^* \to 0 \tag{C.2}$$

where $\mathbb{Z}$ is the constant sheaf, $\mathcal{O}_X$ is the structural sheaf and $\mathcal{O}_X^*$ is the sheaf of multiplicative inverses in $\mathcal{O}_X$.

C.4 Sheaf of Sections of a Sheaf Space

It turns out that every sheaf is a sheaf of sections of a sheaf space. To prove this we start by considering a presheaf $\mathcal{F}$. We denote by $E(\mathcal{F})$ the disjoint union $\sqcup_{x\in X}\mathcal{F}_x$ and we let $\pi\colon E(\mathcal{F}) \to X$ be the map sending every point $p \in \mathcal{F}_x$ to x. The canonical map $\mathcal{F}(U) \to \mathcal{F}_x$, $s \mapsto s_x$ associates to every element $s \in \mathcal{F}(U)$ a map $\widetilde{s}\colon U \to E(\mathcal{F})$, $x \mapsto s_x$ such that $\pi(\widetilde{s}(x)) = x$ for every $x \in U$. We now equip $E(\mathcal{F})$ with the coarsest topology for which the maps $\widetilde{s}$ ($s \in \mathcal{F}(U)$, U open set in X) are continuous. The map π is then a local homeomorphism. We simplify notation by setting $\mathcal{F}^+ := \widehat{E(\mathcal{F})}$ as in Hartshorne (see [Har77, Definition II.1.2 and Exercise II.1.13]).

If the presheaf $\mathcal{F}$ is a presheaf of abelian groups (resp. rings etc.), $E(\mathcal{F})$ has a natural continuous composition law $(p, q) \mapsto p+q$ defined whenever $\pi(p)=\pi(q)$, which induces on every fibre $\mathcal{F}_x$ an abelian group structure. (In the case of rings, for example, there is also a second continuous composition law $(p, q) \mapsto pq$ defined whenever $\pi(p) = \pi(q)$ and these two laws turn $\mathcal{F}_x$ into a ring). If s and t are sections of $(E(\mathcal{F}), \pi)$ over an open set U in X then we denote by $s + t$ the section $x \mapsto s(x) + t(x)$ (resp. by st the section $x \mapsto s(x)t(x)$).

Definition C.4.1 (*Sheaf associated to a presheaf*) The pair $(E(\mathcal{F}), \pi)$ is called the *sheaf space associated to the presheaf* $\mathcal{F}$ and the sheaf $\mathcal{F}^+$ is called the *sheaf associated* to the presheaf $\mathcal{F}$.

The sheafification $\mathcal{F}^+$ of the presheaf $\mathcal{F}$ has a natural morphism

$$\mathcal{F} \longrightarrow \mathcal{F}^+$$

which is universal for morphisms from $\mathcal{F}$ to a sheaf. In other words, any morphism $\alpha\colon \mathcal{F} \to \mathcal{G}$ to a sheaf $\mathcal{G}$ factorises through a unique morphism $\tilde{\alpha}\colon \mathcal{F}^+ \to \mathcal{G}$

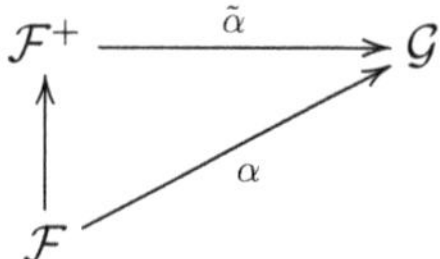

In particular, if $\mathcal{F}$ is a subpresheaf of a sheaf $\mathcal{G}$ then $\mathcal{F}^+$ is a subsheaf of $\mathcal{G}$ determined as follows: for any open set $U \subset X$ an element $f \in \mathcal{G}(U)$ belongs to $\mathcal{F}^+(U)$ if and only if there is an open covering $\{U_i\}_{i \in I}$ of U such that for any $i \in I$, $\rho_{UU_i}(f) \in \mathcal{F}(U_i)$.

Theorem C.4.2 ([God58, Théorème 1.2.1], [Har77, Proposition II.1.2]) *Let X be a topological space and let $\mathcal{F}$ be a presheaf over X. The sheaf $\mathcal{F}^+$ of sections of the sheaf space $E(\mathcal{F})$ is isomorphic to the presheaf $\mathcal{F}$ if and only if $\mathcal{F}$ is a sheaf.*

Corollary C.4.3 *Any sheaf over X is isomorphic to the sheaf of sections of some sheaf space over X and this sheaf space is unique up to canonical isomorphism.*

Remark C.4.4 When $\mathcal{F}$ is a presheaf of K-valued functions on X (see Example C.1.2) the local sections of $\mathcal{F}^+$ over an open subset $U \subset X$ are defined by

$$\mathcal{F}^+(U) = \{f\colon U \to K \mid \forall x \in U, \exists V \subset U \text{ open neighbourhood of } x \text{ and } \exists g \in \mathcal{F}(V) \mid f|_V = g\}\,.$$

Definition C.4.5 Let X be a topological space, let $\mathcal{F}$ be a sheaf on X and let $\mathcal{F}'$ be a subsheaf of $\mathcal{F}$. We then have that $U \mapsto \mathcal{F}(U)/\mathcal{F}'(U)$ is a presheaf over X. The sheafification $\mathcal{F}/\mathcal{F}' := (U \mapsto \mathcal{F}(U)/\mathcal{F}'(U))^+$ is called the *quotient sheaf*.

Proposition C.4.6 *Let X be a topological space, $\mathcal{F}$ a sheaf on X and $\mathcal{F}'$ a subsheaf of $\mathcal{F}$. We have that*

$$\forall x \in X, \quad (\mathcal{F}/\mathcal{F}')_x = \mathcal{F}_x/\mathcal{F}'_x\,.$$

Remark C.4.7 The sheaf space $E(\mathcal{F})$ is not generally Hausdorff even if X is. Indeed, for any sections s, t over an open set U in X, the set of elements $x \in U$ such that $s(x) = t(x)$ is an open subset of U but if $E(\mathcal{F})$ is Hausdorff it is also a closed subset of U: this implies that two sections which are equal at x are equal on the whole of the connected component of U containing x. In other words, the sheaf $\mathcal{F}$ satisfies the *analytic continuation property*. For example, if X is a complex analytic variety and $\mathcal{F}$ is the sheaf of analytic functions then $E(\mathcal{F})$ is Hausdorff, but if $\mathcal{F}$ is the sheaf of continuous functions then $E(\mathcal{F})$ is certainly not Hausdorff.

Definition C.4.8 Let X be a topological space, let $\mathcal{F}$ be a sheaf over X and let $(E(\mathcal{F}), \pi)$ be the associated sheaf space. Let $Y \subset X$ be a topological subspace and set $E|_Y := \pi^{-1}(Y)$. The associated sheaf $\widehat{E|_Y}$, denoted $\mathcal{F}|_Y$, is called the *restricted sheaf* or the *restriction* of $\mathcal{F}$ to Y.

Example C.4.9 If $\mathcal{F}$ is a presheaf on X and $U \subset X$ is open then on setting $\mathcal{F}|_U(V) := \mathcal{F}(V)$ for any open set $V \subset U$, we get a presheaf on U called the *restriction* of the presheaf $\mathcal{F}$ to U. Of course, if $\mathcal{F}$ is a sheaf this is just the restricted sheaf $\mathcal{F}|_U$ defined above.

Proposition C.4.10 *Let X be a topological space, $\mathcal{F}$ a sheaf over X and $Y \subset X$ a topological subspace. For any $x \in Y$ we then have that*

$$(\mathcal{F}|_Y)_x = \mathcal{F}_x \ .$$

Let X and Y be topological spaces, let $\mathcal{F}$ be a sheaf over X, let $\mathcal{G}$ be a sheaf over Y and let $\varphi \colon X \to Y$ be a continuous map. For any open set $V \subset Y$ the set $\varphi^{-1}(V)$ is then an open set of X and we denote by $\varphi_*\mathcal{F}$ the sheaf on Y given by $V \mapsto \mathcal{F}(\varphi^{-1}(V))$.

Definition C.4.11 The sheaf $\varphi_*\mathcal{F}$ on Y is called the *direct image* of $\mathcal{F}$.

We can also define an inverse image sheaf of $\mathcal{G}$, generally denoted $\varphi^{-1}\mathcal{G}$.

Definition C.4.12 The sheaf $\varphi^{-1}\mathcal{G}$ on X associated to the presheaf

$$U \mapsto \varinjlim_{V \supset \varphi(U)} \mathcal{G}(V)$$

where U is an open subset of X and the limit is taken over all open sets V in Y containing $\varphi(U)$ is called the *inverse image* of $\mathcal{G}$.

Remark C.4.13 Warning: in [God58] the inverse image sheaf of $\mathcal{G}$ under φ is denoted $\varphi^*\mathcal{G}$. As in [Har77, Section II.5], we will only use this notation when $\mathcal{G}$ is a sheaf of $\mathcal{O}_Y$-modules (Definition C.5.4 below): $\varphi^{-1}\mathcal{G}$ is then a sheaf of $\varphi^{-1}\mathcal{O}_Y$-modules and $\varphi^*\mathcal{G} := \varphi^{-1}\mathcal{G} \otimes_{\varphi^{-1}\mathcal{O}_Y} \mathcal{O}_X$ is a sheaf of $\mathcal{O}_X$-modules.

Proposition C.4.14 $\forall x \in X$, $(\varphi^{-1}\mathcal{G})_x = \mathcal{G}_{\varphi(x)}$.

Exercise C.4.15 If $i: Z \hookrightarrow Y$ is the canonical injection of a topological subspace of Y then

$$i^{-1}\mathcal{G} = \mathcal{G}|_Z .$$

C.5 Ringed Spaces

When working with algebraic varieties (Definition 1.3.1) we can restrict ourselves to sheaves that are subsheaves of function sheaves. When working with schemes, we use sheaves of local rings, which gives rise to the notion of ringed spaces.

Definition C.5.1 A *ringed space* (which is short for *locally ringed in local rings*) is the data of a topological space X and a sheaf of rings $\mathcal{O}_X$ on X such that $\mathcal{O}_{X,x}$ is a local ring for all $x \in X$. The sheaf $\mathcal{O}_X$ is called the *structural sheaf* of $(X, \mathcal{O}_X)$. Let $(X, \mathcal{O}_X)$ and $(Y, \mathcal{O}_Y)$ be ringed spaces: a *morphism of ringed spaces* is a pair $(\varphi, \varphi^\#)$ where $\varphi: X \to Y$ is a continuous map and $\varphi^\#: \mathcal{O}_Y \to \varphi_*\mathcal{O}_X$ is a morphism of sheaf of rings on Y.

Definition C.5.2 A morphism $(\varphi, \varphi^\#): (X, \mathcal{O}_X) \to (Y, \mathcal{O}_Y)$ is called an *open embedding* (resp. *closed embedding*) if and only if

1. φ is a homeomorphism onto $\varphi(X)$;
2. $\varphi(X)$ is open (resp. closed) in Y ;
3. $\varphi_x^\#$ is an isomorphism (resp. surjective morphism) for all $x \in X$.

Example C.5.3 When $\mathcal{O}_X$ (resp. $\mathcal{O}_Y$) is a subsheaf of the sheaf $\mathcal{F}_X$ (resp. $\mathcal{F}_Y$) of K-valued functions, any continuous map $\varphi: X \to Y$ induces a morphism $\varphi^\#: \mathcal{O}_Y \to \varphi_*\mathcal{F}_X$ of sheaves of rings on Y associated to the pull back map

$$\forall U \text{ open set in } Y,\ f \in \mathcal{O}_Y(U) \mapsto \big(f \circ \varphi : \varphi^{-1}(U) \to K\big) .$$

The pair $(\varphi, \varphi^\#)$ is then a morphism of ringed spaces if and only if $\varphi^\#$ is contained in $\varphi_*\mathcal{O}_X$. In particular, if $(X, \mathcal{O}_X)$ and $(Y, \mathcal{O}_Y)$ are algebraic varieties over the same base field K then $(\varphi, \varphi^\#)$ is a morphism of ringed spaces if and only if φ is a morphism of algebraic varieties over K.

Definition C.5.4 Let $(X, \mathcal{O}_X)$ be a ringed space. A *sheaf of $\mathcal{O}_X$-modules* (also called an *$\mathcal{O}_X$-module*) is a sheaf $\mathcal{F}$ over X such that for any open set $U \subset X$ the group $\mathcal{F}(U)$ is an $\mathcal{O}_X(U)$-module, and for any inclusion of open sets $V \subset U$ the restriction morphism $\mathcal{F}(U) \to \mathcal{F}(V)$ is compatible with the module structure *via* the ring morphism $\mathcal{O}_X(U) \to \mathcal{O}_X(V)$.

Definition C.5.5 Let $(X, \mathcal{O}_X)$ be a ringed space. The *direct sum* of two $\mathcal{O}_X$-modules $\mathcal{F}$ and $\mathcal{G}$ is an $\mathcal{O}_X$-module denoted $\mathcal{F} \oplus \mathcal{G}$. The *tensor product* of two $\mathcal{O}_X$-modules $\mathcal{F}$ and $\mathcal{G}$ is the sheaf denoted $\mathcal{F} \otimes_{\mathcal{O}_X} \mathcal{G}$ associated to the presheaf $U \mapsto \mathcal{F}(U) \otimes_{\mathcal{O}_X(U)} \mathcal{G}(U)$. See Proposition A.4.1 for the definition of the tensor product of two A-modules over a ring A.

Definition C.5.6 (*Locally free sheaf*) Let $(X, \mathcal{O}_X)$ be a ringed space. A $\mathcal{O}_X$-module $\mathcal{F}$ is said to be *free* if and only if it is isomorphic to a direct sum $\mathcal{O}_X \oplus \mathcal{O}_X \oplus \ldots$. It is said to be *locally free* if and only if there exists an open cover of X by sets U such that $\mathcal{F}|_U$ is a free $\mathcal{O}_X|_U$-module. Such an open set U is said to be a *trivialising open set* for $\mathcal{F}$. The *rank* of a locally free $\mathcal{O}_X$-module over a trivialising open set is the (finite or infinite) number of copies of $\mathcal{O}_X$ required. A *sheaf of ideals* over X is an $\mathcal{O}_X$-module $\mathcal{I}$ which is a subsheaf of $\mathcal{O}_X$.

An $\mathcal{O}_X$-module $\mathcal{F}$ is therefore locally free if and only if there is an open cover of X and a set I for each set U of this covering such that

$$\mathcal{F}|_U \simeq \mathcal{O}_X^{(I)}|_U$$

where $\mathcal{O}_X^{(I)}$ is the direct sum of copies of $\mathcal{O}_X$ indexed by I.

Let $r \geqslant 1$ be an integer. An $\mathcal{O}_X$-module $\mathcal{F}$ is locally free of rank r if and only if there is an open cover of X by open sets U such that

$$\mathcal{F}|_U \simeq \mathcal{O}_X^r|_U$$

where $\mathcal{O}_X^r$ denotes the direct sum of r copies of $\mathcal{O}_X$. More generally, as the rank of a locally free sheaf is constant on connected components of X, a locally free sheaf is said to be *of finite type* if and only if its rank is finite on each connected component of X.

Example C.5.7 Let $(X, \mathcal{O}_X)$ be a ringed space and let $\mathcal{F}$ be an $\mathcal{O}_X$-module. We define the *tensor algebra* $T(\mathcal{F})$ (resp. the *symmetric algebra* $S(\mathcal{F})$, resp. the *exterior algebra* $\bigwedge(\mathcal{F})$) of the $\mathcal{O}_X$-module $\mathcal{F}$ by taking the sheafification of the presheaf $U \mapsto T(\mathcal{F}(U))$ (resp. $U \mapsto S(\mathcal{F}(U))$, resp. $U \mapsto \bigwedge(\mathcal{F}(U))$) where the tensor operations are taken with respect to the $\mathcal{O}_X(U)$-module structure on $\mathcal{F}(U)$. See Definition A.4.8 for more details.

If $\mathcal{F}$ is locally free of rank r then $T^k(\mathcal{F})$ (resp. $S^k(\mathcal{F})$, resp. $\bigwedge^k(\mathcal{F})$) is also a locally free sheaf of rank r^k (resp. $\binom{r+k-1}{r-1}$, resp. $\binom{r}{k}$).

Definition C.5.8 (*Invertible sheaf*) An *invertible sheaf* over X is a locally free sheaf of rank 1, by which we mean that there is a covering of X by open sets U such that $\mathcal{F}|_U$ is isomorphic to $\mathcal{O}_X|_U$.

Proposition C.5.9 *Let $(X, \mathcal{O}_X)$ be a ringed space and let $\mathcal{F}$ be a $\mathcal{O}_X$-module. The sheaf $\mathcal{F}$ is then locally free if and only if $\mathcal{F}_x$ is a free $\mathcal{O}_{X,x}$-module for every $x \in X$.*

Proposition C.5.10 (Projection formula) *Let $\varphi\colon (X, \mathcal{O}_X) \to (Y, \mathcal{O}_Y)$ be a morphism of ringed spaces. If $\mathcal{F}$ is an $\mathcal{O}_X$-module and $\mathcal{E}$ is a locally free $\mathcal{O}_Y$-module of finite rank then there is a natural isomorphism*

$$\varphi_*(\mathcal{F} \otimes_{\mathcal{O}_X} \varphi^*\mathcal{E}) \simeq \varphi_*(\mathcal{F}) \otimes_{\mathcal{O}_Y} \mathcal{E} .$$

C.6 Coherent Sheaves

We start with the most general definition of coherent sheaves, and prove later that in the case we are interested in (Example C.6.8) a coherent sheaf is just a sheaf that is isomorphic to a quotient of a locally free sheaf of finite type.

Definition C.6.1 (*Sheaf generated by its global sections*) Let $(X, \mathcal{O}_X)$ be a ringed space and let $\mathcal{F}$ be an $\mathcal{O}_X$-module. We say that $\mathcal{F}$ is *generated by its global sections at* $x \in X$ if and only if the canonical map $\mathcal{F}(X) \otimes_{\mathcal{O}_X(X)} \mathcal{O}_{X,x} \to \mathcal{F}_x$ is surjective. We say that $\mathcal{F}$ is *generated by global sections* if and only if this holds at any point x in X.

Example C.6.2 Let $(X, \mathcal{O}_X)$ be a ringed space. Let I be a set. The sheaf $\mathcal{O}_X^{(I)}$—the direct sum of copies of $\mathcal{O}_X$ indexed by I—is generated by global sections.

Lemma C.6.3 *Let $(X, \mathcal{O}_X)$ be a ringed space. An $\mathcal{O}_X$-module $\mathcal{F}$ is generated by its global sections if and only if there is a set I and a surjective morphism of $\mathcal{O}_X$-modules*

$$\mathcal{O}_X^{(I)} \to \mathcal{F} \to 0\,.$$

Definition C.6.4 (*Quasi-coherent sheaf*) Let $(X, \mathcal{O}_X)$ be a ringed space and let $\mathcal{F}$ be an $\mathcal{O}_X$-module. We say that $\mathcal{F}$ is *quasi-coherent* if and only if for every $x \in X$ there is an open neighbourhood U of x in X and an exact sequence of $\mathcal{O}_X$-modules

$$\mathcal{O}_X^{(J)}|_U \to \mathcal{O}_X^{(I)}|_U \to \mathcal{F}|_U \to 0\,.$$

Example C.6.5 Let $(X, \mathcal{O}_X)$ be a ringed space.

1. The structural sheaf $\mathcal{O}_X$ is quasi-coherent.
2. Any locally free $\mathcal{O}_X$-module is quasi-coherent.
3. Any sheaf of ideals is quasi-coherent.

Definition C.6.6 (*Sheaf of finite type*) Let $(X, \mathcal{O}_X)$ be a ringed space and let $\mathcal{F}$ be an $\mathcal{O}_X$-module. We say that $\mathcal{F}$ is *of finite type* if and only if for every $x \in X$ there is an open neighbourhood U of x in X, an integer $r \geqslant 1$ and a surjective morphism of $\mathcal{O}_X$-modules

$$\mathcal{O}_X^r|_U \to \mathcal{F}|_U \to 0\,.$$

Definition C.6.7 (*Coherent sheaves*) Let $(X, \mathcal{O}_X)$ be a ringed space and let $\mathcal{F}$ be an $\mathcal{O}_X$-module. We say that $\mathcal{F}$ is *coherent* if and only if it is of finite type and for every open subset U in X, every integer r and every morphism $\alpha\colon \mathcal{O}_X^r|_U \to \mathcal{F}|_U$ the kernel $\ker \alpha$ is of finite type.

Example C.6.8 (*Structural sheaf*)

1. The structural sheaf $\mathcal{O}_X$ of an algebraic variety $(X, \mathcal{O}_X)$ over an algebraically closed field K is coherent.

2. The structural sheaf $\mathcal{O}_X$ of a locally Noetherian scheme $(X, \mathcal{O}_X)$ is coherent.
3. The sheaf of germs of holomorphic functions over a non singular complex analytic variety is coherent ([Oka50]).

Definition C.6.9 (*Finitely presented sheaf*) Let $(X, \mathcal{O}_X)$ be a ringed space and let $\mathcal{F}$ be an $\mathcal{O}_X$-module. We say that $\mathcal{F}$ is *finitely presented* if and only if for every $x \in X$ there is an open neighbourhood U of x in X, integers $r \geqslant 1$ and $c \geqslant 1$ and an exact sequence of $\mathcal{O}_X$-modules

$$\mathcal{O}_X^c|_U \to \mathcal{O}_X^r|_U \to \mathcal{F}|_U .$$

Every coherent sheaf is finitely presented. The converse holds whenever the structural sheaf is also coherent.

Proposition C.6.10 *Let $(X, \mathcal{O}_X)$ be a ringed space. If the sheaf $\mathcal{O}_X$ is coherent then an $\mathcal{O}_X$-module is coherent if and only if it is finitely presented.*

Corollary C.6.11 *Let $(X, \mathcal{O}_X)$ be a ringed space with coherent structural sheaf $\mathcal{O}_X$.*

1. *Any locally free $\mathcal{O}_X$-module of finite type (i.e. of finite rank on every trivialising open set) is coherent. Any locally free $\mathcal{O}_X$-module of finite rank and in particular every invertible sheaf is coherent.*
2. *Any finitely generated sheaf of ideals is coherent. In particular, the sheaf of ideals of regular functions vanishing on a closed subvariety of an algebraic variety X with coherent structural sheaf is coherent.*

C.7 Algebraic Varieties over an Algebraically Closed Base Field

Proposition C.7.1 *Let $(X, \mathcal{O}_X)$ be an algebraic variety over an algebraically closed base field K. A sheaf of $\mathcal{O}_X$-modules $\mathcal{F}$ is then coherent if and only if it is of finite type and quasi-coherent.*

Definition C.7.2 (*$\mathcal{O}_X$-module associated to a $\Gamma(X, \mathcal{O}_X)$-module*). Let X be an affine algebraic variety over an algebraically closed base field K, let $A := \mathcal{A}(X) = \Gamma(X, \mathcal{O}_X)$ be the ring of affine coordinates of X and let M be an A-module. We define a $\mathcal{O}_X$-module $\widetilde{M}$ on the principal open sets of X (which form an open basis for X by Exercise 1.3.15(3)) as follows: for any $f \in A$ we set $\widetilde{M}(\mathcal{D}(f)) = M_f = M \otimes_A A_f$. In particular we have that $\widetilde{M}(X) = \Gamma(X, \widetilde{M}) = M$.

Theorem C.7.3 *Let $(X, \mathcal{O}_X)$ be an algebraic variety over an algebraically closed base field K.*

A sheaf of $\mathcal{O}_X$-modules $\mathcal{F}$ is said to be quasi-coherent *if and only if for every open affine subset U of X the $\mathcal{O}_X(U)$-modules $\mathcal{F}|_U$ and $\widetilde{\mathcal{F}(U)}$ are isomorphic.*

It is said to be coherent *if and only if the $\mathcal{O}_X(U)$-modules $\mathcal{F}|_U$ and $\widetilde{\mathcal{F}(U)}$ are isomorphic and finitely generated.*

Theorem C.7.4 *Let* $(X, \mathcal{O}_X)$ *be an irreducible algebraic variety over an algebraically closed base field* K. *We then have that* Ω_X *is a locally free sheaf of dimension* $n = \dim X$ *if and only if* X *is non singular.*

Proof See [Har77, Theorem II.8.15]. □

Proposition C.7.5 *Let* $(X, \mathcal{O}_X)$ *be an algebraic variety over an algebraically closed base field* K *and let* $\mathcal{F}$ *be a coherent sheaf. The sheaf* $\mathcal{F}$ *is then invertible if and only if there is a coherent sheaf* $\mathcal{G}$ *such that* $\mathcal{F} \otimes \mathcal{G} \simeq \mathcal{O}_X$.

See [Per95, III.7] for other properties of quasi-coherent sheaves over an algebraically closed base field.

Appendix D
Analytic Geometry

The first part of this appendix is based on Serre's famous article **GAGA** [Ser56].

D.1 Complex Analytic Spaces and Holomorphic Functions

Definition D.1.1 A subset U of $\mathbb{C}^n$ is said to be *analytic* if and only if it is locally the vanishing locus of a set of holomorphic functions. More formally, U is analytic if and only if for every $x \in U$ there are holomorphic functions $f_1, \dots, f_k$ defined in a neighbourhood W of x such that for any $z \in W$ the point z is in $U \cap W$ if and only if $f_i(z) = 0$ for all $i = 1 \dots k$. The restriction to U of the sheaf of holomorphic functions $\mathcal{H}$ on $\mathbb{C}^n$ is called the sheaf of *holomorphic functions on* U, denoted $\mathcal{H}_U$.

Any analytic subset $U \subset \mathbb{C}^n$ is locally closed in $\mathbb{C}^n$: it is therefore also locally compact for the induced topology. For any $x \in U$, the ring of germs $\mathcal{H}_{U,x}$ is isomorphic to the quotient of $\mathcal{H}_x$ by the ideal of germs of functions whose restriction to U is identically zero in some neighbourhood of x.

Definition D.1.2 A *complex analytic space*[8] is a pair $(X, \mathcal{O}_X)$ where X is a topological space and $\mathcal{O}_X$ is a subsheaf of the sheaf of complex valued functions on X satisfying the following two conditions.

1. There is a covering of the space X by open sets U_i such that $(U_i, \mathcal{O}_X|_{U_i})$ is isomorphic as a ringed space to an analytic subset of $\mathbb{C}^n$ with its sheaf of holomorphic functions.
2. The topology on X is Hausdorff (Definition B.1.1). In practice we will also assume that the topological space has a countable basis of open sets.

The sheaf $\mathcal{O}_X$ is called the *sheaf of holomorphic functions* or *sheaf of analytic functions* on X.

---

[8]The use of the term "space" rather than "manifold" implies that the object under consideration may be singular.

F. Mangolte, *Real Algebraic Varieties*, Springer Monographs in Mathematics,
https://doi.org/10.1007/978-3-030-43104-4

If X and Y are complex analytic spaces then a *morphism* (or *analytic map*) $\varphi\colon X \to Y$ is a continuous map such that for any open set $V \subset Y$ and any analytic function $f\colon V \to K$ the function $f \circ \varphi\colon \varphi^{-1}(V) \to K$ is analytic.

Let X be an analytic space, let x be a point of X and let $\mathcal{O}_x$ be the ring of germs of holomorphic functions on X at x. This ring is a local $\mathbb{C}$-algebra whose unique maximal ring $\mathfrak{m}$ contains exactly the functions f that vanish at x. We have that $\mathcal{O}_x/\mathfrak{m} = \mathbb{C}$. When $X = \mathbb{C}^n$ the algebra $\mathcal{O}_x = \mathcal{H}_x$ is simply the algebra $\mathbb{C}\{z_1, \dots, z_n\}$ of convergent series in n variables and in general, $\mathcal{O}_x$ is isomorphic to a quotient algebra $\mathbb{C}\{z_1, \dots, z_n\}/I$. It follows that $\mathcal{O}_x$ is a Noetherian ring. In particular, X is isomorphic to $\mathbb{C}^n$ in a neighbourhood of x if and only if $\mathcal{H}_x$ is isomorphic to $\mathbb{C}\{z_1, \dots, z_n\}$, or in other words if $\mathcal{O}_x$ is a regular local ring (Definition 1.5.32) of dimension n. An analytic space all of whose points are regular is said to be an analytic *variety*: see Definition D.2.1 and Remark D.2.2 for more details.

Any complex algebraic variety $(X, \mathcal{O}_X)$ has a natural complex analytic space structure $(X^h, \mathcal{O}_X^h)$ where X^h is the underlying set of X with its Euclidean topology and $\mathcal{O}_X^h$ is the sheaf of holomorphic functions associated to $\mathcal{O}_X$. In other words, for any $x \in X$, $\mathcal{O}_{X,x}^h$ is the analytic subring of the ring of germs at x of complex valued functions generated by $\mathcal{O}_{X,x}$. One important property of this construction is that the completions of the local rings $\mathcal{O}_{X,x}$ and $\mathcal{O}_{X,x}^h$ are isomorphic for all $x \in X$ [Ser56, pages 9–11].

Theorem D.1.3 (Cartan–Serre finiteness theorem) *Let $(X, \mathcal{O}_X)$ be a compact analytic space and let $\mathcal{F}$ be a coherent sheaf on X. The $\mathbb{C}$-vector space $H^i(X, \mathcal{F})$ is then finite dimensional for any $i \geqslant 0$ and*

$$H^i(X, \mathcal{F}) = \{0\} \quad \textit{for all} \quad i > \dim_{\mathbb{C}} X\,.$$

Proof See [CS53] or [BHPVdV04, Theorem 8.3]. □

D.2 Complex Analytic Varieties

Definition D.2.1 A *complex analytic variety*[9] is an analytic space $(X, \mathcal{O}_X)$ which is locally isomorphic to an open subset of $\mathbb{C}^n$. If X and Y are complex analytic varieties a *morphism* (or *holomorphic map*) $\varphi\colon X \to Y$ is a continuous map such that for any open set $V \subset Y$ and any holomorphic function $f\colon V \to K$ the function $f \circ \varphi\colon \varphi^{-1}(V) \to K$ is holomorphic.

Remark D.2.2 An analytic space is an analytic variety if and only if it is non singular.

[9] Sometimes called a *holomorphic variety* in the literature.

D.2.1 Stein Manifolds

We refer the interested reader to [GR79] for the original definition of a Stein manifold: we will use an alternative definition which is equivalent to the original definition by the Stein embedding theorem [*Ibid.*].

Definition D.2.3 A complex analytic variety is said to be a *Stein manifold* if and only if it has a proper holomorphic embedding in an affine space $\mathbb{C}^n$.

Recall that an algebraic variety V is said to be *affine* if and only if it is isomorphic to a closed subvariety of an affine space.

Example D.2.4 The complex analytic variety underlying a non singular affine algebraic variety over $\mathbb{C}$ is a Stein manifold.

The converse, however, is false—there are non singular algebraic varieties that are Stein but not affine. See [Nee89] for more details.

D.2.2 Serre Duality

Theorem D.2.5 (Serre duality) *Let X be a non singular complex projective variety of dimension n and let $\mathcal{L}$ be a holomorphic vector bundle on X. We then have that*

$$H^k(X, \mathcal{L}) \simeq H^{n-k}(X, \mathcal{L}^\vee \otimes \mathcal{K}_X) .$$

and in particular

$$\chi(\mathcal{L}) = \chi(\mathcal{L}^\vee \otimes \mathcal{K}_X) .$$

Proof See [Ser55b] for the original proof or [Har77, Chapter III, Corollary 7.7] for an algebro-geometric proof. □

D.3 Kähler Manifolds and Hodge Theory

We refer to [Voi02, Chapitre III] for an in depth study of Hodge theory and Kähler manifolds.

Definition D.3.1 Let X be a complex analytic variety of dimension n. We denote by $\mathcal{T}_X$ the *holomorphic tangent bundle* of X, by which we mean the real tangent bundle $T_{X,\mathbb{R}}$ of the underlying differentiable manifold, equipped with the complex structure inherited from the analytic structure of X.

The bundle $\mathcal{T}_X$ is isomorphic to the subbundle $T_X^{1,0} \subset T_{X,\mathbb{C}} = T_{X,\mathbb{R}} \otimes \mathbb{C}$ generated by holomorphic vector fields. See below for more details.

Definition D.3.2 The dual of the holomorphic tangent bundle $\mathcal{T}_X$ is called the *bundle of holomorphic forms* $\Omega_X := \Omega^1_X = \mathcal{T}^\vee_X$. The *bundle of holomorphic p-forms* on X is defined by $\Omega^p_X := \bigwedge^p \Omega_X$. The *canonical bundle* of X is the complex holomorphic line bundle

$$\mathcal{K}_X := \bigwedge^n \mathcal{T}^\vee_X = \Omega^n_X = \det \Omega_X \, .$$

Theorem D.3.3 *Let n be a strictly positive integer. There is then an exact sequence of sheaves on* $\mathbb{P}^n(\mathbb{C})$*:*

$$0 \to \Omega_{\mathbb{P}^n(\mathbb{C})} \to \overbrace{\mathcal{O}_{\mathbb{P}^n(\mathbb{C})}(-1) \oplus \cdots \oplus \mathcal{O}_{\mathbb{P}^n(\mathbb{C})}(-1)}^{n+1 terms} \to \mathcal{O}_{\mathbb{P}^n(\mathbb{C})} \to 0$$

where $\mathcal{O}_{\mathbb{P}^n(\mathbb{C})}(-1)$ *is the tautological bundle (see Section F.1 or Definition 2.6.14).*

Proof See [Har77, Theorem 2.8.13] for an algebro-geometric proof. □

A *Hermitian metric* on a holomorphic variety X is a $\mathcal{C}^\infty$ family of Hermitian products on each holomorphic tangent bundle, which we can think of as a section $h \in \Gamma^\infty\left(X, \left(T^{1,0}_X \otimes \overline{T^{1,0}_X}\right)^*\right)$ such that

1. $h_x(u, \overline{v}) = \overline{h_x(v, \overline{u})}$ for every $x \in X$ and for every $u, v \in T^{1,0}_x$;
2. $h_x(u, \overline{u}) > 0$ for every non zero vector $u \in T^{1,0}_x$.

A Hermitian metric h on a holomorphic variety X provides a Riemannian metric g on the underlying differentiable manifold, namely the real part of h:

$$g = \frac{1}{2}(h + \overline{h}) \, .$$

The form g is a symmetric bilinear form on the complexification $T_{X,\mathbb{C}} = T_{X,\mathbb{R}} \otimes \mathbb{C}$. As g is equal to its conjugate it is also the complexification of a real symmetric bilinear form on $T_{X,\mathbb{R}}$.

The metric h also determines a $(1, 1)$-form $\omega = -\Im(h) = \frac{i}{2}(h - \overline{h})$. As above, ω is the complexification of a real form on the real tangent bundle $T_{X,\mathbb{R}}$.

Definition D.3.4 (*Kähler varieties*) A *Kählerian* or *Kähler* variety is a (non singular) complex analytic variety with a Hermitian metric h such that the 2-form $\omega = -\Im(h)$ is closed. The metric h is then said to be a *Kähler* or *Kählerian* metric on X and the form ω is said to be a *Kähler form* on X.

Remark D.3.5 As the form ω is non degenerate and closed it is a symplectic form. Any Kähler variety therefore has a natural symplectic manifold structure.

Example D.3.6 The analytic variety underlying a non singular complex projective algebraic variety is always Kähler, since it inherits the Kähler *Fubini-Study* metric from projective space. Consider the Hopf fibration $\mathbb{S}^{n+1} \to \mathbb{CP}^n$ whose fibres are great circles on $\mathbb{S}^{n+1}$. The spherical metric on $\mathbb{S}^{n+1}$ is the restriction of the Euclidean metric on the space $\mathbb{R}^{n+2}$ and it is invariant under rotation. The Fubini-Study metric on $\mathbb{CP}^n = \mathbb{S}^{n+1}/\mathbb{S}^1$ is then the metric induced by the spherical metric on $\mathbb{S}^{n+1}$. We refer the interested reader to [Voi02, 3.3.2] for more information, notably the expression of the Fubini Study metric in coordinates. Any projective complex analytic variety is a compact Kähler variety by restriction of the Fubini-Study metric

Example D.3.7 (*Other examples of Kähler varieties*)

1. Every non singular complex analytic curve—i.e. every Riemann surface—is Kähler (see Appendix E) (and projective if it is compact, see Theorem E.2.28) since in complex dimension 1 every 2-form is closed.
2. Any K3 surface is Kähler by Siu's theorem (see [Siu83] or [X85]).
3. Complex Euclidean space $\mathbb{C}^n$ is Kähler with the standard Hermitian metric.
4. Quotienting the above example, any complex torus of the form $\mathbb{C}^n/\Gamma$ where Γ is a lattice in $\mathbb{R}^{2n}$ is Kähler.

D.3.1 Hodge Theory

We refer the interested reader to the first two chapters of [GH78, Chapitre 0] for a more detailed study of Hodge theory.

Let X be a complex analytic variety of dimension n. For any $a \in X$, we consider a system of analytic coordinates centred at a which we denote by $z = (z_1, z_2, \dots, z_n)$. There are three different tangent spaces to X at the point a.

1. *Real tangent space*. We denote by $T_{\mathbb{R},a}$ the usual real tangent space derived from the $\mathcal{C}^\infty$ manifold structure on X, which we realise as the space of $\mathbb{R}$-linear derivations of the ring of germs of real $\mathcal{C}^\infty$ functions. In other words, writing the coordinates z_j in the form $z_j = x_j + iy_j$ in a neighbourhood of a we have that
$$T_{\mathbb{R},a} = \mathbb{R}\{\frac{\partial}{\partial x_j}, \frac{\partial}{\partial y_j}\} .$$
2. *Complex tangent space*. We denote by $T_{\mathbb{C},a} = T_{\mathbb{R},a} \otimes \mathbb{C}$ the complexified tangent bundle. We can think of it as the space of $\mathbb{C}$-linear derivations of the ring of germs of complex $\mathcal{C}^\infty$ functions
$$T_{\mathbb{C},a} = \mathbb{C}\{\frac{\partial}{\partial x_j}, \frac{\partial}{\partial y_j}\} = \mathbb{C}\{\frac{\partial}{\partial z_j}, \frac{\partial}{\partial \overline{z_j}}\} ;$$
where
$$\frac{\partial}{\partial z_j} = \frac{1}{2}(\frac{\partial}{\partial x_j} - i\frac{\partial}{\partial y_j}) ; \qquad \frac{\partial}{\partial \overline{z_j}} = \frac{1}{2}(\frac{\partial}{\partial x_j} + i\frac{\partial}{\partial y_j}) .$$
3. Decomposition of $T_{\mathbb{C},a}$. We denote by $T_a^{1,0} = \mathbb{C}\{\frac{\partial}{\partial z_j}\} \subset T_{\mathbb{C},a}$ the *holomorphic tangent space* to X at a. It can be characterised as the subspace of $T_{\mathbb{C},a}$ of

derivations which are zero on all anti-holomorphic functions f (i.e. functions such that $\bar{f}$ is holomorphic). The space $T_a^{1,0}$ is therefore independent of the choice of coordinates at a. We denote by $T_a^{0,1} = \mathbb{C}\{\frac{\partial}{\partial \overline{z_j}}\}$ the anti-holomorphic tangent space. There is then a direct sum decomposition:

$$T_{\mathbb{C},a} = T_a^{1,0} \oplus T_a^{0,1} . \tag{D.1}$$

Note that by definition $T_{\mathbb{C},a}$ has a real structure. Conjugation

$$\frac{\partial}{\partial z_j} \mapsto \frac{\partial}{\partial \overline{z_j}}$$

is therefore well defined and

$$T_a^{0,1} = \overline{T_a^{1,0}} .$$

We denote by $A^k(X, \mathbb{R})$ the space of real valued differential k-forms on X, by $Z^k(X, \mathbb{R}) \subset A^k(X, \mathbb{R})$ the subspace of closed forms and by $A(X, \mathbb{R}) := \bigoplus_k A^k(X, \mathbb{R})$ the space of all differential forms. Similarly, $A^k(X, \mathbb{C})$ is the space of complex valued k-forms and $Z^k(X, \mathbb{C})$ is the subspace of closed complex valued forms. We set $A(X, \mathbb{C}) := \bigoplus_k A^k(X, \mathbb{C})$. The De Rham cohomology groups of X are then defined as follows.

$$H^k_{\mathrm{DR}}(X, \mathbb{R}) = \frac{Z^k(X, \mathbb{R})}{dA^{k-1}(X, \mathbb{R})} ;$$
$$H^k_{\mathrm{DR}}(X, \mathbb{C}) = \frac{Z^k(X, \mathbb{C})}{dA^{k-1}(X, \mathbb{C})} .$$

Remark D.3.8 We have that $H^k_{\mathrm{DR}}(X, \mathbb{C}) = H^k_{\mathrm{DR}}(X, \mathbb{R}) \otimes \mathbb{C}$.

By (D.1) there is a decomposition of the cotangent space

$$T^*_{\mathbb{C},a} = (T_a^{1,0})^* \oplus (T_a^{0,1})^*$$

for every $a \in X$. It follows that there is a decomposition of the exterior algebra

$$\bigwedge^k T^*_{\mathbb{C},a} = \bigoplus_{p+q=k} (\bigwedge^p (T_a^{1,0})^* \otimes \bigwedge^q (T_a^{0,1})^*) .$$

We set

$$A^{p,q}(X) := \left\{ \varphi \in A^k(X, \mathbb{C}) \mid \varphi(a) \in \bigwedge^p (T_a^{1,0})^* \otimes \bigwedge^q (T_a^{0,1})^*, \forall a \in X \right\}$$

and we obtain that

$$A^k(X,\mathbb{C}) = \bigoplus_{p+q=k} A^{p,q}(X)\,. \tag{D.2}$$

The fundamental Hodge theorem establishes a corresponding decomposition on cohomology groups when the variety X is compact Kähler.

A differential form $\varphi \in A^{p,q}(X)$ is said to be *of type* (p,q). We denote by $\pi^{p,q}\colon A(X,\mathbb{C}) \to A^{p,q}(X)$ the projection maps. Let φ be a differential form of type (p,q): for every $a \in X$ we then have that

$$d\varphi(a) \in \left(\bigwedge^p (T_a^{1,0})^* \otimes \bigwedge^q (T_a^{0,1})^*\right) \wedge T^*_{\mathbb{C},a}\,,$$

or in other words

$$d\varphi \in A^{p+1,q}(X) \oplus A^{p,q+1}(X)\,.$$

We define operators

$$\bar\partial\colon A^{p,q}(X) \to A^{p,q+1}(X) \quad , \qquad \partial\colon A^{p,q}(X) \to A^{p+1,q}(X)$$

by

$$\bar\partial = \pi^{(p,q+1)} \circ d \quad , \qquad \partial = \pi^{(p+1,q)} \circ d\,.$$

We then have that

$$d = \partial + \bar\partial\,.$$

Let $Z^{p,q}_{\bar\partial}(X)$ be the space of $\bar\partial$-closed forms of type (p,q). Since $\bar\partial^2 = 0$ on $A^{p,q}(X)$, we can define the Dolbeault cohomology groups by

$$H^{p,q}_{\bar\partial}(X) = \frac{Z^{p,q}_{\bar\partial}(X)}{\bar\partial A^{p,q-1}(X)}\,.$$

Let $Z^{p,q}(X)$ be the space of closed complex-valued differential forms of type (p,q). We also define

$$H^{p,q}(X) = \frac{Z^{p,q}(X)}{dA(X,\mathbb{C}) \cap Z^{p,q}(X)}\,.$$

D.3.2 De Rham's Theorem

We recall that $H^*(X;\mathbb{R})$ is the group of singular cohomology of X with coefficients in $\mathbb{R}$. It is also the sheaf cohomology of the constant sheaf $\mathbb{R}$. Let φ be a closed p-form and let σ be the boundary of a $(p+1)$-chain τ. Stokes' theorem then gives us

$$\int_\sigma \varphi = \int_\tau d\varphi = 0 \,.$$

The p-form φ therefore defines a singular p-cocycle. Moreover, for any p-form φ and for any p-cycle σ we have that

$$\forall \eta \in A^{p-1}(X, \mathbb{R}), \quad \int_\sigma \varphi = \int_\sigma \varphi + d\eta \,.$$

This gives us a map $H^*_{\mathrm{DR}}(X, \mathbb{R}) \to H^*(X; \mathbb{R})$ which is in fact an isomorphism by the following theorem.

Theorem D.3.9 (De Rham) *Let X be a $\mathcal{C}^\infty$ manifold. There is an isomorphism*

$$H^*_{DR}(X, \mathbb{R}) \simeq H^*(X; \mathbb{R}) \,.$$

We prove this using a fine resolution of the constant sheaf $\mathbb{R}$. Let $\mathcal{A}^p$ be the sheaf of germs of $\mathcal{C}^\infty$ p-forms on X. The sequence

$$0 \to \mathbb{R} \hookrightarrow \mathcal{A}^0 \xrightarrow{d} \mathcal{A}^1 \xrightarrow{d} \mathcal{A}^2 \xrightarrow{d} \cdots$$

is exact by the Poincaré's lemma which says that every closed form is locally exact.

Lemma D.3.10 *Let α be a degree d $\mathcal{C}^1$ form on X with $d > 0$. If $d\alpha = 0$ then for any contractible open set U in X there is a $\mathcal{C}^1$ form β of degree $d - 1$ on U such that $\alpha|_U = d\beta$.*

Proof See [GH78, Section 0.2, page 25] or [Voi02, Proposition 2.31]. □

This sequence can be broken up into a collection of short exact sequences. (In the sequences below, $d\mathcal{A}^p$ denotes the *sheaf of closed p-forms*).

$$0 \to \mathbb{R} \hookrightarrow \mathcal{A}^0 \xrightarrow{d} d\mathcal{A}^0 \to 0$$
$$0 \to d\mathcal{A}^0 \hookrightarrow \mathcal{A}^1 \xrightarrow{d} d\mathcal{A}^1 \to 0$$
$$\vdots$$
$$0 \to d\mathcal{A}^{p-2} \hookrightarrow \mathcal{A}^{p-1} \xrightarrow{d} d\mathcal{A}^{p-1} \to 0$$

The associated cohomology sequences split because $H^q(X, \mathcal{A}^p) = 0$ for any non zero q and we get that

$$\begin{aligned}
&H^p(X;\mathbb{R})\\
&\simeq H^{p-1}(X, d\mathcal{A}^0)\\
&\simeq H^{p-2}(X, d\mathcal{A}^1)\\
&\quad\vdots\\
&\simeq H^1(X, d\mathcal{A}^{p-2})\\
&\simeq \frac{H^0(X, d\mathcal{A}^{p-1})}{dH^0(X, \mathcal{A}^{p-1})}\\
&= \frac{\Gamma(d\mathcal{A}^{p-1})}{d\Gamma(\mathcal{A}^{p-1})}\\
&= H^p_{\mathrm{DR}}(X, \mathbb{R}).
\end{aligned}$$

D.3.3 Dolbeault's Theorem

Recall that Ω^p is the sheaf of germs of holomorphic p-forms on X and $H^q(X, \Omega^p)$ is the qth cohomology group of this sheaf.

Theorem D.3.11 (Dolbeault) *Let X be a complex analytic variety. We then have that*

$$H^q(X, \Omega^p_X) \simeq H^{p,q}_{\bar{\partial}}(X)$$

We denote by $\mathcal{A}^{p,q}_X$ the sheaf of $\mathcal{C}^\infty$ forms of type (p, q) on X. We have a fine resolution of the sheaf Ω^p_X

$$0 \to \Omega^p_X \hookrightarrow \mathcal{A}^{p,0}_X \xrightarrow{\bar{\partial}} \mathcal{A}^{p,1}_X \xrightarrow{\bar{\partial}} \cdots$$

This sequence is exact by the $\bar{\partial}$ Poincaré lemma which states that any $\bar{\partial}$-closed form is locally $\bar{\partial}$-exact:

Lemma D.3.12 *Let α be a $\mathcal{C}^1$ form of type (p, q) on X with $q > 0$. If $\bar{\partial}\alpha = 0$ then for every contractible open set U in X there is a $\mathcal{C}^1$ form β of type $(p, q - 1)$ on U such that $\alpha|_U = \bar{\partial}\beta$.*

Proof See [GH78, Section 0.2, page 25] or [Voi02, Proposition 2.31]. □

The rest of the proof is exactly the same as the proof given above for the de Rham case.

D.3.4 Hodge Decomposition

It remains to show that the decomposition (D.2) page 373 also holds on the cohomology groups when X is compact Kähler. There is no known algebraic proof of this fact: the proof uses transcendental methods. This theorem is proved using the fact that Hodge's theorem tells us that every cohomology class in $H^*_{\mathrm{DR}}(X, \mathbb{C})$ is represented by a unique harmonic form and since X is Kähler, the space of harmonic forms decomposes as $\mathcal{H}^r = \bigoplus_{p+q=r} \mathcal{H}^{p,q}$.

We start by defining harmonic forms. The Hermitian metric X, yields a Hermitian product $(\cdot, \cdot)$, with turns $A^{p,q}(X)$ into a inner product space. We prove that this space is in fact a Hilbert space and then introduce the Laplacian $\Delta_{\bar\partial}$ in order to answer the following question:

Given a form $\psi \in Z^{p,q}_{\bar\partial}(X)$, can we find a representative of the cohomology class $[\psi] \in H^{p,q}_{\bar\partial}(X)$ of ψ which is of minimal norm?

The operator $\bar\partial$ turns out to be bounded on $A^{p,q}(X)$ so we can introduce its adjoint $\bar\partial^* \colon A^{p,q}(X) \to A^{p,q-1}(X)$, defined by

$$\forall \eta \in A^{p,q-1}(X), \quad (\bar\partial^* \psi, \eta) = (\psi, \bar\partial \eta) .$$

We then prove that ψ is of minimal norm in $\psi + \bar\partial A^{p,q-1}$ if and only if

$$\bar\partial^* \psi = 0 .$$

Elements of the group $H^{p,q}_{\bar\partial}(X)$ are therefore represented by solutions of the second order system

$$\bar\partial \psi = 0 , \quad \bar\partial^* \psi = 0 .$$

The Laplacian enables us to replace this system by a single equation

$$\Delta_{\bar\partial} = \bar\partial \bar\partial^* + \bar\partial^* \bar\partial .$$

On the one hand, $\bar\partial \psi = \bar\partial^* \psi = 0$ clearly implies that $\Delta_{\bar\partial} \psi = 0$: on the other hand, the equation

$$(\Delta_{\bar\partial} \psi, \psi) = (\bar\partial \bar\partial^* \psi, \psi) + (\bar\partial^* \bar\partial \psi, \psi) = |\bar\partial^* \psi|^2 + |\bar\partial \psi|^2$$

proves that the converse also holds.

A form ψ such that $\Delta_{\bar\partial} \psi = 0$ is said to be harmonic. We denote by $\mathcal{H}^r$ the space of harmonic forms of degree r and by $\mathcal{H}^{p,q}(X)$ the space of harmonic forms of type (p, q).

Theorem D.3.13 (Hodge) *Let X be a compact complex analytic variety. We then have that*

1. $\dim \mathcal{H}^{p,q}(X) < \infty$.

2. *The orthogonal projection*

$$\mathcal{H}\colon A^{p,q}(X) \to \mathcal{H}^{p,q}(X)$$

is well defined and there is a unique operator (known as Green's operator)

$$G\colon A^{p,q}(X) \to A^{p,q}(X)$$

such that

$$G(\mathcal{H}^{p,q}(X)) = 0\,, \quad \bar{\partial} G = G\bar{\partial}\,, \quad \bar{\partial}^* G = G\bar{\partial}^*$$

and

$$Id = \mathcal{H} + \Delta G \quad on \quad A^{p,q}(X)\,.$$

This equation can also be written in the form $\forall \psi \in A^{p,q}(X)$,

$$\psi = \mathcal{H}(\psi) + \bar{\partial}(\bar{\partial}^* G\psi) + \bar{\partial}^*(\bar{\partial} G\psi)\,.$$

It follows that for any $\psi \in Z^{p,q}_{\bar{\partial}}(X)$ we have that $\psi = \mathcal{H}(\psi) + \bar{\partial}(\bar{\partial}^* G\psi)$ because $\bar{\partial} G\psi = G\bar{\partial}\psi = 0$ which yields an isomorphism

$$H^{p,q}_{\bar{\partial}}(X) \simeq \mathcal{H}^{p,q}\,.$$

We also introduce the operator $\Delta_d = dd^* + d^*d$ and the Kähler condition then implies that $\Delta_d = 2\Delta_{\bar{\partial}}$, from which the following result follows.

Proposition D.3.14 *If X is Kähler then the complex vector spaces $H^{p,q}_{\bar{\partial}}(X)$ and $H^{p,q}(X)$ are isomorphic.*

For the same reason it follows that if X is Kähler then the Laplacian $\Delta_{\bar{\partial}}$ is a real operator so that $\mathcal{H}^{q,p} = \overline{\mathcal{H}^{p,q}}$. Moreover, as $\Delta_{\bar{\partial}}$ is real it commutes with the projections $\pi^{p,q}$ and we have that

$$\mathcal{H}^r(X) \simeq \bigoplus_{p+q=r} \mathcal{H}^{p,q}(X)\,.$$

Corollary D.3.15 *Let X be a compact Kähler variety. There is then a direct sum decomposition*

$$H^r(X;\mathbb{C}) \simeq \bigoplus_{p+q=r} H^{p,q}(X)$$

such that

$$\overline{H^{p,q}(X)} = H^{q,p}(X)\,.$$

Proof By De Rham's theorem, we know that $H^*_{\mathrm{DR}}(X,\mathbb{C}) \simeq H^*_{\mathrm{DR}}(X,\mathbb{R}) \otimes \mathbb{C} \simeq H^*(X;\mathbb{R}) \otimes \mathbb{C} \simeq H^*(X;\mathbb{C})$. The corollary follows. □

Poincaré duality induces an isomorphism

$$H^{n-k}(X;\mathbb{C}) \simeq H^k(X;\mathbb{C}) .$$

Passing to harmonic forms we see that this isomorphism is compatible with the Hodge decomposition and this gives us an isomorphism (which can also be proved directly using Serre duality, Theorem D.2.5—see Remark D.4.2):

$$H^{n-p,n-q}(X) \simeq H^{p,q}(X) .$$

D.3.5 Consequences

(a) If $q = 0$, $H^{p,0}(X) \simeq H^{p,0}_{\bar\partial}(X) \simeq H^0(X, \Omega^p)$ which is the space of global holomorphic p-forms on X. A holomorphic form is therefore harmonic for any Kähler metric on a compact variety.

(b) Odd degree Betti numbers on Kähler manifolds are even. Indeed, if we denote by $b_k(X) = \dim_{\mathbb{C}} H^k(X;\mathbb{C})$ the Betti numbers of X and by $h^{p,q}(X) = \dim H^{p,q}(X)$ the Hodge numbers of X, then

$$b_k(X) = \sum_{p+q=k} h^{p,q}(X)\ ; \quad h^{p,q}(X) = h^{q,p}(X) .$$

It follows that if $k = 2q+1$, then $b_k(X) = 2\sum_{p=0}^{q} h^{p,2q+1-p}(X)$.

(c) We organise the cohomology groups of X into a diagram called the Hodge diamond as in Figure D.1.

The kth cohomology group of X is the direct sum of all groups in the kth horizontal line. The diagram is symmetric under rotation about its centre $h^{n-p,n-q} = h^{p,q}$ and symmetry in the vertical axis $h^{q,p} = h^{p,q}$.

For any connected surface ($n = 2$) this gives us a diagram where $q = h^{0,1}$ is the irregularity of the surface and $p_g = h^{0,2}$ is its geometric genus (see Definition D.4.1).

$$\begin{array}{ccccc} & & 1 & & \\ & q & & q & \\ p_g & & h^{1,1} & & p_g \\ & q & & q & \\ & & 1 & & \end{array}$$

Example D.3.16 Calculating the cohomology of a compact Riemann surface S of genus g.

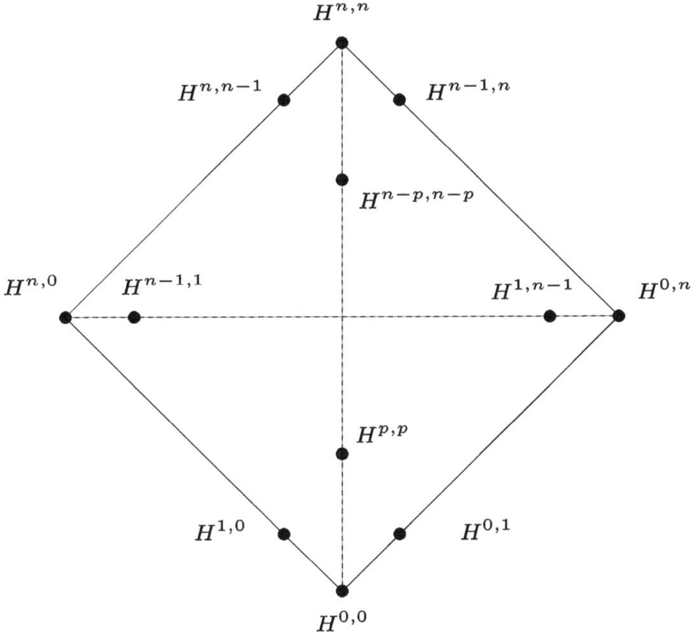

Fig. D.1 Hodge diamond

$$b_0 = b_2 = h^{0,0} = h^{1,1} = 1 \;;$$
$$h^{1,0} = h^{0,1} = \dim H^0(S, \Omega^1) = g \text{ from which it follows that } b_1 = 2g \;.$$

The existence of the Hodge decomposition has an important consequence for $\mathbb{R}$-varieties.

Lemma D.3.17 *Let (X, σ) be a compact Kähler $\mathbb{R}$-variety. If we denote by σ^* the action induced by σ on $H^*(X; \mathbb{C}) = H^*(X; \mathbb{Q}) \otimes_{\mathbb{Q}} \mathbb{C}$ then we have that*

$$\sigma^* H^{p,q}(X) = H^{q,p}(X) \;.$$

Proof See [Sil89, Lemma I.(2.4) page 10]. □

D.4 Numerical Invariants

Definition D.4.1 Let $(X, \mathcal{O}_X)$ be a compact complex analytic variety (such as the underlying analytic space of a non singular complex projective variety) of dimension n.

The *geometric genus* of X is defined to be $p_g(X) := \dim H^n(X, \mathcal{O}_X)$. The *irregularity* of X is defined to be $q(X) := \dim H^1(X, \mathcal{O}_X)$. The *Hodge numbers* of X are defined to be the numbers $h^{p,q}(X) := \dim H^q(X, \Omega^p_X)$.

Remark D.4.2 By Serre duality (Theorem D.2.5) applied to the line bundle $\mathcal{O}_X = \Omega^0_X$ we have that $p_g(X) = \dim H^0(X, \mathcal{K}_X) = \dim H^0(X, \Omega^n_X)$. Noting that $(\Omega^p_X)^\vee \otimes \mathcal{K}_X = \Omega^{n-p}_X$ (see [Har77, II, Exercice 5.16b], for example), Serre duality applied to the vector bundles Ω^p_X gives us the more general result that $h^{p,q}(X) = h^{n-p,n-q}(X)$ for any p, q. If additionally X is Kähler, the Hodge numbers satisfy $h^{p,q}(X) = h^{q,p}(X)$ for any p, q.

Definition D.4.3 (*Chern numbers of a complex surface*) Let $(X, \mathcal{O}_X)$ be a compact complex analytic variety of dimension 2. The *Chern numbers* of X are given by:

$$c_1^2(X) := c_1^2(\mathcal{K}_X) = (K_X^2) \text{ and } c_2(X) := \chi_{\text{top}}(X) = \sum_{k=0}^{4} (-1)^k b_k(X) .$$

Example D.4.4 (*Numerical invariants of a surface in* $\mathbb{P}^3$) Let X be a non singular complex hypersurface of degree d in $\mathbb{P}^3(\mathbb{C})$. We then have the following formulas. See [GH78, pages 601–602] for more details.

$$\begin{aligned} b_1(X) &= 0 ; \\ b_2(X) &= d^3 - 4d^2 + 6d - 2 ; \\ c_2(X) &= d^3 - 4d^2 + 6d ; \\ h^{0,2}(X) &= \frac{1}{6}(d-1)(d-2)(d-3) ; \\ h^{1,1}(X) &= \frac{1}{3}d(2d^2 - 6d + 7) . \end{aligned}$$

Example D.4.5 (*Numerical invariants of a double cover of the plane*) Using [BHPVdV04, V.22, page 237], for example, we can calculate the numerical invariants of a double cover X of $\mathbb{P}^2$ ramified over a non singular irreducible curve of degree $2k$. (Some of the formulas below are also proved in [Wil78, Section 5, page 65–66]):

$$\begin{aligned} q(X) = 0, \quad b_1(X) &= 0 ; \\ h^{0,2}(X) = 1 + \frac{1}{2}k(k-3) &= \frac{1}{2}(k-1)(k-2) ; \\ c_1^2(X) &= 2(k-3)^2 ; \\ c_2(X) = \chi_{top}(X) &= 4k^2 - 6k + 6 . \end{aligned}$$

It follows that

$$h^{1,1}(X) = c_2(X) - 2 - 2h^{0,2}(X) = 3k^2 - 3k + 2 \ .$$

Definition D.4.6 (*Algebraic dimension*) Let X be a compact connected complex analytic variety. The *algebraic dimension* of X is the transcendence degree over $\mathbb{C}$ of the field of meromorphic functions on X:

$$a(X) := \text{trdeg}_{\mathbb{C}} \mathcal{M}(X) \ .$$

This definition makes sense because the field of meromorphic functions on a compact connected complex analytic variety X is a function field over $\mathbb{C}$ (Definition A.5.8). See [BHPVdV04, Section I.7] for more details.

Proposition D.4.7 *The algebraic dimension is a bimeromorphic invariant.*

Definition D.4.8 (*Kodaira dimension*) Let X be a compact connected complex analytic variety. For any integer $m \geqslant 1$, the number $P_m(X) := \dim H^0(X, \mathcal{K}_X{}^{\otimes m})$ is the *m th plurigenus* of X: in particular, $P_1(X) = p_g(X)$. The *Kodaira dimension* of X is defined as follows.

$$\kappa(X) := \begin{cases} -\infty & \text{if and only if } P_m(X) = 0 \text{ for all } m \geqslant 1 \ ; \\ k \geqslant 0 & \text{is the smallest integer such that the sequence } \left\{ \frac{P_m(X)}{m^k} \right\}_m \text{ is bounded.} \end{cases}$$

Proposition D.4.9 *The Kodaira dimension of a variety is a bimeromorphic invariant. If the variety is a projective surface then it is a birational invariant.*

Proof See [*Ibid.*]. □

Remark D.4.10 Let X be a compact connected complex analytic variety. We then have that

$$\kappa(X) \leqslant a(X) \leqslant \dim X \ .$$

See [*Ibid.*] for more details.

In particular, the Kodaira dimension $\kappa(X)$ of a compact complex variety X of dimension n is contained in this list: $-\infty, 0, 1, \ldots, n$.

Definition D.4.11 A compact complex variety X (resp. $\mathbb{R}$-surface (X, σ)) of dimension n is said to be *of general type* if and only if $\kappa(X) = n$ and *of special type* otherwise ($\kappa(X) < n$).

Proposition D.4.12 *Let X and Y be compact connected complex analytic varieties. We then have that*

$$\kappa(X \times Y) = \kappa(X) + \kappa(Y) \ .$$

Proof See [Uen75, page 63]. □

Theorem D.4.13 (Iitaka's conjecture $C_{2,1}$) *Let X be a compact connected complex analytic surface, let Y be a compact connected curve and let $\pi : X \to Y$ be a fibration, by which we mean that π is surjective, holomorphic and proper (this last condition not being necessary for our purposes because X is compact). If X is minimal then*

$$\kappa(X) \geqslant \kappa(Y) + \kappa(\textit{general fibre of } \pi) \ .$$

Proof See [BHPVdV04, Theorem III.18.4]. □

D.5 Projective Varieties

Unlike the compact complex curves discussed in Appendix E, compact complex varieties of dimension $n \geqslant 2$ are not all projective and in fact they are not even all Kähler. On the other hand, Chow's famous theorem tells us that any projective complex analytic variety is algebraic.

Theorem D.5.1 (Chow's theorem) *Let X be a subset of a complex projective space. If X is a closed analytic subspace then X is an algebraic subvariety.*

Proof See [GR65, Section V.D, Theorem 7]. □

Corollary D.5.2 *Let X be a compact complex analytic variety. X can be equipped with a projective algebraic structure variety if and only if there is an analytic embedding $X \hookrightarrow \mathbb{P}^N(\mathbb{C})$ in projective space.*

To any coherent algebraic sheaf $\mathcal{F}$ on a complex algebraic variety X we can associate a natural coherent analytic sheaf $\mathcal{F}^h$ on X^h. See [Ser56, Section 3, 9] for more details. The next three theorems, collectively known as the "GAGA" theorems, state that if X is projective then the theory of coherent analytic sheaves on X^h is essentially the same as the theory of coherent algebraic sheaves on X. These theorems are valid for projective varieties only and in particular they do not hold for affine X. We refer the interested reader to [Ser56, Section 3, 12] for the proofs.

Theorem D.5.3 *Let X be a complex projective algebraic variety and let $\mathcal{F}$ be a coherent algebraic sheaf over X. For any integer $i \geq 0$ there is an isomorphism*

$$H^i(X, \mathcal{F}) \simeq H^i(X^h, \mathcal{F}^h) \ .$$

Theorem D.5.4 *Let X be a complex projective algebraic variety and let $\mathcal{F}$ and $\mathcal{G}$ be coherent algebraic sheaves over X. Any analytic homomorphism from $\mathcal{F}^h$ to $\mathcal{G}^h$ arises from an unique algebraic homomorphism from $\mathcal{F}$ to $\mathcal{G}$.*

Theorem D.5.5 *Let X be a complex projective algebraic variety. For any coherent analytic sheaf $\mathcal{M}$ on X^h there is a coherent algebraic sheaf $\mathcal{F}$ over X such that $\mathcal{F}^h$ is isomorphic to $\mathcal{M}$. The sheaf $\mathcal{F}$ is unique up to isomorphism.*

D.6 Picard and Albanese Varieties

We define Cartier divisors on a complex analytic variety X as in Definition 2.6.7. Let $U \subset X$ be an open subset in the Euclidean topology and let $f \in \mathcal{M}_X(U)$ be a meromorphic function on U. By definition there is a dense open subset $V \subset U$ such that $\forall p \in V$, $f(p) = \frac{g(p)}{h(p)}$ for some $g, h \in \mathcal{O}_X(V)$. Attention: there is generally a subset of codimension 2 where this function is not defined.

Definition D.6.1 The quotient sheaf $\mathcal{D}_X = \mathcal{M}_X^*/\mathcal{O}_X^*$ is the sheaf of divisors of X arising from the exact sequence

$$1 \longrightarrow \mathcal{O}_X^* \longrightarrow \mathcal{M}_X^* \longrightarrow \mathcal{D}_X \longrightarrow 1$$

where $\mathcal{O}_X^*$ is the sheaf of germs of nowhere vanishing holomorphic functions and $\mathcal{M}_X^*$ is the sheaf of germs of non identically zero meromorphic functions. A *Cartier divisor* is a global section of the quotient sheaf $\mathcal{D}_X$. A *principal divisor* is the divisor associated to a global meromorphic function.

Let $U \subset X$ be a Euclidean open set and let $D = (U_i, f_i)_i \in \mathrm{Div}(X) = \Gamma(U, \mathcal{M}_X^*/\mathcal{O}_X^*)$ be a divisor described with respect to an open cover $\{V_i\}_i$ of U. This means that there are germs of holomorphic functions $g_i, h_i \in \mathcal{O}(V_i)$ such that

$$f_i = \frac{g_i}{h_i} \quad \text{and} \quad \frac{g_i}{h_i} \cdot \left(\frac{g_j}{h_j}\right)^{-1} \in \mathcal{O}^*(V_i \cap V_j).$$

Proposition D.6.2 *Let X be a non singular complex projective algebraic variety. The group of divisors modulo linear equivalence is then isomorphic to the Picard group of isomorphism classes of holomorphic line bundles. (Compare with Definition 2.6.11.)*

$$\mathrm{Div}(X)/\mathcal{P}(X) \simeq H^1(X, \mathcal{O}^*) \simeq \mathrm{Pic}(X)\ .$$

Proof See [Hir66, Chapter I] or [GH78, Section 1.1]. □

Proposition D.6.3 *In the long exact sequence*

$$\to H^1(X, \mathcal{O}_X) \to H^1(X, \mathcal{O}_X^*) \xrightarrow{\delta} H^2(X; \mathbb{Z}) \to H^2(X, \mathcal{O}_X) \to$$

associated to the exponential short exact sequence (see Example C.3.15)

$$0 \to \mathbb{Z} \xrightarrow{incl.} \mathcal{O}_X \xrightarrow{\exp(2\pi i \cdot)} \mathcal{O}_X^* \to 0\ ,$$

the coboundary map $\delta\colon H^1(X, \mathcal{O}_X^*) \to H^2(X; \mathbb{Z})$ *can be identified with the first Chern class morphism* $c_1\colon \mathrm{Pic}(X) \to H^2(X; \mathbb{Z})$.

Proof See [Hir66, Chapter I] or [GH78, Section 1.1]. □

Proposition D.6.4 *Let (X, σ) be an $\mathbb{R}$-variety and set $G = \mathrm{Gal}(\mathbb{C}|\mathbb{R})$. The exponential exact sequence*

$$0 \to \mathbb{Z} \xrightarrow{incl.} \mathcal{O}_X \xrightarrow{\exp(2\pi i \cdot)} \mathcal{O}_X^* \to 0$$

gives rise to a sequence of G-sheaves on "twisting" the constant sheaf $\mathbb{Z}$ by the G-action given by $\sigma \cdot n = -n$.

For any $d \in Pic(X)$, we have that

$$c_1(\sigma^*(d)) = -\sigma^*(c_1(d)) .$$

Proof Passing to the long exact sequence

$$\to H^1(X, \mathcal{O}_X) \to H^1(X, \mathcal{O}^*) \xrightarrow{c_1} H^2(X; \mathbb{Z}) \to$$

we get that for any divisor class $d \in H^1(X, \mathcal{O}^*)$, we have that

$$c_1(\sigma^*(d)) = -\sigma^*(c_1(d)) .$$

See [Sil89, I.(4.7)] for more details. □

Proposition D.6.5 *Let (X, σ) be a non singular projective $\mathbb{R}$-variety such that $p_g(X) = 0$ and $q(X) = 0$. The map $c_1\colon \mathrm{Pic}(X) \to H^2(X; \mathbb{Z})$ then induces an isomorphism of $\mathbb{Z}_2$-vector spaces*

$$H^2(G, \mathrm{Pic}(X)) \simeq H^1\left(G, H_2(X; \mathbb{Z})\right) .$$

Proof See [Sil89, I.(4.7–4.12) and III.(3.3–3.4)]. □

D.6.1 Picard Variety

Definition D.6.6 Let X be a compact connected Kähler variety, such as a non singular projective complex variety. The *Picard variety* $\mathrm{Pic}^0(X) \subset \mathrm{Pic}(X)$ of X is the kernel of the morphism $c_1\colon \mathrm{Pic}(X) \to H^2(X; \mathbb{Z})$.

By Proposition 2.6.12 the quotient group $\mathrm{Pic}(X)/\mathrm{Pic}^0(X)$ is therefore isomorphic to a subgroup of $H^2(X; \mathbb{Z})$ known as the *Néron–Severi group*, $\mathrm{NS}(X)$, of X. See Definition 2.6.34.

Proposition D.6.7 *If $q(X) > 0$ then $\mathrm{Pic}^0(X)$ is a complex torus. If X is projective (and non singular) then it is an abelian variety.*

Proof See [BHPVdV04, Section I.13]. □

If X has a real structure then the exact sequence

$$0 \to H^1(X;\mathbb{Z}) \xrightarrow{i^*} H^1(X,\mathcal{O}_X) \to \operatorname{Pic}^0(X) \to 0 \tag{D.3}$$

induces a real structure on the quotient torus

$$\operatorname{Pic}^0(X) = H^1(X,\mathcal{O}_X)/i^*(H^1(X;\mathbb{Z}))$$

on "twisting" by the Galois action on the constant sheaf $\mathbb{Z}$ as in the exponential exact sequence in Proposition D.6.4. See [Sil82, II.8] or [Sil89, IV.1] for more details.

Proposition D.6.8 *Let (X,σ) be a non singular projective $\mathbb{R}$-variety of irregularity $q > 0$. The Picard variety* $\operatorname{Pic}^0(X)$ *then has an induced $\mathbb{R}$-variety structure,* $\operatorname{Pic}^0(X)^G = \operatorname{Pic}^0(X)(\mathbb{R})$ *is a real compact Lie group and*

$$\operatorname{Pic}^0(X)^G = \operatorname{Pic}^0(X)(\mathbb{R}) \simeq (\mathbb{R}/\mathbb{Z})^q \times (\mathbb{Z}/2)^{q-\lambda_1}$$

where $\lambda_1 := \dim_{\mathbb{Z}_2}(1+\sigma_*)H^1(X;\mathbb{Z}_2)$ is the Comessatti characteristic of the involutive module $\big(H^1(X;\mathbb{Z}),\sigma_*\big)$ (Definition 3.1.3).

Proof See [Sil82, II.8] or [Sil89, IV.1]. □

Remark D.6.9 It follows from Poincaré duality that $\dim_{\mathbb{Z}_2}(1+\sigma_*)H^1(X;\mathbb{Z}_2) = \dim_{\mathbb{Z}_2}(1+\sigma_*)\big(H_1(X;\mathbb{Z})_f \otimes \mathbb{Z}_2\big)$ or in other words that the Comessatti characteristic of the involutive modules $\big(H^1(X;\mathbb{Z}),\sigma_*\big)$ and $\big(H_1(X;\mathbb{Z})_f,\sigma_*\big)$ are equal.

D.6.2 Albanese Variety

We refer to [BHPVdV04, Section I.13], amongst others, for more about the Albanese variety. Consider a compact connected Kähler variety X—for example a non singular complex projective algebraic variety—such that $q(X) \neq 0$. Let $\omega_1,\dots,\omega_q$ be holomorphic forms that form a basis of the complex vector space $H^0(X,\Omega_X)$ of global holomorphic forms on X. The family $\omega_1,\dots,\omega_q,\bar\omega_1,\dots,\bar\omega_q$ is then a basis for $H^1(X;\mathbb{C})$ by Theorem D.3.11 and Corollary D.3.15. We denote by $H_1(X;\mathbb{Z})_f := H_1(X;\mathbb{Z})/\operatorname{Tor}(H_1(X;\mathbb{Z}))$ the *free part* of $H_1(X;\mathbb{Z})$ and we consider a basis $\gamma_1,\dots,\gamma_{2q}$ of the free $\mathbb{Z}$-module $H_1(X;\mathbb{Z})_f$. The vectors

$$v_j = \begin{pmatrix} \int_{\gamma_j}\omega_1 \\ \vdots \\ \int_{\gamma_j}\omega_q \end{pmatrix} \in \mathbb{C}^q \quad \text{for } j = 1,\dots,2q$$

are therefore $\mathbb{R}$-linearly independent and generate a *lattice* in $\mathbb{C}^q$. The group morphism $H_1(X;\mathbb{Z})_f \to H^0(X,\Omega_X)^*$, $\gamma \mapsto (\omega \mapsto \int_\gamma \omega)$ is therefore injective.

Definition D.6.10 Let X be a compact connected Kähler variety—for example a non singular complex projective algebraic variety. The *Albanese variety* of X is defined by the exact sequence:

$$0 \to H_1(X;\mathbb{Z})_f \to H^0(X,\Omega_X)^* \to \mathrm{Alb}(X) \to 0\,. \tag{D.4}$$

In other words, $\mathrm{Alb}(X)$ is the *cokernel* of $H_1(X;\mathbb{Z})_f \to H^0(X,\Omega_X)^*$.

Under these hypotheses, if $q(X) \neq 0$ then the variety $\mathrm{Alb}(X)$ is a complex torus of dimension $q(X)$: if moreover X has a real structure σ then the exact sequence (D.4) induces a real structure on $\mathrm{Alb}(X)$. (See [Sil82, II.5] or [Sil89, IV.1] for more details). If X is projective then $\mathrm{Alb}(X)$ is an abelian variety [Voi02, Corollaire 12.12].

Proposition D.6.11 *Let (X,σ) be a Kähler $\mathbb{R}$-variety of irregularity $q > 0$. The Albanese variety $\mathrm{Alb}(X)$ then has an induced $\mathbb{R}$-variety structure, $\mathrm{Alb}(X)^G = \mathrm{Alb}(X)(\mathbb{R})$ is a compact real Lie group and we have that*

$$\mathrm{Alb}(X)^G = \mathrm{Alb}(X)(\mathbb{R}) \simeq (\mathbb{R}/\mathbb{Z})^q \times (\mathbb{Z}/2)^{q-\lambda_1}$$

where $\lambda_1 := \dim_{\mathbb{Z}_2}(1+\sigma_*)\big(H_1(X;\mathbb{Z})_f \otimes \mathbb{Z}_2\big)$ is the Comessatti characteristic of the involutive module $\big(H_1(X;\mathbb{Z})_f, \sigma_*\big)$ (Definition 3.1.3).

Proof See [Sil82, II.5] or [Sil89, IV.1]. □

Remark D.6.12 The complex tori $\mathrm{Alb}(X)$ and $\mathrm{Pic}^0(X)$ associated to the same compact Kähler variety X are isomorphic but in general the $\mathbb{R}$-varieties associated to the same compact Kähler $\mathbb{R}$-variety (X,σ) are not. If we denote by $\sigma_{\mathrm{Alb}(X)}$ and $\sigma_{\mathrm{Pic}^0(X)}$ the real structure associated to σ, then the $\mathbb{R}$-variety $\big(\mathrm{Alb}(X), \sigma_{\mathrm{Alb}(X)}\big)$ is isomorphic to the $\mathbb{R}$-variety $\big(\mathrm{Pic}^0(X), -\sigma_{\mathrm{Pic}^0(X)}\big)$. See [Sil82, II.8, after Lemma 3] for more details.

Definition D.6.13 Let X be a compact connected Kähler variety of irregularity $q(X) \neq 0$ and let P_0 be a point in X. We define the *Albanese map*:

$$\alpha_{P_0}\colon X \to \mathrm{Alb}(X),\ P \mapsto \begin{pmatrix} \int_{P_0}^P \omega_1 \\ \vdots \\ \int_{P_0}^P \omega_q \end{pmatrix} \quad \mathrm{mod}\ \big(v_1,\ldots,v_{2q}\big)\ .$$

If X has a real structure σ and $P_0 \in X(\mathbb{R})$ then α_{P_0} is an $\mathbb{R}$-morphism. See [*Ibid.*] for more details.

D.7 Riemann–Roch Theorem

D.7.1 Riemann–Roch for Curves

The Riemann–Roch theorem on divisors of an abstract curve Theorem E.3.1 can be generalised to rank r bundles on curves embedded in a non singular projective variety X.

Theorem D.7.1 *If C is a curve (which is not assumed non singular, reduced nor irreducible) on a non singular projective variety X and $\mathcal{F}$ is a locally free $\mathcal{O}_C$-module of rank r then*

$$\chi(\mathcal{F}) = \deg(\mathcal{F}) + r\chi(\mathcal{O}_C) ;$$
$$h^0(C, \mathcal{F}) - h^1(C, \mathcal{F}) = \deg(\mathcal{F}) + r(1 - p_a(C)) .$$

If $\mathcal{F}$ is a vector bundle of rank r on a non singular irreducible curve C then we have that

$$h^0(C, \mathcal{F}) - h^1(C, \mathcal{F}) = \int_C c_1(\mathcal{F}) + r(1 - g(C)) .$$

Proof See [BHPVdV04, Theorem II.3.1]. □

D.7.2 Riemann–Roch on Surfaces

See Theorem 4.1.18 for more details.

Theorem D.7.2 *If X is a non singular projective surface and D is a divisor on X then*

$$h^0(D) - h^1(D) + h^0(K_X - D) = \frac{1}{2} D \cdot (D - K_X) + \chi(\mathcal{O}_X)$$

where $\chi(\mathcal{O}_X)$ is the holomorphic Euler characteristic of X.

Proof See [BHPVdV04, Theorem I.5.5] or [Har77, Theorem V.1.6] for an algebro-geometric proof. □

D.8 Vanishing Theorems

We refer the interested reader to [EV92] for a deeper discussion of vanishing theorems, by which we mean theorems that give sufficient conditions for the vanishing for cohomology groups of coherent sheaves.[10] We state two of the most important vanishing theorems below.

Theorem D.8.1 (Serre vanishing theorem) *Let X be a non singular projective algebraic variety over an algebraically closed base field K, let $\mathcal{L}$ be an invertible sheaf on X and let $\mathcal{F}$ be a coherent sheaf over X. If $\mathcal{L}$ is* ample *then there is a natural number m_0 such that*

$$H^i(X, \mathcal{F} \otimes \mathcal{L}^m) = \{0\} \quad \text{for } i > 0 \text{ and } m \geqslant m_0 .$$

In particular on taking $\mathcal{F} = \mathcal{O}_X$ we get that $H^i(X, \mathcal{L}^m)$ vanishes for $i > 0$ and m sufficiently large.

Theorem D.8.2 (Kodaira's vanishing theorem) *Let X be a non singular complex projective variety of dimension n and let $\mathcal{L}$ be an invertible sheaf on X. If $\mathcal{L}$ is* ample *then*

1. $H^i(X, \mathcal{L} \otimes \mathcal{K}_X) = \{0\}$ *for all* $i > 0$;
2. $H^i(X, \mathcal{L}^{-1}) = \{0\}$ *for all* $i < n$.

Note that by Serre duality equations (1) and (2) are equivalent. The original proof is in [Kod53].

D.9 Other Fundamental Theorems

Theorem D.9.1 (Bertini's theorem) *Let $N \geqslant n \geqslant 2$ be strictly positive integers and let X be a complex analytic subvariety of dimension n in $\mathbb{P}^N(\mathbb{C})$. If X is connected then any* general *hyperplane $H \subset \mathbb{P}^N(\mathbb{C})$(i.e. in the complement of some strict algebraic subset of $\mathbb{P}^N(\mathbb{C})^\vee$) meets X transversally and the hypersurface $X \cap H$ in H is non singular.*

Proof See [BHPVdV04, Corollary I.20.3] and [GH78, 1.1, page 137]. □

Theorem D.9.2 (Lefschetz hyperplane theorem) *Let $N \geqslant n \geqslant 2$ be strictly positive integers, let X be a complex analytic subvariety of dimension n in $\mathbb{P}^N(\mathbb{C})$ and let $H \subset \mathbb{P}^N(\mathbb{C})$ be a hyperplane such that $X \cap H$ is a non singular variety. The inclusion morphisms*

$$H_i(X \cap H; \mathbb{Z}) \to H_i(X; \mathbb{Z}) \quad \text{and} \quad \pi_i(X \cap H, \mathbb{Z}) \to \pi_i(X, \mathbb{Z})$$

are then isomorphisms whenever $0 \leqslant i \leqslant n-2$.

Proof See [Mil63a]. □

[10] Sheaf cohomology of coherent sheaves is sometimes called *coherent cohomology*.

Let X be a compact Kähler variety. By Corollary D.3.15 there is then a decomposition $H^2(X;\mathbb{C}) = H^{2,0}(X) \oplus H^{1,1}(X) \oplus H^{0,2}(X)$.

The image of the first Chern class map $c_1 \colon \mathrm{Pic}(X) \to H^2(X;\mathbb{Z})$ is contained in the set of integral classes of type $(1,1)$ [Voi02, Section I.7.1]. This is expressed, slightly abusively, in [GH78, Section 1.2] as meaning that this image is contained in the "intersection" $H^{1,1}(X) \cap H^2(X;\mathbb{Z})_f$, by which they mean that we consider an inclusion map $H^2(X;\mathbb{Z})_f \to H^2(X;\mathbb{C})$ obtained by composing the inclusion $H^2(X;\mathbb{Z})_f \simeq \mathrm{Hom}(H_2(X;\mathbb{Z}),\mathbb{Z})$ (Theorem B.4.1) with the $\mathbb{Z}$-module inclusion $\mathrm{Hom}(H_2(X;\mathbb{Z}),\mathbb{Z}) \hookrightarrow \mathrm{Hom}(H_2(X;\mathbb{Z}),\mathbb{C})$ induced by the unique ring morphism $\mathbb{Z} \hookrightarrow \mathbb{C}$.

Theorem D.9.3 (Lefschetz theorem on (1, 1)-cycles) *Let X be a compact Kähler variety. The first Chern class map $c_1 \colon \mathrm{Pic}(X) \to H^2(X;\mathbb{Z})$ is a surjection onto the intersection $H^{1,1}(X) \cap H^2(X;\mathbb{Z})_f$.*

Proof The original proof uses Poincaré's *normal functions*: we refer the interested reader to [Lef71] which reproduces the famous 1924 article *L'Analysis situs et la géométrie algébrique*. Here is a proof based on the exponential exact sequence (Proposition D.6.3):

$$0 \to \mathbb{Z} \xrightarrow{\mathrm{i}} \mathcal{O}_X \xrightarrow{\exp(2\pi i\cdot)} \mathcal{O}_X^* \to 0$$

whose long exact sequence is

$$\to H^1(X,\mathcal{O}_X) \to H^1(X,\mathcal{O}_X^*) \xrightarrow{c_1} H^2(X;\mathbb{Z}) \xrightarrow{i^*} H^2(X,\mathcal{O}_X) \to \ .$$

In this exact sequence of $\mathbb{Z}$-modules, we use the fact that $H^1(X,\mathcal{O}_X^*)$ is isomorphic to $\mathrm{Pic}(X)$ and $H^2(X,\mathcal{O}_X)$ is isomorphic to the free $\mathbb{Z}$-module $H^{0,2}(X)$. The map i^* vanishes on the torsion subgroup of $H^2(X;\mathbb{Z})$ and factors through the inclusion described above $H^2(X;\mathbb{Z})_f \to H^2(X;\mathbb{C})$ and the projection $H^2(X;\mathbb{C}) \to H^{0,2}(X)$. The restriction of i^* to $H^{1,1}(X) \cap H^2(X;\mathbb{Z})_f$ therefore vanishes, which proves the theorem. We refer the interested reader to [Voi02, I.7.9] for more details. □

Theorem D.9.4 (Kodaira's embedding theorem) *Let $(X,\mathcal{O}_X)$ be a compact complex analytic variety. The variety X is isomorphic to a non singular projective variety (which is algebraic by Chow's theorem) if and only if it has a* Hodge metric, *or in other words a Kähler metric whose class $\omega \in H^2(X;\mathbb{R})$ is integral. In this case, $\omega \in H^2(X;\mathbb{Z}) \cap H^{1,1}(X)$.*

Proof See [Kod54, Theorem 4] or [GH78, 1.4]. □

Appendix E
Riemann Surfaces and Algebraic Curves

This appendix is a summary of important results on Riemann surfaces, central objects in complex geometry.

E.1 Genus and Topological Classification of Surfaces

Topological surfaces appear in two different contexts in this book.

1. The underlying topological space of a complex algebraic curve is a topological surface. In this statement, the word "curve" refers to the algebraic—i.e. complex—dimension of the object, and the word "surface" refers to its real dimension. For example, $\mathbb{R}^2$ is the topological surface underlying the complex curve $\mathbb{C}$. The underlying topological surface of the complex projective line $\mathbb{P}^1(\mathbb{C})$ is the sphere $\mathbb{S}^2$.
2. Topological surfaces can also appear as real loci of algebraic surfaces defined over $\mathbb{R}$. For example, $\mathbb{R}^2$ is the real locus of the algebraic surface $\mathbb{C}^2$. The real locus of the quadric surface $x^2 + y^2 + z^2 = 1$ is the sphere $\mathbb{S}^2$.

   It is important to understand the difference between these two types of surfaces.

Definition E.1.1 A topological surface S is a Hausdorff topological space which is locally homeomorphic to $\mathbb{R}^2$. In other words, for any $x \in S$ there is a pair (U, φ), where U is an open neighbourhood of x in S and $\varphi\colon U \to \mathbb{R}^2$ is a homeomorphism.

Definition E.1.2 The *genus* $g := g(S)$ of a topological surface S is defined to be the maximal number of disjoint simple closed curves (i.e. embedded circles) $\sqcup C_i \subset S$ which can be cut out of S without disconnecting it.

1. $\forall i, j\,,\quad i \neq j \Rightarrow C_i \cap C_j = \varnothing$
2. $S \setminus \sqcup C_i$ is connected.

F. Mangolte, *Real Algebraic Varieties*, Springer Monographs in Mathematics,
https://doi.org/10.1007/978-3-030-43104-4

Proposition E.1.3 *Let S, S' be two topological surfaces. If there is a homeomorphism, i.e. a continuous bijective map with continuous inverse*

$$f : S \to S' ,$$

then $g(S) = g(S')$.

Definition E.1.4 The *Euler-Poincaré characteristic* χ of a polyhedron whose faces, edge and vertices sets are denoted F, E and V respectively, is defined by the formula

$$\chi = \#F - \#E + \#V .$$

See Proposition E.1.10 below for a characterisation of orientability using the differentiable manifold structure on a surface. See Definition B.5.3 for a definition of orientability using only the topological structure.

The key fact is that a surface S is orientable if and only if any closed simple curve in S has a trivial tubular neighbourhood, i.e. a tubular neighbourhood homeomorphic to $\mathbb{S}^1 \times [-1, 1]$. On the other hand, S is non orientable if and only if it contains a simple closed curve which has a tubular neighbourhood homeomorphic to a Möbius band. See Lemma 3.4.4 for more details.

As there are at least two incompatible definitions of the genus of a non orientable surface in the literature, it is useful to explain the relationship between the Euler-Poincaré characteristic of a polyhedron and the genus as defined in E.1.2. (The existence of a polyhedron underlying a surface is guaranteed by a theorem of Radó's proved in 1925. See [Mas67, Chapitre 1] for more details.)

Proposition E.1.5 *Let S be a topological surface with a polyhedral decomposition. The following then hold.*

1. *If S is orientable*
$$\chi(S) = 2 - 2g(S) .$$

2. *If S is non-orientable*
$$\chi(S) = 2 - g(S) .$$

In particular, the Euler-Poincaré characteristic of a surface does not depend on the choice of polyhedral decomposition.

For example, the Klein bottle $\mathbb{K}^2$, which is non orientable and has zero Euler characteristic, is of genus 2 whereas the torus, which also has zero Euler characteristic but is orientable, is of genus 1.

Theorem E.1.6 (Classification of compact surfaces) *Let S, S' be two connected compact topological surfaces without boundary. Assume that both surfaces are orientable (resp. both surfaces are non orientable). We then have that.*

$$g(S) = g(S') \iff S \text{ homeomorphic to } S' .$$

More generally, two compact connected topological surfaces are homeomorphic if and only if they have the same orientability, the same Euler characteristic and the same number of connected components in their boundary.

See [Mas67, Chapitre 1] or [FK80] for a proof of this fact.

Definition E.1.7 A *two dimensional differentiable manifold* is a topological surface S with a maximal atlas $\mathcal{A}$ whose transition functions are diffeomorphisms.

Formally:

1. $\forall x \in S, \exists (U, \varphi) \in \mathcal{A}$, U open neighbourhood of x in S, $\varphi\colon\ U \to \mathbb{R}^2$ is a homeomorphism
2. $\forall (U_1, \varphi_1), (U_2, \varphi_2) \in \mathcal{A}, U_1 \cap U_2 \neq \varnothing \Rightarrow$

$$\varphi_1 \circ \varphi_2^{-1} \text{ is a } \mathcal{C}^\infty \text{ map from } \varphi_2(U_1 \cap U_2) \subset \mathbb{R}^2 \text{ on} \varphi_1(U_1 \cap U_2) \subset \mathbb{R}^2 .$$

$$\Leftrightarrow \varphi_1 \circ \varphi_2^{-1}|_{\varphi_2(U_1 \cap U_2)} \in \mathcal{C}^\infty(\varphi_2(U_1 \cap U_2)).$$

In practice we do not need our atlas to be maximal: any open cover of S by charts satisfying (2) will do.

See the standard reference [Laf96] (English translation [Laf15]) for an introduction to differentiable manifolds.

By convention, unless otherwise specified the transition maps of a differentiable manifold are assumed to be $\mathcal{C}^\infty$, even though the definition makes sense for $\mathcal{C}^k$ functions for any strictly positive integer k. When $k = 0$ the corresponding objects are topological surfaces.

Remark E.1.8 In real dimension 2 any topological manifold has a unique $\mathcal{C}^\infty$ differentiable manifold structure and any homeomorphism between topological manifolds can be approximated by $\mathcal{C}^\infty$ maps. See [Hir76, Chapter 9] for more details.

Exercise E.1.9 We can replace condition (1) of the definition by the following:

1. $\forall x \in S, \exists (U, \varphi) \in \mathcal{A}$, U open neighbourhood of x in S such that $\varphi\colon U \to \mathbb{R}^2$ is a homeomorphism from U to $\varphi(U) \subset \mathbb{R}^2$.

[Hint: any open ball in $\mathbb{R}^n$ is diffeomorphic to $\mathbb{R}^n$.]

Proposition E.1.10 *A differentiable surface is said to be* orientable *if and only if it has an atlas $\mathcal{A}$ whose transition functions preserve orientation, or in other words*

$$\forall (U_1, \varphi_1), (U_2, \varphi_2) \in \mathcal{A}, U_1 \cap U_2 \neq \varnothing \Rightarrow \forall x \in U_1 \cap U_2, \det d_x(\varphi_1 \circ \varphi_2^{-1}) > 0$$

Proof See [Hir76, Section 4.4]. □

We denote by $[S]$ the orientability class of a surface S.

Corollary E.1.11 *Let S, S' be two compact connected differentiable surfaces. The following are equivalent:*

$$S \text{ diffeomorphic to } S' \Leftrightarrow g(S) = g(S') \text{ and } [S] = [S'].$$

Exercise E.1.12 Prove that the product torus $S^1 \times S^1 \subset \mathbb{R}^4$ is diffeomorphic to a revolution torus in $\mathbb{R}^3$ by constructing an explicit diffeomorphism.

E.2 Complex Curves and Riemann Surfaces

For detailed statements and proofs of foundational results on Riemann surfaces we refer to [FK80, Chapitre 1]. If U is an open subset of $\mathbb{C}$ we denote by $\mathcal{H}(U)$ the ring of holomorphic functions $U \to \mathbb{C}$.

Definition E.2.1 (*Compare with Definition* D.1.2) A complex analytic curve or *Riemann surface* X is a Hausdorff topological space, locally homeomorphic to $\mathbb{C}$, equipped with a maximal atlas $\mathcal{A}$ whose transition functions are holomorphic.

Formally

1. $\forall x \in X, \exists (U, \varphi) \in \mathcal{A}$, U open neighbourhood of x in X, $\varphi\colon U \to \mathbb{C}$ is a homeomorphism from U to $\varphi(U) \subset \mathbb{C}$,
2. $\forall (U_1, \varphi_1), (U_2, \varphi_2) \in \mathcal{A}$, $U_1 \cap U_2 \neq \varnothing \Rightarrow$

$$\varphi_1 \circ \varphi_2^{-1} \text{ is holomorphic on } \varphi_2(U_1 \cap U_2) \subset \mathbb{C} \text{ [d'image } \varphi_1(U_1 \cap U_2) \text{ .]}$$

$$\Leftrightarrow \varphi_1 \circ \varphi_2^{-1}|_{\varphi_2(U_1 \cap U_2)} \in \mathcal{H}(\varphi_2(U_1 \cap U_2)).$$

By convention, a Riemann surface is assumed *connected*.

Remark E.2.2 (*Transition functions*)

1. The maps $\varphi_1 \circ \varphi_2^{-1}|_{\varphi_2(U_1 \cap U_2)}$ are therefore biholomorphisms.
2. As any biholomorphism is a $\mathcal{C}^\infty$ diffeomorphism the atlas $\mathcal{A}$ equips X with a differentiable surface structure.
3. Warning. An open disc is not biholomorphic to the plane $\mathbb{C}$, and more generally any bounded open set is not biholomorphic to the plane: by Liouville's theorem, if $f \in \mathcal{O}(\mathbb{C})$ and $|f| < M$ on $\mathbb{C}$ then f is constant. Holomorphic geometry is more rigid than differentiable geometry.
4. Transition functions on a Riemann surface are holomorphic, so their determinant is strictly positive (exercise). The underlying differentiable manifold of a Riemann surface is therefore orientable.
5. In practice we do not need our atlases to be maximal: any covering of X by open charts satisfying condition 2. of the definition will do (exercise).
6. Warning: the term *surface* in the name "Riemann surface" refers to this underlying differentiable manifold structure modeled on $\mathbb{R}^2$ (which is isomorphic to $\mathbb{C}$ as an $\mathbb{R}$-vector space). See Section E.1 for more details.

Exercise E.2.3 (*Examples of Riemann surfaces*)

1. (a) The field $\mathbb{C}$ of complex numbers with its usual topology and one chart, namely the identity.
   (b) Any connected open subset of $\mathbb{C}$ with one chart, namely inclusion.
   (c) Any connected open subset $U \subset X$ of a Riemann surface X with the atlas given by restrictions to U of charts on X.
2. The Riemann sphere $\mathbb{C} \cup \{\infty\}$ with atlas

$$(\mathbb{C} \to \mathbb{C}, z \mapsto z), (\mathbb{C}^* \cup \{\infty\} \to \mathbb{C}, z \mapsto \frac{1}{z}, \infty \mapsto 0) .$$

   Prove that this Riemann surface is diffeomorphic to the usual sphere in $\mathbb{R}^3$.

   [Hint: use stereographic projection as in Proposition 5.3.1.]
3. The tori: consider $\tau \in \mathbb{H}(\Leftrightarrow Im(\tau) > 0)$ and set

$$T := \mathbb{C}/\mathbb{Z} \oplus \tau\mathbb{Z}$$

   where $z_1 \sim z_2 \Leftrightarrow \exists (n, m) \in \mathbb{Z}^2$ such that $z_2 = z_1 + n + m\tau$.

   [Hint: consider charts on a fundamental domain.]

Definition E.2.4 (*Holomorphic*)

1. A continuous function $f : X \to \mathbb{C}$ on a Riemann surface X is said to be a *holomorphic function* if and only if $\forall x \in X$ there is a chart (U, φ) in a neighbourhood of x such that $f \circ \varphi^{-1} : \varphi(U) \to \mathbb{C}$ is a holomorphic function. When this is the case we write $f \in \mathcal{O}_X(X)$. More generally, for any open connected set W in X we will denote by $\mathcal{O}_X(W)$ the ring of holomorphic functions on W.
2. A *holomorphic map* (or *morphism*) between Riemann surfaces

$$g : X \to Y$$

   is a continuous map such that for any open set $V \subset Y$ and any holomorphic function $f : V \to \mathbb{C}$ the function

$$f \circ g : g^{-1}(V) \to \mathbb{C}$$

   is holomorphic.

   We set $g^*(f) := f \circ g|_{g^{-1}(V)}$: g^* is called the *pull back* of f by g. With this notation, g is holomorphic if and only if for any open set $V \subset Y$ we have that

$$f \in \mathcal{O}_Y(V) \Rightarrow g^*(f) \in \mathcal{O}_X(g^{-1}(V)) .$$

A holomorphic map $f: X \to Y$ between Riemann surfaces is said to be *conformal* if and only if it is both injective and surjective.

Exercise E.2.5 The function f is then a biholomorphism by Proposition E.2.9.

Exercise E.2.6 (*Characterisation of holomorphic functions*)

1. Prove that any $\mathcal{C}^\infty$ map between surfaces $g: X \to Y$ is a continuous map such that for any $\mathcal{C}^\infty$ function $f: Y \to \mathbb{R}^2$ the function $f \circ g: X \to \mathbb{R}^2$ is $\mathcal{C}^\infty$.
2. Prove that any continuous map $f: X \to Y$ between Riemann surfaces is holomorphic if and only if for any pair of charts (U, φ) in X and (V, ψ) in Y such that $f(U) \cap V \neq \varnothing$, the expression of f in coordinates $\psi \circ f \circ \varphi^{-1}: \varphi(U \cap f^{-1}(V)) \subset \mathbb{C} \to \psi(V) \subset \mathbb{C}$ is a holomorphic function between open subsets of $\mathbb{C}$.

Remark E.2.7 If a Riemann surface X has an anti-holomorphic involution σ, we say (X, σ) is *separating* if the complement $X \setminus X^\sigma$ is non connected. We refer the interested reader to [Gab06] for a detailed study of this property.

Definition E.2.8 A holomorphic map $g: X \to Y$ is said to be *constant* if and only if the image of X under g is a point.

Proposition E.2.9 (Open image) *Let $f: X \to Y$ be a holomorphic map between Riemann surfaces. If f is non constant then the image of any open connected subset of X under f is an open set in Y. In other words, any non constant holomorphic map between Riemann surfaces is open.*

Exercise E.2.10 Prove the above proposition using the analogous result for holomorphic functions on $\mathbb{C}$.

Theorem E.2.11 *Let X be a* compact *Riemann surface and let Y be a Riemann surface. Any holomorphic map $f: X \to Y$ is either constant or surjective, and in this latter case Y is also compact.*

In particular, any holomorphic function $X \to \mathbb{C}$ is constant and the ring of global holomorphic functions on X satisfies $\mathcal{O}_X(X) = \mathbb{C}$.

Proof If f is non constant then $f(X)$ is open by Proposition E.2.9 and compact because the image of a compact space under a continuous map is compact. It follows that $f(X)$ is a closed subset of Y because Y is Hausdorff. Since X and Y are assumed connected by convention, $f(X) = Y$. □

Exercise E.2.12 The affine complex line $\mathbb{A}^1(\mathbb{C}) \simeq \mathbb{C}$ is a Riemann surface. Prove that the projective complex line $\mathbb{P}^1(\mathbb{C}) \simeq \mathbb{C}^2/\mathbb{C}^*$ is a Riemann surface which is isomorphic (i.e. biholomorphic) to the Riemann sphere.

Proposition E.2.13 (Local expression of a holomorphic map) *Let $f: X \to Y$ be a holomorphic non constant map between Riemann surfaces. Consider a point $x_0 \in X$ and set $y_0 = f(x_0)$. Let ψ be a chart of Y centred on y_0. There is then a local* *coordinate z on X vanishing at x_0—i.e. a chart centred at x_0—and a natural number d such that the expression of f in these charts is*

$$z \mapsto z^d \ .$$

Proof Consider a coordinate $\tilde{z}$ on X centred at x_0 and let $\tilde{f}$ be the expression of f in the charts $\tilde{z}$ and ψ. We then have that $\tilde{f}(0) = 0$. In a neighbourhood of 0 we can develop $\tilde{f}$ as a power series $\tilde{f}(\tilde{z}) = \sum a_k \tilde{z}^k$. Let d be the smallest integer such that $a_d \neq 0$ and choose $c \in \mathbb{C}$ such that $c^d = a_d$. The function $\tilde{f}$ is of the form $\tilde{f}(\tilde{z}) = (c\tilde{z})^d(1 + u(\tilde{z}))$ where u is holomorphic and $u(0) = 0$. As the holomorphic function $w \mapsto \sqrt[d]{w}$ is well defined in a certain neighbourhood of 1, the function $1 + u$ is of the form h^d where h is holomorphic in a neighbourhood of 0 and $h(0) = 1$, so $\tilde{f}(\tilde{z}) = (c\tilde{z} \cdot h(\tilde{z}))^d$. Set $z(x) = c\tilde{z}(x) \cdot h(\tilde{z}(x))$. The implicit function theorem implies that z is a holomorphic coordinate in a neighbourhood of x_0 and we have that $\psi(f(z)) = (z(x))^d$ for any x in some neighbourhood of x_0. □

Exercise E.2.14 Check that the natural number d depends on f and x_0 but is independent of the choice of charts.

Definition E.2.15 The natural number d is called the *ramification index* of f at x_0. We also say that f is equal to $f(x_0)$ with *multiplicity* d at x_0. (In a neighbourhood of $f(x_0)$, the fibre of f meets a neighbourhood of x_0 in d points). The number $b_f(x_0) := d - 1$ is called the *branching number* of f at x_0.

Proposition E.2.16 *Let $f\colon X \to Y$ be a non constant holomorphic map between* compact *Riemann surfaces. There is then an integer m such that every $y \in Y$ has exactly m preimages, counting multiplicities. In other words,*

$$\forall y \in Y, \sum_{x \in f^{-1}(y)} (b_f(x) + 1) = m \ .$$

Proof We refer to [FK80, page 12] for the details.

We set

$$\Sigma_n := \Big\{ y \in Y \,;\ \sum_{x \in f^{-1}(y)} (b_f(x) + 1) \geqslant n \Big\}$$

and we prove that every set of this form is either empty or equal to Y. For any point $y_0 \in Y$ we then set $m := \sum_{x \in f^{-1}(y_0)} (b_f(x) + 1)$. This gives us $0 < m < \infty$ and since $y_0 \in \Sigma_m$ we get that $\Sigma_m = Y$. Since $y_0 \notin \Sigma_{m+1}$, Σ_{m+1} must be empty.

To prove that every set of this form is either empty or Y we prove that Σ_n is both open and closed in the connected set Y. □

Definition E.2.17 The integer m, denoted $\deg(f)$, is called the *degree* of f. We will also say that f is an m-sheeted (ramified) covering of Y by X.

Theorem E.2.18 (Riemann–Hurwitz) *Let $f\colon X \to Y$ be a non constant holomorphic map between compact Riemann surfaces. If m is the degree of f then*

$$g(X) = m(g(Y) - 1) + 1 + \frac{1}{2} \sum_{x \in X} b_f(x) .$$

Proof This result follows from a relationship between Euler characteristics.

$$2 - 2g(X) = m(2 - 2g(Y)) - \sum_{x \in X} b_f(x) .$$

As the set of branching points is finite, we can assume they are contained in the set of vertices of a polyhedral decomposition (F, A, S) of Y. This decomposition can be lifted by f to a decomposition that has $m\#F$ faces, $m\#A$ edges and $m\#S - \sum_{x \in X} b_f(x)$ vertices. □

Example E.2.19 (*Plane cubics and hyperelliptic curves*) Affine cubics of the form $\{y^2 = x^3 + ax + b\} \subset \mathbb{A}^2(\mathbb{C})$ and projective cubics of the form $C := \{y^2 = x^3 + ax + b\} \cup \{\infty\} \subset \mathbb{P}^2(\mathbb{C})$ $(\infty = (1 : 0 : 0))$ are said to be reduced. We calculate the genus of C using the form $\{y^2 = x(x-1)(x-h)\} \cup \{\infty\}$. If the curve is non singular—note that C is then a compact Riemann surface—then C is a torus, $g = 1$. This can be proved by considering the map

$$\pi \colon C \to \mathbb{P}^1(\mathbb{C}), (X : Y : Z) \mapsto (X : Z) , \infty \mapsto (1 : 0) .$$

in homogeneous coordinates.

In affine coordinates this gives us $(x, y) \mapsto x$, so this is a degree 2 morphism with four ramification points. We get the Riemann surface structure on C by pulling back the Riemann surface structure on $\mathbb{P}^1(\mathbb{C})$ as below.

More generally, the same argument shows that hyperelliptic curves $\{y^2 = P_{2g+1}(x)\} \cup \{\infty\} \subset \mathbb{P}^2(\mathbb{C})$, where P_{2g+1} is a degree $2g + 1$ polynomial with simple roots, have genus g. These curves are ramified double covers of the Riemann sphere.

Definition E.2.20 Let X be a Riemann surface.

1. A holomorphic map from X to $\mathbb{C}$ is called a *holomorphic function* on X (see Definition E.2.4).
2. A holomorphic map from X to the Riemann sphere $\mathbb{C} \cup \{\infty\}$ is called a *meromorphic function* on X. For any open set $U \subset X$ we denote by $\mathcal{M}(U)$ the $\mathbb{C}$-algebra of meromorphic functions on U. When U is connected, $\mathcal{M}(U)$ is a field and in particular $\mathcal{M}(X)$ is the field of meromorphic functions on the Riemann surface X.

Remark E.2.21 The reader should be aware that in higher dimension E.2.20 (2) fails because there can be points at which a meromorphic function is not defined, even when including the value ∞.

Exercise E.2.22 1. Any polynomial of degree d can be extended to a meromorphic function on $\mathbb{P}^1(\mathbb{C})$ with a pole of order d at ∞.

2. (Very important!) If f is a non constant meromorphic function on a Riemann surface X then with multiplicity f has the same number of zeros and poles. (Use Proposition E.2.16).
3. Recall that the usual definition of a meromorphic function on X is a function $f : X \setminus D \to \mathbb{C}$ where $D \subset X$ is a discrete closed subset whose expression in any chart of X is meromorphic, by which we mean that it is holomorphic outside of a discrete subset and has a pole at every point in which it is not defined. Extending f to the whole of X by setting $f(x) = \infty$ at every pole x we recover Definition E.2.20.

Proposition E.2.23 *Any meromorphic function on the Riemann sphere is a rational function, by which we mean a function of the form* $\frac{p}{q}$ *where* p *and* q *are polynomials*

Proof Let $f : \mathbb{P}^1(\mathbb{C}) \to \mathbb{C}$ be a meromorphic function. Since $\mathbb{P}^1(\mathbb{C})$ is compact, f has only a finite number of poles. Replacing f by $1/f$, we can assume that $\infty \in \mathbb{P}^1(\mathbb{C})$ is not a pole. Let $(a_1, \ldots, a_n) \in \mathbb{C}$ be the set of poles of f. In a neighbourhood of a_ν, let the polar part of f be

$$h_\nu(z) = \sum_{l=-k_\nu}^{-1} c_l^\nu (z - a_\nu)^l.$$

The function $f - (h_1 + \ldots h_n)$ is then holomorphic on $\mathbb{P}^1(\mathbb{C})$. By Theorem E.2.11 it is therefore constant, so f is rational. □

Note that it follows that a single non constant meromorphic function entirely determines the complex structure on X. Indeed, if $f \in \mathcal{M}(X)$ is non constant, x is a point of X and $n - 1 = b_f(x)$ then f determines a local coordinate centred on x :

$$\begin{cases} (f - f(x))^{\frac{1}{n}} & \text{if } f(x) \neq \infty\,, \\ (f)^{-\frac{1}{n}} & \text{if } f(x) = \infty\,. \end{cases}$$

Exercise E.2.24 Deduce a proof of the D'Alembert-Gauss theorem (also known as the fundamental theorem of algebra): any non constant polynomial has a root in $\mathbb{C}$. This theorem can be generalised as follows "any non constant polynomial has a root in the algebraic closure of its field of coefficients".

Any Riemann surface, and particularly any compact Riemann surface, has a globally defined non constant meromorphic function. (This is a difficult result). On the other hand, there are complex surfaces, such as general tori of complex dimension $\geqslant 2$, which do not have any globally defined non constant meromorphic function.

Theorem E.2.25 *Any Riemann surface has a globally defined non constant meromorphic function.*

Proof See [FK80, Cor. II.5.3]: the key point in the proof is Weyl's lemma.[11] □

---

[11] Hermann Weyl (1885–1955), not to be confused with André Weil (1906–1998)

Corollary E.2.26 *Any Riemann surface is triangulable.*

Remark E.2.27 1. Any such non constant function $f \in \mathcal{M}(X)$ gives us a holomorphic map $f: X \to \mathbb{P}^1$ which by Proposition E.2.16 is an $m = \deg(f)$-sheeted ramified covering map.
2. More generally, any non constant holomorphic map $f: X \to Y$ between compact Riemann surfaces is also a ramified covering map. Restricting this covering to the complement of the branch points we get a degree m non ramified covering. It can be proved that this non ramified covering determines and is determined by f. (See [Dol90, 5.(6.3.1) and 5.(6.3.4)] for more details).

Theorem E.2.28 *Any compact Riemann surface is projective.*

More precisely, if X is a compact Riemann surface then there is a natural number N and a holomorphic embedding

$$\Phi: X \to \mathbb{P}^N(\mathbb{C}) .$$

Moreover, $\Phi(X) \subset \mathbb{P}^N(\mathbb{C})$ *is a complex projective algebraic curve.*

Sketch Proof Theorem E.2.25 implies the existence of an ample line bundle (Definition 2.6.20) $\mathcal{L}$ on X. The existence of such a line bundle implies the existence of a morphism $\varphi_{\mathcal{L}}$ (see Definition 2.6.20 or [Dol90, page 182 (8.7.3)]) and we then simply apply Chow's Theorem D.5.1. We refer the interested reader to [Jos06, Theorem 5.7.1] for a full proof. □

Example E.2.29 Weierstrass's elliptic $\wp$ function (see [Car61, V.2.5], [FK80, page 4], [Sil09, VI.3]) gives a direct proof—i.e. a proof which does not depend on the above theorem—of the fact that all complex tori of dimension 1 of type $T_\tau := \mathbb{C}/\mathbb{Z} \oplus \tau\mathbb{Z}$ seen in Exercise E.2.3(3) are algebraic.

For any $\tau \in \mathbb{H}$ we set

$$\wp(\tau; z) = \frac{1}{z^2} + \sum_{\substack{(n,m)\neq(0,0)\\(n,m)\in\mathbb{Z}^2}} \left(\frac{1}{(z-n-m\tau)^2} - \frac{1}{(n+m\tau)^2} \right).$$

The function thus defined is a meromorphic function on the plane which is doubly periodic with respect to the lattice $\mathbb{Z} \oplus \tau\mathbb{Z}$ and therefore defines a meromorphic function on the torus T_τ. It can be proved that the derivative $\wp'$ of $\wp$ satisfies an algebraic equation in $\wp$:

$$(\wp')^2 = 4(\wp - e_1)(\wp - e_2)(\wp - e_3) \tag{E.1}$$

where $e_1 = \wp(\frac{1}{2})$, $e_2 = \wp(\frac{\tau}{2})$ and $e_3 = \wp(\frac{1+\tau}{2})$.

The function $\wp'$ is also meromorphic on the torus. The Riemman surface T_τ can be thought of as the plane algebraic curve determined by the equation

$$y^2 = 4(x - e_1)(x - e_2)(x - e_3) .$$

E.3 The Riemann–Roch Theorem for a Curve

Theorem E.3.1 (Riemann–Roch theorem) *Let X be a non singular projective curve and let D be a divisor on X: recall that $h^k(D) = \dim H^k(X, \mathcal{O}_X(D))$. We then have that*

$$h^0(D) - h^0(K_X - D) = \deg D + 1 - g(X)\ .$$

Proof See [Jos06, Theorem 5.4.1], for example. □

E.4 Jacobian Variety Associated to a Curve

Let X be a compact Riemann surface of non zero genus g and let $(\omega_1, \dots, \omega_g)$ be a basis of the complex vector space $H^0(X, \Omega_X)$ of global holomorphic differentiable forms on X. Let $\gamma_1, \dots, \gamma_{2g}$ be a basis of the free $\mathbb{Z}$-module $H_1(X; \mathbb{Z})$: the $2g$ vectors

$$v_j = \begin{pmatrix} \int_{\gamma_j} \omega_1 \\ \vdots \\ \int_{\gamma_j} \omega_g \end{pmatrix} \in \mathbb{C}^g \quad j = 1, \dots, 2g$$

are then $\mathbb{R}$-linearly independent (see [ACGH85, Section I.3] for example) and generate a *lattice* in $\mathbb{C}^g$. It follows that integration of holomorphic differentiable forms of a compact Riemann surface X along 1-cycles gives us an injection

$$H_1(X; \mathbb{Z}) \hookrightarrow H^0(X, \Omega_X)^*, \quad \gamma \mapsto (\omega \mapsto \int_\gamma \omega)$$

which enables the following definition.

Definition E.4.1 Let X be a compact Riemann surface of non zero genus. The complex torus

$$\mathrm{Jac}(X) := H^0(X, \Omega_X)^* / H_1(X; \mathbb{Z})$$

is called the *Jacobian variety* of the curve X.

Remark E.4.2 The Jacobian $\mathrm{Jac}(X)$ of a Riemann surface X is a special case of an Albanese variety $\mathrm{Alb}(X)$ (Definition D.6.10).

Proposition E.4.3 *Let X be a compact Riemann surface of non zero genus g with a real structure σ. Let $s = \#\pi_0(X(\mathbb{R}))$ be the number of connected components of $X(\mathbb{R})$. The Jacobian* $\mathrm{Jac}(X)$ *then has an induced $\mathbb{R}$-variety structure and* $\mathrm{Jac}(X)^G = \mathrm{Jac}(X)(\mathbb{R})$ *is a real compact Lie group which is isomorphic to*

1. $(\mathbb{R}/\mathbb{Z})^g \times (\mathbb{Z}/2)^{s-1}$ *if $s \neq 0$;*

2. $(\mathbb{R}/\mathbb{Z})^g$ *if* $s = 0$ *and* g *is even;*
3. $(\mathbb{R}/\mathbb{Z})^g \times \mathbb{Z}/2$ *if* $s = 0$ *and* g *is odd.*

Proof If $X(\mathbb{R})$ is non empty then $\lambda_1 = g + 1 - s$ (see Exercise 3.3.12 and Example 3.6.9). By Remark E.4.2, when $s \neq 0$ the result below is a special case of Proposition D.6.11. We refer to [Sil82, Proposition 10] for more details. □

Theorem E.4.4 (Abel-Jacobi theorem) *Let X be a non singular complex projective algebraic curve. The Abel Jacobi map*

$$\pi_{P_0}\colon X \to \mathrm{Jac}(X),\ P \mapsto \begin{pmatrix} \int_{P_0}^{P} \omega_1 \\ \vdots \\ \int_{P_0}^{P} \omega_q \end{pmatrix} \quad \bmod\ \left(v_1, \ldots, v_{2q}\right)$$

then induces a group isomorphism.

$$\mathrm{Pic}^0(X) \to \mathrm{Jac}(X)\ .$$

Proof The injectivity of this map is simply *Abel's theorem* and its surjectivity is equivalent to *Jacobi's inversion theorem*. Voir [ACGH85, Section I.3]. □

Appendix F
Blow Ups

Blow ups are one of the main technical tools in this book. We summarise their main "algebraic" and "differentiable" properties in this section.

F.1 Blowing Up $\mathcal{C}^\infty$ Manifolds

This section is based on [Mik97, 2.1].

F.1.1 Tautological Bundle

We denote by $B_n \to \mathbb{RP}^n$ the tautological bundle—often denoted $\mathcal{O}_{\mathbb{P}^n}(-1)$ in algebraic geometry: see Definition 2.6.14 for more details—over projective space $\mathbb{RP}^n$. The fibre of this bundle at the point $L \in \mathbb{RP}^n$ is just the line passing through zero in $\mathbb{R}^{n+1}$ represented by L. It is a real rank 1 bundle. We recall how to construct a local trivialisation of this bundle. Let v be a non zero vector in $\mathbb{R}^{n+1}$ and let $L \in \mathbb{RP}^n$ be the line generated by v. Let $H \subset \mathbb{R}^{n+1}$ be a hyperplane which is a linear complement to L. Denote by $A \subset \mathbb{RP}^n$ the set of lines not contained in H. Every line L' contained in A contains exactly one vector of the form $v + w(L')$ with $w(L') \in H$. This yields a homeomorphism

$$A \times \mathbb{R} \to B_n|_A, \quad (L', t) \mapsto (L', tw(L'))$$

linear on each fibres.

B_n is a submanifold of the product $\mathbb{R}^{n+1} \times \mathbb{RP}^n$ by construction and the tautological bundle morphism is the restriction of the projection map

$$\mathbb{R}^{n+1} \times \mathbb{RP}^n \to \mathbb{RP}^n.$$

F. Mangolte, *Real Algebraic Varieties*, Springer Monographs in Mathematics,
https://doi.org/10.1007/978-3-030-43104-4

We denote by $\pi\colon B_n \to \mathbb{R}^{n+1}$ the restriction of projection to the first factor. The map π then induces a diffeomorphism

$$B_n \setminus E_P \xrightarrow{\approx} \mathbb{R}^{n+1} \setminus \{P\}$$

where $P = (0, \ldots, 0) \in \mathbb{R}^{n+1}$ and $E_P := \pi^{-1}(P)$.

We say that $\pi\colon B_n \to \mathbb{R}^{n+1}$ is the *blow up* of $\mathbb{R}^{n+1}$ at P. The submanifold E_P of codimension 1 in B_n is called the *exceptional divisor* of this blow up. It follows immediately from the definition that E_P is diffeomorphic to $\mathbb{RP}^n$.

Remark F.1.1 The tautological bundle is also the universal bundle over $\mathbb{RP}^n = \mathbb{G}_{n+1,1}(\mathbb{R})$. See Definition 5.2.11 for more details.

F.1.2 Projectivisation of the Normal Bundle

Consider a compact submanifold without boundary C of codimension r in a smooth manifold M: for simplicity, we equip M with a Riemannian metric. Let $\mathcal{N}_{M|C} \to C$ be the normal bundle to C in M: this is a vector bundle of rank r. We denote by

$$\pi_1\colon E_C \to C$$

the projectivisation of the bundle $\mathcal{N}_{M|C} \to C$. By definition, the fibre $\pi_1^{-1}(P)$ over $P \in C$ is the projective space of lines in the vector space $\mathcal{N}_{M|C,P}$ and E_C is therefore the total space of an $\mathbb{RP}^{r-1}$-bundle over C.

F.1.3 Blowing Up a Manifold Along a Submanifold

As C is embedded in M, there is an injective $\mathcal{C}^\infty$ map $j\colon \mathcal{N}_{M|C} \hookrightarrow M$ identifying $\mathcal{N}_{M|C}$ with an open neighbourhood $U = j(\mathcal{N}_{M|C})$ of C in M. The injection j is called a *tubular neighbourhood* of C in M. (The open subset U is also often called a tubular neighbourhood). The map j then identifies C with the zero section of $\mathcal{N}_{M|C}$. By abuse of notation we write $C \subset \mathcal{N}_{M|C}$ and j then induces a diffeomorphism $\mathcal{N}_{M|C} \setminus C \xrightarrow{\approx} U \setminus C$. We denote by $\widetilde{U}$ the total space of the tautological bundle over E_C and we identify E_C with the zero section $E_C \subset \widetilde{U}$. The space $\widetilde{U}$ is then a manifold of the same dimension as M by construction and we have a natural diffeomorphism

$$\mu\colon \widetilde{U} \setminus E_C \xrightarrow{\approx} U \setminus C$$

which extends to a $\mathcal{C}^\infty$ map

$$f\colon \widetilde{U} \to U \subset M$$

such that $f|_{E_C} = \pi_1$.

Ignoring the various choices involved in this construction, we have the following definition.

Definition F.1.2 The *blow up* $\widetilde{M}$ of M along C is constructed by gluing together $\widetilde{U}$ and $M \setminus C$ by the diffeomorphism μ. The $\mathcal{C}^\infty$ map

$$\pi \colon \widetilde{M} \to M$$

defined by $\pi|_{M\setminus C} = \mathrm{id}$ and $\pi|_{\widetilde{U}} = f$ is called the *topological blow up* of M along C.

The submanifold $C \subset M$ is called the *centre* of the blow up and the codimension 1 submanifold E_C in $\widetilde{M}$ is called the *exceptional divisor*. We often denote the blow up by $B_C M := \widetilde{M}$.

If $L \subset M$ is a closed suset then we say that a subset $\widetilde{L} \subset \widetilde{M}$ is the *strict transform* of L if and only if

- $\pi(\widetilde{L}) = L$,
- $\widetilde{L}$ is closed in $\widetilde{M}$,
- $\widetilde{L} \setminus E_C$ is dense in $\widetilde{L}$.

We refer the interested reader to [AK85, Section 2] for more details.

F.2 Blow Ups of Algebraic Varieties

F.2.1 Strict Transform

Consider an algebraic subvariety $W \subset \mathbb{P}^N$ given by r equations $\{f_1 = 0, \ldots, f_r = 0\}$.

Definition F.2.1 The *blow up* of $\mathbb{P}^N$ along W is the subvariety $B_W\mathbb{P}^N$ of $\mathbb{P}^N_{x_0:\cdots:x_N} \times \mathbb{P}^{r-1}_{y_1:\cdots:y_r}$ given by the $r-1$ equations

$$\begin{cases} y_1 f_2(x_0, \ldots, x_N) - y_2 f_1(x_0, \ldots, x_N) = 0, \\ y_2 f_3(x_0, \ldots, x_N) - y_3 f_2(x_0, \ldots, x_N) = 0, \\ \qquad \vdots \\ y_{r-1} f_r(x_0, \ldots, x_N) - y_r f_{r-1}(x_0, \ldots, x_N) = 0. \end{cases}$$

The *blow up map* $\pi_W \colon B_W\mathbb{P}^N \to \mathbb{P}^N$ is given by

$$((x_0 : \cdots : x_N),\ (y_1 : \cdots : y_r)) \mapsto (x_0 : \cdots : x_N).$$

If $\operatorname{codim} W = r$, we recover the previous interpretation of a blow up in terms of the normal bundle at every smooth point of W.

For any subvariety $V \subset \mathbb{P}^N$, we denote by $\widetilde{V}$ the Zariski closure of $\pi_W^{-1}(V \setminus W \cap V)$ in $B_W \mathbb{P}^N$.

Definition F.2.2 The subvariety $\widetilde{V}$ is called the *strict transform* of V under π_W.

It is possible to prove that the variety $\widetilde{V}$ does not depend on the embeddings $V \subset \mathbb{P}^N$ and $W \subset \mathbb{P}^N$ but only on the embedding $W \cap V \subset V$ (see [Har77, II.7]). We denote by π_W the restriction $\widetilde{V} \to V$ of π_W to $\widetilde{V}$.

Definition F.2.3 Let V be a projective variety and let $W \subset V$ be a subvariety. The restriction $\pi_W : \widetilde{V} \to V$ is called the *blow up* of V of *centre* W. We denote by $B_W V := \widetilde{V}$ the *blow up of* V *along* W. The divisor $\pi_W^*(W)$ is called the *exceptional divisor* of the blow up.

We denote the blown up variety by $B_W V := \widetilde{V}$.

Proposition F.2.4 (Universal property of blow ups on surfaces) *Let $f : Y \to X$ be a birational morphism of non singular projective complex surfaces. If $P \in X$ is a point at which the inverse rational map $f^{-1} : X \dashrightarrow Y$ is not well defined then f factorises uniquely as a map*

$$f : Y \xrightarrow{g} B_P X \xrightarrow{\pi_P} X$$

where g is a birational map and π_P is the blow up of X at P.

Proof See [Bea78, Proposition II.8]. □

Remark F.2.5 A more general statement which also holds in higher dimension is given in [Har77, Proposition II.7.14]: this result is weaker than the above in the two dimensional case. See [Har77, Remark V.5.4.1] for more details.

Corollary F.2.6 *Let X be a non singular projective complex surface and let $f : X \dashrightarrow X$ be a birational map. Let P be a point of X fixed by f. There is then a unique birational endomorphism $g : B_P X \to B_P X$ such that $f \circ \pi_P = \pi_P \circ g$:*

$$\begin{array}{ccc} B_P X & \xrightarrow{g} & B_P X \\ \downarrow{\scriptstyle \pi_P} & & \downarrow{\scriptstyle \pi_P} \\ X & \xrightarrow{f} & X \end{array}$$

Proof Simply apply the previous proposition to the birational morphism $f \circ \pi_P$: $B_P X \to X$, noting that the inverse map $(f \circ \pi_P)^{-1}$ is not defined at $P = f(P)$. Indeed, if f^{-1} were well defined at P then $f^{-1}(P) = P$ would be a point at which π_P^{-1} is not defined. □

When V and W are smooth there is a diffeomorphism between the topological and algebraic blow ups that commutes with morphisms over V.

F.3 Topology of Blow Ups

The Definition B.5.12 of the connected sum of two varieties will be needed in this section.

Proposition F.3.1 *Consider a point $P \in \mathbb{R}^n$. The blow up $B_P\mathbb{R}^n$ is then diffeomorphic to $\mathbb{R}^n \# \mathbb{RP}^n$. More generally, if $C \subset M$ is a non singular submanifold of codimension r with trivial normal bundle then $B_C M$ is diffeomorphic to $M \# C \times \mathbb{RP}^r$.*

Corollary F.3.2 *Let X be a real surface and let $B_P X$ be the blow up of X at a point P in X. The differentiable manifold $B_P X$ of real dimension 2 is then diffeomorphic to*

$$X \# \mathbb{RP}^2 .$$

In particular, if $P \in \mathbb{RP}^2$ then the blow up $B_P\mathbb{RP}^2$ is diffeomorphic to the Klein bottle $\mathbb{K}^2$.

Proof Indeed, the boundary $\partial(U_{\mathbb{RP}^2|\mathbb{RP}^1})$ of a tubular neighbourhood of $\mathbb{RP}^1$ in $\mathbb{RP}^2$ is diffeomorphic to a circle $\mathbb{S}^1$ and the tubular neighbourhood itself is a Möbius band. □

Proposition F.3.3 *Let X be a complex surface and let $B_P X$ be the blow up of X at a point P in X. For example, $B_{(0,0)}\mathbb{C}^2$ is the complex subvariety in $\mathbb{C}^2{}_{x,y} \times \mathbb{CP}^1_{u:v}$ given by the equation $xv = yu$. The differentiable manifold $B_P X$ of real dimension 4 is then diffeomorphic in the Euclidean topology to*

$$X \# \overline{\mathbb{CP}^2}$$

where $\overline{\mathbb{CP}^2} = -\mathbb{CP}^2$ is the complex projective plane with the inverse orientation.

Proof The boundary $\partial(U_{\mathbb{CP}^2|\mathbb{CP}^1})$ of a tubular neighbourhood of $\mathbb{CP}^1$ in $\mathbb{CP}^2$ is diffeomorphic to the sphere $\mathbb{S}^3$ and the circle bundle induced by projection

$$\mathbb{S}^3 \approx \partial(U_{\mathbb{CP}^2|\mathbb{CP}^1}) \to \mathbb{CP}^1 \approx \mathbb{S}^2$$

is the Hopf fibration (see Proposition B.8.3). The blow up of a point is therefore equivalent to the surgery that replaces the tubular neighbourhood $U_{X|P}$ of a point in X, diffeomorphic to the unit ball of dimension 4, by $U_{\mathbb{CP}^2|\mathbb{CP}^1}$, gluing them together along their boundaries spheres by a orientation reversing diffeomorphism. See [KM61] or [Hir51] for more details. □

Example F.3.4 We now illustrate the above with a worked example due to Kollár. See [Kol99a, Example 1.4] for more details.

Let X be a non singular real algebraic variety of dimension 3 and let $D \subset X$ be a real curve with a unique real point $\{0\} = D(\mathbb{R})$. Suppose moreover that close to 0 this curve is given by equations $\{z = x^2 + y^2 = 0\}$. Let $Y_1 = B_D X$ be the variety obtained by blowing up Y in D. This new variety is real and has a unique singular point P. Consider $Y := B_P Y_1$, the variety obtained by blowing up Y_1 at P, which is a non singular real algebraic variety. Let $\pi\colon Y \to X$ be the composition of blow ups. We will prove that the connected component $M \subset X(\mathbb{R})$ containing P satisfies

$$\pi^{-1}M \approx M\#(\mathbb{S}^2 \times \mathbb{S}^1),$$

or in other words

$$B_P(B_D M) \approx M\#(\mathbb{S}^2 \times \mathbb{S}^1).$$

As our aim is to calculate the topology of the blow up, it is reasonable to use the $\mathcal{C}^\infty$ blow up, which enables us to work in an open set U which is "arbitrarily small", such as, for example, an open neighbourhood of the unique real point of the singular curve D. We will make free us of the identification $U \approx \mathbb{R}^3$: such an identification does not exist either in the analytic category (since $\mathbb{C}^3$ is not biholomorphic to any of its strict open subsets) or in the algebraic category (because the Zariski topology is too weak). See [Sha94, Chapter VI Section 2.2] for a more complete explanation.

Consider the curve D whose equations are $(z = x^2 + y^2 = 0)$ in $X = \mathbb{R}^3$. We will calculate $\pi_1\colon Y_1 = B_D\mathbb{R}^3 \to \mathbb{R}^3$ and $\pi\colon Y = B_P Y_1 \to \mathbb{R}^3$, where $P \in Y_1$ is the unique singular point of Y_1.

By definition, $B_D\mathbb{R}^3 \subset \mathbb{R}^3_{x,y,z} \times \mathbb{P}^1_{\alpha:\beta}$ is determined by the equation

$$\alpha(x^2 + y^2) - \beta z = 0$$

which has a unique singular point P in the affine chart $\alpha \neq 0$. Restricting to this chart, Y_1 is the affine hypersurface of equation $x_1^2 + y_1^2 - z_1 t_1 = 0$ where $x_1{=}x$, $y_1{=}y$, $z_1 = z$, $t_1 = \beta$ and $P = (0, 0, 0, 0)$.

We calculate $B_P Y_1$ by blowing up $\pi_P\colon \widetilde{\mathbb{R}^4} \to \mathbb{R}^4$ at the point P and considering the strict transform $Y = B_P Y_1$ of Y_1.

The four equations of $\pi_P^{-1}(Y_1) \subset \mathbb{R}^4_{x_1,y_1,z_1,t_1} \times \mathbb{P}^3_{a:b:c:d}$ are

$$x_1^2 + y_1^2 - z_1 t_1 = a y_1 - b x_1 = b z_1 - c y_1 = c t_1 - d z_1 = 0.$$

Restricting to the chart $c \neq 0$, $\pi_P^{-1}(Y_1)$ is the affine variety of equation $z_2^2(x_2^2 + y_2^2 - t_2) = 0$ in the affine subspace of $\mathbb{R}^7_{x_2,y_2,z_2,t_2,a,b,d}$ whose equations are

$$\begin{cases} x_2 = a, \\ y_2 = b, \\ t_2 = d. \end{cases}$$

where $z_2 = z_1$, $x_2 = x_1/z_1$, $y_2 = y_1/z_1$, $t_2 = t_1/z_1$.

The trace of the exceptional divisor in this chart is given by $z_2 = 0$ and it follows that the equation of Y is $x_2^2 + y_2^2 - t_2 = 0$ in the affine subspace $x_2 - a = y_2 - b = t_2 - d = 0$. The real locus is therefore the product variety of a paraboloid of revolution and the real line $\mathbb{R}$. In the chart $d \neq 0$ the topological situation is the same and simply need to check that the gluing is diffeomorphic to

$$\mathbb{R}^3 \# \mathbb{S}^2 \times \mathbb{S}^1 \approx \mathbb{S}^2 \times \mathbb{S}^1 \setminus \mathbb{D}^3$$

where $\mathbb{D}^3$ is the ball of dimension 3.

Glossary of Notations

$\simeq$	Isomorphisms of structures, varieties, $\mathbb{R}$-varieties and so on, 15
$\sim$	Linear equivalence of divisors, 110
$\approx$	Diffeomorphism, 190
$\lfloor x \rfloor$	Round down of x, 188
$\lceil x \rceil$	Round up of x, 188
$\smile$	Singular cohomology cup-product, 338
$\frown$	Cap-product in singular (co)homology, 341
$\Gamma(U, \mathcal{F}) = \mathcal{F}(U)$	Sections of the sheaf $\mathcal{F}$ on the open set U, 10
$\Gamma(V)$	Quotient group $H^2(V; \mathbb{Z})/H^2_{\mathbb{C}-\mathrm{alg}}(V; \mathbb{Z})$, 269
Ω_X	Bundle of (regular or holomorphic) differential forms, 370
Ω_X^p	Bundle of (regular or holomorphic) differential p-forms, 370
χ	Euler characteristic, 392
$\chi_{\mathrm{top}}(X)$	Topological Euler characteristic of X, 180
$\chi(\mathcal{O}_X)$	Holomorphic Euler characteristic of X, 175
$\gamma_{n,k}$	Universal bundle over $\mathbb{G}_{n,k}(K)$, 261
λ, λ_σ	Comessatti characteristic, 126
$\nu\colon \widetilde{X} \to X$	Normalisation of X, 44
$\sigma_{\mathbb{A}} := \sigma_{\mathbb{A}^n}$	Standard complex conjugation on $\mathbb{C}^n$, 66
$\sigma_{\mathbb{P}} := \sigma_{\mathbb{P}^n}$	Standard complex conjugation on $\mathbb{P}^n(\mathbb{C})$, 66
${}^\sigma\mathcal{L}$	Conjugate sheaf of $\mathcal{L}$, 77
${}^\sigma\psi$	Conjugate map of ψ, 71
${}^\sigma f$	Conjugate function of f, 67
$(A \cdot B)$	Intersection number of divisors A and B, 177
(A^2)	Self-intersection number of a divisor A, 177
$\mathbb{A}^n(K)$	Affine space of dimension n over K, 3
$\mathcal{A}(F)$	K-algebra of affine coordinates on F, 13
$A_{\mathfrak{p}}$	Localisation of the ring A at the prime ideal $\mathfrak{p}$, 316
A_f	Localisation of A by f, 316
$\mathrm{Alb}(X)$	Albanese variety of X, 386
$B_P X$	Blow up of X at P, 182
$\mathcal{C}^\infty(V, W)$	Space of maps from V to W, 258

F. Mangolte, *Real Algebraic Varieties*, Springer Monographs in Mathematics,
https://doi.org/10.1007/978-3-030-43104-4

$\widehat{C}$ Projective completion of C, 28
$\widetilde{C}$ Normalisation of C, 44
C^{Φ} Curve of equation f^{Φ}, 49
$\mathrm{CaCl}(X)$ Group of linear divisor classes of Cartier divisors on X, 103
$\mathrm{Cl}(X)$ Group of linear divisor classes on X, 102
$\mathcal{D}(f)$ Non-vanishing locus of f, 5
$|D|$ Linear system associated to D, 106
$\mathrm{Div}(X)$ Group of Cartier divisors on X, 103
E_P Exceptional line of the blow up of P, 84
$\mathrm{Ext}(H, A)$ Group of extension classes of H by A, 331
$F_{\mathbb{C}}$ Complexification of an algebraic set F, 87
$\mathrm{Frac}\, A$ Fraction ring of A, 317
$\mathbb{G}_{n,k}(K)$ Grassmannian of the k-subspaces of K^n, 165
$\mathrm{Gal}(\mathbb{C}|\mathbb{R})$ Galois group of the extension $\mathbb{C}|\mathbb{R}$, 72
$H_{2k}^{\mathbb{R}-\mathrm{alg}}(X; \mathbb{Z})$ Classes represented by invariant complex cycles, 166
$\mathbb{H}$ Field of quaternions, 260
$H^2_{\mathbb{C}-\mathrm{alg}}(V; \mathbb{Z})$ Classes of algebraic rank 1 $\mathbb{C}$-vector bundles on V, 269
$H^k(G, M)$ kth Galois cohomology group of the module M, 126
$H^k(X, L; A)$ kth singular cohomology group of the pair (X, L) with coefficients in A, 124
$H^k_{\mathrm{alg}}(X(\mathbb{R}); \mathbb{Z}_2)$ Group of fundamental classes of algebraic subvarieties, 164
$H_k(X, L; A)$ kth singular homology group of the pair (X, L) with coefficients in A, 124
$H_k^{\mathrm{alg}}(X(\mathbb{R}); \mathbb{Z}_2)$ Group of fundamental classes of algebraic subvarieties, 164
$\mathcal{I}(U)$ Homogeneous ideal of polynomials vanishing on U, 3
$\mathcal{I}(U)$ Ideal of polynomials vanishing on U, 3
$\sqrt{I}$ Radical of I, 315
I_L Ideal in $L[X_1, \ldots, X_n]$ generated by I, 6
$I_{\mathbb{C}}$ Ideal $I \otimes_{\mathbb{R}[X_1,\ldots,X_n]} \mathbb{C}[X_1, \ldots, X_n]$, 6
$\mathrm{Jac}(X)$ Jacobian of X, 401
$K(U)$ Field of rational functions of U over K, 21
K_X Canonical divisor on X, 175
$\mathcal{K}_X$ Canonical bundle of X, 370
$\mathcal{K}_X$ Canonical sheaf of X, 175
$\overline{\mathcal{L}}$ Anti-sheaf of a sheaf $\mathcal{L}$, 68
M^G, M^{σ} Submodule of invariants, 124
$M^{-\sigma}$ Submodule of anti-invariants, 124
$[M]$ Fundamental homology class of M, 335
$\mathcal{M}_X$ Sheaf of rational functions on X, 32
$M_{\mathfrak{p}}$ Localisation of the module M at the prime ideal $\mathfrak{p} \subset A$, 316
M_f Localisation of the A-module M by $f \in A$, 316
$\mathcal{N}_{M|C}$ Normal bundle of C in M, 404
$\mathrm{NS}(X)$ Néron–Severi group of X, 109
$N_{X|X(\mathbb{R})}$ Normal bundle of $X(\mathbb{R})$ in X, 144

$\mathrm{Num}(X)$	Group of divisors up to numerical equivalence, 180
$(\mathcal{O}_X)^G_{X(\mathbb{R})}$	Invariant subset of the restricted sheaf on the real locus, 78
$\mathcal{O}_X$	Sheaf of regular functions on X, 10
$\mathcal{O}_X(D)$	Invertible sheaf associated to a divisor D on X, 105
$\mathcal{O}_X(D)$	Line bundle associated to a divisor D on X, 105
$\mathcal{O}_x = \mathcal{O}_{U,x}$	Algebra of germs of regular functions at x in U, 10
$\mathcal{O}_{\mathbb{P}^n}(-1)$	Tautological bundle, 105
$\mathcal{O}_{\mathbb{P}^n}(1)$	Serre's twisting sheaf, 105
$\mathbb{P}^n(K)$	Projective space of dimension n over K, 5
$\mathrm{Pic}(X/B)$	Relative Picard group of $\pi\colon X \to B$, 197
$\mathrm{Pic}(X/\pi)$	Relative Picard group of $\pi\colon X \to B$, 197
$\mathcal{P}(U)$	K-algebra of polynomial functions on U, 8
$\widehat{P}$	Homogenisation of the polynomial P, 11
$\mathcal{P}(X)$	Group of principal divisors on X, 102
$\mathrm{Pic}(X)$	Picard group of X, 105
$\mathrm{Pic}^0(X)$	Picard variety of X, 110
$\mathcal{R}(U)$	K-algebra of regular functions on U, 9
$\mathcal{R}(V, W)$	Space of regular maps from V to W, 259
$\mathrm{Reg}X$	Regular locus of X, 44
$\mathcal{S}(F)$	K-algebra of homogeneous coordinates on F, 4
$S^{-1}M$	Localisation of the module M at S, 316
$\mathrm{Sing}X$	Singular locus of X, 44
$T_X^{0,1}$	Anti-holomorphic tangent bundle, 371
$T_X^{1,0}$	Holomorphic tangent bundle, 369
T_X	Tangent bundle, 129
$T_{X,\mathbb{C}}$	Complex tangent bundle, 369
$T_{X,\mathbb{R}}$	Real tangent bundle, 369
$\mathcal{T}_X$	Holomorphic tangent bundle, 371
T_aF	Usual tangent space to F at a point $a \in F$, 39
$T_a^{Zar}F$	Zariski tangent space to F at the point $a \in F$, 39
$\mathrm{Tor}(H, A)$	Tor group of A and H, 330
$U(A)$	Multiplicative set of invertible elements of a ring A, 314
$U_{\mathbb{C}}$	Complexification of a quasi-algebraic set U, 87
$\overline{X} := (X, \overline{\mathcal{O}_X})$	Conjugate variety of $(X, \mathcal{O}_X)$, 68
$\mathbb{Z}_m$	Cyclic group of order m, 124
$\mathcal{Z}(I)$	Zero-set of the ideal I, 5
$\mathcal{Z}(f)$	Set of zeros of the function f, 5
$\mathcal{Z}_L(I)$	Zero-set in $\mathbb{A}^n(L)$ of the ideal I, 5
$Z^1(X)$	Group of Weil divisors on X, 102
$a(X)$	Algebraic dimension of X, 381
$b_*(X)$	Total Betti number of X, 135
$b_*(X; \mathbb{Z}_2)$	Total Betti number of X with coefficients in $\mathbb{Z}_2$, 135
$b_k(X)$	kth Betti number of X, 135
$b_k(X; \mathbb{Z}_2)$	kth Betti number of X with coefficients in $\mathbb{Z}_2$, 135

$c_k(X)$	kth Chern class X, 129
f^Φ	Pull back of f by Φ, 49
$g(C)$	Geometric genus of a curve C, 184
$h^{a,b}(X)$	Hodge numbers of X, 175
$\kappa(X)$	Kodaira dimension of X, 176
$\mathfrak{m}_x$	Maximal ideal of functions vanishing at x, 10
$\mathfrak{m}_x^{\mathcal{P}}$	Maximal ideal of polynomial functions vanishing at x, 11
$\mathfrak{m}_x^{\mathcal{R}}$	Maximal ideal of regular functions vanishing at x, 11
$\mathrm{mult}_A(D)$	Multiplicity of a Cartier divisor D along a divisor A, 103
$\mathrm{mult}_A(D)$	Multiplicity of a Weil divisor D along a divisor A, 101
$\mathrm{mult}_A(f)$	Multiplicity of a rational function f along a divisor A, 102
$p_a(C)$	Arithmetic genus of a curve C, 185
$p_g(X)$	Geometric genus of a variety X, 379
$p_v(D)$	Virtual genus of a divisor D, 187
$q(X)$	Irregularity of a variety X, 379
$r_k(X)$	dimension of the invariant subspace of $H_k(X;\mathbb{Q})$, 146
$\mathrm{trdeg}_K L$	Transcendence degree of an extension $L\vert K$, 321
$\mathrm{v}_k(X)$	kth Wu class of X, 142
$\mathrm{w}_k(X)$	kth Stiefel–Whitney class of X, 129

References

[A'C80] N. A'Campo – " Sur la première partie du seizième problème de Hilbert", in *Séminaire Bourbaki (1978/79)*, Lecture Notes in Math., vol. 770, Springer, Berlin, 1980, p. Exp. No. 537, pp. 208–227.

[AC02] M. Alberich-Carramiñana – *Geometry of the plane Cremona maps*, Lecture Notes in Mathematics, vol. 1769, Springer-Verlag, Berlin, 2002.

[ACGH85] E. Arbarello, M. Cornalba, P. A. Griffiths & J. Harris – *Geometry of algebraic curves. Vol. I*, Grundlehren der Mathematischen Wissenschaften [Fundamental Principles of Mathematical Sciences], vol. 267, Springer-Verlag, New York, 1985.

[AK85] S. Akbulut & H. King – " Submanifolds and homology of nonsingular real algebraic varieties", *Amer. J. Math.* **107** (1985), no. 1, p. 45–83.

[AK91] S. Akbulut & H. King – " Rational structures on 3-manifolds", *Pacific J. Math.* **150** (1991), no. 2, p. 201–214.

[AK03] C. Araujo & J. Kollár – " Rational curves on varieties", in *Higher dimensional varieties and rational points (Budapest, 2001)*, Bolyai Soc. Math. Stud., vol. 12, Springer, Berlin, 2003, p. 13–68.

[AM65] M. Artin & B. Mazur – " On periodic points", *Ann. of Math. (2)* **81** (1965), p. 82–99.

[AM08] M. Akriche & F. Mangolte – " Nombres de Betti des surfaces elliptiques réelles", *Beiträge Algebra Geom.* **49** (2008), no. 1, p. 153–164.

[AM15] M. Akriche & S. Moulahi – " Fibre singulière d'un pinceau réel en courbes de genre 2", *Ann. Fac. Sci. Toulouse Math. (6)* **24** (2015), no. 3, p. 427–482.

[And05] M. T. Anderson – " Géométrisation des variétés de dimension 3 via le flot de Ricci", *Gaz. Math.* (2005), no. 103, p. 24–40, Translated from Notices Amer. Math. Soc. 51 (2004), no. 2, 184–193 by Zindine Djadli.

[Arn71] V. I. Arnol'd – " The situation of ovals of real plane algebraic curves, the involutions of four-dimensional smooth manifolds, and the arithmetic of integral quadratic forms", *Funkcional. Anal. i Priložen.* **5** (1971), no. 3, p. 1–9.

[Art62] M. Artin – " Some numerical criteria for contractability of curves on algebraic surfaces", *Amer. J. Math.* **84** (1962), p. 485–496.

[Art66] M. Artin, " On isolated rational singularities of surfaces", *Amer. J. Math.* **88** (1966), p. 129–136.

[AS68] M. F. Atiyah & I. M. Singer – " The index of elliptic operators. III", *Ann. of Math. (2)* **87** (1968), p. 546–604.

[Ati66] M. F. Atiyah – " K-theory and reality", *Quart. J. Math. Oxford Ser. (2)* **17** (1966), p. 367–386.

F. Mangolte, *Real Algebraic Varieties*, Springer Monographs in Mathematics,
https://doi.org/10.1007/978-3-030-43104-4

[Băd01] L. Bădescu – *Algebraic surfaces*, Universitext, Springer-Verlag, New York, 2001, Translated from the 1981 Romanian original by Vladimir Maşek and revised by the author.

[Bal91] E. Ballico – " An addendum on: "Algebraic models of smooth manifolds" [Invent. Math. **97** (1989), no. 3, 585–611] by J. Bochnak and W. Kucharz", *Geom. Dedicata* **38** (1991), no. 3, p. 343–346.

[BB06] B. Bertrand & E. Brugallé – " A Viro theorem without convexity hypothesis for trigonal curves", *Int. Math. Res. Not.* (2006), p. Art. ID 87604, 33.

[BBK89] J. Bochnak, M. Buchner & W. Kucharz – " Vector bundles over real algebraic varieties", *K-Theory* **3** (1989), no. 3, p. 271–298.

[BBK90] J. Bochnak, M. Buchner & W. Kucharz , " Erratum: "Vector bundles over real algebraic varieties" [K-Theory **3** (1989), no. 3, 271–298", *K-Theory* **4** (1990), no. 1, p. 103.

[BBM+10] L. Bessières, G. Besson, S. Maillot, M. Boileau & J. Porti – *Geometrisation of 3-manifolds*, EMS Tracts in Mathematics, vol. 13, European Mathematical Society (EMS), Zürich, 2010.

[BCM+16] F. Bogomolov, I. Cheltsov, F. Mangolte, C. Shramov & D. Testa – " Spitsbergen volume [Editorial]", *Eur. J. Math.* **2** (2016), no. 1, p. 1–8.

[BCP11] I. Bauer, F. Catanese & R. Pignatelli – " Surfaces of general type with geometric genus zero: a survey", in *Complex and differential geometry*, Springer Proc. Math., vol. 8, Springer, Heidelberg, 2011, p. 1–48.

[BCR87] J. Bochnak, M. Coste & M.-F. Roy – *Géométrie algébrique réelle*, Ergebnisse der Mathematik und ihrer Grenzgebiete (3) [Results in Mathematics and Related Areas (3)], vol. 12, Springer-Verlag, Berlin, 1987.

[BCR98] J. Bochnak, M. Coste & M.-F. Roy, *Real algebraic geometry*, Ergebnisse der Mathematik und ihrer Grenzgebiete (3), vol. 36, Springer-Verlag, Berlin, 1998, Translated from the 1987 French original.

[BDIM19] E. Brugallé, A. Degtyarev, I. Itenberg & F. Mangolte – " Real algebraic curves with large finite number of real points", *Eur. J. Math.* **5** (2019), p. 686–711.

[Bea78] A. Beauville – *Surfaces algébriques complexes*, Société Mathématique de France, Paris, 1978, Astérisque, No. 54.

[Bea14] A. Beauville, " Some surfaces with maximal Picard number", *J. Éc. polytech. Math.* **1** (2014), p. 101–116.

[Ben16a] M. Benzerga – " Real structures on rational surfaces and automorphisms acting trivially on Picard groups", *Math. Z.* **282** (2016), no. 3-4, p. 1127–1136.

[Ben16b] M. Benzerga, " Structures réelles sur les surfaces rationnelles", Thèse, Université d'Angers, 2016, `https://tel.archives-ouvertes.fr/tel-01471071`.

[Ben17] M. Benzerga – " Finiteness of real structures on KLT Calabi-Yau regular smooth pairs of dimension 2", Preprint, `arXiv:1702.08808 [math.AG]`, 2017.

[Ben18] O. Benoist – " Sums of three squares and Noether-Lefschetz loci", *Compos. Math.* **154** (2018), no. 5, p. 1048–1065.

[Ben19] O. Benoist, " The period-index problem for real surfaces", *Publ. Math. Inst. Hautes Études Sci.* **130** (2019), p. 63–110.

[Bes05] L. Bessières – " Conjecture de Poincaré: la preuve de R. Hamilton et G. Perelman", *Gaz. Math.* (2005), no. 106, p. 7–35.

[Bes13] G. Besson – " La conjecture de Poincaré", *Gaz. Math.* (2013), no. 135, p. 5–16.

[BH59] A. Borel & F. Hirzebruch – " Characteristic classes and homogeneous spaces. II", *Amer. J. Math.* **81** (1959), p. 315–382.

[BH61] A. Borel & A. Haefliger – " La classe d'homologie fondamentale d'un espace analytique", *Bull. Soc. Math. France* **89** (1961), p. 461–513.

[BH75] E. Bombieri & D. Husemoller – " Classification and embeddings of surfaces", in *Algebraic geometry (Proc. Sympos. Pure Math., Vol. 29, Humboldt State Univ., Arcata, Calif., 1974)*, Amer. Math. Soc., Providence, R.I., 1975, p. 329–420.

[BH07] I. Biswas & J. Huisman – " Rational real algebraic models of topological surfaces", *Doc. Math.* **12** (2007), p. 549–567.

[BHPVdV04] W. P. Barth, K. Hulek, C. A. M. Peters & A. Van de Ven – *Compact complex surfaces*, second ed., Ergebnisse der Mathematik und ihrer Grenzgebiete. 3. Folge., vol. 4, Springer-Verlag, Berlin, 2004.

[Bih01a] F. Bihan – " Betti numbers of real numerical quintic surfaces", in *Topology, ergodic theory, real algebraic geometry*, Amer. Math. Soc. Transl. Ser. 2, vol. 202, Amer. Math. Soc., Providence, RI, 2001, p. 31–38.

[Bih01b] F. Bihan, " Une sextique de l'espace projectif réel avec un grand nombre d'anses", *Rev. Mat. Complut.* **14** (2001), no. 2, p. 439–461.

[BK89] J. Bochnak & W. Kucharz – " Algebraic models of smooth manifolds", *Invent. Math.* **97** (1989), no. 3, p. 585–611.

[BK91] J. Bochnak & W. Kucharz, "Nonisomorphic algebraic models of a smooth manifold", *Math. Ann.* **290** (1991), no. 1, p. 1–2.

[BK99] J. Bochnak & W. Kucharz, "The Weierstrass approximation theorem for maps between real algebraic varieties", *Math. Ann.* **314** (1999), no. 4, p. 601–612.

[BK10] J. Bochnak & W. Kucharz, "Algebraic approximation of smooth maps", *Univ. Iagel. Acta Math.* (2010), no. 48, p. 9–40.

[BKS82] J. Bochnak, W. Kucharz & M. Shiota – " The divisor class groups of some rings of global real analytic, Nash or rational regular functions", in *Real algebraic geometry and quadratic forms (Rennes, 1981)*, Lecture Notes in Math., vol. 959, Springer, Berlin-New York, 1982, p. 218–248.

[BKS97] J. Bochnak, W. Kucharz & R. Silhol – " Morphisms, line bundles and moduli spaces in real algebraic geometry", *Inst. Hautes Études Sci. Publ. Math.* (1997), no. 86, p. 5–65 (1998).

[BKS00] J. Bochnak, W. Kucharz & R. Silhol, " Erratum to: "Morphisms, line bundles and moduli spaces in real algebraic geometry" [Inst. Hautes Études Sci. Publ. Math. No. 86, (1997), 5–65 (1998)", *Inst. Hautes Études Sci. Publ. Math.* (2000), no. 92, p. 195 (2001).

[BKVV13] M. Bilski, W. Kucharz, A. Valette & G. Valette – " Vector bundles and regulous maps", *Math. Z.* **275** (2013), no. 1-2, p. 403–418.

[BM76] E. Bombieri & D. Mumford – " Enriques' classification of surfaces in char. p. III", *Invent. Math.* **35** (1976), p. 197–232.

[BM77] E. Bombieri & D. Mumford, " Enriques' classification of surfaces in char. p. II", in *Complex analysis and algebraic geometry*, Iwanami Shoten, Tokyo, 1977, p. 23–42.

[BM92] R. Benedetti & A. Marin – " Déchirures de variétés de dimension trois et la conjecture de Nash de rationalité en dimension trois", *Comment. Math. Helv.* **67** (1992), no. 4, p. 514–545.

[BM07] F. Bihan & F. Mangolte – " Topological types and real regular Jacobian elliptic surfaces", *Geom. Dedicata* **127** (2007), p. 57–73.

[BM11] J. Blanc & F. Mangolte – " Geometrically rational real conic bundles and very transitive actions", *Compos. Math.* **147** (2011), no. 1, p. 161–187.

[BM14] J. Blanc & F. Mangolte, " Cremona groups of real surfaces", in *Automorphisms in Birational and Affine Geometry*, Springer Proceedings in Mathematics & Statistics, vol. 79, Springer, 2014, p. 35–58.

[BMP03] M. Boileau, S. Maillot & J. Porti – *Three-dimensional orbifolds and their geometric structures*, Panoramas et Synthèses [Panoramas and Syntheses], vol. 15, Société Mathématique de France, Paris, 2003.

[Bom73] E. Bombieri – " Canonical models of surfaces of general type", *Inst. Hautes Études Sci. Publ. Math.* (1973), no. 42, p. 171–219.

[Bor60] A. Borel – *Seminar on transformation groups*, With contributions by G. Bredon, E. E. Floyd, D. Montgomery, R. Palais. Annals of Mathematics Studies, No. 46, Princeton University Press, Princeton, N.J., 1960.

[BR90] R. Benedetti & J.-J. Risler – *Real algebraic and semi-algebraic sets*, Actualités Mathématiques. [Current Mathematical Topics], Hermann, Paris, 1990.

[Bre72] G. E. Bredon – *Introduction to compact transformation groups*, Academic Press, New York-London, 1972, Pure and Applied Mathematics, Vol. 46.

[Bru06] E. Brugallé – " Real plane algebraic curves with asymptotically maximal number of even ovals", *Duke Math. J.* **131** (2006), no. 3, p. 575–587.

[BS64] A. Borel & J.-P. Serre – " Théorèmes de finitude en cohomologie galoisienne", *Comment. Math. Helv.* **39** (1964), p. 111–164.

[BW69] G. E. Bredon & J. W. Wood – " Non-orientable surfaces in orientable 3-manifolds", *Invent. Math.* **7** (1969), p. 83–110.

[BW18a] O. Benoist & O. Wittenberg – " On the integral hodge conjecture for real varieties, I", Preprint, `arXiv:1801.00872 [math.AG]`, 2018.

[BW18b] O. Benoist & O. Wittenberg, " On the integral hodge conjecture for real varieties, II", *Journal de l'École Polytechnique* **7** (2020), p. 373–429.

[Cam92] F. Campana – " Connexité rationnelle des variétés de Fano", *Ann. Sci. École Norm. Sup. (4)* **25** (1992), no. 5, p. 539–545.

[Car61] H. Cartan – *Théorie élémentaire des fonctions analytiques d'une ou plusieurs variables complexes*, Avec le concours de Reiji Takahashi, Enseignement des Sciences. Hermann, Paris, 1961.

[Cas01] G. Castelnuovo – " Le trasformazioni generatrici del gruppo cremoniano nel piano", *Atti della R. Accad. delle Scienze di Torino* **36** (1901), p. 861–874.

[Cat03] F. Catanese – " Moduli spaces of surfaces and real structures", *Ann. of Math. (2)* **158** (2003), no. 2, p. 577–592.

[Cat08] F. Catanese, " Differentiable and deformation type of algebraic surfaces, real and symplectic structures", in *Symplectic 4-manifolds and algebraic surfaces*, Lecture Notes in Math., vol. 1938, Springer, Berlin, 2008, p. 55–167.

[CD13] D. Cerveau & J. Déserti – *Transformations birationnelles de petit degré*, Cours Spécialisés [Specialized Courses], vol. 19, Société Mathématique de France, Paris, 2013.

[CF03] F. Catanese & P. Frediani – " Real hyperelliptic surfaces and the orbifold fundamental group", *J. Inst. Math. Jussieu* **2** (2003), no. 2, p. 163–233.

[Che78] A. Chenciner – *Courbes algébriques planes*, Publications Mathématiques de l'Université Paris VII, vol. 4, Université de Paris VII U.E.R. de Mathématiques, Paris, 1978, Ré-édité en 2008 chez Springer-Verlag.

[CL98] A. Chambert-Loir – *Algèbre commutative et introduction à la géométrie algébrique*, Cours de troisième cycle, Paris VI, 1998.

[CLO15] D. A. Cox, J. Little & D. O'Shea – *Ideals, varieties, and algorithms*, fourth ed., Undergraduate Texts in Mathematics, Springer, Cham, 2015, An introduction to computational algebraic geometry and commutative algebra.

[CM08] F. Catanese & F. Mangolte – " Real singular del Pezzo surfaces and 3-folds fibred by rational curves. I", *Michigan Math. J.* **56** (2008), no. 2, p. 357–373.

[CM09] F. Catanese & F. Mangolte, " Real singular del Pezzo surfaces and 3-folds fibred by rational curves. II", *Ann. Sci. Éc. Norm. Supér. (4)* **42** (2009), no. 4, p. 531–557.

[Com12] A. Comessatti – " Fondamenti per la geometria sopra le suerficie razionali dal punto di vista reale", *Math. Ann.* **73** (1912), no. 1, p. 1–72.

[Com14] A. Comessatti, " Sulla connessione delle superfizie razionali reali", *Annali di Math.* **23** (1914), no. 3, p. 215–283.

[Com25] A. Comessatti, " Sulle varietà abeliane reali", *Ann. Mat. Pura Appl.* **2** (1925), no. 1, p. 67–106.

[Com26] A. Comessatti, " Sulle varietà abeliane reali", *Ann. Mat. Pura Appl.* **3** (1926), no. 1, p. 27–71.

[Com28] A. Comessatti, " Sulla connessione delle superficie algebriche reali", *Ann. Mat. Pura Appl.* **5** (1928), no. 1, p. 299–317.

[Cos92] M. Coste – " Épaississement d'une hypersurface algébrique réele", *Proc. Japan Acad. Ser. A Math. Sci.* **68** (1992), no. 7, p. 175–180.

[Cos02] M. Coste, " An introduction to semialgebraic geometry", `https://perso.univ-rennes1.fr/michel.coste/polyens/SAG.pdf` 2002.

[CS53] H. Cartan & J.-P. Serre – " Un théorème de finitude concernant les variétés analytiques compactes", *C. R. Acad. Sci. Paris* **237** (1953), p. 128–130.

[CS10] D. I. Cartwright & T. Steger – " Enumeration of the 50 fake projective planes", *C. R. Math. Acad. Sci. Paris* **348** (2010), no. 1-2, p. 11–13.

[Deb01] O. Debarre – *Higher-dimensional algebraic geometry*, Universitext, Springer-Verlag, New York, 2001.

[Deg12] A. Degtyarev – *Topology of algebraic curves. An approach via dessins d'enfants*, De Gruyter Studies in Mathematics, vol. 44, Walter de Gruyter & Co., Berlin, 2012.

[Del73] P. Deligne – " Le Théorème de Noether", in *SGA 7 II*, Lecture Notes in Mathematics, vol. 340, Springer, Berlin, 1973, p. 328–340.

[DFM18] A. Dubouloz, G. Freudenburg & L. Moser-Jauslin – " Algebraic vector bundles on the 2-sphere and smooth rational varieties with infinitely many real forms", Preprint, arXiv:1807.05885 [math.AG], 2018.

[Die70] J. Dieudonné – *Éléments d'analyse. Tome III: Chapitres XVI et XVII*, Cahiers Scientifiques, Fasc. XXXIII, Gauthier-Villars Éditeur, Paris, 1970.

[DIK00] A. Degtyarev, I. Itenberg & V. Kharlamov – *Real Enriques surfaces*, Lecture Notes in Mathematics, vol. 1746, Springer-Verlag, Berlin, 2000.

[DIK08] A. Degtyarev, I. Itenberg & V. Kharlamov, " On deformation types of real elliptic surfaces", *Amer. J. Math.* **130** (2008), no. 6, p. 1561–1627.

[Dim85] A. Dimca – " Monodromy and Betti numbers of weighted complete intersections", *Topology* **24** (1985), no. 3, p. 369–374.

[DIZ14] A. Degtyarev, I. Itenberg & V. Zvonilov – " Real trigonal curves and real elliptic surfaces of type I", *J. Reine Angew. Math.* **686** (2014), p. 221–246.

[DK81] H. Delfs & M. Knebusch – " Semialgebraic topology over a real closed field. II. Basic theory of semialgebraic spaces", *Math. Z.* **178** (1981), no. 2, p. 175–213.

[DK96a] A. Degtyarev & V. Kharlamov – " Halves of a real Enriques surface", *Comment. Math. Helv.* **71** (1996), no. 4, p. 628–663.

[DK96b] A. Degtyarev & V. Kharlamov, " Topological classification of real Enriques surfaces", *Topology* **35** (1996), no. 3, p. 711–729.

[DK00] A. Degtyarev & V. Kharlamov, " Topological properties of real algebraic varieties: Rokhlin's way", *Uspekhi Mat. Nauk* **55** (2000), no. 4(334), p. 129–212.

[DK02] A. Degtyarev & V. Kharlamov, " Real rational surfaces are quasi-simple", *J. Reine Angew. Math.* **551** (2002), p. 87–99.

[DM16] A. Dubouloz & F. Mangolte – " Real frontiers of fake planes", *Eur. J. Math.* **2** (2016), no. 1, p. 140–168.

[DM17] A. Dubouloz & F. Mangolte, " Fake real planes: exotic affine algebraic models of $\mathbb{R}^2$", *Selecta Math. (N.S.)* **23** (2017), p. 1619–1668.

[DM18] A. Dubouloz & F. Mangolte, " Algebraic models of the line in the real affine plane", *Geom Dedicata* (2020). https://doi.org/10.1007/s10711-020-00539-1.

[DO19] T.-C. Dinh & K. Oguiso – " A surface with discrete and nonfinitely generated automorphism group", *Duke Math. J.* **168** (2019), no. 6, p. 941–966.

[Dol90] P. Dolbeault – *Analyse complexe*, Collection Maîtrise de Mathématiques Pures. [Collection of Pure Mathematics for the Master's Degree], Masson, Paris, 1990.

[DPT80] M. Demazure, H. C. Pinkham & B. Teissier (eds.) – *Séminaire sur les Singularités des Surfaces*, Lecture Notes in Mathematics, vol. 777, Springer, Berlin, 1980, Held at the Centre de Mathématiques de l'École Polytechnique, Palaiseau, 1976–1977.

[Duc14] A. Ducros – *Introduction à la théorie des schémas*, Preprint, 2014, arXiv:1401.0959 [math.AG].

[Edm81] A. L. Edmonds – " Orientability of fixed point sets", *Proc. Amer. Math. Soc.* **82** (1981), no. 1, p. 120–124.

[EGH00] Y. Eliashberg, A. Givental & H. Hofer – " Introduction to symplectic field theory", *Geom. Funct. Anal.* (2000), no. Special Volume, Part II, p. 560–673, GAFA 2000 (Tel Aviv, 1999).

[Ehr51] C. Ehresmann – " Les connexions infinitésimales dans un espace fibré différentiable", in *Colloque de topologie (espaces fibrés), Bruxelles, 1950*, Georges Thone, Liège; Masson et Cie., Paris, 1951, p. 29–55.

[Ehr95] C. Ehresmann, " Les connexions infinitésimales dans un espace fibré différentiable", in *Séminaire Bourbaki, Vol. 1*, Soc. Math. France, Paris, 1995, p. Exp. No. 24, 153–168.

[Eis95] D. Eisenbud – *Commutative algebra*, Graduate Texts in Mathematics, vol. 150, Springer-Verlag, New York, 1995, With a view toward algebraic geometry.

[EV92] H. Esnault & E. Viehweg – *Lectures on vanishing theorems*, DMV Seminar, vol. 20, Birkhäuser Verlag, Basel, 1992.

[FHMM16] G. Fichou, J. Huisman, F. Mangolte & J.-P. Monnier – " Fonctions régulues", *J. Reine Angew. Math.* **718** (2016), p. 103–151.

[FK80] H. M. Farkas & I. Kra – *Riemann surfaces*, Graduate Texts in Mathematics, vol. 71, Springer-Verlag, New York, 1980.

[Flo60] E. E. Floyd – " Periodic maps via smith theory", in *Seminar on transformation groups* (A. Borel, ed.), Princeton University Press, Princeton, N.J., 1960, p. 35–48.

[FM94] R. Friedman & J. W. Morgan – *Smooth four-manifolds and complex surfaces*, Ergebnisse der Mathematik und ihrer Grenzgebiete (3) [Results in Mathematics and Related Areas (3)], vol. 27, Springer-Verlag, Berlin, 1994.

[FMQ17] G. Fichou, J.-P. Monnier & R. Quarez – " Continuous functions on the plane regular after one blowing-up", *Math. Z.* **285** (2017), no. 1-2, p. 287–323.

[FQ90] M. H. Freedman & F. Quinn – *Topology of 4-manifolds*, Princeton Mathematical Series, vol. 39, Princeton University Press, Princeton, NJ, 1990.

[Fuj82] T. Fujita – " On the topology of noncomplete algebraic surfaces", *J. Fac. Sci. Univ. Tokyo Sect. IA Math.* **29** (1982), no. 3, p. 503–566.

[Ful89] W. Fulton – *Algebraic curves*, Advanced Book Classics, Addison-Wesley Publishing Company Advanced Book Program, Redwood City, CA, 1989, Notes written with the collaboration of Richard Weiss, Reprint of 1969 original.

[Ful98] W. Fulton, *Intersection theory*, second ed., Ergebnisse der Mathematik und ihrer Grenzgebiete. 3. Folge. A Series of Modern Surveys in Mathematics [Results in Mathematics and Related Areas. 3rd Series. A Series of Modern Surveys in Mathematics], vol. 2, Springer-Verlag, Berlin, 1998.

[Gab00] A. Gabard – " Topologie des courbes algébriques réelles: une question de Felix Klein", *Enseign. Math. (2)* **46** (2000), no. 1-2, p. 139–161.

[Gab04] A. Gabard, " Sur la topologie et la géométrie des courbes algébriques réelles", 2004, Thèse de l'Université de Genève `http://archive-ouverte.unige.ch/unige:273`.

[Gab06] A. Gabard, " Sur la représentation conforme des surfaces de Riemann à bord et une caractérisation des courbes séparantes", *Comment. Math. Helv.* **81** (2006), no. 4, p. 945–964.

[GH78] P. Griffiths & J. Harris – *Principles of algebraic geometry*, Wiley-Interscience [John Wiley & Sons], New York, 1978, Pure and Applied Mathematics.

[God58] R. Godement – *Topologie algébrique et théorie des faisceaux*, Actualit'es Sci. Ind. No. 1252. Publ. Math. Univ. Strasbourg. No. 13, Hermann, Paris, 1958.

[GP99] R. V. Gurjar & C. R. Pradeep – " **Q**-homology planes are rational. III", *Osaka J. Math.* **36** (1999), no. 2, p. 259–335.

[GPS97] R. V. Gurjar, C. R. Pradeep & A. R. Shastri – " On rationality of logarithmic **Q**-homology planes. II", *Osaka J. Math.* **34** (1997), no. 3, p. 725–743.

[GR65] R. C. Gunning & H. Rossi – *Analytic functions of several complex variables*, Prentice-Hall, Inc., Englewood Cliffs, N.J., 1965.

[GR79] H. Grauert & R. a. Remmert – *Theory of Stein spaces*, Grundlehren der Mathematischen Wissenschaften [Fundamental Principles of Mathematical Sciences], vol. 236, Springer-Verlag, Berlin-New York, 1979, Translated from the German by Alan Huckleberry.

[Gre67] M. J. Greenberg – *Lectures on algebraic topology*, W. A. Benjamin, Inc., New York-Amsterdam, 1967.

[Gro57] A. Grothendieck – " Sur quelques points d'algèbre homologique", *Tôhoku Math. J. (2)* **9** (1957), p. 119–221.

[Gro95] A. Grothendieck, " Technique de descente et théorèmes d'existence en géometrie algébrique. I. Généralités. Descente par morphismes fidèlement plats", in *Séminaire Bourbaki, Vol. 5*, Soc. Math. France, Paris, 1995, p. Exp. No. 190, 299–327.

[Gud69] D. A. Gudkov – " Complete topological classification of the disposition of ovals of a sixth order curve in the projective plane", *Gor/kov. Gos. Univ. Učen. Zap. Vyp.* **87** (1969), p. 118–153.

[Gud71] D. A. Gudkov, " Construction of a new series of M-curves", *Dokl. Akad. Nauk SSSR* **200** (1971), p. 1269–1272.

[Haa95] B. Haas – " Les multilucarnes: nouveaux contre-exemples à la conjecture de Ragsdale", *C. R. Acad. Sci. Paris Sér. I Math.* **320** (1995), no. 12, p. 1507–1512.

[Har74] V. M. Harlamov – " A generalized Petrovskiĭ inequality", *Funkcional. Anal. i Priložen.* **8** (1974), no. 2, p. 50–56.

[Har76] V. M. Harlamov, " Topological types of nonsingular surfaces of degree 4 in $\mathbf{R}P^3$", *Funkcional. Anal. i Priložen.* **10** (1976), no. 4, p. 55–68.

[Har77] R. Hartshorne – *Algebraic geometry*, Springer-Verlag, New York, 1977, Graduate Texts in Mathematics, No. 52.

[Hat02] A. Hatcher – *Algebraic topology*, Cambridge University Press, Cambridge, 2002.

[Hem76] J. Hempel – *3-Manifolds*, Princeton University Press, Princeton, N. J., 1976, Ann. of Math. Studies, No. 86.

[Hir51] F. Hirzebruch – " Über eine Klasse von einfachzusammenhängenden komplexen Mannigfaltigkeiten", *Math. Ann.* **124** (1951), p. 77–86.

[Hir64] H. Hironaka – " Resolution of singularities of an algebraic variety over a field of characteristic zero. I, II", *Ann. of Math. (2) 79 (1964), 109–203; ibid. (2)* **79** (1964), p. 205–326.

[Hir66] F. Hirzebruch – *Topological methods in algebraic geometry*, Third enlarged edition. New appendix and translation from the second German edition by R. L. E. Schwarzenberger, with an additional section by A. Borel. Die Grundlehren der Mathematischen Wissenschaften, Band 131, Springer-Verlag New York, Inc., New York, 1966.

[Hir69] F. Hirzebruch, " The signature of ramified coverings", in *Global Analysis (Papers in Honor of K. Kodaira)*, Univ. Tokyo Press, Tokyo, 1969, p. 253–265.

[Hir75] H. Hironaka – " Triangulations of algebraic sets", in *Algebraic geometry (Proc. Sympos. Pure Math., Vol. 29, Humboldt State Univ., Arcata, Calif., 1974)*, Amer. Math. Soc., Providence, R.I., 1975, p. 165–185.

[Hir76] M. W. Hirsch – *Differential topology*, Springer-Verlag, New York, 1976, Graduate Texts in Mathematics, No. 33.

[HM05a] J. Huisman & F. Mangolte – " Every connected sum of lens spaces is a real component of a uniruled algebraic variety", *Ann. Inst. Fourier (Grenoble)* **55** (2005), no. 7, p. 2475–2487.

[HM05b] J. Huisman & F. Mangolte, " Every orientable Seifert 3-manifold is a real component of a uniruled algebraic variety", *Topology* **44** (2005), no. 1, p. 63–71.

[HM09] J. Huisman & F. Mangolte, " The group of automorphisms of a real rational surface is n-transitive", *Bull. Lond. Math. Soc.* **41** (2009), no. 3, p. 563–568.

[HM10] J. Huisman & F. Mangolte, " Automorphisms of real rational surfaces and weighted blow-up singularities", *Manuscripta Math.* **132** (2010), no. 1-2, p. 1–17.

[Hor75] E. Horikawa – " On deformations of quintic surfaces", *Invent. Math.* **31** (1975), no. 1, p. 43–85.

[HP52] W. V. D. Hodge & D. Pedoe – *Methods of algebraic geometry. Vol. II. Book III: General theory of algebraic varieties in projective space. Book IV: Quadrics and Grassmann varieties*, Cambridge, at the University Press, 1952.

[HR96] B. Hughes & A. Ranicki – *Ends of complexes*, Cambridge Tracts in Mathematics, vol. 123, Cambridge University Press, Cambridge, 1996.

[Hu59] S.-t. Hu – *Homotopy theory*, Pure and Applied Mathematics, Vol. VIII, Academic Press, New York, 1959.

[Hui94] J. Huisman – " Cycles on real abelian varieties", Prépublication de l'Institut Fourier 271, Grenoble, 1994.

[Hui95] J. Huisman, " On real algebraic vector bundles", *Math. Z.* **219** (1995), no. 3, p. 335–342.

[Hui11] J. Huisman, " Topology of real algebraic varieties; some recent results on rational surfaces", in *Real Algebraic Geometry, Rennes : France (2011)*, Prépublication, 2011, `hal-00609687`, p. 51–62.

[IMS09] I. Itenberg, G. Mikhalkin & E. Shustin – *Tropical algebraic geometry*, second ed., Oberwolfach Seminars, vol. 35, Birkhäuser Verlag, Basel, 2009.

[Isk65] V. A. Iskovskih – " On birational forms of rational surfaces", *Izv. Akad. Nauk SSSR Ser. Mat.* **29** (1965), p. 1417–1433.

[Isk67] V. A. Iskovskih, " Rational surfaces with a pencil of rational curves", *Mat. Sb. (N.S.)* **74 (116)** (1967), p. 608–638.

[Ite93] I. Itenberg – " Contre-examples à la conjecture de Ragsdale", *C. R. Acad. Sci. Paris Sér. I Math.* **317** (1993), no. 3, p. 277–282.

[Ite95] I. Itenberg, " Counter-examples to Ragsdale conjecture and T-curves", in *Real algebraic geometry and topology (East Lansing, MI, 1993)*, Contemp. Math., vol. 182, Amer. Math. Soc., Providence, RI, 1995, p. 55–72.

[JO69] K. Jänich & E. Ossa – " On the signature of an involution", *Topology* **8** (1969), p. 27–30.

[Jos06] J. Jost – *Compact Riemann surfaces*, third ed., Universitext, Springer-Verlag, Berlin, 2006, An introduction to contemporary mathematics.

[JP00] N. Joglar-Prieto – " Rational surfaces and regular maps into the 2-dimensional sphere", *Math. Z.* **234** (2000), no. 2, p. 399–405.

[JPM04] N. Joglar-Prieto & F. Mangolte – " Real algebraic morphisms and del Pezzo surfaces of degree 2", *J. Algebraic Geom.* **13** (2004), no. 2, p. 269–285.

[Kam75] T. Kambayashi – " On the absence of nontrivial separable forms of the affine plane", *J. Algebra* **35** (1975), p. 449–456.

[Kas77] A. Kas – " On the deformation types of regular elliptic surfaces", in *Complex analysis and algebraic geometry*, Iwanami Shoten, Tokyo, 1977, p. 107–111.

[Kaw92] Y. Kawamata – " Boundedness of $\mathbf{Q}$-Fano threefolds", in *Proceedings of the International Conference on Algebra, Part 3 (Novosibirsk, 1989)* (Providence, RI), Contemp. Math., vol. 131, Amer. Math. Soc., 1992, p. 439–445.

[KB32] B. O. Koopman & A. B. Brown – " On the covering of analytic loci by complexes", *Trans. Amer. Math. Soc.* **34** (1932), no. 2, p. 231–251.

[KI96] V. Kharlamov & I. Itenberg – " Towards the maximal number of components of a nonsingular surface of degree 5 in $\mathbf{RP}^3$", in *Topology of real algebraic varieties and related topics*, Amer. Math. Soc. Transl. Ser. 2, vol. 173, Amer. Math. Soc., Providence, RI, 1996, p. 111–118.

[KK02] V. S. Kulikov & V. M. Kharlamov – " On real structures on rigid surfaces", *Izv. Ross. Akad. Nauk Ser. Mat.* **66** (2002), no. 1, p. 133–152.

[KK16] W. Kucharz & K. Kurdyka – " Some conjectures on continuous rational maps into spheres", *Topology Appl.* **208** (2016), p. 17–29.

[KK18] W. Kucharz & K. Kurdyka, " Stratified-algebraic vector bundles", *J. Reine Angew. Math.* **745** (2018), p. 105–154.

[KKK18] J. Kollár, W. Kucharz & K. Kurdyka – " Curve-rational functions", *Math. Ann.* **370** (2018), no. 1-2, p. 39–69.

[Kle82] F. Klein – " Ueber Riemann's Theorie der algebraischen Functionen und ihrer Integrale.", Leipzig. Teubner (1882)., 1882.

[Kle66] S. L. Kleiman – " Toward a numerical theory of ampleness", *Ann. of Math. (2)* **84** (1966), p. 293–344.

[KM61] M. A. Kervaire & J. W. Milnor – " On 2-spheres in 4-manifolds", *Proc. Nat. Acad. Sci. U.S.A.* **47** (1961), p. 1651–1657.

[KM09] J. Kollár & F. Mangolte – " Cremona transformations and diffeomorphisms of surfaces", *Adv. Math.* **222** (2009), no. 1, p. 44–61.

[KM12] K. Kuyumzhiyan & F. Mangolte – " Infinitely transitive actions on real affine suspensions", *J. Pure Appl. Algebra* **216** (2012), no. 10, p. 2106–2112.

[KM16] J. Kollár & F. Mangolte – " Approximating curves on real rational surfaces", *J. Algebraic Geom.* **25** (2016), p. 549–570.

[KMM92] J. Kollár, Y. Miyaoka & S. Mori – " Rational connectedness and boundedness of Fano manifolds", *J. Differential Geom.* **36** (1992), no. 3, p. 765–779.

[KN15] J. Kollár & K. Nowak – " Continuous rational functions on real and p-adic varieties", *Math. Z.* **279** (2015), p. 85–97.

[Kne76a] M. Knebusch – " On algebraic curves over real closed fields. I", *Math. Z.* **150** (1976), no. 1, p. 49–70.

[Kne76b] M. Knebusch, " On algebraic curves over real closed fields. II", *Math. Z.* **151** (1976), no. 2, p. 189–205.

[Kod53] K. Kodaira – " On a differential-geometric method in the theory of analytic stacks", *Proc. Nat. Acad. Sci. U. S. A.* **39** (1953), p. 1268–1273.

[Kod54] K. Kodaira, " On Kähler varieties of restricted type (an intrinsic characterization of algebraic varieties)", *Ann. of Math. (2)* **60** (1954), p. 28–48.

[Kod64] K. Kodaira, " On the structure of compact complex analytic surfaces. I", *Amer. J. Math.* **86** (1964), p. 751–798.

[Kol96] J. Kollár – *Rational curves on algebraic varieties*, Ergebnisse der Mathematik und ihrer Grenzgebiete. 3. Folge. A Series of Modern Surveys in Mathematics [Results in Mathematics and Related Areas. 3rd Series. A Series of Modern Surveys in Mathematics], vol. 32, Springer-Verlag, Berlin, 1996.

[Kol97] J. Kollár, " Real algebraic surfaces", `arXiv:alg-geom/9712003`, 1997.

[Kol98a] J. Kollár, " The Nash conjecture for threefolds", *Electron. Res. Announc. Amer. Math. Soc.* **4** (1998), p. 63–73 (electronic).

[Kol98b] J. Kollár, " Real algebraic threefolds. I. Terminal singularities", *Collect. Math.* **49** (1998), no. 2-3, p. 335–360, Dedicated to the memory of Fernando Serrano.

[Kol99a] J. Kollár, " Real algebraic threefolds. II. Minimal model program", *J. Amer. Math. Soc.* **12** (1999), no. 1, p. 33–83.

[Kol99b] J. Kollár, " Real algebraic threefolds. III. Conic bundles", *J. Math. Sci. (New York)* **94** (1999), no. 1, p. 996–1020, Algebraic geometry, 9.

[Kol00] J. Kollár, " Real algebraic threefolds. IV. Del Pezzo fibrations", in *Complex analysis and algebraic geometry*, de Gruyter, Berlin, 2000, p. 317–346.

[Kol01a] J. Kollár, " The topology of real algebraic varieties", in *Current developments in mathematics, 2000*, Int. Press, Somerville, MA, 2001, p. 197–231.

[Kol01b] J. Kollár, " The topology of real and complex algebraic varieties", in *Taniguchi Conference on Mathematics Nara '98*, Adv. Stud. Pure Math., vol. 31, Math. Soc. Japan, Tokyo, 2001, p. 127–145.

[Kol01c] J. Kollár, " Which are the simplest algebraic varieties?", *Bull. Amer. Math. Soc. (N.S.)* **38** (2001), no. 4, p. 409–433.

[Kol02] J. Kollár, " The Nash conjecture for nonprojective threefolds", in *Symposium in Honor of C. H. Clemens (Salt Lake City, UT, 2000)*, Contemp. Math., vol. 312, Amer. Math. Soc., Providence, RI, 2002, p. 137–152.

[Kol07] J. Kollár, *Lectures on resolution of singularities*, Annals of Mathematics Studies, vol. 166, Princeton University Press, Princeton, NJ, 2007.

[Kol17] J. Kollár, " Nash's work in algebraic geometry", *Bull. Amer. Math. Soc. (N.S.)* **54** (2017), no. 2, p. 307–324.

[Kra83] V. A. Krasnov – " Harnack-Thom inequalities for mappings of real algebraic varieties", *Izv. Akad. Nauk SSSR Ser. Mat.* **47** (1983), no. 2, p. 268–297.

[Kra06] J. Kollár, " Rigid isotopy classification of real three-dimensional cubics", *Izv. Ross. Akad. Nauk Ser. Mat.* **70** (2006), no. 4, p. 91–134.

[Kra09] J. Kollár, " On the topological classification of real three-dimensional cubics", *Mat. Zametki* **85** (2009), no. 6, p. 886–893.

[KS04] J. Kollár & F.-O. Schreyer – " Real Fano 3-folds of type V_{22}", in *The Fano Conference*, Univ. Torino, Turin, 2004, p. 515–531.

[Kuc96] W. Kucharz – " Algebraic equivalence and homology classes of real algebraic cycles", *Math. Nachr.* **180** (1996), p. 135–140.

[Kuc99] J. Kollár, " Algebraic morphisms into rational real algebraic surfaces", *J. Algebraic Geom.* **8** (1999), no. 3, p. 569–579.

[Kuc09] J. Kollár, " Rational maps in real algebraic geometry", *Adv. Geom.* **9** (2009), no. 4, p. 517–539.

[Kuc13] J. Kollár, " Regular versus continuous rational maps", *Topology Appl.* **160** (2013), no. 12, p. 1375–1378.

[Kuc14a] J. Kollár, " Approximation by continuous rational maps into spheres", *J. Eur. Math. Soc. (JEMS)* **16** (2014), no. 8, p. 1555–1569.

[Kuc14b] J. Kollár, " Continuous rational maps into the unit 2-sphere", *Arch. Math. (Basel)* **102** (2014), no. 3, p. 257–261.

[Kuc16a] J. Kollár, " Continuous rational maps into spheres", *Math. Z.* **283** (2016), no. 3-4, p. 1201–1215.

[Kuc16b] J. Kollár, " Stratified-algebraic vector bundles of small rank", *Arch. Math. (Basel)* **107** (2016), no. 3, p. 239–249.

[Laf96] J. Lafontaine – *Introduction aux variétés différentielles.*, Grenoble: Presses Universitaires de Grenoble; Les Ulis: EDP Sciences, 1996 (French).

[Laf15] J. Lafontaine, *An introduction to differential manifolds*, second ed., Springer, Cham, 2015.

[Lau71] H. B. Laufer – *Normal two-dimensional singularities*, Princeton University Press, Princeton, N.J., 1971, Annals of Mathematics Studies, No. 71.

[Laz04] R. Lazarsfeld – *Positivity in algebraic geometry. I*, Ergebnisse der Mathematik und ihrer Grenzgebiete. 3. Folge. A Series of Modern Surveys in Mathematics [Results in Mathematics and Related Areas. 3rd Series. A Series of Modern Surveys in Mathematics], vol. 48, Springer-Verlag, Berlin, 2004, Classical setting: line bundles and linear series.

[Lef71] S. Lefschetz – *Selected papers*, Chelsea Publishing Co., Bronx, N.Y., 1971.

[Les18] J. Lesieutre – " A projective variety with discrete, non-finitely generated automorphism group", *Invent. Math.* **212** (2018), no. 1, p. 189–211.

[Liu02] Q. Liu – *Algebraic geometry and arithmetic curves*, Oxford Graduate Texts in Mathematics, vol. 6, Oxford University Press, Oxford, 2002, Translated from the French by Reinie Erné, Oxford Science Publications.

[LM89] H. B. Lawson, Jr. & M.-L. Michelsohn – *Spin geometry*, Princeton Mathematical Series, vol. 38, Princeton University Press, Princeton, NJ, 1989.

[Łoj64] S. Łojasiewicz – " Triangulation of semi-analytic sets", *Ann. Scuola Norm. Sup. Pisa (3)* **18** (1964), p. 449–474.

[LV06] Y. Laszlo & C. Viterbo – " Estimates of characteristic numbers of real algebraic varieties", *Topology* **45** (2006), no. 2, p. 261–280.

[LW33] S. Lefschetz & J. H. C. Whitehead – " On analytical complexes", *Trans. Amer. Math. Soc.* **35** (1933), no. 2, p. 510–517.

[Mal67] B. Malgrange – *Ideals of differentiable functions*, Tata Institute of Fundamental Research Studies in Mathematics, No. 3, Tata Institute of Fundamental Research, Bombay; Oxford University Press, London, 1967.

[Man67] Y. I. Manin – " Rational surfaces over perfect fields. II", *Mat. Sb. (N.S.)* **72 (114)** (1967), p. 161–192.

[Man86] Y. I. Manin, *Cubic forms*, second ed., North-Holland Mathematical Library, vol. 4, North-Holland Publishing Co., Amsterdam, 1986, Algebra, geometry, arithmetic, Translated from the Russian by M. Hazewinkel.

[Man94] F. Mangolte – " Une surface réelle de degré 5 dont l'homologie est entièrement engendrée par des cycles algébriques", *C. R. Acad. Sci. Paris Sér. I Math.* **318** (1994), no. 4, p. 343–346.

[Man97] F. Mangolte, " Cycles algébriques sur les surfaces $K3$ réelles", *Math. Z.* **225** (1997), no. 4, p. 559–576.

[Man00] F. Mangolte, " Surfaces elliptiques réelles et inégalité de Ragsdale-Viro", *Math. Z.* **235** (2000), no. 2, p. 213–226.

[Man01] M. Manetti – " On the moduli space of diffeomorphic algebraic surfaces", *Invent. Math.* **143** (2001), no. 1, p. 29–76.

[Man03] F. Mangolte – " Cycles algébriques et topologie des surfaces bielliptiques réelles", *Comment. Math. Helv.* **78** (2003), no. 2, p. 385–393.

[Man04] F. Mangolte, " Real algebraic geometry of some 2-dimensional and 3-dimensional varieties", Habilitation à diriger des recherches, Université de Savoie, June 2004, https://tel.archives-ouvertes.fr/tel-00006900/file/tel-00006900.pdf.

[Man06] F. Mangolte, " Real algebraic morphisms on 2-dimensional conic bundles", *Adv. Geom.* **6** (2006), no. 2, p. 199–213.

[Man14] F. Mangolte, " Topologie des variétés algébriques réelles de dimension 3", *Gaz. Math.* **139** (2014), p. 5–34.

[Man17a] F. Mangolte, " Real rational surfaces", in *Real Algebraic Geometry*, vol. 51, Panoramas et synthèses, 2017, p. 1–26.

[Man17b] F. Mangolte, *Variétés algébriques réelles*, Cours Spécialisés [Specialized Courses], vol. 24, Société Mathématique de France, Paris, 2017, viii + 484 pages.

[Mar80] A. Marin – " Quelques remarques sur les courbes algébriques planes réelles", in *Seminar on Real Algebraic Geometry (Paris, 1977/1978 and Paris, 1978/1979)*, Publ. Math. Univ. Paris VII, vol. 9, Univ. Paris VII, Paris, 1980, p. 51–68.

[Mas67] W. S. Massey – *Algebraic topology: An introduction*, Harcourt, Brace & World, Inc., New York, 1967.

[Maz86] B. Mazur – " Arithmetic on curves", *Bull. Amer. Math. Soc. (N.S.)* **14** (1986), no. 2, p. 207–259.

[MH73] J. Milnor & D. Husemoller – *Symmetric bilinear forms*, Springer-Verlag, New York-Heidelberg, 1973, Ergebnisse der Mathematik und ihrer Grenzgebiete, Band 73.

[Mik97] G. Mikhalkin – " Blowup equivalence of smooth closed manifolds", *Topology* **36** (1997), no. 1, p. 287–299.

[Mil62] J. Milnor – " A unique decomposition theorem for 3-manifolds", *Amer. J. Math.* **84** (1962), p. 1–7.

[Mil63a] J. Milnor, *Morse theory*, Based on lecture notes by M. Spivak and R. Wells. Annals of Mathematics Studies, No. 51, Princeton University Press, Princeton, N.J., 1963.

[Mil63b] J. Milnor, " Spin structures on manifolds", *Enseignement Math. (2)* **9** (1963), p. 198–203.

[Mil64] J. Milnor, " On the Betti numbers of real varieties", *Proc. Amer. Math. Soc.* **15** (1964), p. 275–280.

[Mil04] J. Milnor, " Vers la conjecture de Poincaré et la classification des variétés de dimension 3", *Gaz. Math.* (2004), no. 99, p. 13–25, Translated from Notices Amer. Math. Soc. **50** (2003), no. 10, 1226–1233.

[Moĭ66a] B. G. Moĭšezon – " On n-dimensional compact complex manifolds having n algebraically independent meromorphic functions. I", *Izv. Akad. Nauk SSSR Ser. Mat.* **30** (1966), p. 133–174.

[Moĭ66b] B. G. Moĭšezon, " On n-dimensional compact complex manifolds having n algebraically independent meromorphic functions. II", *Izv. Akad. Nauk SSSR Ser. Mat.* **30** (1966), p. 345–386.

[Moĭ66c] B. G. Moĭšezon, " On n-dimensional compact complex manifolds having n algebraically independent meromorphic functions. III", *Izv. Akad. Nauk SSSR Ser. Mat.* **30** (1966), p. 621–656.

[Moi67] B. Moishezon – " On n-dimensional compact varieties with n independent meromorphic functions", *Amer. Math. Soc. Translations* **63** (1967), p. 51–177.

[Mon18] J.-P. Monnier – " Semi-algebraic geometry with rational continuous functions", *Math. Ann.* **372** (2018), no. 3-4, p. 1041–1080.

[MS74] J. W. Milnor & J. D. Stasheff – *Characteristic classes*, Princeton University Press, Princeton, N. J., 1974, Annals of Mathematics Studies, No. 76.

[MT86] R. Mneimné & F. Testard – *Introduction à la théorie des groupes de Lie classiques*, Collection Méthodes. [Methods Collection], Hermann, Paris, 1986.

[MT19] L. Moser-Jauslin & R. Terpereau – " Real structures on symmetric spaces", Preprint, `arXiv:1904.10723 [math.AG]`, 2019.

[MTB18] L. Moser-Jauslin, R. Terpereau & M. Borovoi – " Real structures on horospherical varieties", Preprint, `arXiv:1808.10793 [math.AG]`, 2018.

[Mum69] D. Mumford – " Enriques' classification of surfaces in char p. I", in *Global Analysis (Papers in Honor of K. Kodaira)*, Univ. Tokyo Press, Tokyo, 1969, p. 325–339.

[Mum79] D. Mumford – " An algebraic surface with K ample, $(K^2) = 9$, $p_g = q = 0$", *Amer. J. Math.* **101** (1979), no. 1, p. 233–244.

[Mun84] J. R. Munkres – *Elements of algebraic topology*, Addison-Wesley Publishing Company, Menlo Park, CA, 1984.

[MvH98] F. Mangolte & J. van Hamel – " Algebraic cycles and topology of real Enriques surfaces", *Compositio Math.* **110** (1998), no. 2, p. 215–237.

[MW12] F. Mangolte & J.-Y. Welschinger – " Do uniruled six-manifolds contain Sol Lagrangian submanifolds?", *Int. Math. Res. Not. IMRN* (2012), no. 7, p. 1569–1602.

[Nas52] J. Nash – " Real algebraic manifolds", *Ann. of Math. (2)* **56** (1952), p. 405–421.

[Nat99] S. M. Natanzon – " Moduli of real algebraic curves and their superanalogues. Spinors and Jacobians of real curves", *Uspekhi Mat. Nauk* **54** (1999), no. 6(330), p. 3–60.

[Nee89] A. Neeman – " Ueda theory: theorems and problems", *Mem. Amer. Math. Soc.* **81** (1989), no. 415, p. vi+123.

[Nik83] V. V. Nikulin – " Involutions of integer quadratic forms and their applications to real algebraic geometry", *Izv. Akad. Nauk SSSR Ser. Mat.* **47** (1983), no. 1, p. 109–188.

[Nik96] V. V. Nikulin, " On the topological classification of real Enriques surfaces. I", in *Topology of real algebraic varieties and related topics*, Amer. Math. Soc. Transl. Ser. 2, vol. 173, Amer. Math. Soc., Providence, RI, 1996, p. 187–201.

[Now17] K. J. Nowak – " Some results of algebraic geometry over Henselian rank one valued fields", *Selecta Math. (N.S.)* **23** (2017), no. 1, p. 455–495.

[Oka50] K. Oka – " Sur les fonctions analytiques de plusieurs variables. VII. Sur quelques notions arithmétiques", *Bull. Soc. Math. France* **78** (1950), p. 1–27.

[Ore01] S. Y. Orevkov – " Real quintic surface with 23 components", *C. R. Acad. Sci. Paris Sér. I Math.* **333** (2001), no. 2, p. 115–118.

[Ore03] S. Y. Orevkov – " Riemann existence theorem and construction of real algebraic curves", *Ann. Fac. Sci. Toulouse Math. (6)* **12** (2003), no. 4, p. 517–531.

[Per81] U. Persson – " Chern invariants of surfaces of general type", *Compositio Math.* **43** (1981), no. 1, p. 3–58.

[Per82] U. Persson, " Horikawa surfaces with maximal Picard numbers", *Math. Ann.* **259** (1982), no. 3, p. 287–312.

[Per95] D. Perrin – *Géométrie algébrique*, Savoirs Actuels. [Current Scholarship], InterEditions, Paris, 1995, Une introduction. [An introduction].

[Pet33] I. G. Petrovsky – " Sur la topologie des courbes réelles et algébriques", *C. R. Acad. Sci. Paris* **197** (1933), p. 1270–1272.

[Pet38] I. G. Petrovsky, " On the topology of real plane algebraic curves", *Ann. of Math. (2)* **39** (1938), no. 1, p. 189–209.

[PP12] P. Popescu-Pampu – " La dualité de poincaré", *Images des Mathématiques* (2012).

[PP17] A. Parusiński & L. Păunescu – " Arc-wise analytic stratification, Whitney fibering conjecture and Zariski equisingularity", *Adv. Math.* **309** (2017), p. 254–305.

[PY07] G. Prasad & S.-K. Yeung – " Fake projective planes", *Invent. Math.* **168** (2007), no. 2, p. 321–370.

[PY10] G. Prasad & S.-K. Yeung, " Addendum to "Fake projective planes" Invent. Math. 168, 321–370 (2007)", *Invent. Math.* **182** (2010), no. 1, p. 213–227.

[Rag06] V. Ragsdale – " On the Arrangement of the Real Branches of Plane Algebraic Curves", *Amer. J. Math.* **28** (1906), no. 4, p. 377–404.

[Ram71] C. P. Ramanujam – " A topological characterisation of the affine plane as an algebraic variety", *Ann. of Math. (2)* **94** (1971), p. 69–88.

[Ren15] A. Renaudineau – " A real sextic surface with 45 handles", *Math. Z.* **281** (2015), no. 1-2, p. 241–256.

[Ren17] A. Renaudineau, " A tropical construction of a family of real reducible curves", *J. Symbolic Comput.* **80** (2017), no. part 2, p. 251–272.

[Ris85] J.-J. Risler – " Type topologique des surfaces algébriques réelles de degré 4 dans $\mathbf{RP}^3$", *Astérisque* (1985), no. 126, p. 153–168, Geometry of $K3$ surfaces: moduli and periods (Palaiseau, 1981/1982).

[Ris93] J.-J. Risler, " Construction d'hypersurfaces réelles (d'après Viro)", *Astérisque* (1993), no. 216, p. Exp. No. 763, 3, 69–86, Séminaire Bourbaki, Vol. 1992/93.

[Rob16] M. F. Robayo – " Prime order birational diffeomorphisms of the sphere", *Ann. Sc. Norm. Super. Pisa Cl. Sci.* **16** (2016), p. 909–970.

[Rus02] F. Russo – " The antibirational involutions of the plane and the classification of real del Pezzo surfaces", in *Algebraic geometry*, de Gruyter, Berlin, 2002, p. 289–312.

[RV05] F. Ronga & T. Vust – " Birational diffeomorphisms of the real projective plane", *Comment. Math. Helv.* **80** (2005), no. 3, p. 517–540.

[RZ18] M. F. Robayo & S. Zimmermann – " Infinite algebraic subgroups of the real Cremona group", *Osaka J. Math.* **55** (2018), no. 4, p. 681–712.

[Sam67] P. Samuel – *Méthodes d'algèbre abstraite en géométrie algébrique*, Seconde édition, corrigée. Ergebnisse der Mathematik und ihrer Grenzgebiete, Band 4, Springer-Verlag, Berlin-New York, 1967.

[Sch11] M. Schütt – " Quintic surfaces with maximum and other Picard numbers", *J. Math. Soc. Japan* **63** (2011), no. 4, p. 1187–1201.

[Sch15] M. Schütt, " Picard numbers of quintic surfaces", *Proc. Lond. Math. Soc. (3)* **110** (2015), no. 2, p. 428–476.

[Sco83] P. Scott – " The geometries of 3-manifolds", *Bull. London Math. Soc.* **15** (1983), no. 5, p. 401–487.

[Ser55a] J.-P. Serre – " Faisceaux algébriques cohérents", *Ann. of Math. (2)* **61** (1955), p. 197–278.

[Ser55b] J.-P. Serre, " Un théorème de dualité", *Comment. Math. Helv.* **29** (1955), p. 9–26.

[Ser56] J.-P. Serre, " Géométrie algébrique et géométrie analytique", *Ann. Inst. Fourier, Grenoble* **6** (1955–1956), p. 1–42.

[Ser77] J.-P. Serre, *Cours d'arithmétique*, Presses Universitaires de France, Paris, 1977, Deuxième édition revue et corrigée, Le Mathématicien, No. 2. Version anglaise : A course in arithmetic, Graduate Texts in Mathematics, No. 7, Springer-Verlag, New York-Heidelberg, 1973.

[Ser94] J.-P. Serre, *Cohomologie galoisienne*, fifth ed., Lecture Notes in Mathematics, vol. 5, Springer-Verlag, Berlin, 1994.

[Sha94] I. R. Shafarevich – *Basic algebraic geometry*, second ed., Springer-Verlag, Berlin, 1994, Two volumes. Translated from the 1988 Russian edition and with notes by Miles Reid.

[Shi71] G. Shimura – *Introduction to the arithmetic theory of automorphic functions*, Publications of the Mathematical Society of Japan, No. 11. Iwanami Shoten, Publishers, Tokyo, 1971, Kanô Memorial Lectures, No. 1.

[Shi72a] J.-P. Serre, " On the field of rationality for an abelian variety", *Nagoya Math. J.* **45** (1972), p. 167–178.

[Shi72b] T. Shioda – " On elliptic modular surfaces", *J. Math. Soc. Japan* **24** (1972), p. 20–59.

[Shi81] T. Shioda, " On the Picard number of a complex projective variety", *Ann. Sci. École Norm. Sup. (4)* **14** (1981), no. 3, p. 303–321.

[Sie55] C. L. Siegel – " Meromorphe Funktionen auf kompakten analytischen Mannigfaltigkeiten", *Nachr. Akad. Wiss. Göttingen. Math.-Phys. Kl. IIa.* **1955** (1955), p. 71–77.

[Sil82] R. Silhol – " Real abelian varieties and the theory of Comessatti", *Math. Z.* **181** (1982), no. 3, p. 345–364.

[Sil84] T. Shioda, " Real algebraic surfaces with rational or elliptic fiberings", *Math. Z.* **186** (1984), no. 4, p. 465–499.

[Sil89] T. Shioda, *Real algebraic surfaces*, Lecture Notes in Mathematics, vol. 1392, Springer-Verlag, Berlin, 1989.

[Sil92] T. Shioda, " Compactifications of moduli spaces in real algebraic geometry", *Invent. Math.* **107** (1992), no. 1, p. 151–202.

[Sil09] J. H. Silverman – *The arithmetic of elliptic curves*, second ed., Graduate Texts in Mathematics, vol. 106, Springer, Dordrecht, 2009.

[Siu83] Y. T. Siu – " Every $K3$ surface is Kähler", *Invent. Math.* **73** (1983), no. 1, p. 139–150.

[Slo80] P. Slodowy – *Simple singularities and simple algebraic groups*, Lecture Notes in Mathematics, vol. 815, Springer, Berlin, 1980.

[Spa66] E. H. Spanier – *Algebraic topology*, McGraw-Hill Book Co., New York, 1966.

[Sta62] J. Stallings – " The piecewise-linear structure of Euclidean space", *Proc. Cambridge Philos. Soc.* **58** (1962), p. 481–488.

[Sti92] J. Stillwell – *Geometry of surfaces*, Universitext, Springer-Verlag, New York, 1992, Corrected reprint of the 1992 original.

[Suw69] T. Suwa – " On hyperelliptic surfaces", *J. Fac. Sci. Univ. Tokyo Sect. I* **16** (1969), p. 469–476 (1970).

[Tho54] R. Thom – " Quelques propriétés globales des variétés différentiables", *Comment. Math. Helv.* **28** (1954), p. 17–86.

[Tho65] R. Thom, " Sur l'homologie des variétés algébriques réelles", in *Differential and Combinatorial Topology (A Symposium in Honor of Marston Morse)*, Princeton Univ. Press, Princeton, N.J., 1965, p. 255–265.

[Tog73] A. Tognoli – " Su una congettura di Nash", *Ann. Scuola Norm. Sup. Pisa (3)* **27** (1973), p. 167–185.

[Tou72] J.-C. Tougeron – *Idéaux de fonctions différentiables*, Springer-Verlag, Berlin-New York, 1972, Ergebnisse der Mathematik und ihrer Grenzgebiete, Band 71.

[Tro98] M. Troyanov – " L'horizon de SOL", *Exposition. Math.* **16** (1998), no. 5, p. 441–479.

[Uen73] K. Ueno – " Classification of algebraic varieties. I", *Compositio Math.* **27** (1973), p. 277–342.

[Uen75] K. Ueno, *Classification theory of algebraic varieties and compact complex spaces*, Lecture Notes in Mathematics, Vol. 439, Springer-Verlag, Berlin-New York, 1975, Notes written in collaboration with P. Cherenack.

[vH00] J. van Hamel – *Algebraic cycles and topology of real algebraic varieties*, CWI Tract, vol. 129, Stichting Mathematisch Centrum Centrum voor Wiskunde en Informatica, Amsterdam, 2000, Dissertation, Vrije Universiteit Amsterdam, Amsterdam.

[Vir80] O. J. Viro – " Curves of degree 7, curves of degree 8 and the Ragsdale conjecture", *Dokl. Akad. Nauk SSSR* **254** (1980), no. 6, p. 1306–1310.

[Vit99] C. Viterbo – " Symplectic real algebraic geometry", Unpublished, 1999.

[Voi02] C. Voisin – *Théorie de Hodge et géométrie algébrique complexe*, Cours Spécialisés [Specialized Courses], vol. 10, Société Mathématique de France, Paris, 2002.

[Wae30] B. v. d. Waerden – " Topologische begründung des kalküls der abzählenden geometrie", *Mathematische Annalen* **102** (1930), p. 337–362.

[Wal35] R. J. Walker – " Reduction of the singularities of an algebraic surface", *Ann. of Math. (2)* **36** (1935), no. 2, p. 336–365.

[Wei56] A. Weil – " The field of definition of a variety", *Amer. J. Math.* **78** (1956), p. 509–524.

[Wei94] C. A. Weibel – *An introduction to homological algebra*, Cambridge Studies in Advanced Mathematics, vol. 38, Cambridge University Press, Cambridge, 1994.

[Wel02] J.-Y. Welschinger – " Forme d'intersection tordue et extension de la congruence d'Arnol' d", *Math. Z.* **242** (2002), no. 3, p. 589–614.

[Wel03] J.-Y. Welschinger, " Real structures on minimal ruled surfaces", *Comment. Math. Helv.* **78** (2003), no. 2, p. 418–446.

[Wil78] G. Wilson – " Hilbert's sixteenth problem", *Topology* **17** (1978), no. 1, p. 53–73.

[Wit34] E. Witt – " Zerlegung reeller algebraischer Funktionen in Quadrate. Schiefkörper über reellem Funktionenkörper", *J. Reine Angew. Math.* **171** (1934), p. 4–11.

[X85] *Géométrie des surfaces $K3$: modules et périodes* – Société Mathématique de France, Paris, 1985, Papers from the seminar held in Palaiseau, October 1981–January 1982, Astérisque No. 126 (1985).

[Yas16] E. Yasinsky – " Subgroups of odd order in the real plane Cremona group", *J. Algebra* **461** (2016), p. 87–120.

[Zeu74] H. G. Zeuthen – " Sur les différentes formes des courbes planes du quatrième ordre.", *Math. Ann.* **7** (1874), p. 410–432 (French).

[Zim18] S. Zimmermann – " The Abelianization of the real Cremona group", *Duke Math. J.* **167** (2018), no. 2, p. 211–267.

Index

F. Mangolte, *Real Algebraic Varieties*, Springer Monographs in Mathematics,
https://doi.org/10.1007/978-3-030-43104-4

List of Examples

F. Mangolte, *Real Algebraic Varieties*, Springer Monographs in Mathematics,
https://doi.org/10.1007/978-3-030-43104-4

GPSR Compliance
The European Union's (EU) General Product Safety Regulation (GPSR) is a set of rules that requires consumer products to be safe and our obligations to ensure this.

If you have any concerns about our products, you can contact us on

ProductSafety@springernature.com

In case Publisher is established outside the EU, the EU authorized representative is:

Springer Nature Customer Service Center GmbH
Europaplatz 3
69115 Heidelberg, Germany

www.ingramcontent.com/pod-product-compliance
Ingram Content Group UK Ltd.
Pitfield, Milton Keynes, MK11 3LW, UK
UKHW021832270726
14058UKWH00001B/114

9783030431068